Lochnagar: The Natural History of a Mountain Lake

Developments in Paleoenvironmental Research

VOLUME 12

Lochnagar

The Natural History of a Mountain Lake

Edited by

Neil L. Rose

Environmental Change Research Centre, University College London,
London, UK

 Springer

A C.I.P. Catalogue record for this book is available from the Library of Congress.

ISBN-10 1-4020-3900-X (HB)
ISBN-13 978-1-4020-3900-3 (HB)
ISBN-10 1-4020-3986-7 (e-book)
ISBN-13 978-1-4020-3986-7 (e-book)

Published by Springer,
P.O. Box 17, 3300 AA Dordrecht, The Netherlands.

www.springer.com

Printed on acid-free paper

Cover painting: Lochnagar from below Meikle Pap, Watercolour by Ray Rose.

DEDICATION

This book is dedicated to Jo Porter. Without the c. 270 visits (so far) he has made to Lochnagar in support of the various sampling programmes there, this book could not have been written.

TABLE OF CONTENTS

Part IV: Future impacts

PREFACE

Previous volumes in this 'Developments in Paleoenvironmental Research' (DPER) series have focussed on providing in-depth descriptions of palaeoenvironmental techniques or have described the applications of these approaches on various regional bases. The former of these now provide an invaluable series of standard text books for scientists and students, while the latter show how the application of palaeo-techniques can be used across broad geographical scales. In this current volume, we have attempted something a little different. Not only are a variety of palaeo-techniques applied to a single, small lake, but we have tried to show how these methods, and the data derived from them, can be integrated synergistically with contemporary monitoring and predictive modelling.

The acidification and metals research provide two good examples of this. Along with other upland lakes across the UK, the early research work at Lochnagar was based on assessing the competing hypotheses for the causes of surface water acidification. As a result, palaeolimnological techniques were used to assess the timing and extent of pH changes over hundreds of years. The subsequent establishment of the UK Acid Waters Monitoring Network (UK AWMN) then allowed a range of biological and chemical parameters to be assessed routinely in order to determine the rate at which the lakes and streams, including Lochnagar, were recovering following emissions reductions. Furthermore, the use of the dynamic model MAGIC (Modelling of Acidification of Groundwater in Catchments) at all the UK AWMN sites provided predictions for how the sites might be expected to respond under a variety of emission reduction scenarios as well as a retrodictive comparison with the palaeolimnological data. Current research now uses all these approaches to assess how the role of nitrogen, dissolved organic carbon and a changing climate may have, and may continue to, confound recovery.

The work on trace metals at Lochnagar began as a spin-off from the acidification research as the sediment record provided a means of showing increasing metal deposition concomitant with the acidification of the loch. This has now expanded to the monitoring of trace metals in many ecological compartments in both aquatic and terrestrial environments as well as a ten year record (so far) of metals in fortnightly rain and loch water samples. Data from contemporary measurements can be compared with temporal trends derived from sediment core data and are also used within the models WHAM (Windermere Humic Aqueous Model) and CHUM (Chemistry of the Uplands Model) to consider how biogeochemical processes control metal behaviour as well as a means by which to predict future changes to soil, water and sediment metal chemistry. Furthermore, the role of climate change in the release of previously deposited metals stored in catchment soils is the subject of research ongoing at the time of writing and this process could well enhance metal availability to freshwater biota despite dramatic reductions in the release of metals to the atmosphere over recent decades. There is therefore a further need for monitoring as an empirical 'ground truthing' for model predictions. Conversely, the sediment record allows these contemporary data to be placed in an historical context, and provides an indication of natural variability as a base-line against which to gauge future impacts.

As a consequence of considering these various data-types (i.e., palaeolimnological, monitoring, modelling) this volume is not solely focussed on the palaeoenvironment and I offer apologies to those who think a book in this DPER series should be so directed. However, neither should this volume be considered a 'final report' on scientific work at Lochnagar, but rather a 'work in progress'. During the compilation of the volume more intriguing research has been developed at the site particularly with stable isotopes and seasonal patterns in methyl mercury which there has been neither time nor space to include. In particular though, one theme has recurred time and again during the compilation of this book and that is the unique status of datasets and research at Lochnagar, at least within the UK. While interesting data and time-series provide an insight into the contamination and biological status of the loch, it is unknown whether this is typical of upland sites across the UK. Considerable added value would be achieved if similar datasets were available for other comparable upland lakes and lochs and there is, therefore, a need for a national and international mountain lake monitoring network.

Many people have aided our work at Lochnagar and contributed to the production of this volume. Ongoing research at the site would not be possible without the support of the Balmoral Estates Office and, in particular, Mr Peter Ord, the Factor. I would also like to thank Jo Porter for his many visits to Lochnagar in support of our monitoring programmes and for providing me with a wealth of background information on Lochnagar. I am also grateful to the past and present members of the Environmental Change Research Centre, University College London for their help in the field. Funding (as explained in Chapter 1) has mainly come from the EU and DEFRA but other smaller studies, including PhD research, have been funded from a number of different sources. I am sure all the Chapter authors would join me in acknowledging their funding sources and the support of their respective Institutions during the research for, and the writing of, this volume.

 On a personal note, I would like to thank all the chapter authors and reviewers for their time and energies; Judith Terpos at Springer for dealing with my questions and editorial struggles; the series editors John Smol and Bill Last; and Rick Battarbee for his inspiration.

 I would also like to thank my parents, Marion and Ray Rose, for their support and encouragement - for as long as I can remember - and for the watercolour on the cover! This book is for them.

 Last, but by no means least, my thanks go to Cath, Ellen and Peter Rose for putting up with my frequent absences in the name of 'science'.

Neil Rose
September 2006

THE EDITOR

Neil Rose is a Principal Research Fellow in the Environmental Change Research Centre at University College London.

Major areas of his research include: the morphology, chemistry and source apportionment of fly-ash particles and the relationships between the components of fly-ash (spheroidal carbonaceous particles (SCPs) and inorganic ash spheres) with other atmospherically deposited pollutants such as acidifying compounds, trace metals and persistent organic pollutants; the spatial and temporal distributions of SCPs around the world using lake sediments as a natural archive; the use of SCP temporal profiles to provide lake sediment chronologies. More information is available on the CARBYNET (http://www.ecrc.ucl.ac.uk/content/view/299/17/) and Environmental Change Research Centre (http://www.ecrc.ucl.ac.uk/) websites. Recent research has been undertaken in Svalbard, China, Japan, Uganda, USA, Greenland and at mountain lakes throughout Europe including, of course, Lochnagar.

AIMS AND SCOPE OF *DEVELOPMENTS IN PALEOENVIRONMENTAL RESEARCH* BOOK SERIES

Paleoenvironmental research continues to enjoy tremendous interest and progress in the scientific community. The overall aims and scope of the *Developments in Paleoenvironmental Research* book series are to capture this excitement and document these developments. Volumes related to any aspect of paleoenvironmental research, encompassing any time period, are within the scope of the series. For example, relevant topics include studies focussed on terrestrial, peatland, lacustrine, riverine, estuarine, and marine systems, ice cores, cave deposits, palynology, isotopes, geochemistry, sedimentology, paleontology, etc. Methodological and taxonomic volumes relevant to paleoenvironmental research are also encouraged. The series will include edited volumes on a particular subject, geographic region or time period, conference and workshop proceedings, as well as monographs. Prospective authors and/or editors should consult the series editors for more details. The series editors also welcome any comments or suggestions for future volumes.

PUBLISHED AND FORTHCOMING TITLES IN THE
DEVELOPMENTS IN PALEOENVIRONMENTAL RESEARCH BOOK SERIES

Series Editors:
John P. Smol, smolj@biology.queensu.ca
Department of Biology, Queen's University Kingston, Ontario, Canada
William M. Last, wm_last@umanitoba.ca
Department of Geological Sciences, University of Manitoba, Winnipeg, Manitoba, Canada

For more information on this series, please visit:

www.springeronline.com

http://home.cc.umanitoba.ca/~mlast/paleolim/dper.html

LIST OF CONTRIBUTORS

Rick W. Battarbee
Environmental Change Research Centre
University College London
Pearson Building, Gower Street
London WC1E 6BT
United Kingdom
r.battarbee@geog.ucl.ac.uk

John S. Bell
Macaulay Land Use Research Institute
Craigiebuckler
Aberdeen AB15 8QH
United Kingdom
j.bell@macaulay.ac.uk

Anna C. Benedictow
Air Pollution Section
Norwegian Meteorological Institute
P.O.Box 43 Blindern
0313 Oslo
Norway
anna.benedictow@met.no

H. John B. Birks
Department of Biology   and   *Environmental Change Research Centre*
University of Bergen   *University College London*
Allégaten 41   *Pearson Building, Gower Street*
N-5007 Bergen   *London WC1E 6BT*
Norway   *United Kingdom*
John.Birks@bio.uib.no

Muriel Bonjean
Centre for Ecology and Hydrology
Wallingford
Oxfordshire OX10 8BB
United Kingdom

Peter Collen
FRS Freshwater Laboratory
Faskally, Pitlochry,
Perthshire, PH16 5LB
United Kingdom
p.collen@marlab.ac.uk

Catherine Dalton
Department of Geography
Mary Immaculate College
University of Limerick
South Circular Road
Limerick
Ireland
catherine.dalton@mic.ul.ie

Chris D. Evans
Centre for Ecology and Hydrology
Orton Building,
Deiniol Road
BangorLL57 2UP
United Kingdom
cev@ceh.ac.uk

Arne Fjellheim
Stavanger Museum
Musegata 16
N-4010 Stavanger
Norway
arne.fjellheim@stavanger.museum.no

Roger J. Flower
Environmental Change Research Centre
University College London
Pearson Building, Gower Street
London WC1E 6BT
United Kingdom
r.flower@geog.ucl.ac.uk

Sally Goodman
Department of Earth and Planetary Sciences
McGill University, 3450 University
Montreal H3A 2A7
Quebec
Canada
sgoodman@eps.mcgill.ca

Joan O. Grimalt
Department of Environmental Chemistry,
Institute of Chemical and Environmental Research (CSIC)
Jordi Girona, 18
08034-Barcelona
Spain
jgoqam@cid.csic.es

Adrian M. Hall
Fettes College
Edinburgh EH4 1QX
United Kingdom
am.hall@fettes.com

Ron Harriman
FRS Freshwater Laboratory
Faskally, Pitlochry
Perthshire, PH16 5LB
United Kingdom

Rachel C. Helliwell
Macaulay Land Use Research Institute
Craigiebuckler
Aberdeen, AB15 8QH
United Kingdom
r.helliwell@macaulay.ac.uk

Rudolf Hofer
Institute of Zoology, University of Innsbruck
Technikerstr. 25
A-6020 Innsbruck
Austria
rudolf.hofer@uibk.ac.at

Michael Hughes
Environmental Change Research Centre
University College London
Pearson Building, Gower Street
London WC1E 6BT
United Kingdom
m.hughes@ucl.ac.uk

Mike Hutchins
Centre for Ecology and Hydrology
Wallingford
Oxfordshire OX10 8BB
United Kingdom
mihu@ceh.ac.uk

Alan Jenkins
Centre for Ecology and Hydrology
Wallingford
Oxfordshire OX10 8BB
United Kingdom
jinx@ceh.ac.uk

Helen Kettle
School of GeoSciences
The University of Edinburgh
Crew Building, Kings Buildings,
West Mains Rd,
Edinburgh EH9 3JN
United Kingdom
h.kettle@ed.ac.uk

Reinhard Lackner
Institute of Zoology, University of Innsbruck
Technikerstr. 25
A-6020 Innsbruck
Austria
reinhard.Lackner@uibk.ac.at

Alan J. Lawlor
Centre for Ecology and Hydrology (Lancaster)
Bailrigg
Lancaster LA1 4AP
United Kingdom

Katrin Layer
School of Biological and Chemical Sciences
Queen Mary University of London
Mile End Road
London E1 4NS
United Kingdom.
k.layer@qmul.ac.uk

Martin Lees
Centre for Ecology and Hydrology
Wallingford
Oxfordshire OX10 8BB
United Kingdom

Allan Lilly
Macaulay Land Use Research Institute
Craigiebuckler
Aberdeen AB15 8QH
United Kingdom
a.lilly@macaulay.ac.uk

Jean-Charles Massabuau
Laboratoire d'Ecophysiologie et Ecotoxicologie des Systèmes Aquatiques
UMR 5805, CNRS & Univ. Bordeaux 1
Place du Dr Peyneau
33 120 Arcachon
France
massabuau@lnpc.u-bordeaux.fr

Sarah E. Metcalfe
School of Geography,
The University of Nottingham,
University Park,
Nottingham NG7 2RD
United Kingdom
sarah.metcalfe@nottingham.ac.uk

Donald T. Monteith
Environmental Change Research Centre
University College London
Pearson Building, Gower Street
London WC1E 6BT
United Kingdom
dmonteit@geog.ucl.ac.uk

Derek C.G. Muir
Water Science and Technology Directorate
Environment Canada
Burlington, Ontario L7R 4A6
Canada
Derek.Muir@ec.gc.ca

Jim Nicholson
jpgn@seawynd.fsnet.co.uk

Benjamin Piña
Department of Molecular and Cellular Biology,
Institute of Chemical and Environmental Research (CSIC)
Jordi Girona, 18
08034-Barcelona
Spain
bpcbmc@cid.csc.es

Sergi Pla
Department of Geography
Loughborough University
Loughborough
Leicestershire LE11 3TU
United Kingdom
S.Pla-Rabes@lboro.ac.uk

Gunnar Raddum
Institute of Biology, University of Bergen
Thormøhlensgt. 49
N-5006 Bergen
Norway
gunnar.raddum@zoo.uib.no

Nick Reynard
Centre for Ecology and Hydrology
Wallingford
Oxfordshire OX10 8BB
United Kingdom

Sigurd Rognerud
Norwegian Institute of Water Research, NIVA Branch Office East
Sandvikaveien 4,
N-2312 Ottestad
Norway
sigurd.rognerud@niva.no

Neil L. Rose
Environmental Change Research Centre
University College London
Pearson Building, Gower Street
London WC1E 6BT
United Kingdom
nrose@geog.ucl.ac.uk

Bjørn Olav Rosseland
Department of Ecology and Natural Resource Management
Norwegian University of Life Sciences
P.O. Box 5003, N-1432 Aas
Norway
bjorn.rosseland@umb.no

Ewan Shilland
Environmental Change Research Centre
University College London
Pearson Building, Gower Street
London WC1E 6BT
United Kingdom
e.shillan@geog.ucl.ac.uk

Laura Shotbolt
Department of Geography,
Queen Mary, University of London
Mile End Road
London E1 4NS
United Kingdom
l.shotbolt@qmul.ac.uk

Roy Thompson
School of GeoSciences
The University of Edinburgh
Edinburgh EH9 3JW
United Kingdom
roy@ed.ac.uk

Edward Tipping
Centre for Ecology and Hydrology (Lancaster)
Bailrigg
Lancaster LA1 4AP
United Kingdom
et@ceh.ac.uk

Martin Todd
Department of Geography
University College London
Pearson Building,Gower Street,
London WC1E 6BT
United Kingdom
m.todd@geog.ucl.ac.uk

Jonathan Tyler
Environmental Change Research Centre
University College London
Pearson Building, Gower Street
London WC1E 6BT
United Kingdom
j.tyler@ucl.ac.uk

Ingrid Vives
Department of Environmental Chemistry,
Institute of Chemical and Environmental Research (CSIC)
Jordi Girona, 18
08034-Barcelona
Spain
ivrqam@cid.csic.es

Guy Woodward
School of Biological and Chemical Sciences
Queen Mary University of London
Mile End Road
London E1 4NS
United Kingdom.
g.woodward@qmul.ac.uk

Handong Yang
Environmental Change Research Centre
University College London
Pearson Building, Gower Street
London WC1E 6BT
United Kingdom
hyang@geog.ucl.ac.uk

1. AN INTRODUCTION TO LOCHNAGAR

NEIL L. ROSE (nrose@geog.ucl.ac.uk)
Environmental Change Research Centre
University College London
Pearson Building, Gower Street
London WC1E 6BT
United Kingdom

Key words: Lochnagar, lochs, monitoring, mountain lakes, Scotland

The importance of mountains and their lakes

Mountainous areas are now recognised as important regions of biodiversity and endemism. Steep gradients and high relief compress 'latitudinal' climatic zones into small spatial scales such that ecological zones are narrow (Pauli et al. 2003; Hofer 2005) and often isolated (Greenwood 2005). Further, mountain regions are a principal source of fresh water, giving rise to many of the world's major rivers and thus fundamental to the provision of this resource to downstream ecosystems, agriculture and human populations. They also sustain many economic activities such as forestry, mining, and hydro-electric power generation (Beniston 2000).

Due to the range of climatic zones present within mountain regions, sensitivity to climatic change is high with the potential for loss of alpine species and considerable impact to hydrology across broad areas. However, it was only in the 1990s that mountain issues finally received international attention on the global environmental agenda with a specific chapter in 'Agenda 21' at the United Nations Conference on Environment and Development in Rio de Janeiro, and inclusion in Paragraph 8(g) of Article 4 of the United Nations Framework Convention on Climate Change (UNFCCC), where they were identified as requiring special consideration with regard to changes in climate.

The lakes of mountain regions often possess fragile ecosystems and this is due to a number of factors including low weathering rates of underlying geologies, thin soils and often harsh meteorological conditions. Therefore, despite their isolation, impacts from additional anthropogenic stressors such as atmospheric pollutant deposition often result in detectable chemical and/or biological changes (e.g., Jones et al. 1993; Battarbee et al. 2005; Lotter 2005). Remote mountain lakes can therefore act as 'early warning' indicators for less sensitive sites, and the wider environment, and they have become a

1

N.L. Rose (ed.), Lochnagar: The Natural History of a Mountain Lake
Developments in Paleoenvironmental Research, 1–25.
© 2007 *Springer.*

useful tool in monitoring environmental change, where they have several advantages. First, they are generally less disturbed by direct human influence thereby allowing indirect impacts to be identified where these might be masked in lakes at lower altitudes and in more accessible areas. Second, via lake sediments, a natural archive of environmental change is stored. This can record both the natural development of the lake and its catchment, possibly over millennia, as well as a record of more recent anthropogenic impact and any corresponding biological response, often at a sub-decadal resolution (Lamoureux and Gilbert 2004; Battarbee et al. 2005). Recent studies in Europe have shown that deposited pollutants have been impacting remote lakes for hundreds of years (Bindler et al. 2001; Yang et al 2002) and that this deposition can result in the accumulation of both trace metals and organochlorine compounds in biota sometimes to significant levels (Rognerud et al. 2002; Vives et al. 2005). Critical load exceedance, resulting from both sulphur and nitrogen deposition, is also known to be widespread in mountain lakes throughout Europe (Curtis et al. 2002).

   The UK does not possess the classic alpine ranges seen across Europe and with only nine lakes with surface areas larger than 1 ha at altitudes higher than 1000 m above sea level (a.s.l.) (UKlakes database; http://www.uklakes.net/) it is an arguable point whether it possesses any mountain lakes at all. However, the tree-line in Scotland, where the majority of the UK's uplands lie, is only at an altitude of 700 – 800 m, and as low as 500 m in the north and west (Birks 1988) compared with 2600 m in the Alps and 1200 m in central Norway (Wightman 1996). Therefore, the Scottish montane zone, defined as being above the local tree-line, and including sub-, low- and mid-alpine zones and consisting of moss and lichen heaths, blanket bog and dwarf shrubs, covers approximately 12% of Scotland's land surface (Wightman 1996). Allowing for this lower altitude definition, Lochnagar (Figure 1), at 788 m a.s.l. and by virtue of a quarter of a century of multi-disciplinary scientific study, has become known as the "UK's mountain lake" [1]. Indeed, as a key site in both national (e.g., Monteith and Evans 2005) and international (e.g., Straškrabová et al. 1999; MOLAR Water Chemistry Group 1999; Marchetto and Rogora 2004) research projects and monitoring networks, it has become one of the most studied freshwater bodies in Europe.

   However, Lochnagar is more than just a focus of research for European freshwater scientists. Its location in an area of outstanding, although harsh, natural beauty has made it a favourite destination for hill-walkers and climbers for almost 200 years (McConnochie 1891) whilst the "gem of corrie and loch, buttress and crag, precipice and plain, towering like a proud giant over lovely Deeside" (Skakle 1934) has also made it irresistible to artists and poets for almost as long. It is also a rare UK habitat for many alpine species of birds, animals and plants attracting both domestic and migrant bird-watchers and botanists. In recognition of its importance from these various points-of-view, Lochnagar lies within the Lochnagar and Loch Muick Nature Reserve (on the Balmoral Estate) and within the boundary of the Cairngorm National Park although outwith the Cairngorms Special Area of Conservation (SAC). The Lochnagar area is also designated a UK 'Important Bird Area' (IBA) as a breeding site for the Eurasian

---

[1] Aware of the consternation that the term 'lake' causes to Scottish friends and colleagues, the word 'loch' is used in its place throughout this volume where the waterbody at Lochnagar is specifically referred to. Elsewhere, where lakes are referred to more generally, the English term is retained.

Figure 1. Lochnagar. A "gem of corrie and loch" (Skakle 1934) in summer and winter plumage. Photographs: Neil Rose.

Dotterel (*Eudromias morinellus*) (Birdlife International 2005) and part of the Deeside and Lochnagar National Scenic Area (NSA).

The aim of this book is to bring together the knowledge gained over 25 years of multi-disciplinary scientific study to show how the loch has changed and developed both naturally and as a result of human impact. In particular there is an emphasis on the role that global climate change may play on this fragile ecosystem both directly (i.e., increasing air temperatures and changing precipitation patterns) and indirectly (e.g., loss of ice cover, biological changes, enhanced distribution and release of toxic compounds, etc.). However, while Lochnagar is well-known amongst the freshwater scientific community for this research, the name is more famous to the general public for its connections to the Royal Estate of Balmoral, to students of literature for the writings of Byron, and to whisky lovers for the Royal Lochnagar single malt. A brief introduction from some of these perspectives is therefore required.

Location and site description

Lochnagar (56° 57' 29" N; 3° 13' 5" W) lies in a small mountainous area of northeast Scotland to the southeast of the Cairngorm Mountains from which it is separated by the River Dee valley (Figure 2). The area forms part of the Grampian Mountains, a series of ranges lying north of the Highland Boundary Fault and south of Strathspey. Aberdeen, the nearest city, is 50 km to the east whilst Braemar and Ballater, the nearest small towns in the Dee valley (or 'Deeside') lie 12 km to the northwest and 15 km to the northeast respectively. Although the name Lochnagar is often used to refer to the mountain, it rightly only belongs to the loch below (Alexander 1928; Watson 1992) and the summits around the loch each have their own individual names (Table 1).

Table 1. The summits of the Lochnagar massif, their heights (metres above sea level) and translations of the names. References refer to the source of the meaning.

Summit name	Altitude	Meaning of the name	Reference
Cac Càrn Beag	1155	Little cairn of excrement	Watson 1975
Cac Càrn Mòr	1150	Big cairn of excrement	Watson 1975
Càrn a' Choire Bhoidheach	1110	Cairn of the beautiful corrie	Cook et al. 1998
Cuidhe Cròm	1082	Crooked snow wreath	Alexander 1946
Càrn an t-Sagairt Mòr	1047	Big cairn of the priest	Alexander 1946
Càrn an t-Sagairt Beag	1044	Little cairn of the priest	Alexander 1946
Meikle Pap† *(Cioch mhor)*	980	Large breast	Alexander 1946
Creag a' Ghlas-uillt	975	Crag of the grey burn	Cook et al. 1998
Meall Coire na Saobhaidhe	973	Corrie of the foxes den	Alexander 1946
Little Pap† *(Cioch beag)*	952	Little breast	Alexander 1946

(† Translated from the Gaelic term in italics: Alexander 1968).

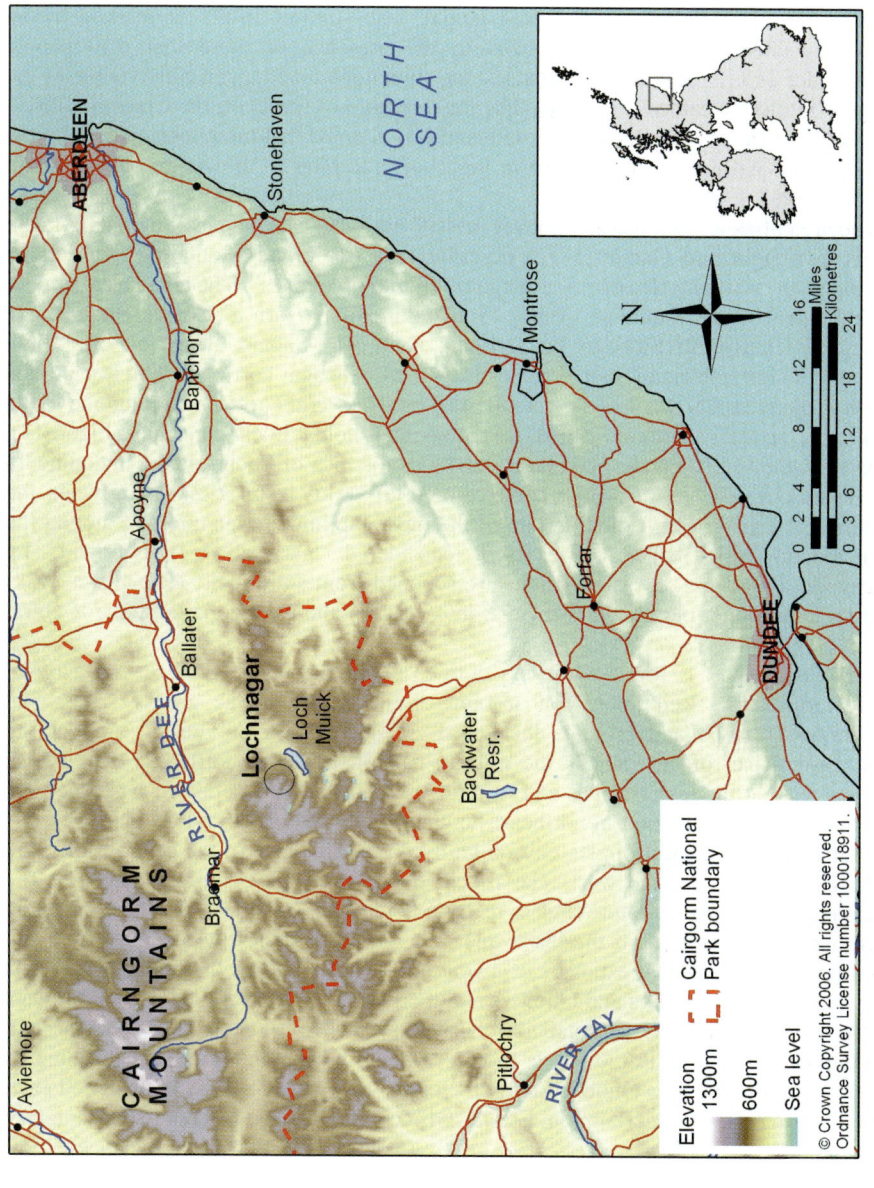

Figure 2. Location map of Lochnagar.

The highest point is Cac Càrn Beag which reaches 1155 m above sea level (a.s.l.) just to the west of the main corrie (Figure 3) making it Scotland's 20th highest mountain (Alexander 1928). The first record of an ascent is from 1810 (McConnochie 1891) made by the Rev. George Keith, a keen surveyor, who measured the height at 1158 m, remarkably close to today's accepted height. The summit itself is marked by an indicator, unveiled on July 12th 1924 in the presence of more than 140 people (Alexander 1924), and shows distances and directions to the main hills visible at the summit in clear weather (Figure 4). The most remote of these are the Cheviot Hills in England 108 miles (174 km) to the south and it is claimed that the Lochnagar summit is visible from Arthur's Seat, the 250 m high volcanic plug in the centre of Edinburgh (Parker 1923).

The area often termed the 'Lochnagar massif' has been called a small mountain range in its own right and Geikie (1887) describes it as a "broad undulating moorland... ending in the north at the edge of a range of granite precipices". The area has a number of individual named summits (Table 1) including four (Càrn a' Choire Bhoidheach; Creag a' Ghlas-uillt; Stob an Dubh Loch; Càrn an t-Sagairt Beag) on the White Mounth, a high 'tableland' that stretches from the Lochnagar corrie to Dubh Loch to the south (Figure 3). The Lochnagar 'massif' also holds several corries including one to the south of the main summit called the Stuic that holds Loch nan Eùn (Figure 3). However, the largest corrie, and that which holds the loch of Lochnagar, is undoubtedly the feature which has made the name "immortal" (Hutchison 1927) mainly via the writings of Byron, who called it "one of the most sublime and picturesque among the 'Caledonian Alps'". The corrie itself, considered "unequalled in Scotland" by Alexander (1968), is "a vast amphitheatre scooped from solid rock" (Hutchison 1927) and its 2 km upper perimeter, is divided into western and eastern parts by the 'Central Buttress' (Figure 3).

The eastern part is lower, with less defined cliffs, the main feature being a broad gully termed the 'Red Spout' on account of the colour of the granite screes that dominate its slope and produce the coarse sand found in the littoral area of the southeastern part of the loch. In the western part of the corrie, cliffs rise over 200 m from screes on the lower slopes and it is for these cliffs that Lochnagar has become a major climbing destination in northeast Scotland. The two most obvious gullies in this western section are the Douglas-Gibson Gully on the east side of the central cliffs and 'The Black Spout' to the west. The Black Spout is the first name to appear in the Lochnagar climbing literature and is really a scree shoot, known since the 1830s by gem hunters as a source of the black quartz 'Cairngorm stones' (Alexander 1950) on account of the occasional exposure of a vein in the gully. Today, the Lochnagar area is very popular with walkers and climbers and there are over 60 recorded climbing routes (Watson 1992) and these are detailed in Nisbet (1995). However, as might be expected, the number of visitors to the site is highly seasonal with monthly figures reaching 3-4000 in summer months (June – August) while c.500 visitors have been recorded each month during the winter (Balmoral Estates Office data; Rose and Battarbee: Chapter 19 this volume).

Both mountain and loch form part of the 18660 hectare Balmoral Estate, situated within the Balmoral Forest, and now also within the boundary of the Cairngorms National Park. This National Park was established in 2003 and at 3800 km^2 is Britain's

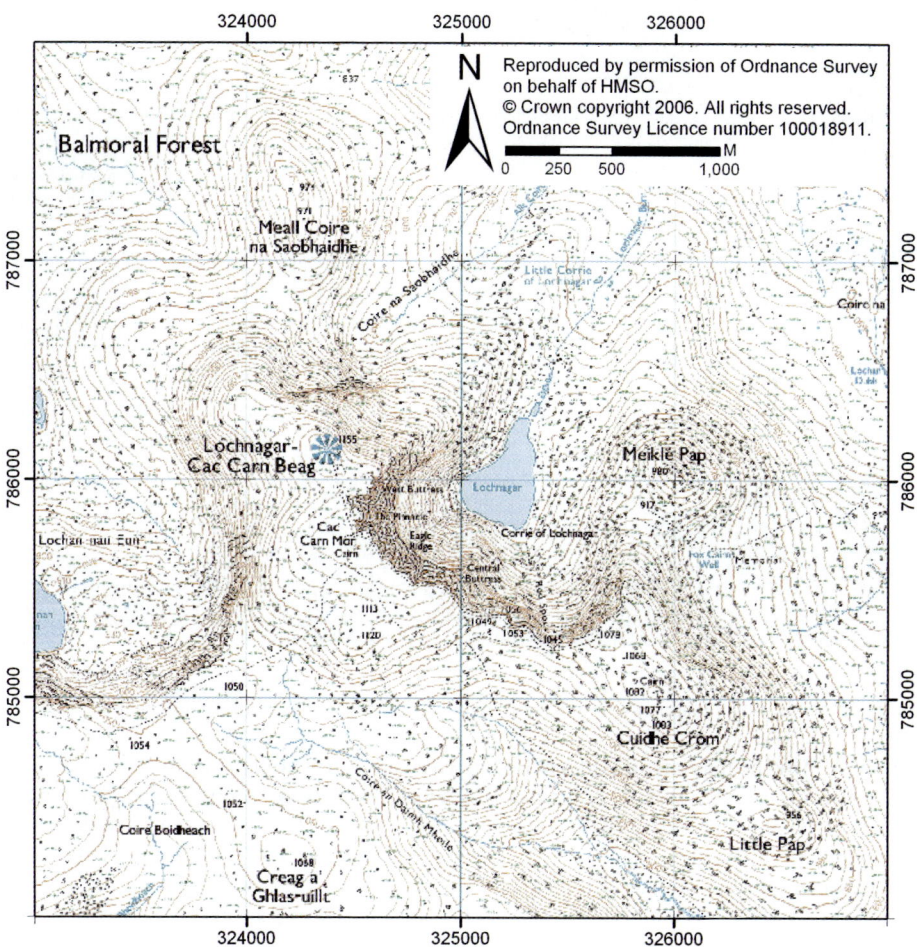

Figure 3. Lochnagar and surrounding area (Digimap 1: 25000 raster). Reproduced by permission of Ordnance Survey on behalf of HMSO.

Figure 4. The Lochnagar summit indicator unveiled in 1924. Photograph by Michael Hughes

largest. The heather moorland on Lochnagar and surrounding area supports healthy populations of red grouse (*Lagopus lagopus*), ptarmigan (*Lagopus mutus*), snow bunting (*Plectrophenax nivalis*) and mountain hares (*Lepus timidus*) whilst other montane bird species such as the dotterel (*Chardrius morinellus*), for which the Lochnagar area is one of Britain's main nesting sites, and raptors including golden eagles (*Aquila chrysaetos*) and ospreys (*Pandion haliaetus*) are also represented. The Lochnagar plateau has long been associated with an importance for birds as the names of lochs and lochans to the west of the summit show, i.e., Loch nan Eùn ("Loch of the birds"), Lochan na Feadaige ("Little loch of the plovers") and Lochan an Tarmachain ("Little loch of the ptarmigan") (Alexander 1946), and in 1998 the Lochnagar plateau was designated a Special Protection Area under the European Birds Directive (79/409/EEC). The importance of the upland heath itself is also recognised and is a UK Priority Habitat based on the habitat classification of the Cairngorms Local Biodiversity Action Plan. The catchment and more extensive massif also support a surprisingly high number of nationally rare and scarce species of terrestrial flowering plants and bryophytes (e.g., *Cicerbita alpina*; *Gnaphalium norvegicum*) (Chapter 7: Birks this volume) many of which are associated with long-lasting snow-beds. Other plants present at the site are 'Red Data Book' species and have 'threatened' or 'vulnerable' status, while species of both fauna and flora in the area are considered UK Priority Species and UK Species of Conservation Concern by the UK Biodiversity Group.

Inflow streams to the loch are generally small and almost without exception originate from the steeper parts of the catchment and enter the loch at its southern end. Although a few flow all year round, many of these streams are ephemeral and only after periods of rain do they flow under and between the boulder fields and screes which dominate large areas of the catchment. After intense rainfall, water often flows across the surface of the moss and grass areas in the southeast of the catchment directly into the loch, or down gullies between eroded peat hags especially along the northeastern shore. All the lochs and streams in the Lochnagar area, including Loch Muick at the head of Glen Muick, ultimately drain into the River Dee. Lochnagar, itself drains to the north via a series of outflow pools and then via the Lochnagar and Gelder Burns to the south side of the River Dee where they flow eastwards. The waters lost from Lochnagar therefore ultimately enter the North Sea, with the River Dee, at Aberdeen (Figure 2).

A brief history

There is a long history of human settlement in Deeside. During the Roman occupation of Britain, the area was inhabited by Pictish tribes and, in the Braemar area, particularly by the Vacomagi. Roman activity in the area is recorded in 81 AD and 138 AD, but following the Roman withdrawal there is little mention until 700 – 900 AD when the Picts were converted to Christianity (Franklin 2003). Lands including Lochnagar came under the rule of the Mormaor of Mar and later the Earldom of Mar (Burke's Peerage and Gentry 2004) before the Bisset family became landlords of the Glen Muick area as the Crown Vassels to the Scottish Kings in the 13th century (Cook et al. 1998). The lands then passed through various ownerships including that of Sir William Keith, Great Marshall of Scotland, in 1351, the Earl of Huntly and Aboyne and eventually to Sir James Mackenzie who bought the land as a sporting estate in 1863. These, and the area around the White Mounth, were added to the Balmoral Estate between 1947 and 1951 by King George VI (Cook et al. 1998).

Lochnagar first appears on a map, as Loch Garr, on Robert Gordon of Straloch's map of 1640 (Figure 5) and is clearly shown in the hills above Loch Muick outflowing via the "Yelder" (today's Gelder Burn) to join the River Dee. By 1826, the name had changed to Loch-na-gar and had become associated with the name of the mountain (Figure 6). There have been a remarkable number of variations on the name over the centuries and this has led to a great deal of confusion as to its meaning. Watson and Allan (1984) provide an excellent summary of the various names and their origins which, apart from those mentioned above, include Lochan na Gàire, Lochan a' ghair, Lochan-a-gharr, Lochan gharr, Lochan ghàir (and Lochĕn i gyàr in Gaelic) for the loch and Laghin y gair and Lochnagarbh for the mountain. The original name for the hill was Beinn nan Ciochan (Watson and Allan 1984) and this name explains the location of Binchichins on Gordon's 1640 map. The origins of this name are obscure although Michie (1896) suggests that it is an attempt at an English pronunciation of Beinn-Cichean or Cichin or "mountain with the paps". Although Gordon's map places this to the south of Loch Muick rather than to the west, the names of two summits to the east

Figure 5. Section of Robert Gordon of Straloch's map of 1640, showing Lochnagar as Loch Garr. Reproduced by permission of the Trustees of the National Library of Scotland.

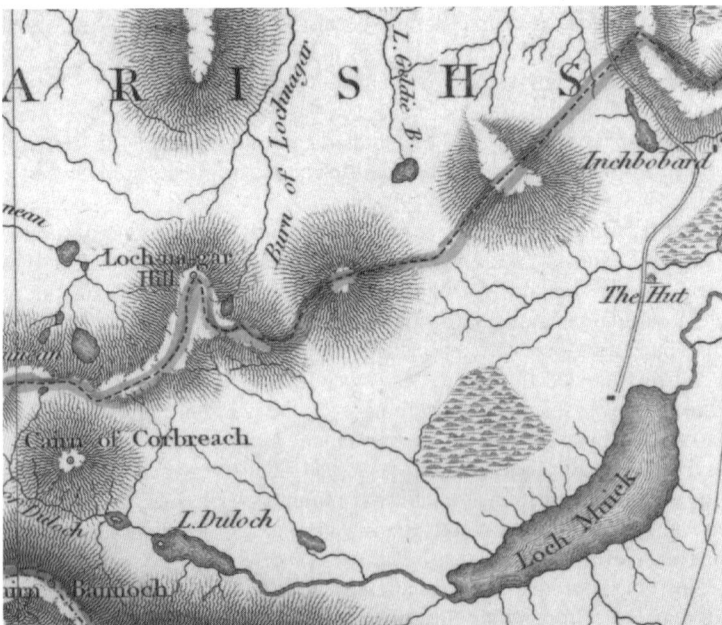

Figure 6. Section from John Thomson's 1826 map of "Northern part of Aberdeen and Banff Shires". Reproduced by permission of the Trustees of the National Library of Scotland.

and southeast of Lochnagar, 'Meikle Pap' and 'Little Pap' (Figure 3), retain a connection to this original name.

A debate on the meaning of the name Lochnagar appeared in the Cairngorm Club Journal in 1913 (Anon 1913) where the Hon. Ruaidhri Erskine proposed that Loch-na-gearra referred to the 'loch of the hare', Mr James MacDonald was "disposed to think the most plausible suggestion yet offered is that the root may be 'gair' or 'gaoir', 'wailing, moaning, shouting, confused noise' applying to the wild howling of the wind on the face of the crags", while Dr John Milne thought it a corruption of "Lochan Gearr" meaning 'short loch'. McConnochie (1891) and Watson and Allan (1984) both prefer the 'noisy' derivation suggesting "the loch of sobbing and wailing" and "the tarn of the noisy sound" or "the noisy little loch" respectively, while Cook et al. (1998) in the Balmoral Estate Guide to the area suggest "the loch of the goats or the rough loch". Today, although the original meaning may be lost, it is generally accepted that the name of Lochnagar rightly refers to the loch alone and individual names should be used for the summits of the mountain above. However, despite this, the name is still widely used for both loch and mountain probably as a result of its popularisation by Byron in the early 19th century.

The Royal loch

Queen Victoria bought the lease for the Balmoral Estate in 1848 and it was purchased for her by Prince Albert four years later. The building of the new castle began in 1852 and was completed in 1856 using granite taken from Glen Gelder, through which the Lochnagar outflow passes on its way to the River Dee. The Queen is said to have had a great affection for the area and wrote extensively about it in her *"Leaves from the Journal of our Life in the Highlands"*. Her first visit to Lochnagar was on September 16th 1848, but the cold and misty start turned to thick fog as they ascended "so as to hide everything not within one hundred yards" and leading the Queen to describe it as "cold, and wet, and cheerless" (1868). However, the Royal party were regular visitors to Lochnagar and in *'Leaves'* for September 6th 1850, the Queen described it as "the jewel of all mountains here" while Prince Albert is said to have likened the corrie to the crater of Vesuvius.

The Queen's Diamond Jubilee in 1897 was celebrated on the summit by members of the Cairngorm Club with a bonfire and fireworks, a "splendid show that had never been seen on the top of Lochnagar before" (McHardy 1897) and her death, four years later, was felt keenly on Deeside. William Allan, Member of Parliament for Gateshead, wrote the following:

> " *Frae Aberdeen to Lochnagar,*
> *Frae glens an' corries roon' Braemar,*
> *This cry o' grief is heard afar –*
> *'Oor Queen will come nae mair "*

Curiously, and rather reminiscent of today's press reportage, prominence was given to a "well faked up" (Anon 1901) story concerning Lochnagar at this time. It was stated that "There is a tradition which avers that when snow appears on a certain part of

Lochnagar the laird of Balmoral dies", and that "No snow has been seen on this part of the mountain since the late Queen became land-lord until the present year". Unfortunately, the part of the mountain for which this was claimed was never actually identified nor, for those interested in climate change impacts, the regular snow cover observations that must surely have led to the establishment of this 'legend'! Today, while winter snow cover is certainly less permanent, Balmoral Castle remains the Scottish home of the Royal Family, and Lochnagar the subject of watercolour paintings by HRH Prince Charles as well as his children's book "The Old Man of Lochnagar".

From a scientific point-of-view, the Royal connection to Lochnagar has been greatly beneficial. First, and most importantly, the facilitation of research at the site, by the Balmoral Estates Office, has been such that without it the work contained within this book could not have been achieved. The 'protection' that this affords equipment left *in situ* to allow high resolution monitoring to take place (e.g., the automatic weather station) should also not be taken for granted, as we know to our cost from experiences at other monitoring locations across the UK. Second, there is an increased visibility for scientific research. Figure 7 shows an article produced by the Scottish edition of a UK national newspaper on the publication of the 10-year report of the United Kingdom Acid Waters Monitoring Network (UK AWMN) (Monteith and Evans 2000) while Figure 8 appeared on the BBC website following the pre-publication of an article on flame retardants in European mountain lakes (Vives et al. 2004) of which Lochnagar was one. Although these raise the profile of the work undertaken at the site, the emphasis on the Royal connection causes frustration and annoyance to both the Estate and the scientists involved.

The study on flame retardants is probably the best example of this. The analysis of PBDEs in fish from Lochnagar was part of the EU-funded EMERGE project (Chapter 12: Rosseland et al. this volume) and while it is true that the PBDE concentrations in the fish from Lochnagar were the highest in the dataset, there is no mention of the fact that there were no other UK data with which to compare the Lochnagar results. Thus, the important point, that high concentrations were found in a remote upland water body in the UK, with the 'obvious' follow-up question, 'what does this mean for freshwater fish elsewhere in the UK?' is lost, and the emphasis is placed instead on the fact that the site is located on the Balmoral Estate, as if this were the reason for the high PBDE content.

"Steep frowning glories": The Lochnagar of artists and poets

Attempts to describe the "majesty and grandeur" of the Lochnagar corrie (Symmers 1931), even in articles outlining the practicalities of a new climbing route, seem to bring out the poet in people. For example Bain (1947) in describing his first visit to the site wrote "Here was a hill to fill the imagination – here was a name to dream of and to conjure up pictures of towering precipices, black crevice and dark, wet rock, dully, malevolently glistening", whilst Skea (1894) likened the view of the corrie from Meikle Pap to a scene from Dante's "Inferno" and Davidson (1922) imagined grinning gargoyles on the corrie walls. However, it is undoubtedly the words of George (later Lord) Byron (1788 – 1824) in his famous poem that for many people epitomise the

Figure 7. Article that appeared in the Scottish edition of News of the World on June 25th 2000 following the publication of the UK AWMN 10 year report. Reproduced with permission from News of the World.

Figure 8. How the pre-publication of the Vives et al. (2004) paper was reported on the BBC News website. The 'story' was treated in a similar fashion in many national newspapers. Image from BBC News at bbcnews.co.uk.

scenery, especially the final verse:

> *Years have roll'd on, Lochnagar, since I left you!*
> *Years must elapse ere I tread you again.*
> *Though nature of verdure and flow'rs has bereft you,*
> *Yet still are you dearer than Albion's plain.*
> *England, thy beauties are tame and domestic*
> *To one who has roamed over mountains afar*
> *Oh! for the crags that are wild and majestic,*
> *The steep frowning glories of dark Lochnagar.*

As James Brown, writing in 1866 in "The New Deeside Guide" (quoted from Smith 1947) put it, "Opposite is the great mountain of Lochnagar, a most renowned and celebrated mountain, not so much for its own sake as for a poem Lord Byron made upon it." The names of Byron and Lochnagar are thus inextricably linked and an internet search for a combination of the two terms results in almost 1200 websites. However, the scenery that Byron described as "wild and majestic" has also proved an inspiration for other poets such as MacGillivray (1915):

> *From mountain brow and where, in corries gaunt,*
> *Dark waters lie asleep; the while we dare*
> *A doubtful path by half-lost cairns to where*
> *Blue Lochnagar's proud peak leaps like a taunt.*

and also artists such as James Giles. Giles produced a series of watercolours around the Balmoral Estate including Lochnagar between 1848 and 1850. Figure 9 shows a reproduction of his "Lochnagar, the loch in the corries of the Mountain" from November 1849 and this view from the outflow looking south to the corrie wall is immediately recognisable to anyone visiting the loch today. A photograph taken in 2002 (Figure 10) shows the scene has changed little in more than 150 years, and the similarity between this recent photograph and an etching by Archibald Geikie (1887) (Figure 11) as well as the first photograph taken of Lochnagar by George Washington Wilson in September 1855 (see p65 in Clark 1981) shows a remarkable lack of change. It is exactly this stability, over a time-scale of centuries that made the loch of interest in studies on the causes of lake acidification, and continues to make it of such great value in ongoing environmental change research.

The scientific 'snowball effect'

The earliest reported studies of surface water acidification in the Cairngorm region were those of Gorham (1957) who considered the influence of environmental factors on the water quality of high lochs in the area. Many were found to be acidic (pH <5.5) and this was thought to be due to slow weathering and acidic precipitation. In the 1970s the Cairngorms were "rediscovered" as an acid sensitive region (Harriman and Morrison 1989) and the Review Group on Acid Rain (RGAR) began monitoring rainfall at a site they called "Lochnagar" in November 1977. This site, in use for over ten years (Cape

Figure 9. "Lochnagar, the loch in the corries of the mountain". Watercolour by James Giles from 1849. Reproduced by kind permission of The Royal Collection © 2006, Her Majesty Queen Elizabeth II.

Figure 10. Lochnagar from the outflow looking south. 2002. Photograph: Neil Rose.

Figure 11. Etching of Lochnagar from the outflow, looking south, c. 1880 by Sir Archibald
Geikie. Reproduced with permission from Geikie (1887).

and Paterson 1989; RGAR 1990), was located near the Allt-na-giubhsaich on the
Lochnagar track, but 5 km from the loch and at an altitude c. 320 m below it. However,
between 1977 and 1987, a clear decline in sulphate in deposition was observed there.
Water samples from the outflow at Lochnagar itself were taken in 1980 (June and
December), by the Freshwater Fisheries Laboratory in Faskally (now Fisheries
Research Services) as part of an assessment into the extent of surface water
acidification in the area, but further sampling was sporadic and only one more water
sample was taken, in 1983, before regular monitoring began in 1988.

   The first sediment cores were taken from Lochnagar in 1986 as part of the Surface
Water Acidification Project (SWAP). One core was radiometrically dated and the pH
reconstructed from the fossil diatom assemblage (Jones et al. 1993; Chapter 14:
Monteith et al. this volume). The resulting inferred acidification, with supporting
evidence of atmospherically deposited contamination from spheroidal carbonaceous
particles (SCPs), magnetic susceptibility and trace metals data from the same core
(Jones et al. 1993), showed that there had been considerable impact at the site from acid
deposition since the mid-19th century. Similar results were found at other sites in
the region. As a consequence of this work, Lochnagar became one of the 22 sites
included within the UK Acid Waters Monitoring Network (UK AWMN; see
http://www.ukawmn.ucl.ac.uk/) at its instigation in 1988. This major initiative, funded
by DEFRA (then the Department of the Environment), included lakes and streams
across the UK, although Lochnagar was the only standing water included from the
northeast of Scotland and the only mountain lake. Inclusion within the UK AWMN

meant that water samples from the outflow and epilithic (rock-dwelling) diatom samples were taken every three months and surveys of aquatic macrophytes, invertebrates and fish (in the outflow stream) were undertaken annually. This monitoring has continued to the present and the UK AWMN has become one of the longest running integrated chemical and biological monitoring networks in Europe. In 1990 annual deep water sediment trapping was included within the UK AWMN in order to supplement the initial coring work and also to provide, for the first time, a means of monitoring atmospheric pollutant deposition in the form of SCPs. Although this trapping programme has occasionally been interrupted as a result of ice on the loch dragging the traps, on the whole it has been very successful and continues to the present. The UK AWMN chemical and biological monitoring has led to the inclusion of Lochnagar in a number of other projects and networks. The loch featured in DEFRA funded projects on critical loads of acidification and recovery 'Critical Loads of Acidity and Metals' (CLAM; CLAM2; Freshwater Umbrella see http://www.freshwaters.org.uk/ research/) and is also now included in the Environmental Change Network (ECN) and Acid Deposition Network while data feed into the UNECE International Co-operative Programme on Assessment and Monitoring of Acidification of Rivers and Lakes.

Interest in the sediment record of atmospherically deposited contaminants as supporting data for acidification led to Lochnagar being included in the EU funded research project AL:PE (Acidification of mountain lakes: Palaeolimnology and Ecology; 1990 – 1993), which also included mountain lakes in Norway, Italy and France. This project was expanded to include mountain lakes in many other European countries in the follow-up projects AL:PE 2 (1993-1996) and MOLAR (Measuring and modelling the dynamic response of remote mountain lake ecosystems to environmental change. A programme of mountain lake research; 1996 – 1999). Lochnagar was not included in AL:PE 2, but its inclusion in MOLAR increased the scale of monitoring at the site enormously. This involved, in 1996, the installation of an automatic weather station in the catchment and thermistors to monitor lake water temperature. The monitoring of bulk deposition and lake waters for major ions, trace metals and SCPs was also initiated. In 1997, the monitoring of trace metals in deposition was expanded to include Hg on a monthly basis. This monitoring, due to end after 18 months of the MOLAR project, was able to continue with funding from DEFRA under the CLAM and CLAM2 projects and in 2001 methyl Hg was also introduced. This now continues as part of the UK AWMN. The trace metals work was expanded still further during the PhD studies of Handong Yang (Yang 2000), where trace metals in many ecological compartments were assessed, including terrestrial and aquatic plants as well as a multi-core survey of the sediment basin and catchment soils. Under CLAM, this biological sampling and analysis became part of the metals monitoring programme and macroinvertebrates and zooplankton have since been added.

After MOLAR the final EU funded mountain lake project EMERGE (European mountain lake ecosystems: Regionalisation, diagnostics and socio-economic evaluation; 2000 – 2003) began to shift emphasis towards climate issues and this has been taken further in Euro-limpacs (2004 – 2009) which is concerned with the direct and indirect effects of climate on the whole range of European freshwaters. With its long-term datasets, Lochnagar was a key site in both of these projects and, while monitoring of other parameters has continued, some additional monitoring has also been started. For example, the fortnightly collection of cloud water for major ions and SCPs started in 2001, in order to provide additional information on the input of atmospheric pollutants at the site by orographic cloud (Chapter 17: Rose and Yang this volume) and more recently, in 2004, the measurement of persistent organic pollutants (POPs) in fortnightly deposition samples was started as part of Euro-limpacs.

Climate research now dominates environmental change studies, and while fortnightly observations have been made since 1996, snow and ice cover at Lochnagar has been monitored daily since 2002 using an automatic camera (see http://ecrc.geog.ucl.ac.uk/ lochnagar/digcam/) The EU funded CHILL-10,000 project also aimed to identify a climate signal, over a far longer time-scale, by using a range of biological and sedimentological proxies from a c. 8000 year sediment core (Dalton et al. 2005). More recently, the monitoring of stable isotopes (O and H) in bulk deposition, lake waters and epilithic diatoms is in progress in order to try and explore ways in which these can be used to reconstruct climate from the lake sediment record (Jonathan Tyler, University College London pers. commun.)

Indirect effects of climate such as potential impacts on the recovery from acidification and the enhanced release of toxic pollutants from catchment soils are also major issues at Lochnagar and the ongoing monitoring and research programmes reflect this. The development of research at Lochnagar over 25 years has therefore resulted in a wealth of data from all areas of the loch and catchment. Furthermore, most of the monitoring programmes have been continued beyond the length of their original research projects and this has provided a unique long-term data series for a number of parameters. The current Lochnagar monitoring programme is summarised in Table 2.

Scope of this book

With 25 years of research work already completed at Lochnagar and with a broad range of monitoring programmes established at the site, it is an appropriate time to bring together data, results and current thinking into a single volume. The aim of this book is to summarise this work so that an assessment can be made about how the contemporary chemical and biological status determined by monitoring compares with historical trends, ascertained by palaeolimnological work, and a modelled future over a range of time-scales (Figure 12). Only by using such a combination of scientific approaches can we determine whether the pollutant status is improving, whether the ecosystem is showing any recovery and ascertain the steps required to both enhance recovery from past impacts and protect from future threats such as global climate change.

Following two introductory chapters (Chapter 2 by Hughes deals with physical characteristics and the light regime at Lochnagar), the book is divided into four, broadly chronological, sections. The first deals with the longer time-scale to provide an historical context. Each chapter deals separately with a fundamental area that has shaped the loch and catchment and consequently influenced the work that has been undertaken there. In Chapter 3, Goodman discusses the formation and composition of the underlying geology and the characteristics that have led the site to be sensitive to, for example, acid deposition. In Chapter 4, Hall considers the geomorphology of the area and the generations of landforms from pre- to periglacial, including the development of the Lochnagar corrie itself, while in Chapter 5 Thompson et al.,

Table 2. Monitoring activities at Lochnagar in 2005. (* denotes current research programme; CEH = Centre for Ecology and Hydrology, Wallingford; NIGL = NERC Isotope Geosciences Laboratory, Keyworth; ECRC = Environmental Change Research Centre, London; FU = Freshwater Umbrella).

	Determinand(s)	Frequency	Start	Programme *
Automatic Weather Station	Air temperature; Wind speed; Wind direction; Air pressure; Relative humidity	30 mins	1996	UK AWMN
	Rainfall	1 day	1996	UK AWMN
	Solar radiation; Net radiation; Soil temperature	30 mins	2003	UK AWMN
Thermistor chain (7 depths)	Water temperature	2 hours	1996	UK AWMN
Outflow	Water flow	15 mins	1999	CEH/ECRC
Snow	% catchment cover	2 weeks	1996	Euro-limpacs
Ice	% loch cover (obs.)	2 weeks	1996	Euro-limpacs
	% loch cover (camera)	1 day	2002	Euro-limpacs
Bulk deposition	pH conductivity NH_4^+ Ca^{2+} Mg^{2+} Na^+ K^+ SO_4^{2-} NO_3^- Cl^-	2 weeks	1996	Acid Deposition Network
	Ni Zn Cu Cd Pb	2 weeks	1996	UK AWMN
	SCPs	2 weeks	1996	UK AWMN
	Hg	1 month	1997	UK AWMN
	MeHg	2 weeks	2001	UK AWMN
	Stable isotopes	2 weeks	2001	NIGL/PhD
	Persistent organic pollutants	2 weeks	2004	Euro-limpacs
Cloud water	pH conductivity NH_4^+ Ca^{2+} Mg^{2+} Na^+ K^+ SO_4^{2-} NO_3^- Cl^-	2 weeks	2001	ECRC
	SCPs	2 weeks	2001	ECRC
Lake water	pH conductivity NH_4^+ Ca^{2+} Mg^{2+} Na^+ K^+ SO_4^{2-} NO_3^- Cl^-	2 weeks	1996	FU / UK AWMN
	Ni Zn Cu Cd Pb	2 weeks	1996	UK AWMN
	Hg	1 month	1997	UK AWMN
	Stable isotopes	2 weeks	2001	NIGL/PhD
'Inflow' water	Stable isotopes	2 weeks	2001	NIGL/PhD
Sediment traps	Diatoms, sediment flux; SCPs;	1 year	1990	UK AWMN
	Hg Pb Cd Ni Cu Zn	1 year	1997	UK AWMN
Epilithic diatoms	Species composition	3 months	1988	UK AWMN
Aquatic invertebrates	Species composition	1 year	1988	UK AWMN
	Hg Pb Cd Ni Cu Zn	1 year	2001	UK AWMN
Aquatic macrophytes	Spatial distribution change	2 years	1988	UKAWMN
	Hg Pb Cd Ni Cu Zn	1 year	1997	UK AWMN
Fish (outflow stream)	Species; age; size class	1 year	1988	UK AWMN
Terrestrial plants	Hg Pb Cd Ni Cu Zn	1 year	1997	UK AWMN
Zooplankton	Hg Pb Cd Ni Cu Zn	1 year	2001	UK AWMN

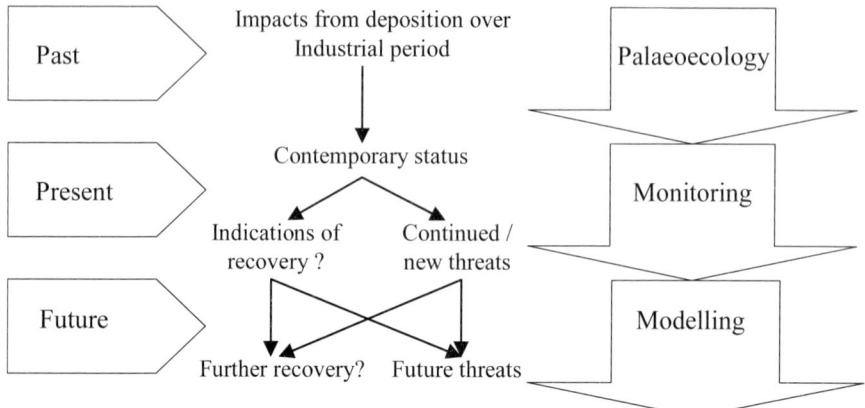

Figure 12. The roles of palaeoecology, monitoring and modelling in understanding historical impacts, contemporary biological and chemical status and the potential for future recovery or further environmental threats.

describe the various lines of evidence for the reconstruction of climate parameters, the sensitivity of mountain sites to changes in climate, and contemporary meteorology. Helliwell et al. discuss the formation, distribution, physico-chemical properties and sensitivity of the catchment soils in Chapter 6, and Birks considers the vegetational history of the site, the present flora and the possible effects of climate warming on alpine plants in Chapter 7.

The second Section deals with the contemporary physical and biological status of Lochnagar. In Chapter 8, Rose describes sediment distribution, accumulation and composition while Jenkins et al., in Chapter 9, discuss the hydrology and hydrochemistry and, in particular, seasonality and short-term events and how these may respond to a changing climate. The next three chapters are concerned with the contemporary biology of the loch itself. So, Flower et al. discuss the aquatic flora from micro-phytobenthos to macrophytes in Chapter 10, Woodward et al. consider the structure and dynamics of the invertebrate community and its status within the broader aquatic ecosystem in Chapter 11, and Rosseland et al. discuss the population and contamination of Lochnagar's brown trout in Chapter 12.

In Section 3 pollution issues are covered, from historical contamination, contemporary loading and future recovery perspectives. In Chapter 13 Rose et al., describe the evidence for the various sources of contamination at Lochnagar, from Scotland, through UK and Europe to global sources. They also consider the various levels of environmental protection legislation. Chapters 14 to 17 then each deal with a separate contamination issue. In Chapter 14, Monteith et al. discuss the acidification of Lochnagar, its history, current status and the prospects for recovery, particularly with respect to possible confounding factors. Chapter 15 is concerned with trace metals and Tipping et al. describe data from all ecological compartments and the modelling of the movement of trace metals from catchment to loch. In Chapter 16, Muir et al. discuss the historical profiles of persistent organic pollutants (POPs), evidence for biodegradation and the status of Lochnagar as an indicator loch for new pollutants, given the potential for increased distribution of semi-volatile POPs under a changing climate, while in Chapter 17, Rose and Yang describe the record of SCPs over various time-scales, from the historical sediment record to atmospheric inputs determined from monitoring bulk deposition.

Finally, Section 4 is concerned with future changes. In Chapter 18, Kettle and Thompson describe the modelling of future climate with respect to Lochnagar and the best current predictions of air and water temperatures, rainfall, storminess, etc. and how these will impact on snow and ice cover of the loch and its catchment. Lastly, in the final chapter of the book, Rose and Battarbee consider the future for Lochnagar, how the various stressors will change as a result of interactions with a changing climate, and the impacts that these will have on both physical aspects and the terrestrial and aquatic biota.

And finally…

A great deal of information has been gathered, regarding Lochnagar, over the last quarter of a century, but perhaps the greatest threat to the mountain loch and its ecology is still to come. Although emissions reductions have caused widespread improvement to the European environment since the 1970s, at present we can only speculate on the extent and rate of direct and indirect impacts that will result from global climate change. At Lochnagar, there is a wealth of background data providing a baseline against which to assess further change and the monitoring tools are now in place with which to make that assessment. The value of monitoring lies in its longevity as trends and patterns can only be accurately determined with a sufficient time-series, while new scientific questions undoubtedly still wait to be answered at remote, sensitive mountain lakes.

There is, therefore, plenty of work still to be done at Lochnagar and this volume should perhaps best be seen as a 'progress report' rather than a final summary. Indeed, Bain was quite correct when, in 1947, he wrote that Lochnagar had "an air of mystery, a suggestion of new and great discoveries which was infinitely compelling". But, I will leave the last word to another Rose who could as well have been thinking about scientists as mountaineers when he wrote in 1895, "The student of the forces of nature, the lover of mountain scenery, and the climber will ever find something new to wonder at, to admire, or to conquer in these grim precipices; and when the ever-changing conditions of weather are taken into account, he will be a very easily satisfied mountaineer and scarcely worthy of the name who can fancy that he has exhausted the attractions of Lochnagar" (Rose 1895)

Acknowledgements

I would like to thank Mike Hughes for producing the maps used in this Chapter and to Rick Battarbee and John Smol for comments on earlier versions of the manuscript.

Ongoing research at Lochnagar would not be possible without the support of the Balmoral Estates Office and, in particular, Mr Peter Ord, the Factor. I would also like to thank Jo Porter for his many visits to Lochnagar in support of our monitoring programmes and to members of the Environmental Change Research Centre, University College London for their help in the field.

Financial support for this research has principally come from the EU via the following projects:

- AL:PE (Acidification of mountain lakes: Palaeolimnology and Ecology. Contract No. STEP-CT-90-0079);
- MOLAR (Measuring and modelling the dynamic response of remote mountain lake ecosystems to environmental change. A programme of mountain lake research. EU contract ENV4-CT95-0007);
- EMERGE (European mountain lake ecosystems: Regionalisation, diagnostics and socio-economic evaluation. EU Contract No. EVK1-CT-1999-00032);
- CHILL-10,000 (Climate History as recorded by ecologically sensitive Arctic and Alpine lakes in Europe during the last 10,000 years: A multi-proxy approach. EU Contract No. ENV4-CT97-0642);
- Euro-limpacs (Integrated Project to Evaluate the Impacts of Global Change on European Freshwater Ecosystems. EU project 505540);

and the UK's Department for Environment, Food and Rural Affairs via:

- CLAM (Critical Loads of Acidity and Metals. Contract No. EPG 1/3/117);
- CLAM2 (Recovery of Acidified Waters in the UK. Contract No. EPG/1/3/183);
- UK Acid Waters Monitoring Network (Contract No. EPG 1/3/160).
- Freshwater Umbrella (Contract No. CPEA17).

References

Alexander H. 1924. The Lochnagar indicator: It's building and unveiling. Cairngorm Club J. 11: 53–67.

Alexander H. 1928. The Cairngorms. The Scottish Mountaineering Club Guide. 1st edition. Scottish Mountaineering Club. Edinburgh. 218 pp.

Alexander H. 1950. The Cairngorms. The Scottish Mountaineering Club Guide. 3rd edition. Scottish Mountaineering Club. Edinburgh. 298 pp.

Alexander H. 1968. The Cairngorms. The Scottish Mountaineering Club District Guide Books. 4th edition. Scottish Mountaineering Trust. Edinburgh. 251 pp.

Alexander W.M. 1946. The place-names on and around Lochnagar. Cairngorm Club J. 16: 60–64.

Anon. 1901. Queen Victoria and Lochnagar. Cairngorm Club J. 3: 310–311.

Anon. 1913. The meaning of "Lochnagar". Cairngorm Club J. 7: 242–243.

Bain R. 1947. Lochnagar – A first visit. Cairngorm Club J. 15: 112–114.

Battarbee R.W., Patrick S.T., Kernan M., Psenner R., Thies H., Grimalt J., Rosseland B.O., Wathne B., Catalan J., Mosello R., Lami A., Livingstone D., Stuchlik E., Straskrabova V. and Raddum G. 2005. In: Huber U., Bugmann H.K.M. and Reasoner M.A. (eds.) Global change and mountain regions. An overview of current knowledge. Springer. Dordrecht. pp 113–121.

Beniston M. 2000. Environmental change in mountains and uplands. Key issues in Environmental Change. Arnold. London. 172 pp.

Bindler R., Olofsson C., Renberg I. and Frech W. 2001.Temporal trends in mercury accumulation in lake sediments in Sweden. Wat. Air Soil Pollut.: Focus 1: 343–355.

Birdlife International. 2005. Birdlife's online World Bird Database: The site for bird conservation. Version 2.0. Cambridge, UK. Birdlife International. Available at: http://birdlife.org (accessed 27th February 2006).

Birks H.J.B. 1988. Long-term ecological change in the British uplands. In: Usher M.B. and Thompson D.B.A. (eds.). Ecological Change in the Uplands. Blackwell Scientific Publications. Special Publication No. 7. British Ecological Society. pp. 37–56.

Birks H.J.B. This volume. Chapter 7. Flora and vegetation of Lochnagar – past, present and future. In: Rose N.L. (ed.) 2007. Lochnagar: The natural history of a mountain lake. Springer. Dordrecht.

Burke's Peerage and Gentry. 2004. The ancient Earldom of Mar. Extract from 'The great historic families of Scotland' by James Taylor, published in 1887. http://www.burkes-peerage.net/sites/scotland/esnews/es0302.asp. Accessed 11th October 2004.

Cape J.N. and Paterson I.S. 1989. Ten years of rainfall chemistry measurements at Lochnagar. In: Environmental Research on the Balmoral Estate. Aberdeen Centre for Land Use. Aberdeen. 35 pp.

Clark R.W. 1981. Balmoral. Queen Victoria's highland home. Thames and Hudson Ltd. London. 114 pp.

Cook N., Freeman P. and Bird C. 1998. Loch Muick and Lochnagar. Balmoral Estate. 28 pp.

Curtis C.J., Barbieri A. Camarero L., Gabathuler M., Galas J., Hanselmann K., Kopáček J., Mosello R., Nickus U., Rose N.L., Stuchlik E., Thies H., Ventura M. and Wright R. 2002. Application of static critical load models for acidity to high mountain lakes in Europe. Wat. Air Soil Pollut.: Focus 2: 115–126.

Dalton C., Birks H.J.B., Brooks S.J., Cameron N.G., Evershed R.P., Peglar S.M., Scott J.A. and Thompson R. 2005. A multi-proxy study of lake-development in response to catchment changes during the Holocene at Lochnagar, north-east Scotland. Palaeogeog. Palaeoclimat. Palaeoecol. 221: 175–201

Davidson C. 1922. Lochnagar in mid-December. Cairngorm Club J. 10: 193–195.

Franklin M. 2003. A brief history of Braemar, Royal Deeside, Scotland. http://www.lochnagar.org/history/brochhist.htm. Accessed 17th June 2005.

Geikie A. 1887. The Scenery of Scotland viewed in connection with its physical geology. 2nd edition. Macmillan and Co. London. 540pp.

Gorham E. 1957. The chemical composition of some natural waters in the Cairn Gorm – Strath Spey district of Scotland. Limnol. Oceanog. 2: 143–154.

Greenwood G.B. 2005. Climate science in the American west. Mountain Res. Develop. 25: 80–81.

Harriman R. and Morrison B.R.S. 1989. Acidification studies in the Cairngorm region. In: Environmental Research on the Balmoral Estate. Aberdeen Centre for Land Use. Aberdeen. 35 pp.

Her Majesty the Queen Victoria. 1868. First ascent of Loch-na-gar. In: Helps, A. (ed.). Leaves from the Journal of our life in the Highlands from 1848 to 1861, to which are prefixed and added extracts from the same Journal giving an account of earlier visits to Scotland and tours in England and Ireland and yachting excursions. Smith, Elder and Co. London pp 67–70.

Hofer T. 2005. Introduction: The International Year of Mountains challenge and opportunity for mountain research. In: Huber U., Bugmann H.K.M. and Reasoner M.A. (eds.) Global change and mountain regions. An overview of current knowledge. Springer. Dordrecht. pp 1–8.

Hutchison A.G. 1927. Lochnagar. The Deeside Field. 3: 15–18

Jones V.J., Flower R.J., Appleby P.G., Natkanski J., Ricahrdson N., Rippey B., Stevenson A.C. and Battarbee R.W. 1993. Palaeolimnological evidence for the acidification and atmospheric contamination of lochs in the Cairngorm and Lochnagar areas of Scotland. J. Ecol. 81: 3–24.

Lamoureux S.F. and Gilbert R. 2004. Physical and chemical properties and proxies of high latitude lake sediments. In: Pienitz R., Douglas M.S.V. and Smol J.P. (eds.) Long-term environmental change in Arctic and Antarctic lakes. Springer. Dordrecht. pp 53–87.

Lotter A.F. 2005. Palaeolimnological investigations in the Alps: The long-term development of mountain lakes. . In: Huber U., Bugmann H.K.M. and Reasoner M.A. (eds.) Global change and mountain regions. An overview of current knowledge. Springer. Dordrecht. pp105–112.

Macgillivray P. 1915. Lochnagar. Cairngorm Club J. 8: 43.

Marchetto A. and Rogora M. 2004. Measured and modelled trends in European mountain lakes: results of fifteen years of cooperative studies. J. Limnol. 63: 55–62.

McConnochie A.I. 1891. Lochnagar. Wylie and Son. Aberdeen. 188 pp.

McHardy J.A. 1897. Lochnagar. Cairngorm Club J. 2: 176–177.

Michie Rev. J.G. 1896. The Benchinnans. Cairngorm Club J. 2: 34–37.

MOLAR Water Chemistry Group. 1999. The MOLAR project: Atmospheric deposition and lake water chemistry. J. Limnol. 58: 88–106.

Monteith D.T. and Evans C.D. (eds.) 2000. UK Acid Waters Monitoring Network: 10 year report. Analysis and interpretation of results April 1988–March 1998. ENSIS Publishing. London 364 pp.

Monteith D.T. and Evans C.D. 2005. The United Kingdom Acid Waters Monitoring Network: A review of the first 15 years and introduction to the special issue. Environ. Pollut. 137: 3–13.

Monteith D.T., Evans C.D. and Dalton C. This volume. Chapter 14. Acidification of Lochnagar and prospects for recovery. In: Rose N.L. (ed.) 2007. Lochnagar: The natural history of a mountain lake. Springer. Dordrecht.

Nisbet A. 1995. The Cairngorms Rock and Ice Climbs (Volume 2). Scottish Mountaineering Club. 214 pp.

Parker J.A. 1923. The horizon from Lochnagar. Cairngorm Club J. 10: 259–261.

Pauli H., Gottfried M., Hohenwallner D., Reiter K. and Grabherr G. (eds.). 2003. The Global Observation Research Initiative in Alpine Environments (GLORIA) Field Manual – Multi-summit approach. University of Vienna. 45pp.

RGAR (United Kingdom Review Group on Acid Rain) 1990. Acid deposition in the United Kingdom 1986–1988. Third report. Warren Spring Laboratory. 124 pp.

Rognerud S., Grimalt J.O., Rosseland B.O., Fernández P., Hofer R., Lackner R., Lauritzen B., Lien L., Massabuau J.C. and Ribes A. 2002. Mercury and organochlorine contamination in brown trout (*Salmo trutta*) and Arctic charr (*Salvelinus alpinus*) from high mountain lakes in Europe and the Svalbard archipelago. Wat. Air Soil Pollut.: Focus. 2: 209–232.

Rose J. 1895. The Lochnagar corrie. Cairngorm Club J. 1: 297–300.

Rose N.L. and Battarbee R.W. This volume. Chapter 19. Past and future environmental change at Lochnagar and the impacts of a changing climate. In: Rose N.L. (ed.) 2007. Lochnagar: The natural history of a mountain lake. Springer. Dordrecht.

Rose N.L. and Yang H. This volume. Chapter 17. Temporal and spatial patterns of spheroidal carbonaceous particles (SCPs) in sediments, soils and deposition at Lochnagar. In: Rose N.L. (ed.) 2007. Lochnagar: The natural history of a mountain lake. Springer. Dordrecht.

Rosseland B.O., Rognerud S., Collen P., Grimalt J.O., Vives I., Massabuau J-C., Lackner R., Hofer R., Raddum G.G., Fjellheim A., Harriman R. and Piña B. This volume. Chapter 12. Brown trout in Lochnagar: Population and contamination by metals and organic micropollutants. In: Rose N.L. (ed.) 2007. Lochnagar: The natural history of a mountain lake. Springer. Dordrecht.

Skakle M. 1934. A spring traverse of Lochnagar. Cairngorm Club J. 13: 155–157.

Skea W. 1894. Two days in Glen Muick. Cairngorm Club J. 1: 65–72.

Smith E.W. 1947. On lochs and lochans. Cairngorm Club J. 15: 35–38.

Straškrabová V., Callieri C., Carrillo P., Cruz-Pizarro L., Fott J., Hartman P., Macek M., Medina-Sánchez J.M., Nedoma J. and Šimek K. Investigations on pelagic food webs in mountain lakes – aims and methods. J. Limnol. 58: 77–87.

Symmers G.R. 1931. Some rock-climbs on Lochnagar. Cairngorm Club J. 12: 186–206.

Vives I., Grimalt J.O., Lacorte S., Guillamón M., Barceló D, and Rosseland B.O. 2004. Polybromodiphenyl ether flame retardants in fish from lakes in European high mountains and Greenland. Environ. Sci. Technol. 38: 2338–2344.

Vives I., Grimalt J.O., Ventura M., Catalan J. and Rosseland B.O. 2005. Age dependence of the accumulation of organochlorine pollutants in brown trout (*Salmo trutta*) from a remote high mountain lake (Redó, Pyrenees). Environ. Pollut. 133: 343–350.

Watson A. 1975. The Cairngorms: The Cairngorms, the Mounth, Lochnagar. The Scottish Mountaineering Club District Guide Books. 5th edition. Scottish Mountaineering Trust. Edinburgh. 304 pp.

Watson A. 1992. The Cairngorms: The Cairngorms, Lochnagar and the Mounth. The Scottish Mountaineering Club District Guide Books. 6th edition. Scottish Mountaineering Trust. Edinburgh. 272 pp.

Watson A. and Allan E. 1984. The place names of Upper Deeside. Aberdeen University Press. 192 pp.

Wightman A. 1996. Scotland's mountains. An agenda for sustainable development. Scottish Wildlife and Countryside Link. Perth. 24 pp.

Yang H. 2000. Trace metal storage in lake systems and its relationship with atmospheric deposition with particular reference to Lochnagar, Scotland. Unpublished PhD thesis. University College London. 340 pp.

Yang H., Rose N.L., Battarbee R.W. and Boyle J.F. 2002. Mercury and lead budgets for Lochnagar, a Scottish mountain lake and its catchment. Environ. Sci. Technol. 36: 1383–1388.

2. PHYSICAL CHARACTERISTICS OF LOCHNAGAR

MICHAEL HUGHES (m.hughes@ucl.ac.uk)
Environmental Change Research Centre
University College London
Pearson Building, Gower Street
London WC1E 6BT
United Kingdom

Key words: bathymetry, Geographic Information System (GIS), insolation, light regime, slope analysis, topography

Introduction

The physical characteristics of a lake affect every aspect of its chemistry, biology and hydrology. In this chapter, the morphological and light characteristics of Lochnagar are described in order to provide background information for subsequent chapters.

Physical parameters

Lochnagar has a surface area of 10.4 hectares (25.7 acres), a maximum length of 500 m in a north-south direction and is 360 m across at its widest point (Figure 1) (Data derived from OS 1:10000 LandLine®.) The shoreline is not very complex and has a total length of about 1470m. A common measure of shoreline complexity is the 'shoreline development index' (SDI) which relates the shoreline length of a water body to the circumference of a circle with the same surface area. Thus, a perfectly circular lake would have an index of 1 while a higher index indicates a more complex shoreline. Lochnagar has an SDI of 1.218.

A detailed bathymetric survey was carried out using a combined echo sounder and global positioning system (GPS). The system was mounted on a small inflatable boat and combined soundings and GPS fixes were taken every second as the boat traversed the loch. Over 12000 data points were collected across Lochnagar and these were used in the construction of a bathymetric model for the loch basin (Figure 1). The water level during the survey was recorded from a stageboard at the outflow although the lake level does not usually vary by more than ± 10 cm. A geographic information system (GIS) was used to store and manipulate the bathymetric data and interpolation was used to

N.L. Rose (ed.), Lochnagar: The Natural History of a Mountain Lake
Developments in Paleoenvironmental Research, 27–35.
© 2007 *Springer.*

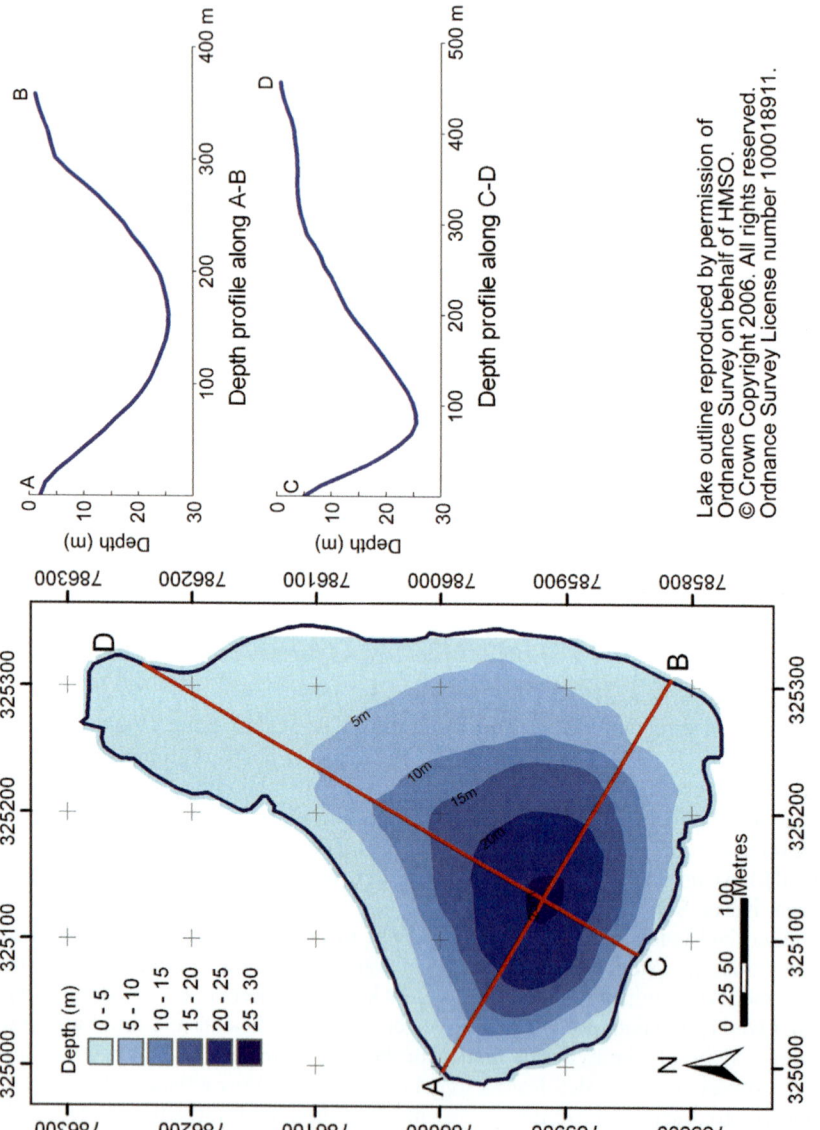

Figure 1. The bathymetry of the Lochnagar lake basin based on more than 12000 measurements taken using an echo sounder linked to a GPS

construct a gridded depth 'surface' at a grid resolution of 10 m. The maximum recorded depth during the survey was 26.4 m offset from the centre of the loch towards the backwall as shown. The gridded depth surface was used to derive the mean depth (8.9m) and thus the volume (925600 m^3) of the loch. A fluctuation of 10 cm around the normal loch level would result in a variation of approximately ± 1% of the loch volume. The hypsographic curve for Lochnagar (percentage of area below a given depth) and an equivalent for lake volume are shown in Figure 2.

(a)

(b)

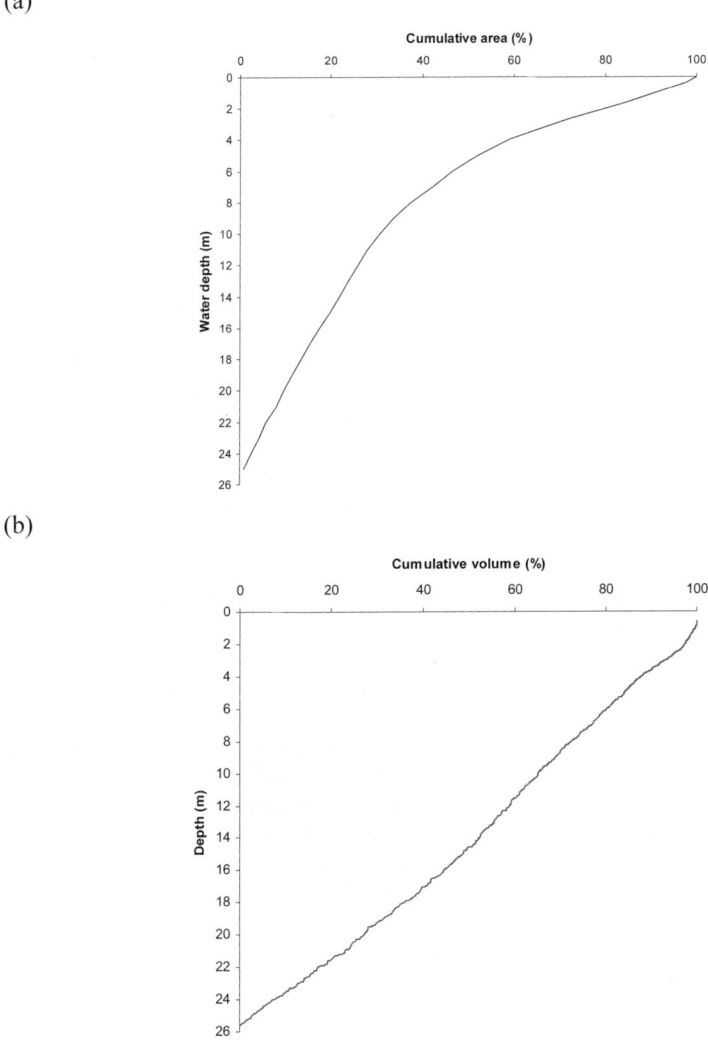

Figure 2. Hypsographic curves based on data in Figure 1. (a) cumulative area versus depth and (b) cumulative volume versus depth.

The loch lies at an altitude of 788 m a.s.l. and the catchment area (108.5 hectares or 286 acres, including the loch area) extends up to the summit on the Lochnagar massif at 1155 m (Chapter 1: Rose this volume) making the range of relief in the catchment 367 m. The mean elevation of the catchment area (including the loch itself) is 925 m. The catchment predominantly faces northeast and is dominated by boulder fields, scree and the steep cliffs of the backwall. Figure 3a shows catchment slope and comparison with Figure 3b highlights the local influence of catchment slope on loch basin slope. To the southwest of the loch catchment slopes exceed 50% and there is a corresponding area of high slope adjacent to it within the loch. The mean slope of the catchment is 25%. The slope within the loch basin does not exceed 30% with the steepest areas formed by a continuation of the backwall at the southern end (Figure 3b). Other slopes exceeding 20%, to the west and east, enclose a central area where slope varies from 0 to c. 15%. The flattest areas of the loch lie towards the north, near the outflow where the basin slope is typically less than 5%. The mean slope for the loch basin is 8.4%.

Light regime

Sunlight is the primary source of energy for any ecosystem and the varying light climate throughout the year is highly influential in driving biological processes, particularly in mountain lakes. The amount of solar radiation reaching the surface of Lochnagar depends on sun-earth geometry but is affected by shading from surrounding hills. The amount of solar radiation (sunlight) can be computed using known relationships between the position of the sun relative to the earth. In order to take the effects of hill shading into account we can employ a digital elevation model combined with the solar radiation data to derive information about how long the loch surface is theoretically exposed to sunlight on any day of the year.

Computing routines in a geographical information system (ESRI's ArcView with Spatial Analyst) were used to carry out a detailed assessment of Lochnagar's light climate. First, a theoretical model was created using the Solar Analyst extension (Fu and Rich 1999) to calculate the total annual incoming solar radiation (direct and indirect solar radiation) for the whole catchment for the year 2005. For the model parameters a 'generally clear sky' is assumed with an average value for atmospheric transmissivity of 0.5 (expressed as the proportion of exoatmospheric radiation transmitted as direct radiation along the shortest atmospheric path). The model produces a map (Figure 4) of incident solar radiation for 2005 in kWh m^{-2} (kilowatt hours per square metre). For comparison, a value of 1 kWh m^{-2} is the equivalent of ten 100W light bulbs illuminating 1 square metre for 1 hour. For the surface of the loch values computed for 2005 range from about 680 to 860 kWh m^{-2} (mean value 789 kWh m^{-2}) although within the catchment the range is much larger (between 100 and 1125 kWh m^{-2}). The highest values are at the northern end of the loch which has a larger sky exposure and is furthest away from the steep cliffs to the southwest which have a significant shading effect on the loch especially during the winter months. Some areas of the catchment on the corrie wall are completely shaded and receive no direct sunlight at any time of the year. Note that the model includes non-direct (scattered) light and this accounts for the fact that incident solar radiation is never less than 100 kWh m^{-2} anywhere in the catchment.

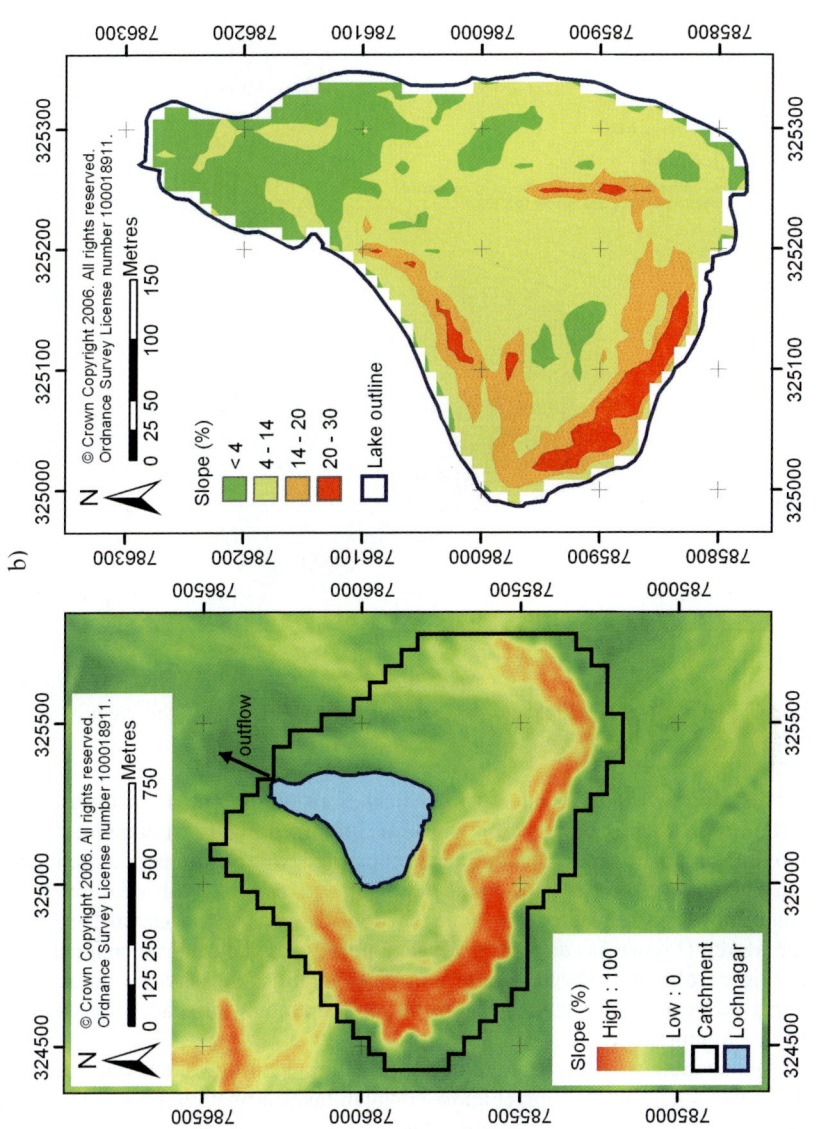

Figure 3. Maps of (a) catchment slope and (b) loch basin slope. The latter is divided into classes (<4%; 4 – 14%, 14 – 20% and 20 – 30%) according to Hilton et al. (1986) in order to provide information on accumulation and transportation zones (see text for explanation)

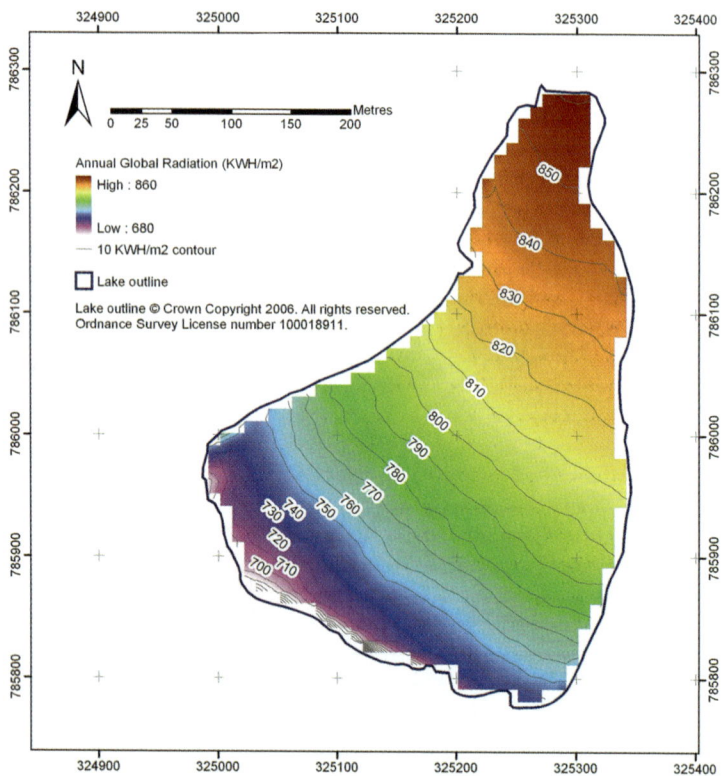

Figure 4. Annual global radiation (2005) across Lochnagar in kWh m^{-2}

Shading of the loch was investigated using calculations of solar azimuth (bearing of the sun) and altitude (angle above the horizon) for 2005 (see Figure 5) combined with the ArcInfo hillshading model. Shading was computed for local noon at ten day intervals throughout the year. Shade-limits for each ten day interval are plotted in Figure 6 and show the retreat of the noon shadow towards the backwall between January and March (Figure 6a) and its re-advance northwards between September and November (Figure 6b). As expected there is a clear relationship between amount of radiation received at the loch surface (Figure 4) and the amount of shading caused by the surrounding topography.

The modelled shading was compared to actual photographs of the site taken from the automated digital camera installation (Figure 7) which takes a photograph every day at 1200 UTC. There was a good agreement between modelled shade limits and actual shade limits although the angle of the photograph and the water surface make it quite difficult to attempt any more than a qualitative analysis.

(a)

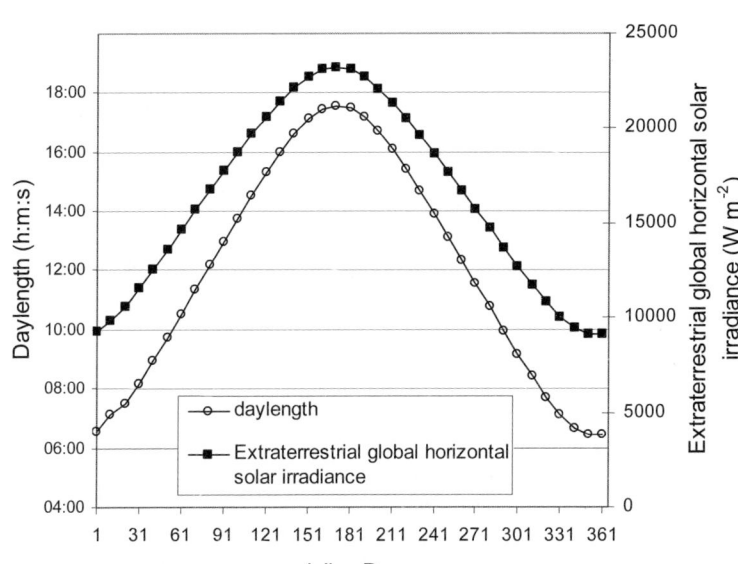

(b)

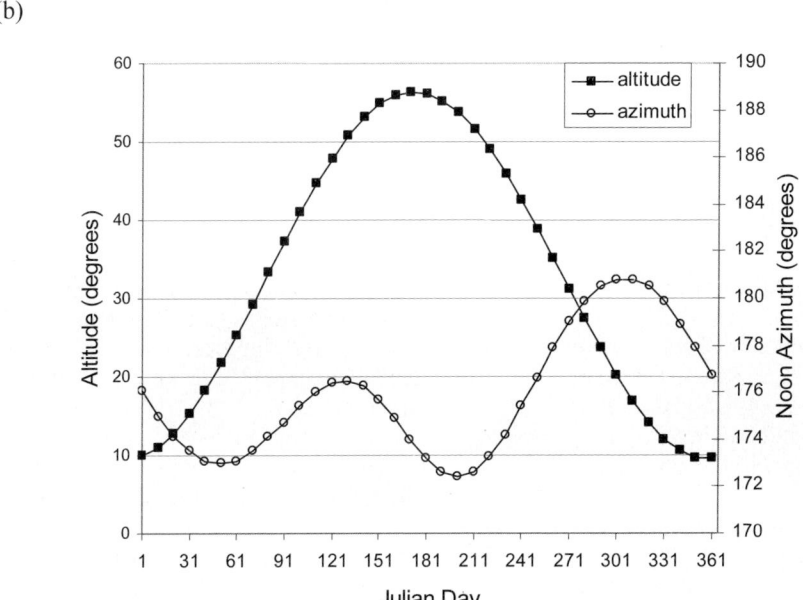

Figure 5. Variation of (a) daylength and extraterrestrial global solar irradiance and (b) sun azimuth and altitude, throughout the year at Lochnagar.

(a)   (b)

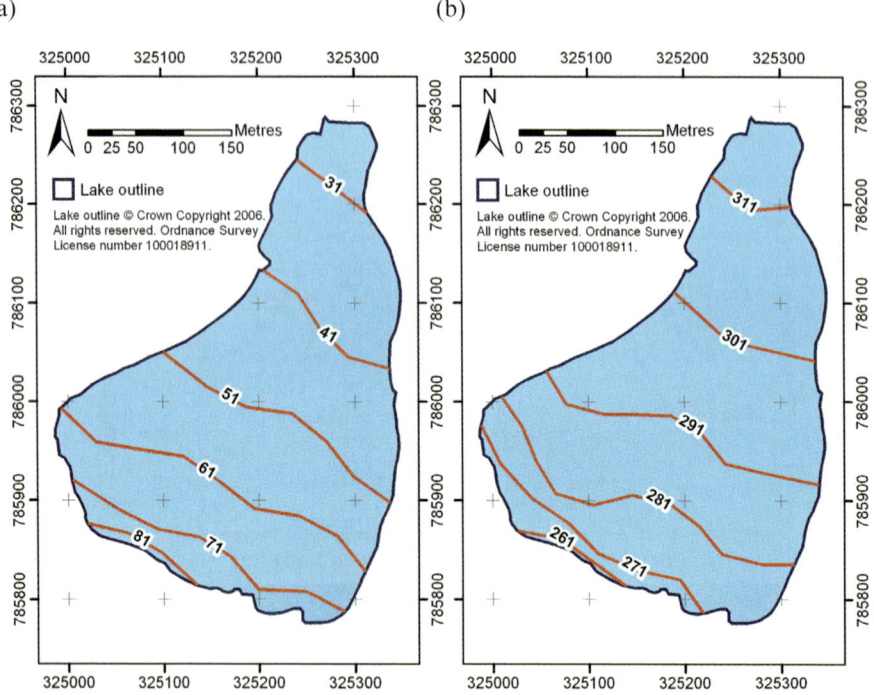

Figure 6. Position of shade limit across Lochnagar at noon on Julian days in (a) January to March and (b) September to November

(a)   (b)

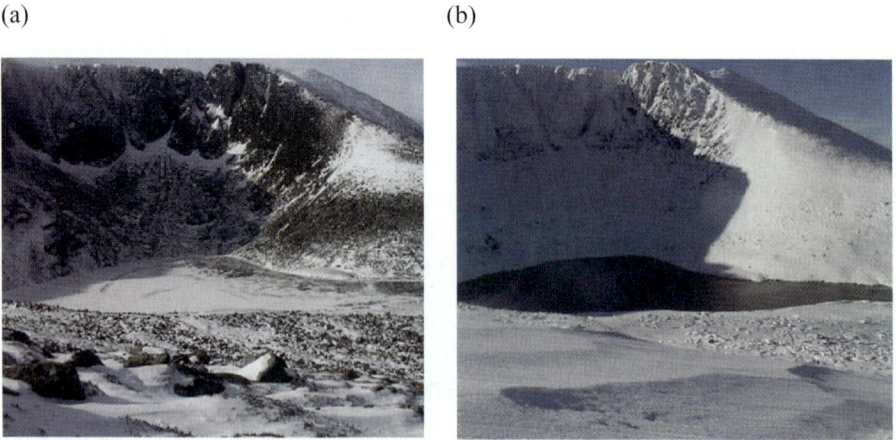

Figure 7. Photographs from the automatic camera showing the shadow across Lochnagar on (a) Julian day 61 (2nd March 2005) and (b) Julian day 301 (28th October 2002)

References

Fu P. and Rich P.M. 1999. Design and implementation of the Solar Analyst: An ArcView extension for modeling solar radiation at landscape scales. Accessed from: http://gis.esri.com/library/userconf/proc99/proceed/papers/pap867/p867.htm Proceedings of ESRI Annual User Conference 1999.

Hilton J. Lishman J.P. and Allen P.V. 1986. The dominant processes of sediment distribution and focusing in a small, eutrophic, monomictic lake. Limnol. Oceanog. 31: 125–133.

Rose N.L. This volume. Chapter 1. An introduction to Lochnagar. In: Rose N.L. (ed.) 2007. Lochnagar: The natural history of a mountain lake. Springer. Dordrecht.

Rose N.L. This volume. Chapter 8. The sediments of Lochnagar: Distribution, accumulation and composition.   In: Rose N.L. (ed.) 2007. Lochnagar: The natural history of a mountain lake. Springer. Dordrecht.

PART I: THE ENVIRONMENTAL LANDSCAPE
          OF LOCHNAGAR

3. GEOLOGY OF LOCHNAGAR AND SURROUNDING REGION

SALLY GOODMAN (sgoodman@eps.mcgill.ca)
Department of Earth and Planetary Sciences,
McGill University,
3450 University,
Montreal H3A 2A7,
Quebec, Canada

Keywords: Caledonides, Dalradian, diorite, geochemistry, granite, Lochnagar, metamorphism

Introduction

The Lochnagar granite massif, situated at the core of Scotland's Caledonian mountains, dominates the high ground to the east of Braemar and south of Crathie, Aberdeenshire, between Glen Muick to the east and Glen Callater to the west (Figure 1). Long before glaciers carved the granite into ridges, corries and glens, indeed long before the granite hardened from molten rock, this area was at the edge of an ancient ocean, the evidence for which lies in the country rocks which surround the granites. Beneath the heather and peat of Glen Muick, Glen Clunie and Glen Doll, lie rocks of the Dalradian succession ('country rocks' on Figure 1), which were once sands, muds and volcanic rocks on the continental shelf at the edge of an ocean. Similar rocks are found from Scandinavia to the US Appalachians, and show that during the Neoproterozoic, some 800 million years ago (800Ma) these regions were united as part of a continental margin on the scale of the current eastern seaboard of the Americas. The ocean lapping these ancient shores, which has been given the name Iapetus Ocean, was probably not as wide as the present-day Atlantic Ocean, but was a significant global feature over a long period of time. The Dalradian rocks in the Lochnagar area were deposited over a time span of approximately 200Ma; an even greater time span is recorded on the Iapetus margin elsewhere.

In the same way that current coastal sediments are brought to the ocean as the result of erosion of the landscape, the Dalradian sediments were eroded from an even older continental area. The continental rocks are not visible in this region today, having been worn down by erosion and buried by generations of younger sediments, though they were probably similar to the ancient gneiss exposed in the North West Highlands of Scotland. Such Archean gneiss forms the core of all continents, and are the oldest rocks on earth. The nature of the Dalradian sediments deposited in the Iapetus Ocean depended on the nature of rocks being eroded in the source area, the processes of weathering and transport, and the way in which the sediment was deposited (Figure 2a).

N.L. Rose (ed.), Lochnagar: The Natural History of a Mountain Lake
Developments in Paleoenvironmental Research, 39–50.
© 2007 *Springer.*

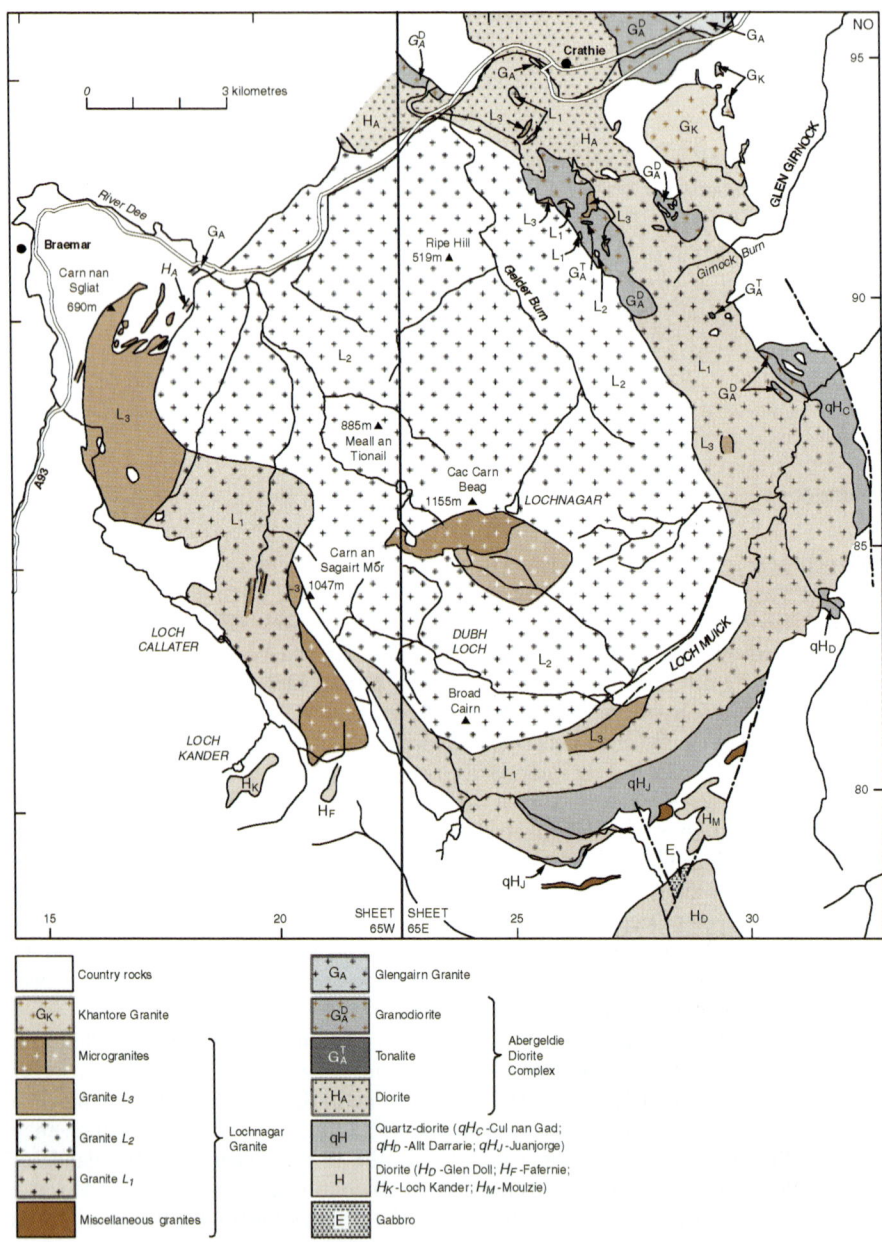

Figure 1: Geological sketch map of the Lochnagar granite massif and surrounding area; Quaternary and Holocene deposits omitted. Note that abundant microgranitic minor intrusions are omitted due to limitations of scale. From Smith et al. (2002) IPR/68-19C BGS 8 NERC All rights reserved.

The resulting sediment ranged from beach sands cleaned by the frequent washing of the waves (which in places even preserve a rippled surface like that seen on beach sands today), through dirtier sands and silts, to thick black muds, deposited further offshore, far away from the winnowing property of the waves. Glacial deposits such as boulder clay are present in the middle part of the Dalradian sequence, indicative of the presence of ice-sheets extending offshore. Carbonate sediments probably formed by evaporation of seawater in shallow areas with limited connection to the open ocean, like the evaporites accumulating off the Arabian Gulf states today.

There are limitations to the use of analogy with the present day; at the time the Dalradian sediments were being deposited there was no vegetation on land, and no higher life in the ocean. The youngest of the Dalradian rocks were formed at the time of the diversification of life in the oceans which defines the start of the Cambrian period, and where they are best preserved they may contain evidence of life such as worm burrows (Kneller 1987). In the Lochnagar area, evidence for ocean life is more subtle, for example the graphite in rocks that were originally black muds was probably carbon of organic origin, but there are none of the plant or animal fossils that mark the rock record in younger sediments.

In addition to oceanic sediments, the Dalradian sequence contains minor amounts of volcanic rocks, and Dalradian sediments are intercalated with layers of basalt lava and tuff (volcanic ash). Movement of he Earth's tectonic plates resulted in stretching and sagging of the Archean continental crust, and decompression of the underlying mantle allowed basaltic magma to well up from below. Some of the lava has a characteristic pillow shape, caused by the hot lava, at temperatures up to 900°C, flowing into cool ocean waters, where pillows were podded off the front of the flow, to accumulate in a pile on the sea floor. The same process has been filmed off Hawaii, where basalt lavas flow into the ocean today.

The Dalradian sequence of sediments and volcanic rocks exposed in the Lochnagar area is only part of the total accumulation over the lifetime of the ocean; evidence from elsewhere suggests that the total sequence was many kilometres in thickness. Deep within such a volcano-sedimentary pile, pressure increases as sediment accumulates at the surface, water is driven out, and new minerals precipitate to cement the sediment grains together. Over the millennia, the soft, wet sediment from the Iapetus Ocean hardened into solid rock; beach sands became white sandstones, muds became mudstones, carbonate sediments became limestones. So the Dalradian sediment, derived from the rocks of an ancient Archean continent, weathered, eroded, swept out to sea and deposited as soft sediment, became hardened into rock once more, and a cycle was complete. The Earth has many such cycles of differing scale, duration and complexity, and the ocean into which the sediment poured had its own cycle. Formed by stretching and swamping of the Archean crust, growing through volcanism in the ocean depths, and accumulating sediment over hundreds of millions of years, the Iapetus eventually closed by the process of subduction in early Palaeozoic times.

The Earth's tectonic plates, the three-dimensional jigsaw that makes up the surface of the Earth, are constantly in motion over a partly molten layer that lies near the top of the mantle. The continental areas, thick, ancient, granitic and buoyant, survive through time, coming together, breaking apart, reassembled in a new order, while the thin, young, dense ocean floors, built by basaltic volcanism as the oceans grow, are eventually consumed. When changing plate motion pushes oceanic crust against the

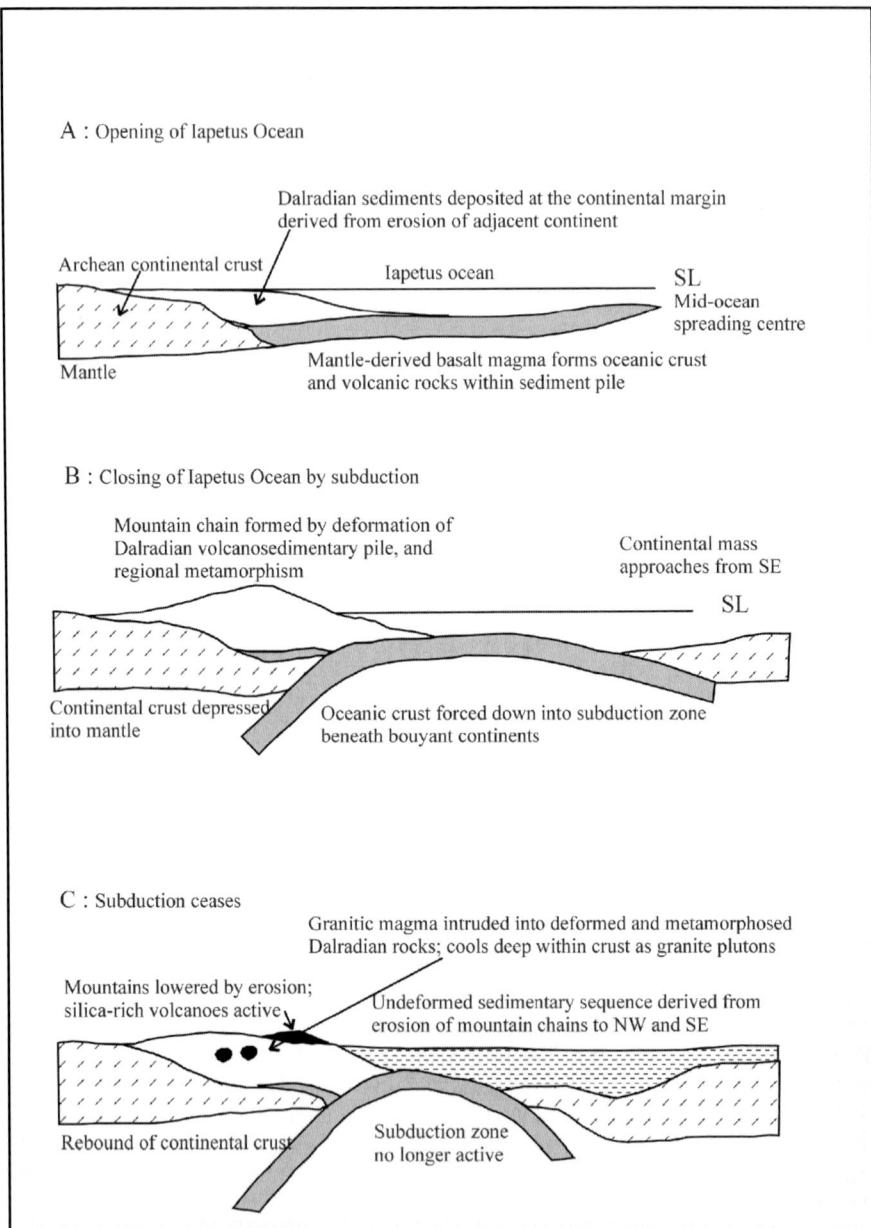

Figure 2: Schematic diagrams to show the geological development of northern Britain, including the Lochnagar area. SL = contemporary sea level.

     A: Deposition of Dalradian sediments and volcanic rocks

     B: Deformation of Dalradian sequence during Caledonian mountain-building episode

     C: Intrusion of the Lochnagar granites

adjacent continent, the continent overrides the ocean floor, which sinks back into the mantle along a subduction zone, to be ultimately reabsorbed and recycled. At the present time, the Pacific Ocean is slowly sliding beneath the western seaboard of the Americas, along a subduction zone that is slowly but surely consuming the ocean floor. Over great periods of time, whole oceans can be consumed by subduction, until the continents that once formed its coasts come into collision. As the Americas override the Pacific plate, and the western margin of the Pacific is consumed by subduction off Japan and Indonesia, so eventually the Pacific Ocean will cease to exist, and the Americas will be sutured to Asia. Where continents collide, subduction eventually ceases, as neither continent can be forced down into the subduction zone.

The immense forces which brought the continents together continue to act for some time after subduction ceases, however, and along the suture line a mountain range rises, its peaks being thrust into the sky, and, deep beneath, equivalent roots being forced down into the mantle. The Himalayas mark the geologically recent collision between Asia and India, and the Caledonian mountains mark the subduction of the Iapetus and the collision between its ancient continental margins. (The relatively recent opening of the Atlantic Ocean split the Caledonian mountain chain, leaving the Scottish and Scandinavian Caledonides sutured to the European continent, while the American Caledonides departed with the Archean heartland).

Deep ocean floor is consumed entirely by subduction. However continental shelves escape subduction, to bear the brunt of compressive forces as the continents collide. At the closure of the Iapetus, the Dalradian shelf sequence was at the core of a mountain chain as impressive in its time as the Himalayas are today (Figure 2b). Under immense pressure, extreme strain and temperatures up to $700^{\circ}C$, the rocks were twisted and folded as the mountains rose, folding that is preserved in rocks currently exposed in the glens and moorland surrounding the Lochnagar massif. Tearing of the crust allowed more mantle melt rock to well up along a fault zone now exposed east of Glen Girnock Figure 1). Unlike the basaltic lavas, these molten rocks did not reach the Earth's surface, but cooled and crystallised within the crust, as intrusions of ultramafic rock that now form the Coyles of Muick.

Many of the minerals that formed the sedimentary and volcanic rocks at the Earth's surface were no longer stable under the high temperatures and pressures reached during mountain building, and were replaced by new minerals in a process known as metamorphism. Sandstones, formed largely of silica, and limestones, dominantly calcium carbonate, had little potential for reaction though their texture was modified, from rounded sediment grains cemented together, to interlocking grains which give the rock an indurated texture. Sandstones hardened into quartzite, and limestone became marble. Rocks with more complex compositions, mudstones, muddy sandstones, or muddy limestones, for example, had greater potential for change, and developed many new minerals (Table 1). The minerals chlorite – biotite - garnet – staurolite – kyanite - sillimanite developed sequentially in metamorphosed mudstones (known as 'pelitic' rocks) as temperature and pressure increased, and their preservation in the rocks give an indication of the intensity of metamorphism in different parts of the Caledonides. The intercalated volcanic rocks developed a different characteristic sequence of minerals: chlorite – biotite – amphibole – epidote – garnet – pyroxene. To the east of the Lochnagar massif, metamorphosed sedimentary rocks contain sillimanite, and basaltic rocks contain pyroxene, indicating that this area suffered extremes of metamorphism at

Table 1: Representative modal analyses of metamorphosed Dalradian country rocks in percent; T = trace. Based on unpublished data, summarised in Smith at al. (2002) and Crane et al. (2002).

	Quartzite	Limestone	Black Schist	Pelitic Schist	Metabasalt
Quartz	68	28	52	22	12
Plagioclase	9	T	17	12	31
K-feldspar	4			14	
Calcite		45			
Biotite	5	5	10	25	11
Muscovite	8		11		
Garnet	3			18	8
Sillimanite				10	
Graphite			10		
Hornblende		T			28
Diopside					4
Epidote		22			3
Apatite	T			T	
Titanite		T	T		2
Tourmaline			T		
Zircon	T			T	T
Oxides	3	T	T	T	T

the heart of the Caledonian mountain chain (Smith et al. 2002), with temperatures as high as 700°C, and pressures up to 7 kbars.

The texture of the rocks also changed during metamorphism, as the new minerals grew under directed stress, and produced platy surfaces within the rock, forming biotite schists and sillimanite gneisses from the mudstones, and amphibole schists from the volcanic rocks. Metamorphism therefore affected the properties of the rocks in two ways, changing both the mineralogy and the texture of the rock, so that, for example, massive mudstones became fissile pelitic schists, studded with hard garnets. Apart from the loss of volatile components (water, carbon dioxide) the chemistry of the rocks was little affected by metamorphism, and largely reflects the chemistry of the original sediments (Table 2). The original sediment distribution, with sandstones, limestones and organic-rich mudstones dominant to the northwest of the area, and silty mudstones and metabasalts to the southeast, is reflected in the landscape to the present day. The combination of original composition and imposed textural changes controlled the way the rocks broke down under weathering processes at the Earth's surface millions of years later when tectonic processes brought them to the surface once more.

Apart from mountain building and metamorphism, another major feature of subduction zones is igneous activity, as evidenced by the abundance of volcanoes, for example in the 'Pacific Ring of Fire', the band of volcanoes at the margins of continents bordering the Pacific Ocean at the present time. Such magmatism differs from the basaltic eruptions discussed previously; water is released by the subducting slab and triggers wet melting in the mantle, at a lower temperature than melting caused by decompression alone. When the melts rise through the overriding tectonic plate they become contaminated by continental crust, which is melted and assimilated en route. As a result, subduction zone magmas are relatively cool (600°C or so) and rich in silica, which makes them viscous and not readily erupted. Evidence for subduction zone magmatism in the Caledonian mountain chain is provided by the diorites and granites which are the deep-level equivalent of silica-rich volcanic rocks, and whose roughly circular outcrop shapes cut through the linear trends of the Dalradian country rocks. The melts, or crystal mushes, forced their way along deep fracture zones and ate their way into the deformed and metamorphosed Dalradian rocks, accumulating as plutons, commonly several kilometres in diameter. The Lochnagar granite massif and nearby Abergeldie complex are examples of large plutons (Figure 1); both are complex intrusions formed by multiple pulses of dioritic and granitic magma. Around the perimeter of the Lochnagar pluton, the granitic rock contains rafts of Dalradian rock, which were broken off into the granite mush as it shouldered its way in. Such rafts within the pluton suggest that the current land surface is close to the original upper contact or roof of the pluton. It seems to have a near vertical outer margin, so is roughly cylindrical in form, but the depth to which it extends has not been determined.

*Table 2.*Chemical analyses of Dalradian metamorphic rocks, major elements in weight percent (Wt %), trace elements in parts per million (ppm); all Fe as Fe_2O_3. Data from Wright (1998); Goodman and Winchester (1993). na=not analysed; nd = not detected.

Wt%	Quartzite	Limestone	Black Schist	Pelitic Schist	Metabasalt
SiO_2	91.94	7.5	56.24	62.6	47.38
TiO_2	0.15	0.16	0.61	0.79	1.41
Al_2O_3	4.39	1.6	23.66	21.2	16.17
Fe_2O_3	0.51	0.75	7.08	5.4	9.40
MnO	0.01	0.02	0.02	0.08	0.17
MgO	0.49	0.76	3.63	2.1	6.85
CaO	0.29	48.20	0.80	1.1	13.85
Na_2O	1.27	0.31	1.20	2.1	2.38
K_2O	1.34	0.3	3.99	4.3	0.58
P_2O_5	0.02	0.05	0.04	0.13	0.25
ppm					
Ba	301	93	594	848	143
Ce	13	16	97	na	11
Cr	na	13	93	na	286
La	5	nd	50	na	1
Nb	na	na	12	20	18
Pb	7	10	23	na	4
Rb	17	12	158	213	18
Sr	19	1954	93	167	580
Th	3	1	15	na	na
V	na	2	na	na	208
Y	4	4	31	56	23
Zr	(40-2406)	20	191	331	102

By the time the diorites and granites were intruded, the Dalradian rocks had cooled somewhat after peak metamorphic conditions, and had become less plastic as a result, hence their tendency to fracture into blocks that could be incorporated into the magma body. Larger crustal fractures, or fault zones, probably guided the emplacement of the granites (Figure 1); smaller fracture zones cut the country rocks and the granites, and may contain veins of quartz, precipitated from hot fluids derived from the cooling granite, or sheets of fine grained microgranitic rock. Intrusion of the hot granite raised the temperature of the surrounding Dalradian rocks in an aureole up to a few hundred metres wide, causing a second phase of metamorphism locally, with growth of minerals such as andalusite and cordierite in pelitic schists, and pyroxenes in the metamorphosed basaltic volcanic rocks (Goodman and Lappin 1996). The minerals formed during the peak of regional metamorphism indicate that the rocks were some 18-20 km below the mountainous surface of the Earth at the apogee of mountain building. When plate collision eventually ceased, and the mountains were no longer growing, erosion took over to reduce the height of the mountains, and the mountain roots, protruding deep down into the mantle, rose up in compensation, and through this process, the deep heart of the mountain chain rose slowly toward the surface. The new metamorphic minerals formed in the aureole of the Lochnagar granite indicate the area was around 8km below the surface when the granite was intruded (Figure 2c), too deep to have been a subvolcanic magma chamber, though still related to the same subduction zone processes.

The Lochnagar massif is in fact not a single granite body, but a series of nested intrusions, which can be placed in time sequence based on their cross-cutting relations. Oldest of all are the quartz diorites that rim the pluton to the east, and send out numerous veins into the surrounding Dalradian country rock. These quartz diorites are in turn veined by the oldest phase of granite (known as L1), which formed an intrusion of roughly circular outline, about 15 km across. The L1 granite was subsequently cored out by a 12 x 9 km pluton of L2 granite, which breached the L1 granite in the northwest. L3 granites cut the L1 granite, but are not observed within the outcrop of L2 granite. All these rock types are formed from an interlocking mosaic of coarse-grained crystals, indicative of the long period of time over which the magma cooled, deep within the crust. All of them consist largely of the minerals quartz and feldspar, with lesser amounts of biotite and hornblende, and traces of muscovite, titanite, apatite, zircon and opaque oxide minerals. They are distinguished on the basis of the relative proportions of the minerals (Table 3), which is a reflection of subtle variations in chemistry as the magmas evolved over time.

The quartz diorites are dominantly formed of plagioclase feldspar, with lesser amounts of quartz, and only minor potassium feldspar; biotite and hornblende are important constituents of the rock. The L1 and L2 granites contain more quartz and much more potassium feldspar; however biotite is less abundant, and hornblende present in very small amounts. In the L3 granite, potassium feldspar is the dominant mineral, with quartz and plagioclase being less abundant, biotite forming only a few percent, and hornblende being absent. This change in the mineral proportions is typical of a chemical evolution in the magma, known as magmatic fractionation, due to precipitation of certain mineral phases and consequent changing composition of the remaining melt. In common with other intrusive rocks in the Caledonian belt in Scotland, there is an evolution in chemistry with time, with enrichment of later phases in silicon, potassium and rubidium, and depletion in titanium, manganese, magnesium, calcium, strontium and phosphorus (Table 4). Comparison of the chemical analyses in Table 4 with the mineral proportions and mineral composition in Table 3 will readily show the control of magma chemistry on the mineral composition of the resulting rock. The evolutionary trend in rock compositions from quartz diorites to L1, L2 and L3 granites is characteristic of the evolution from a single parent magma, a single subduction zone melt event giving rise to a suite of genetically related, though distinctive, rock types (Smith et al. 2002).

Radioactive minerals that precipitated from the granitic melt mark the time that the granite began to cool. Minerals such as zircon contain a small amount of radioactive uranium, which decays in a predictable way over time to form lead. The ratio of parent uranium to daughter lead in zircon measures the time over which decay has taken place. Although the theory is simple enough, in practice it can be difficult to find enough of the tiny zircon grains to provide a robust date for an intrusive rock. In the Lochnagar area, diorites of the Abergeldie Complex (Figure 1) have been dated at 423.4 ± 2.7Ma; these are chemically and mineralogically similar to the quartz diorites fringing the Lochnagar Complex, which are probably of similar age. The Lochnagar L1 granite has been dated at 426.2 ± 2.5Ma, i.e., the age of intrusion of the quartz diorites and L1 granites are within analytical error, and therefore inseparable by our current dating methods. The L3 granite seems to be significantly younger, being dated at 417 ± 1Ma; no date has been obtained from the L2 granite. The age dating is therefore consistent with the evidence from cross-cutting field relations and from chemical evolution that the quartz diorites were intruded first, quickly followed by the L1 granites, the L2 granites, and, some time later, by the L3 granites which represent the products of the final crystallisation of this batch of magma (Smith et al. 2002).

The isotopic age of the granites put the time of intrusion in the Silurian Period, at a time when detritus from the Caledonian mountain chain was filling the adjacent trench, as shown by evidence from the lowlands of the Midland Valley to the south (Cameron and Stephenson 1985). Subduction ceased; erosion of the mountains continued,

Table 3. Representative modal analyses of quartz diorites (Juanjorge, Cul nan Gad) and granites (Lochnagar L1, L2, L3) in percent; T=trace. See Figure 1 for distribution of intrusive phases. From Smith et al. (2002).

Mineral	Mineral composition	Juanjorge	Cul nan Gad	L1	L2	L3
Quartz	SiO_2	16	10	23	29	25
Plagioclase	$(Na,Ca)[Al_{1-2}Si_{3-2}O_8]$	52	38	39	41	33
K-feldspar	$(K,Na)[AlSi_3O_8]$	2	7	28	26	40
Biotite	$K_2(Mg,Fe)_{6-4}(Fe,Al,Ti)_{0-2}$ $[Si_{6-5}Al_{2-3}O_{20}](OH)_4$	13	6	8	4	2
Hornblende	$(Na,K)Ca_2(Mg,Fe,Al)_5$ $[Si_{6-7}Al_{2-1}O_{22}](OH)_2$	14	38	1	0	0
Muscovite	$K_2Al_4[Si_6Al_2O_{10}](OH)_4$	0	0	T	0	0
Titanite	$CaTi[SiO_4](OH)$	0	T	T	T	T
Apatite	$Ca_5(PO_4)_3(OH)$	0	T	T	T	T
Zircon	$Zr[SiO_4]$	0	0	T	0	T
Oxides	$Fe_3O_4, FeTiO_2, TiO_2$	1	T	T	T	T

Table 4. Chemical analyses of quartz diorites (Juanjorge, Cul nan Gad) and granites (Lochnagar L1, L2, L3), major elements in weight percent (Wt %), trace elements in parts per million (ppm). All Fe as Fe_2O_3; LOI = loss on ignition. See Figure 1 for distribution of intrusive phases. From Smith et al. (2002).

Wt %	Juanjorge	Cul nan Gad	L1	L2	L3
SiO_2	55.39	53.78	70.50	73.24	76.34
TiO_2	1.30	1.35	0.42	0.29	0.09
Al_2O_3	17.50	17.12	15.00	14.54	12.56
Fe_2O_3	7.34	7.20	2.46	1.77	0.65
MnO	0.15	0.16	0.10	0.08	0.08
MgO	3.95	5.02	0.95	0.61	0.05
CaO	6.22	8.51	1.93	1.74	0.40
Na_2O	3.85	3.49	3.83	3.78	3.74
K_2O	1.96	1.35	4.13	4.14	4.66
P_2O_5	0.36	0.39	0	0.05	0
LOI	1.20	1.02	0.69	0.41	0.3
Total	99.23	99.39	100.01	100.65	98.87
ppm					
Ba	379	553	560	438	10
Ce	52	71	73	55	36
Cr	62	73	26	21	9
La	31	31	31	23	9
Nb	12	14	14	12	15
Pb	7	10	21	29	19
Rb	27	45	132	171	124
Sr	784	747	305	239	96
Th	5	4	18	17	13
U	5	1	5	8	3
V	136	119	30	20	2
Y	19	26	15	10	9
Zr	208	232	188	136	74

sediment accumulated and the seas receded, until by the subsequent Devonian Period this area formed part of a continental landmass. Red sandstones attest to the arid, desert conditions prevailing at this time, with the continent lying about 30° south of the equator (Cameron and Stephenson 1985). Conditions at the Earth's surface had changed since Dalradian sediment deposition; vegetation had colonised the land, complex life had evolved in the oceans, and creatures had begun to crawl upon the land.

After intrusion of the granites, the geological story of the Lochnagar area is one of gradual exhumation; the region was distant enough from any plate margin to escape the influence of later cycles of ocean formation, subduction and mountain building that affected other parts of the British Isles. Surface processes took over from geological forces, and erosion over many millions of years caused exposure of deep crustal rocks at the land surface, as we see them today. The variation in the chemical and physical properties of the rock types which make up this area has had an impact on the landforms produced, through their varying resistance to erosion, on the soils developed, both *in situ* and on glacial debris smeared across the bedrock, on the hydrology of the area, and hence has affected the flora and fauna which colonised the land over time.

There is no direct evidence for when this area was first exposed at the surface, but one small area 5 km to the southeast of Lochnagar has outcrop of a red silcrete of the kind typically developed in desert areas under highly arid conditions (Smith et al. 2002), and is most likely the same age as red desert sediments of Devonian age in the Midland Valley. It is interesting to speculate that the Lochnagar area may have been at or near the Earth's surface since that time, and witness to the evolution of plant and animal life and the vagaries of climate as this piece of the Caledonian chain was shunted around the globe by plate tectonics, from the low southern latitudes in the Devonian, across the equator, and ever northward to our present-day latitude. How fitting, then, that this witness to the past can also allow us to peer into the future.

Summary

- The district is underlain by regionally metamorphosed Dalradian rocks, which were originally a thick sequence of marine sediments and volcanic rocks, deposited during the Neoproterozoic.
- The Dalradian rocks were deformed and metamorphosed during a mountain building episode in the early Palaeozoic, and intruded by granitic rocks in the Silurian Period.
- The Lochnagar massif is composed of a suite of genetically related granitic rocks, forming a broadly cylindrical intrusion emplaced into the Dalradian rocks.
- The present landscape, water regime, soils and vegetation have all been strongly influenced by the bedrock geology of the region.

Acknowledgements

Much of this chapter draws on experience gained during NERC-funded mapping contracts, held at Aberdeen University, 1986-1996. A full account of the geology of the area is contained in the British Geological Survey memoir for Sheet 65E, "Geology of the Ballater district", which should be consulted for any further information, and for details of primary references. Figure 1 is reproduced from the Ballater Memoir by permission of the British Geological Survey. 8 NERC. All rights reserved. IPR/86-19C. Don Mallick (BGS) is thanked for his constructive comments on an earlier version.

References

Cameron I.B. and Stephenson D. 1985. British Regional Geology: The Midland Valley of Scotland (3rd Edition). London: HMSO for British Geological Survey

Crane A., Goodman S., Krabbendam M., Leslie A.G., Robertson, S. and Rollin K. 2002. Geology of the Glen Shee district, Scotland (Sheet 56W and adjacent areas). Memoir of the British Geological Survey.

Goodman S. and Lappin M.A. 1996. The thermal aureole of the Lochnagar Complex: mineral reactions and implications from thermal modelling. Scot. J. Geol. 32: 159–172.

Goodman S. and Winchester J.A. 1993. Geochemical variations within metavolcanic rocks of the Dalradian Farragon Beds and adjacent formations. Scot. J. Geol. 29: 131–141.

Kneller B.C. 1987. A geological history of north-east Scotland. In: Trewin N.H., Kneller B.C., and Gillen C. (eds.) Excursion guide to the Aberdeen area. Scottish Academic Press for the Geological Society of Aberdeen, pp 1–50.

Smith C.G., Goodman S. and Robertson S. 2002. Geology of the Ballater district, Scotland (Sheet 65E). Memoir of the British Geological Survey.

Wright A.E. 1988. The Appin Group. In: Winchester J.A. (ed.) Later Proterozoic stratigraphy of the Northern Atlantic regions. Blackie, London. pp 177–199.

4. THE SHAPING OF LOCHNAGAR: PRE-GLACIAL, GLACIAL AND POST-GLACIAL PROCESSES

ADRIAN M. HALL (am.hall@fettes.com)
Fettes College
Edinburgh
EH4 1QX
United Kingdom

Key words: corrie, glaciation, granite, Lochnagar, moraine, pre-glacial landforms

Introduction

The Lochnagar massif dominates the skyline of southern Deeside. The granite batholith that underlies Balmoral Forest forms a rolling summit terrain which reaches 1155 m above sea level at Cac Càrn Beag and rises from the high plateau of The Mounth, developed largely in metamorphic and basic igneous rocks (Chapter 3: Goodman this volume). Set into the edge of the massif are the spectacular corries which scallop the north face of the mountain, including the great amphitheatre that holds Lochnagar. Incised also into the plateau is the broad glacial trough occupied by Loch Muick, curving with the margin of the granite ring complex (Figure 1). The geomorphology of the area is also of considerable interest for the detail of its glacial and periglacial landforms (Gordon and Ballantyne 1993). The international significance of the corrie lochan of Lochnagar for studies of Holocene environmental change means that it is important to first set the basin in the context of the processes which have shaped the surrounding terrain and which continue to operate today.

The scenery of Lochnagar and its environs is in many ways a microcosm of the wider Cairngorms, where three generations of relief conventionally are recognised (Gordon and Ballantyne 1993; Sugden 1968):

1. Major non-glacial landforms which predate the onset of regional glaciation in the Pleistocene around 1 million years ago.
2. Major landforms of glacial erosion, including corries and troughs.
3. Minor landforms produced by processes operating under glacial, periglacial and cool temperate processes which ornament the relief and which have developed largely over the past 15 ka.

N.L. Rose (ed.), Lochnagar: The Natural History of a Mountain Lake
Developments in Paleoenvironmental Research, 51–62.
© 2007 *Springer.*

This chapter examines the origins of these relief forms, together with the development of the lake basin and its surroundings through the Lateglacial and Holocene periods.

Ancient landforms

The preglacial relief of Lochnagar and its surroundings developed during the Tertiary period (65-2.5 Ma) under climates probably warmer than today but continued to evolve

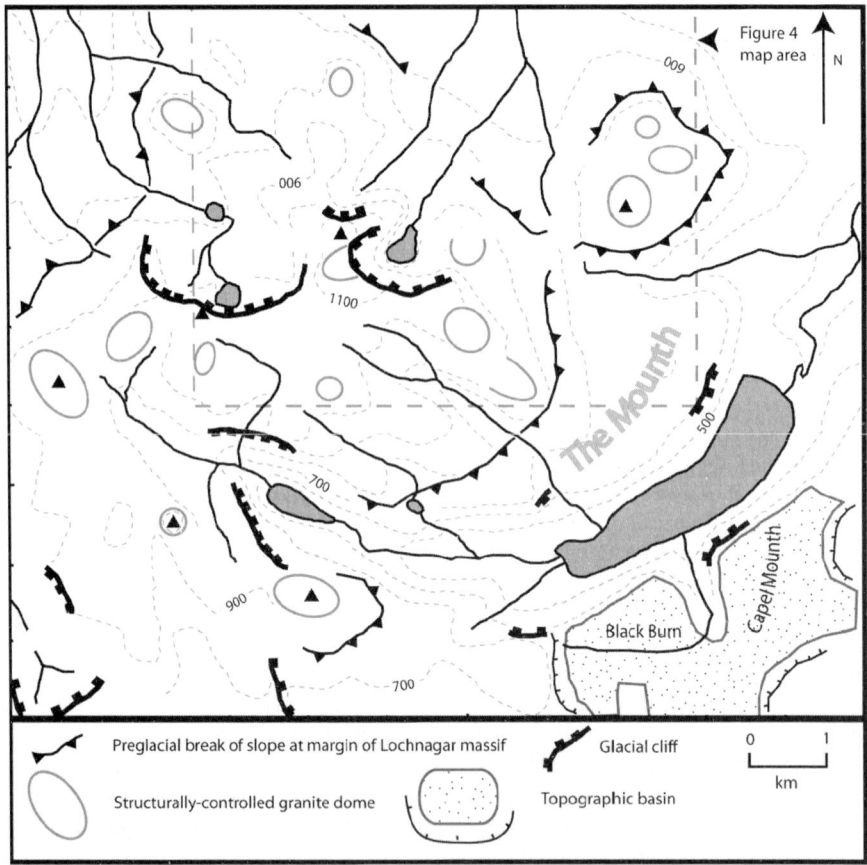

Figure 1. The Lochnagar massif and its pre-glacial and glacial landforms

through erosion throughout the Quaternary. The main elements are residual hill masses, erosion surfaces, topographic basins and valleys. The Lochnagar massif and Mount Keen form the largest areas of residual relief, rising from a marked break of slope at 730 - 750 m. The Mounth plateau abuts the massif and is one of a staircase of erosion surfaces that dominate the relief of northeast Scotland (Fleet 1938; Hall 1983; 1991). The remarkably gentle relief is probably a product of a prolonged phase of erosion close to sea level prior to uplift in the Late Oligocene, 28-23 million years ago (Hall and Bishop 2002). Set into the erosion surface are shallow basins of differential weathering and erosion, notably the diorite basin now occupied by the Black Burn. The Mounth carries a dendritic system of shallow headwater valleys which represent a little-modified preglacial drainage network (Bremner 1919). Incision of the drainage network into the uplifted surface commenced well before glaciation but has yet to penetrate fully the Lochnagar massif.

The Lochnagar massif displays typical granite terrain. Although the L3 granite forms low ground (Chapter 3: Goodman this volume), it is structural rather than lithological controls which dominate the relief. At least two vertical joint sets define cuboidal blocks, whilst alteration zones produced by hydro-fracturing at a late stage of cooling of the granite provide other major lines of weakness. Erosion of the granite and release of overburden leads to the opening of sheet joints that curve parallel to the slopes of the massif. Weathering and erosion have exploited these weaknesses to give a landscape with both curved and linear elements. The domes and rounded spurs are a reflection of the control of sheet structures over slope evolution (Figure 1). The valleys and cols often align with zones of closely-spaced jointing or of hydrothermal alteration. The linear zones of reddened, disintegrating granite in the headwall and gullies of the corrie of Lochnagar represent rocks weakened by late-stage alteration during granite cooling. Small tors rise from several domes (Figure 2), with a rectilinear geometry that displays the orientation of crossing joint sets.

Tors are minor landforms formed by non-glacial processes of weathering and erosion acting on uneven joint densities. The tors emerge from areas of low joint density. In the Cairngorms, recent work indicates that small tors emerged during the last few hundreds of thousands of years and that many have been modified by glacial erosion (Phillips et al. 2006). The subdued form of many tors in the vicinity of Lochnagar (Figure 3), including the summit tor (Addy 2005), leaves little doubt that these tors have also lost superstructure to flowing glacier ice (Hall and Phillips 2006b).

Major glacial landforms

The drama of the scenery around Lochnagar stems from a juxtaposition of the gentle slopes of the massif with the cliffs that define the margins of the corries and glacial valleys. This combination is typical of landscapes of selective linear glacial erosion (Sugden 1968). On the plateau, glacial erosion has been limited to the removal of a few metres of rock whereas in the corries and valleys there has been deep erosion. For Cairngorm corries, Gordon (2001) estimates rates of glacial deepening rates of 600 m Ma^{-1} and the scale of the Lochnagar amphitheatre requires similar rates of erosion. Glacial excavation of the over-deepened basin of Loch Muick has removed a depth of >250 m of rock. Both corries and glacial valleys must be the products of many cycles of glacial erosion during the Pleistocene.

The selectivity of glacial erosion is superbly illustrated on the Capel Mounth, the path that crosses the high plateau from Loch Muick to Glen Clova. The steep slopes which

Figure 2. The granite dome of Meikle Pap, with glacially-modified summit tor and boulder fields.

Figure 3. Glacially-modified tor on the eastern spur above Lochnagar. The tor has lost superstructure to flowing glacier ice.

rise from the valley floor are plastered with glacial deposits left by ice moving out of the Muick trough during the last ice sheet glaciation. Walking up towards the break of slope between the valley side and plateau, the till thins and pockets of weathered diorite appear and steadily increase in extent. The valley of the Black Burn carries many large diorite erratics, and sections show that these are derived in part from the excavation of unweathered rock kernels from within weathering profiles. On the plateau, the diorite is extensively weathered to granular sand and forms an extensive, shallow basin set within the surrounding schists. The Capel Mounth reveals an inverse correlation between erosion and altitude which is typical of landscapes of selective linear glacial erosion.

Such selectivity stems from the former basal thermal regimes of the glaciers that occupied the valley and covered the plateau. To be highly erosive, ice must slide across its bed and so the basal ice must be at the pressure melting point. The presence of meltwater not only promotes sliding and abrasion but also aids plucking of debris from protuberances on the glacier beds. In contrast, ice below the pressure melting point does not slide at the ice-rock interface but deforms internally. Although capable of the removal of loose blocks, as on tors, cold-based ice is essentially non-erosive and so protects the buried terrain from erosion. In general terms, the massif has been covered by cold-based ice whereas the lower ground has been scoured by thicker, warm-based glacier ice. The actual patterns are more complex, with parts of the plateau showing roughening due to glacial erosion by sliding ice, a reflection in part of the role of topography in channelling ice flow.

The last ice sheet and its landforms

The Lochnagar massif forms the highest ground in the southeast Grampians and is important for the evidence it provides for the reconstruction of the dynamics of the last ice sheet. The last (Late Devensian) ice sheet in Scotland built up from 32 k cal yr (Whittington and Hall 2002), reached its maximum extent by 26 ka cal yr (Sejrup et al. 1994), then retreated, re-advanced and finally disappeared from low ground by around 15 ka cal yr (McCabe and Clark 2003). Clapperton (1986) used the distribution of metamorphic erratics and meltwater channels to identify the flow of external ice around the northern and eastern slopes of the Lochnagar massif. Schist erratics reach elevations of 850 m on Conachcraig and channel systems incise the col between this hill and Lochnagar to 825 m (Figure 4). These features probably relate mainly to the last glacial maximum (LGM) when ice from Glen Clunie and Strath Dee deformed around the massif. The absence of schist erratics from higher elevations does not imply that the summit of Cac Càrn Beag was ice-free. Granite erratics comprising large granite blocks rest in areas of different granite lithology or in zones of high joint density at high elevations on the plateau. Complete ice cover is indicated also by glacially-modified summit tors. Many modified tors show only shallow weathering pits on exposed surfaces indicating loss of blocks to or local abrasion by the last ice sheet. Glacially-exposed rock surfaces on the tors at Meikle Pap and Conachraig carry pits up to 35 cm deep (Addy 2005). Comparisons with Cairngorm tors suggest that such pits developed during the last interglacial or earlier and so glacial modification predates the LGM (Hall and Phillips 2006a). The tors and erratics thus provide evidence of an ice sheet flowing to the northeast over the mountain during at least two stages of the Pleistocene. The northern face of the mountain shows a striking sequence of ice-marginal landforms, including boulder lines, moraine ridges and meltwater channels (Brown 1993; Clapperton 1986). The highest channels in the col east of Meikle Pap reach over 800 m OD (Figure 4), requiring that the ice sheet was at that time flowing across the mouth of the corrie of Lochnagar. Other ice-marginal ridges form a roughly parallel sequence down to around 400 m (Figure 4) and formed as the ice sheet retreated from the flanks of the mountain and towards Braemar (Brown 1993).

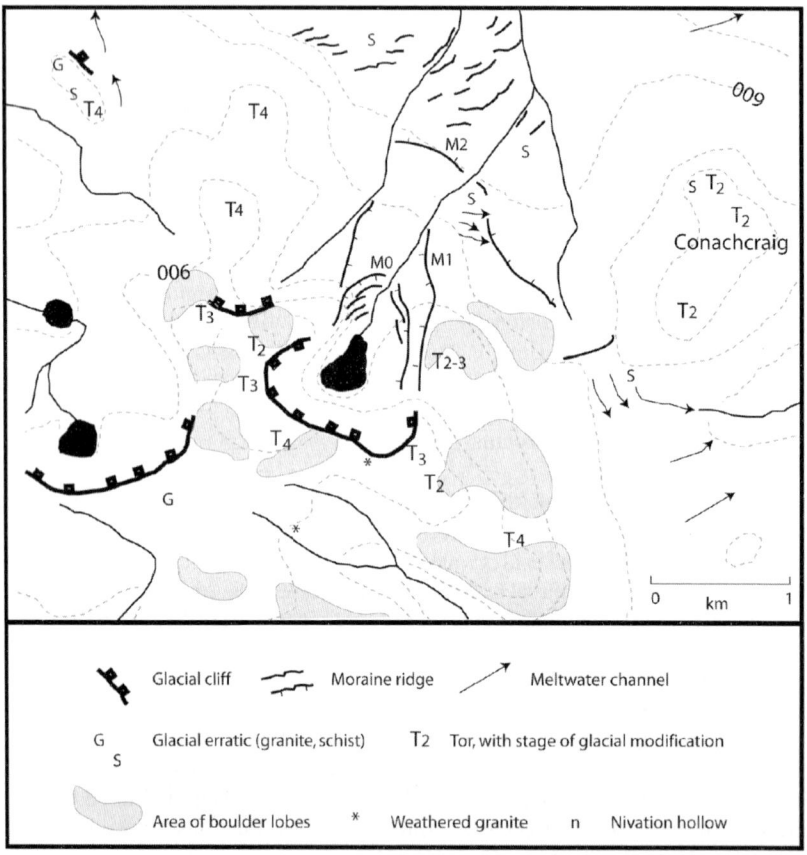

Figure 4. Glacial and periglacial landforms around Lochnagar. Data on meltwater channels, erratics and M0-M2 moraines from Clapperton (1986). Data on moraines on northern slopes from Brown (1993) and for boulder lobes from Gordon and Ballantyne (1993). The stages of glacial modification of tors follow the model of Hall and Phillips (2006b). Unmodified tors (T1) show a delicate superstructure of perched and rounded blocks. Progressive glacial modification involves firstly the removal of the superstructure (T2), followed by the entrainment of tor blocks along open horizontal sheet joints (T3). Many small tors have been reduced to plinths or slabs (T4) by the wholesale removal of upstanding joint blocks. When modified by wet-based ice, tors are shaped by block removal, abrasion, and lee-side plucking into roche moutonnées or whalebacks (T5). T1 and T5 are not represented in the map area.

Periglacial landforms

Large parts of the plateau are mantled by frost-weathered debris, with excellent examples of boulder lobes and terraces on Cuidhe Cròm and Cac Càrn Beag (Galloway 1958). A detailed survey by Shaw (1977) revealed that lobes occur on slopes of 10-34° from 640-1110 m OD. Many are structurally-controlled and developed from the steps between sheet joints. The lobes are stone-banked and up to 6 m high and 33 m across slope and comprise openwork boulders. The boulder terraces are found on gentler slopes (14 - 22°) but have comparable composition and thickness. Shaw considered both features to be inactive currently, but their age is uncertain. The absence of these features from within the limits of Loch Lomond Stadial glaciers suggests that the boulder lobes were actively forming under the intense periglacial conditions of that period (Sissons and Grant 1972) but the manner in which boulder mantles are absent from areas of the plateau last covered by warm-based ice at the LGM suggests that some plateau regolith has a longer residence time.

The northeast corrie of Lochnagar

This is an imposing example of a corrie, with a headwall up to 350 m high which overshadows Lochnagar. The loch occupies a rock basin over 25 m deep (Chapter 2: Hughes this volume). The corrie is the product of many periods of occupation by glaciers. Its northerly orientation ensured the trapping of snow blown from the plateau and ablation was suppressed by its elevation and shade. Well-defined boulder ridges (M1 on Figure 4) extend well beyond the confines of the corrie and appear to be the lateral moraines of a former glacier sourced in the corrie. These ridges terminate against boulder lines which run across the mouth of the corrie (M2) and which probably relate mainly to the margin of a glacier flowing down the Dee valley (Clapperton 1986). A series of nine arcuate boulder ridges (M0), each less than 3 m high, sits inside these former ice limits and terminate abruptly down valley, suggesting separate and later deposition from ice in the corrie. Radial ridges on the east side of the corrie floor may represent fluted moraine from this late phase of corrie glaciation (Clapperton 1986).

These delicate depositional landforms almost certainly relate to the last phases of glaciation in the corrie and its environs. There is uncertainty whether the moraines belong to the period of the retreat of the last ice sheet, when a small glacier could have remained in the corrie, or to the Loch Lomond Stadial, a short period of intense cooling in which renewed glacier growth occurred extensively in the Scottish Highlands (Table 1). Uncorrected ^{10}Be cosmogenic exposure duration times (Phillips et al. 2003) indicate that the M1 moraine ridges date from 10.4-15.2 ka. These dates imply that the M1 moraine dates from the Loch Lomond Stadial, as proposed by Sissons and Grant (1972). The M2 boulder ridge has provided a range of exposure ages from 24.9-13.4 ka. This implies that the M2 moraine relates to ice sheet deglaciation but incorporates boulders reworked from earlier glacial phases. Recent work on the edge-rounding of granite boulders demonstrates a clear difference between boulders from the M1 and M0 moraines (Kirkbride 2006). The greater edge rounding on the boulders from the M1 moraine suggests that the boulders were either exposed to a longer period of weathering spanning several thousand years or that the outer M1 moraine was exposed to intense weathering after it had retreated from its maximum position in the Loch Lomond Stadial (Kirkbride 2006). Alternatively, many boulders in the M1 moraine may have been resident on the corrie floor before the Loch Lomond Stadial and then entrained during the final advance of ice from the corrie.

Table 1. Late Quaternary events and timescales in the Lochnagar area.

Age (cal. ka BP)	OI Stage	Period				Geomorphic environments
11.5 —	1	Holocene				Weathering, soil formation and lake sedimentation. Mass movement, debris flows and avalanches. Cool temperate interglacial climate.
13 —		Loch Lomond Stadial	Late-glacial	Late Devensian		Intense cold. Corrie glacier reforms. Periglacial activity renewed on surrounding slopes.
15 —	2	Windermere Interstadial				Rapid warming to near present levels. Vegetation colonisation of recently deglaciated slopes.
26 —		Dimlington Stadial				Ice sheet glaciation after 32 ka reaching a maximum at ~27 ka. Onset of final deglaciation after 15 ka.
59 —	3	Middle Devensian				Rapidly changing cool to cold temperatures. Long periods of ice cover, with corrie and valley glaciers and probable larger ice masses.
71 —	4	Early Devensian				Probable renewed ice sheet glaciation.
116 —	5a-5d					Two cool to warm intervals separated by cold phases when corrie glaciation probably recurred.
128 —	5e	Ipswichian				Cool temperate interglacial climate, with similar processes operating to those in the Holocene.
	6	Wolstonian				Ice sheet glaciation.

Pollen analysis of peat cores taken from within the arcuate moraines demonstrate that sedimentation commenced before 9700 radiocarbon years BP (Rapson 1985). Samples of peat and a pine stump in the same area gave radiocarbon ages of 7170 ± 80 BP (SRR-2272) and 6080 ± 50 BP (SRR-1808), respectively (Rapson 1985). Cores from Lochnagar indicate that organic sedimentation commenced before 8430 ± 80 (Hela-403) radiocarbon years (Dalton et al. 2005).

Since the final disappearance of glacier ice at or after the end of the Loch Lomond Stadial, substantial volumes of debris have accumulated at the base of the headwall, a sign of recent and continuing mass movement (Figure 5). Many different processes are involved, of which the most important are rockfall and debris flows. Continuing rockfall from cliffs is indicated by the presence of fresh, angular clasts on debris slopes and on late-lying snow patches. A substantial rock slope failure has occurred in recent years at the foot of the Eagle Ridge (Figure 5). The debris slopes extend in places from the base of the cliffs down to the shoreline of the loch and include both unvegetated, active slopes and vegetated, currently inactive slopes. The main activity of snow avalanches in corries such as this lies in the erosion the upper parts of talus slopes (Ballantyne and Harris 1994) but such modification has been modest in scale at

Figure 5. The corrie headwall above Lochnagar. Note the gullies acting as debris chutes and the overlapping debris cones at their exits. A rockfall scar from 1995 is visible on the Eagle Ridge. The deep gulley at the angle of the corrie (centre) is the Black Spout: it feeds a debris flow that extends to the loch shore.

Figure 6. Sand plume on the south-east margin of Lochnagar from recent debris flows. Note the vegetated boulder slopes in the background, with few signs of recent contributions of debris to the loch.

Lochnagar. Ward (1985) surveyed the debris slopes to find slope profiles virtually indistinguishable from those of unmodified rockfall debris. Although snow avalanches occur frequently in this corrie (Ward 1984), the majority are small in volume, confined to the snow layers and travel over beds protected by old snow (Ward 1985). The evidence for recent transport by debris flows is more compelling. Each of the main gullies in the headwall has been largely swept clear of rockfall debris which has accumulated on overlapping talus cones on the middle and lower slopes. The presence of unvegetated talus suggests recent activity and this is in accordance with an increase in debris flow activity throughout the Scottish Highlands over the last millennium (Ballantyne 2002; Brazier and Ballantyne 1989).

Sediment input to Lochnagar is largely restricted to its rim and may have come from a variety of sources. The corrie receives wind-blown snow and sand-sized and smaller particles from the adjacent plateau during southerly and south-westerly gales. Whilst it is conceivable that some of this material may reach the lochan directly, the corrie headwall is probably the dominant source of debris. The distribution of rockfall debris, which mainly stops short of the loch shore, suggests that few clasts enter the loch directly by rolling or sliding. Avalanche debris may have reached the loch in the past but this was probably confined largely to the organic and inorganic materials held within the snow pack. Debris flows have reached the southwest and southeast edges of Lochnagar to leave plumes of grit and sand (Figure 6) but the debris is found only a short distance offshore. On the northeastern side of the loch there is continuing and extensive peat erosion, dating from the last 100-150 years (Chapter 6: Helliwell et al. this volume), and this organic debris must now lie largely on the loch floor. Cores from some areas of the loch show that 1.55 m of largely organic debris has accumulated over the last 9470 years, with higher organic productivity during warmer climate phases (Dalton et al. 2005).

Summary

The landforms of the area around Lochnagar belong to three generations:
- Preglacial, major non-glacial features produced by long-term differential weathering and erosion, and including erosion surfaces, residual hills, topographic basins and granite domes.
- Glacial features formed during the Pleistocene by ice sheets, valley and corrie glaciers. The corrie of Lochnagar and its surroundings display moraines which record the retreat of the last ice sheet from the flanks of the massif and the subsequent build-up and decay of a small glacier in the corrie now occupied by Lochnagar.
- Post-glacial features formed since the final disappearance of ice from the corrie 11500 years ago. Sediment input to Lochnagar has been limited, as most debris sourced from the corrie headwall appears to be retained on debris slopes that extend from the base of the headwall to the lake shore.

References

Addy S. 2005. The tors of Lochnagar. B. Sc. Thesis. Univ. Aberdeen.

Ballantyne C.K. 2002. Debris flow activity in the Scottish Highlands: Temporal trends and wider implications for dating. Studia Geom. Carpatho-Balcanica 36: 7–27.

Ballantyne C.K. and Harris C. 1994. The periglaciation of Great Britain. Cambridge University Press, Cambridge. 330 pp.

Brazier V. and Ballantyne C.K. 1989. Late Holocene debris cone evolution in Glen Feshie, western Cairngorm Mountains, Scotland. Trans. Roy. Soc. Edinburgh: Earth Sci. 80: 17–24.

Bremner A. 1919. A geographical study of the high plateau of the south-east Highlands. Scot. Geog. Mag. 35: 331–351.

Brown I.M. 1993. Pattern of deglaciation of the last (Late Devensian) Scottish ice sheet: Evidence from ice-marginal deposits in the Dee valley, northeast Scotland. J. Quat. Sci. 8: 235–250.

Clapperton C.M. 1986. Glacial geomorphology of northeast Lochnagar. In: Ritchie W., Stone J.C. and Mather A.S. (eds.) Essays for Professor R. E. H. Mellor. University of Aberdeen. Aberdeen, pp. 390–396.

Dalton C., Birks H.J.B., Brooks S.J., Cameron N.G., Evershed R.P., Peglar S.M., Scott, J.A. and Thompson R. 2005. A multi-proxy study of lake-development in response to catchment changes during the Holocene at Lochnagar, north-east Scotland. Palaeogeog. Palaeoclim. Palaeoecol. 221: 175–201.

Fleet H. 1938. Erosion surfaces in the Grampian Highlands of Scotland. Rapport Comm. Cartographique des surface d'applanisement tertiare. Union géographique Internationale. pp. 91–94.

Galloway R.W. 1958. Periglacial phenomena in Scotland. Ph.D Thesis. Univ. Edinburgh.

Goodman S. This volume. Chapter 3. Geology of Lochnagar and surrounding region. In: Rose N.L. (ed.) 2007. Lochnagar: The natural history of a mountain lake. Springer. Dordrecht.

Gordon J.E. 2001. The corries of the Cairngorm Mountains. Scot. Geogr. J. 117: 49–63.

Gordon J.E. and Ballantyne C.K. 1993. Lochnagar. In: Gordon J.E. and Sutherland D.G. (eds.) Quaternary of Scotland. Chapman and Hall, London. pp. 276–280.

Hall A.M. 1983. Weathering and landform evolution in north-east Scotland. Ph.D thesis. Univ. St. Andrews.

Hall A.M. 1991. Pre-Quaternary landscape evolution in the Scottish Highlands. Trans. Roy. Soc. Edinburgh: Earth Sci. 82: 1–26.

Hall A.M. and Bishop P. 2002. Scotland's denudational history: An integrated view of erosion and sedimentation at an uplifted passive margin. In: Doré A.G., Cartwright J., Stoker M.S., Turner J.P. and White N. (eds.) Exhumation of the North Atlantic Margin: Timing, mechanisms and implications for petroleum exploration. Geological Society, Special Publication, London. pp. 271–290.

Hall A.M. and Phillips W.M. 2006a. Weathering pits as indicators of the relative age of granite surfaces in the Cairngorm Mountains, Scotland. Geografiska Annaler 88A: 135–150.

Hall A.M. and Phillips W.M. 2006b. Glacial modification of tors in the Cairngorm Mountains, Scotland. J. Quat. Sci. 21: 811–830.

Helliwell R.C., Lilly A. and Bell J.S. This volume. Chapter 6. The development, distribution and properties of soils in the Lochnagar catchment and their influence on soil water chemistry. In: Rose N.L. (ed.) 2007. Lochnagar: The natural history of a mountain lake. Springer. Dordrecht.

Hughes M. This volume. Chapter 2. Physical characteristics of Lochnagar. In: Rose N.L. (ed.) 2007. Lochnagar: The natural history of a mountain lake. Springer. Dordrecht.

Kirkbride, M.P., 2006. Boulder edge-roundness as an indicator of relative age: a Lochnagar case study. Scot. Geog. J. 121: 219–236.

McCabe A.M. and Clark P.U. 2003. Deglacial chronology from County Donegal, Ireland: Implications for deglaciation of the British-Irish ice sheet. J. Geol. Soc. 160: 847–856.

Phillips W.M., Stewart M.A. and Kubik P.W. 2003. Cosmogenic ^{10}Be deglaciation chronology for Lochnagar, southern Cairngorm Mountains, Scotland. www.ipp.phys.ethz.ch/research/experiments/tandem/Annual/2003/18.pdf

Phillips W.M., Hall A.M., Mottram R., Fifield K. and Sugden D.E. 2006. Cosmogenic exposure ages of tors and erratics on the Cairngorm plateau, Scotland: Timescales for the development of a classic landscape of selective linear glacial erosion. Geomorph. 73: 222–245.

Rapson S.C. 1985. Minimum age of corrie moraine ridges in the Cairngorm Mountains. Boreas 14: 155–159.

Sejrup H.P., Haflidason H., Aarseth I., King E., Forsberg C.F., Long D. and Rokoengen K. 1994. Late Weichselian glaciation history of the northern North Sea. Boreas 23: 1–13.

Shaw R. 1977. Periglacial features in part of the south-east Grampian Highlands of Scotland. Ph.D. Thesis. University of Edinburgh.

Sissons J.B. and Grant A.J.H. 1972. The last glaciers in the Lochnagar area, Aberdeenshire. Scot. J. Geol. 8: 85–93.

Sugden D.E. 1968. The selectivity of glacial erosion in the Cairngorm Mountains, Scotland. Trans. Inst. Brit. Geogr. 45: 79–92.

Ward R.G.W. 1984. Avalanche prediction in Scotland: 1. A survey of avalanche activity. App. Geog. 4: 43–62.

Ward R.G.W. 1985. Geomorphological evidence of avalanche activity in Scotland. Geogr. Ann. 67A: 247–256.

Whittington G. and Hall A.M. 2002. The Tolsta Interstadial, Scotland: Correlation with D-O cycles GI-8 to GI-5? Quat. Sci. Rev. 21: 901–915.

5. LOCHNAGAR WATER-TEMPERATURES, CLIMATE AND WEATHER

ROY THOMPSON (roy@ed.ac.uk) and HELEN KETTLE
School of Geosciences
The University of Edinburgh
Edinburgh EH9 3JW
United Kingdom

and

DONALD T. MONTEITH and NEIL L. ROSE
Environmental Change Research Centre
University College London
Pearson Building, Gower Street
London WC1E 6BT
United Kingdom

Key words: air, ice-cover, instrumental, lake, temperature, thermistor, water, wind-speed

Introduction

The climate of Britain is characterised by mild summers, rain at anytime, strong downpours and overcast winters (Table 1). At low elevations Britain's marine west coast type of climate is ideal for evergreen forest growth. The marine west coast climatic regime is particularly extensive in western Europe where it stretches from central Norway to northern Portugal and as far east as Bulgaria; but is also found as a narrow strip along much of the west coast of North America and southern Chile; in New Zealand, South Africa and Southeast Australia. All these regions are dominated by prevailing winds which blow in from the ocean and give rise to a low annual range in temperature. The ocean's influence also keeps daily temperature fluctuations low. Modest seasonal changes in precipitation result in low clouds, fog and drizzle prevailing during much of the year. Moving equatorwards the influence of the subtropical highs becomes greater and the climate gradually changes to that of the Mediterranean zone, with rain during the mild winters (snow in higher elevations), and dry summers. Poleward lies the cold west coast climate, with average temperatures of below zero in the coldest month, long winters and short summers. Such sub-arctic maritime regions are places of greater extremes; coniferous (boreal) forests dominate

N.L. Rose (ed.), Lochnagar: The Natural History of a Mountain Lake
Developments in Paleoenvironmental Research, 63–91.
© 2007 *Springer*.

their lowlands. To the east the marine west coast climate is replaced by humid continental climates, which on account of their remoteness from the oceans exhibit lower precipitation totals and very distinct seasons. With future climate change the climate of Britain, while remaining of the marine west coast type, is expected to become more Mediterranean. Many aspects of the climate of the English lowlands will match today's climate in upland areas of northern Greece, central Italy, or Sardinia (Broadmeadow et al. 2005), while Scottish headwater catchments will experience conditions similar to those of the present day in the uplands of Brittany or even the western Pyrenees.

Although the Scottish mountains only rise to just over 1300m they generate all the characteristic elements of mountain climates, including orographic enhancement of precipitation, valley winds, cold temperatures, and a closer association with the free atmosphere. Hourly observations made during the period 1891 to 1903 at the Ben Nevis observatory (1344 m), when compared with observations from Fort William at sea level (Roy 1997) allow these effects to be quantified. Strong changes in temperature with height are revealed (Table 2). In the summer months maximum temperatures decline particularly rapidly with increasing elevation. The diurnal temperature fluctuations are much reduced at altitude, especially in the winter when they only amount to 0.5 °C (Roy 1997). Orographic enhancement of precipitation over the mountains, and the generation of rain shadows in their lee, is a further major influence of the Scottish uplands on the local weather particularly as much of the rain is associated with active mid-latitude cyclones (depressions) which sweep in from the Atlantic. As Scotland lies in, or close to, the path of many Atlantic depressions it is very windy, especially in the west. Being remote from the industrial and populated areas of Britain and mainland Europe it often experiences excellent visibility. However, low cloud bases frequently shroud the higher ground with hill fog (Weston 1992). Roy (1997) reports 78% of November to January days in fog, compared to 55% in May and June at Ben Nevis summit. Bad winter visibility is similarly recorded by the Lochnagar Digicam (http://ecrc.geog.ucl.ac.uk/lochnagar/digcam/) (Hughes: Chapter 2 this volume) which shows the loch, 500m distant, to be invisible 38% of the time (on 98 out of the 261 winter (Nov, Dec, Jan) days monitored between 2002 and 2005). By contrast on early-summer (May, June) days the lake was visible 89% of the time (on 109 out of 122 days monitored). It follows that occult deposition, the scavenging of cloud and fog droplets by the land surface and vegetation, can form a significant contribution to the total wet deposition of chemical species and particulates at elevated sites such as Lochnagar (Figure 4 in Weston 1992; Chapter 17: Rose and Yang, this volume).

Spatial variations in the British climate

While the UK as a whole experiences a maritime climate there are, nevertheless, differences in the degree of oceanicity or continentality of the climate in particular areas. Eastern and central Britain experiences greater continentality as it is less influenced by maritime tropical airstreams which impinge on western regions. In summer the highest temperatures are found in the south due to the latitudinal increase in solar radiation (Chapter 2: Hughes this volume). The general pattern of precipitation in Britain also varies spatially.

Table 1. 1971 - 2000 averages for Braemar (57° 00' 23''N; 3° 23'53''W; 339 metres above sea level). (Data from http://www.met-office.gov.uk/climate/uk/stationdata/braemardata.txt).

Month	Max. Temp.	Min. Temp.	Days of air frost	Sunshine	Rainfall	Days of Rainfall (>= 1 mm)
	[deg C]	[deg C]	[days]	[hours]	[mm]	[days]
Jan	4.1	-1.8	18.6	27	113	16.3
Feb	4.4	-1.8	17.7	55	68	12.5
Mar	6.4	-0.4	16.0	96	77	15.0
Apr	9.3	1.0	10.7	131	55	11.2
May	13.1	3.5	5.0	174	60	11.2
Jun	15.9	6.5	0.5	169	56	10.5
Jul	18.1	8.7	0.1	167	54	10.0
Aug	17.4	8.2	0.3	153	61	11.0
Sep	14.2	6.2	1.7	113	83	12.1
Oct	10.6	3.7	5.6	69	102	14.4
Nov	6.8	0.7	11.4	36	92	14.6
Dec	4.8	-0.9	15.9	20	95	15.0
Year	10.5	2.8	103	1210	913	154

Table 2. Change of temperature with elevation (°C) change for every 1000m (1891-1903 Ben Nevis).

	Jan	Feb	Mar	Apr	May	Jun	Jul	Aug	Sep	Oct	Nov	Dec
Max	6.5	6.5	7.5	8.3	8.4	8.0	7.7	7.7	7.4	7.1	6.3	6.3
Min	6.1	5.6	6.0	5.8	5.6	5.2	5.6	5.6	5.5	5.6	5.6	5.6

There is a strong precipitation gradient from west to east. In Scotland rainfall decreases from over 2200 mm yr^{-1} in the western highlands to under 600 mm yr^{-1} near the east coast. Lochnagar has a more easterly setting and so the cooler winters give rise to extended snow cover duration, while precipitation amounts are considerably less than to the west.

A sector analysis of air mass trajectories (at 900m elevation) over the Cairngorms during 1990 (Weston and Fowler 1991) again indicates Lochnagar's easterly climate. The bulk of Cairngorm precipitation was found to be received from the south-west sector. However, the percentage contribution from this sector was distinctly less than for west-coast sites on account of the orographic shelter offered by the Scottish highlands. The pattern of wind trajectories which give rise to rainfall has important implications for the mean acidity of the wet deposition, because quite different acidities are found to be associated with various geographic sectors. In particular, at this time (1990) Lochnagar saw only a fifth of the highly polluted (93-101 μeq H$^+$ l^{-1}) south easterlies responsible for the high annual-mean acidity of east-coast stations. Conversely the influence of clean (7-12 μeq H$^+$ l^{-1}) southwesterlies and northwesterlies, which dominate in the west of Scotland (Weston and Fowler 1991) was much reduced at Lochnagar. The net effect is that at Lochnagar both the annual mean acidity of deposition and its annual mean concentration are intermediate between the extremes of our west- and east-coasts.

Present climate

The 1971-2000 reference period

Although Scotland was well populated with active meteorological stations which operated throughout the 1971-2000 reference period, most of the stations were sited at low elevations. The highest of the stations, Braemar, at 339 m elevation, is conveniently located just 12 km to the northwest of Lochnagar. Weather recording at Braemar was established as long ago as 1855 by the Prince Consort. Braemar has a claim to being the coldest village in Scotland, having twice recorded temperatures of -27.2 °C, the first occasion being on 11 February 1895. The mean temperature is 6.4 °C, with an annual range of 10 °C from January and February, the coldest months, to July, the warmest month (Figure 1; Table 1). The spread of temperatures from year to year is depicted in Figure 1a by the height of each boxplot. Unusual years are denoted by open circles. The summer months can be seen to have the lowest inter-annual variability. The winter months display much greater variability, and an asymmetric distribution with a number of unusually cold years. Such bitterly cold spells are caused by rare incursions of Arctic air masses, in combination with strong air-temperature inversions. A succession of clear, calm nights leads to heat loss by radiation and the build up of a pool of very cold air in the valley bottom. However localities a few 100 m above the valley bottom generally lie above the temperature inversion and so are protected from the severe frost damage experienced in the valley.

Snow typically lies at 09:00 hours (GMT), in Braemar, on 63 days during the year. On the Scottish mountain tops sleet or snow falls on over 100 days in the year. On the highest mountain peaks such as Ben Nevis snow cover persists for around six to seven months of the year. Braemar's inland setting gives rise to its annual total of some 900 mm yr^{-1}, rain falling on almost one day in two (42% of all days; Table 1). Braemar sunshine averages 1200 hour yr^{-1}, which is typical of highland valleys, although around 17% less than that enjoyed at coastal stations. April to July tend to be the driest months, October to January the wettest (Figure 1b and Table 1). In comparison to the Braemar statistics of Table 1, Lochnagar being 446 m higher is likely to be 3 °C cooler, 600 mm yr^{-1} wetter, and about 20% less sunny (unpublished calculations by the authors).

Teleconnections

The North Atlantic Oscillation (NAO) is a phenomenon associated with winter fluctuations in temperatures, rainfall and storminess over much of Europe. One of the simplest definitions of the NAO index is that it is the difference in pressure at sea-level between the Azores and Iceland. Long distance connections are formed by way of changes in synoptic weather patterns, particularly those associated with changes in the frequency and magnitude of the westerlies. When the NAO index is 'positive' there is a steeper than usual pressure gradient between the Azores and Iceland, and as a consequence westerly winds are stronger or more persistent and so northern Europe

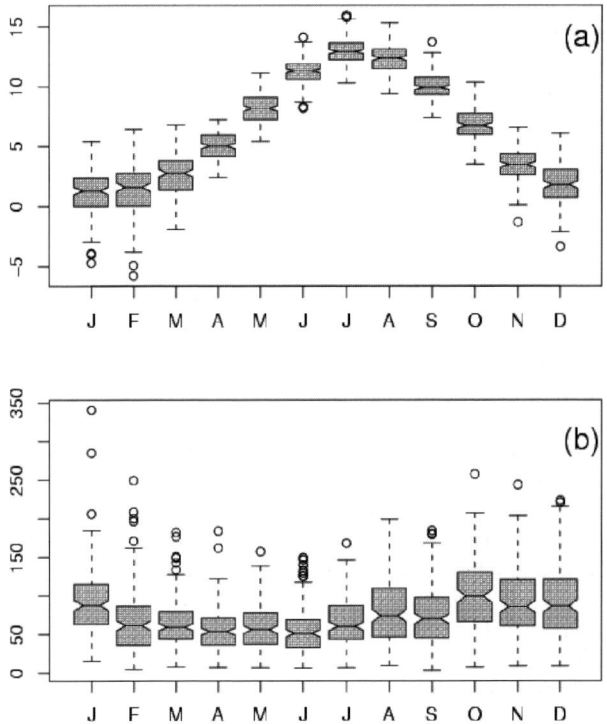

Figure 1. Boxplots summarising the variability of the present day climate (1857 to 2004) at Braemar. a) Mean air-temperature (°C), b) Precipitation (mm). Boxplots summarise a great deal of information very succinctly. The centre of the box shows the median for each month. The bottom and top of the box show the 25 and 75 percentiles respectively (i.e., the location of the middle 50% of the data). The horizontal lines joined to the box by the dashed line (or 'whisker') shows 1.5 times the interquartile range of the data. Points (outliers) beyond the whiskers are drawn individually. Boxplots not only show the location and spread of data but also indicate skewness (asymmetry). For example, in January the range of lower precipitation amounts is much less than the range of higher amounts.

tends to be warmer and wetter than average. When the NAO index is 'negative', westerly winds are weaker or less persistent, northern Europe is colder and drier and southern Europe warmer and wetter than average. In the west of Britain, wintertime (Dec - Feb) precipitation shows a strong correlation (r = +0.75) with the North Atlantic Oscillation (Hurrell 1995). Walker and Bliss (1932) reported a correlation of +0.84 between Stornoway (Dec - Feb) temperatures and the NAO index. The correlations found between mean monthly temperatures at Braemar (1857 to the present day) and the NAO index can be taken as being typical for Lochnagar. The correlations are, as expected, highest in the winter months, rising to +0.77 in January. Precipitation correlations are only significant in the winter, when they rise to +0.37. NAO correlations of this magnitude are typical in easterly settings in Britain. Edinburgh, for example, has January NAO correlations of +0.74 and +0.34, for air temperature and precipitation respectively.

Records from the instrumental period

Reconstructing air-temperature at Lochnagar using instrumental data

Britain is well provided with long instrumental records. The longest is the well-known Central England series of Manley (1953; 1974) beginning in 1659. Although it derives from the English lowlands, 600 km to the south, the central England monthly air-temperatures show respectable correlations with Braemar temperatures. The mean monthly correlation is 0.82. Other instrumental records, dating back to the late 1700s, of air-temperature, precipitation and pressure are available from Inverness 80 km to the north, Aberdeen 90 km to the east and Edinburgh 120 km to the southwest. Correlations of Edinburgh and Braemar mean monthly air temperatures reach to over 0.94 in March, and average 0.89 through the year. The long English and Scottish series can thus be readily used to reconstruct the Braemar record, and so provide estimates of change at Lochnagar back several centuries (Agusti-Panareda et al. 2000; Agusti-Panareda and Thompson 2002). Figure 2 plots the reconstructed air temperatures for winter (DJF), spring (MAM), summer (JJA) and autumn (SON). Following cold decades around the 1780s and 1810s, annual temperatures for Lochnagar are found to have generally warmed at an average linear rate of +0.3 degrees century^{-1} in the 1800s and +0.5 degrees century^{-1} in the 1900s (Table 6 in Agusti-Panareda and Thompson 2002) . The 20th century rise is mostly due to increases during autumn and spring (Figure 2).

Recent years have seen many more recordings being made in mountain areas. For example an automatic weather station (AWS) has been operating on the summit of Cairngorm (1245 m, 30 km to the northwest of Lochnagar) since 1977, recording wind speed, direction and temperature every 30 minutes. The severe icing and wind regime of the summit of Cairngorm is overcome by exposing the meteorological sensors only briefly to the free atmosphere. For the most part the sensors remain protected in a warmed housing (Baker et al. 1979).

The automatic weather station at Lochnagar

An automatic weather station (AWS) was established within the Lochnagar catchment in August 1996 as part of the EU's MOLAR research programme and has been maintained since this time by staff at the Environmental Change Research Centre (ECRC). The AWS was set-up to measure air temperature, wind speed, wind direction, air pressure and relative humidity at 30 minute intervals whilst rainfall was measured daily (see Box 1). Following a series of power failures, the AWS was replaced in 2003 using funding from the Department for Environment, Food and Rural Affairs (DEFRA). Monitoring of these six parameters was continued at the same frequency along with measurement of solar radiation and soil temperature (at a depth of 20cm). All these data

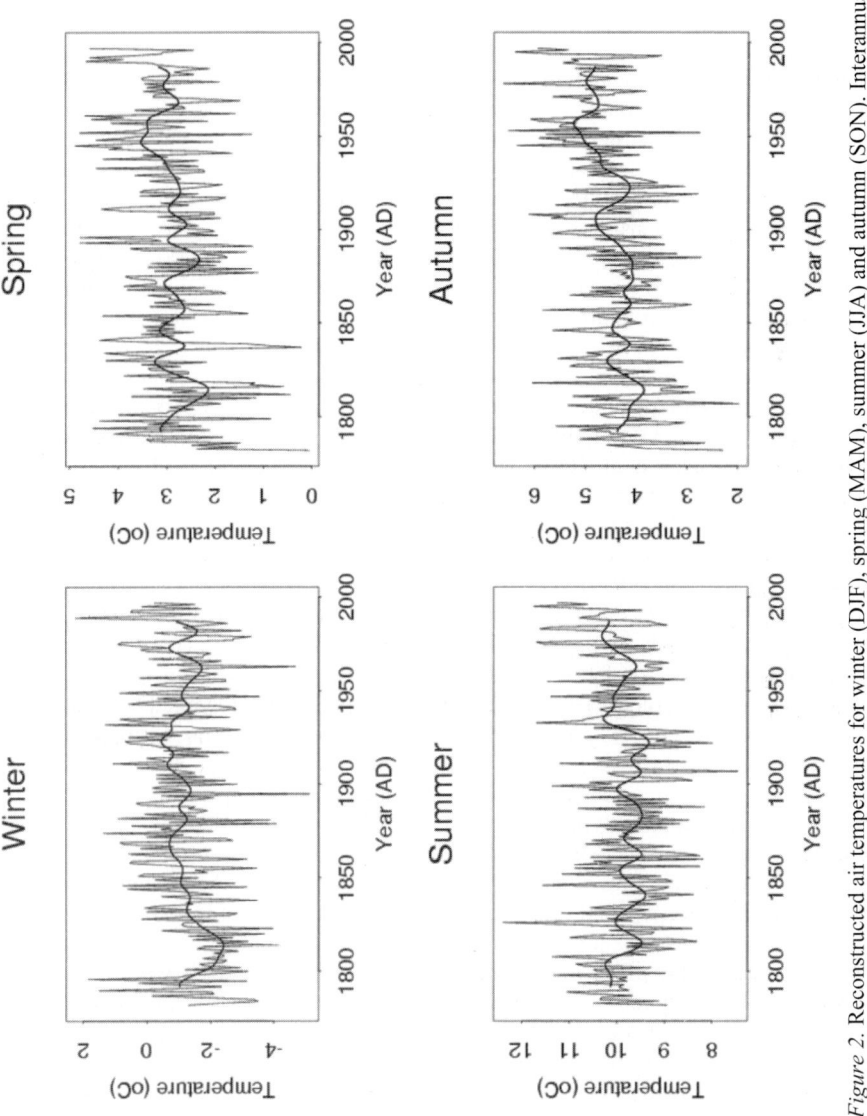

Figure 2. Reconstructed air temperatures for winter (DJF), spring (MAM), summer (JJA) and autumn (SON). Interannual variations are shown by the thin lines.

are now downloadable directly from the logger via a modem link. The climate predictions described in Chapter 18 (Kettle and Thompson this volume) have used these data in their models. The meteorological data from the first AWS, covering the period 1996 – 2000 is given in Rose (2001). The summary below covers the measured data from both AWSs for the 9-year, 1996 – 2004, period.

Automatic weather station data summary

Monthly mean, minimum and maximum air temperatures are shown in Figure 3a. Monthly mean air temperatures generally range between 0 °C and 12 °C. Since the AWS logging began in August 1996, monthly means have only fallen below 0 °C on two occasions and have not done so since December 1999. However, this may in part be an artefact of winter power failures preventing the full temperature record being measured during some winter months. Monthly minima range between -7 °C (Nov 1999) and 10.2 °C (Aug 2003); and maxima between 2.5 °C (Feb 2002) and 17.2 °C (Aug 2002). August has tended to be the warmest month with maximum temperatures being logged during this month in six of the eight years (1997 – 2004). The highest recorded temperature at the Lochnagar AWS (to the end of 2004) was 24.4 °C at 1700 hrs GMT (Greenwich Mean Time) on the 7th August 2003. This was a period of exceptionally high temperature both at Lochnagar, with the highest 23 measurements ever recorded by the AWS falling on the 6th and 7th of that month, and indeed over much of Europe (Schär et al. 2004). However during the 20th century, July has tended to be slightly warmer than August over much of eastern and central Scotland (Roy 1997, Harrison 1997). By contrast, monthly minima and lowest annual recorded temperatures at Lochnagar occur in a number of months from December (1999, 2000, 2001) through to April (1998). The lowest temperature recorded at Lochnagar by the AWS is -9.4 °C at 0400 GMT 8th January 1997 and also at 2200 GMT on 9th April 1998. These were both by far the coldest periods for these years with the coldest 20 recorded temperatures for 1997 recorded between 2230 and 0730 on the 7th and 8th January 1997 and the coldest 12 recorded temperatures for 1998 recorded between 2030 and 0200 on the 9th and 10th April 1998. Since this time the lowest temperatures for each year have ranged between -8.1 °C (2000) and – 6.8 °C (2002). Again, power loss during winter months may have resulted in the loss of the complete record of lowest temperatures. No particularly extreme cold-spells have occurred during the nine-year operation of the Lochnagar AWS. For example, on the two coldest nights experienced by the AWS (early morning 8th January 1997 and late night 9th April 1998) when the minimum temperature reached -9.4 °C the minimum temperature at Braemar was -4.0 and -4.5 °C respectively.

Monthly mean and monthly maximum recorded wind speeds are shown in Figure 3b. In general, windiest months at Lochnagar fall between December and February whilst calm periods (0 m s^{-1}) occur throughout the year but not usually for more than a few hours. Highest recorded wind speed was 16.6 m s^{-1} at 0700 on 27th December 1998. There has been a slightly declining trend in both monthly mean and maximum wind speeds over the recorded period and since the highest recorded speed in December 1998, annual maximum recorded wind speed has decreased each year to only 10.7 m s^{-1} in 2004 (Insert in Figure 3b). However, once again the loss of winter data must be borne in mind.

The direction of wind as recorded by the Lochnagar AWS is undoubtedly greatly influenced by the morphology of the corrie. Figure 4 shows the monthly frequency of wind directions at Lochnagar using 1999 as an example. The dominant wind directions are seen to be north-south which is also the aspect alignment of the corrie basin. For 1999, 28% of the wind directions are within the 45° centred on due north and 24% for the 45° centred on due south. This is in contrast to only 15% for the 45° centred on the southwest, the direction generally considered to be the prevailing wind-direction for the UK and the prevailing direction found for low-lying areas with minimal local topographic effects such as the island of Tiree (Roy 1997). The wind directions within the corrie therefore probably bear little relation to the wind direction beyond the catchment. Nevertheless, whilst only significant on a small scale, these wind directions within the corrie are important with regard to the movement of rafting ice on the loch during ice break-up, and the local wind strength is of paramount importance in determining the state of stratification of the lake in the open-water season.

Monthly rainfall data (Figure 3c) shows that rain is typically recorded on between 15 and 30 days a month and this does not include precipitation via snowfall or occult deposition via cloud water inputs. The wettest season during the last nine years appears to have been the autumn with the highest daily rainfall for each year falling between September and November for every year of AWS operation except 2002 when 64 mm of rain fell on 2nd February. The highest measured daily rainfall occurred on 21st September 1999 when 102 mm of rain fell, accounting for 37% of the total monthly rainfall. No trends are apparent in this short dataset.

Thermistor chain data

VEMCO TR 8-bit thermistor loggers are deployed at ten depths (surface, 1.5, 2.5, 3.5, 4.5, 6, 7, 9, 14 and 19 m) along a 'chain' in the deep area of Lochnagar. These are anchored at the base and buoyed at the surface to maintain their position. The 'surface' logger is located 5 – 10 cm below the buoy and some shading will inevitably occur. The first thermistor chain was installed in September 1999 without a surface logger and this was added in August 2001. However, temperature differences between this surface thermistor and that at 1.5 m are mostly negligible and only occur during ice cover or in exceptionally warm and calm periods. The temperature resolution of the loggers is 0.1 °C with a ± 0.2 °C accuracy over the temperature range – 4 to 20 °C. The maximum water temperature recorded was 18.5 °C at 1800 hrs on the 7th August 2003 corresponding to the period of highest ever recorded air temperature (see above) of 24.4 °C one hour earlier. Wind speeds at this time were low (<1 m s^{-1}) causing the surface thermistor to read up to 4 °C higher than the 1.5 m thermistor. However, by the next afternoon, wind speeds had increased to c. 4 m s^{-1} and this surface stratification had broken down. Periods of stratification and mixing are easily discernible from this dataset.

Figure 5 shows a summary isotherm diagram for the water temperature data for 2000 for the 1.5 – 19 m section covered by the thermistor chain. It is apparent that the water column is well mixed for most of the year with stratification mainly in July to August and inverse stratification in January and February during the main period of ice cover. This is a typical pattern for the loch and similar periods of stratification occur in all other years since the thermistors were first deployed. They are in good agreement with

BOX 1 The Lochnagar AWS logger and sensors

The AWS and all sensors sited at Lochnagar are made, or supplied by Delta-T Devices Ltd of Cambridge, UK. Data are logged on a Delta-T DL2e logger. The specifications for the instruments used at the site are as follows (Data from Delta-T manual, with specific calibrations for individual sensors):

Air temperature:
Delta-T sensor AT1. Accuracy = ± 0.1°C over the range 0 – 70 °C. Range -20 – 80 °C. Measurements are taken every minute and the average over each half-hour recorded.

Wind speed:
Delta-T anemometer AN1. Threshold: 0.2 m s^{-1}. Maximum speed: 75 m s^{-1}. Linearity: 2%. Accuracy: 1% at ± 0.1 m s^{-1}. Calibration: 0.795 revolutions Hz m^{-1} s^{-1}. Temperature range: -30 - +55 °C. The number of anemometer revolutions is counted over a half-hour period and an average wind speed for that time period calculated.

Wind direction:
Delta-T wind vane WD1. Accuracy: ± 2 degrees in steady winds over 5 m s^{-1}. Resolution: 0.2 degrees. Maximum speed: 75 m s^{-1}. Temperature range –50 – +70 °C. Life: 5 x 10^7 cycles, equivalent to 10 year's typical exposure. Measurements are taken each half-hour. No averaging is undertaken.

Relative humidity:
Delta-T sensor RHT2. Operating temperature: –20 °C to 80 °C. Sensitivity: 10mV per %RH; 0 – 1 V. Time response: < 10s between 10 – 75% RH. Accuracy: ± 2% RH (5 – 95% RH), ± 2.5% (RH < 5% and >95%). Measurements are taken every minute and the average over each half-hour recorded.

Air pressure:
Delta-T barometric pressure sensor BS4. Pressure range: 600 – 1060 hPa (mbar). Temperature range: –40 °C to 60 °C. Linearity: ± 0.45 hPa. Accuracy at 20 °C: ± 0.5 hPa; 0 – 40 °C = ± 2 hPa; -20 °C – 40 °C = 4 hPa. Resolution: 0.1 hPa. Measurements are taken every minute and the average over each half-hour recorded.

Rainfall:
Delta-T Tipping bucket rain gauge RG2. Funnel internal diameter: 159.6 mm (giving area of 200 cm^2). 0.200 mm rain per tip, individually calibrated. Accuracy: ± 2%. Capacity: 2.4 mm per minute. Rainfall calculated on a 24 hour basis.

Solar energy:
Delta-T energy sensor ES2 (silicon photodiode). Sensitivity: 10.0 mV per kW m^{-2} of solar radiation in the 300 – 3000 nm waveband. Accuracy: ± 3% at 20°C. Linearity: ± 1% 0 – 2 kW m^{-2}. Azimuth error: ± 1% over 360°. Stability: typically better than ± 2% per year. Response time: 10 µs. Temperature limits: -10 – 60 °C. Measurements are taken every minute and the average over each half-hour recorded.

Soil temperature:
 Delta-T soil temperature probe ST1. Accuracy: ± 0.2 °C over 0 – 70°C. Range: -20 – 80 °C. Response time: 6 secs. Drift: ± 0.02 °C in 8 years at 25 °C. Measurements are taken every minute and the average over each half-hour recorded.

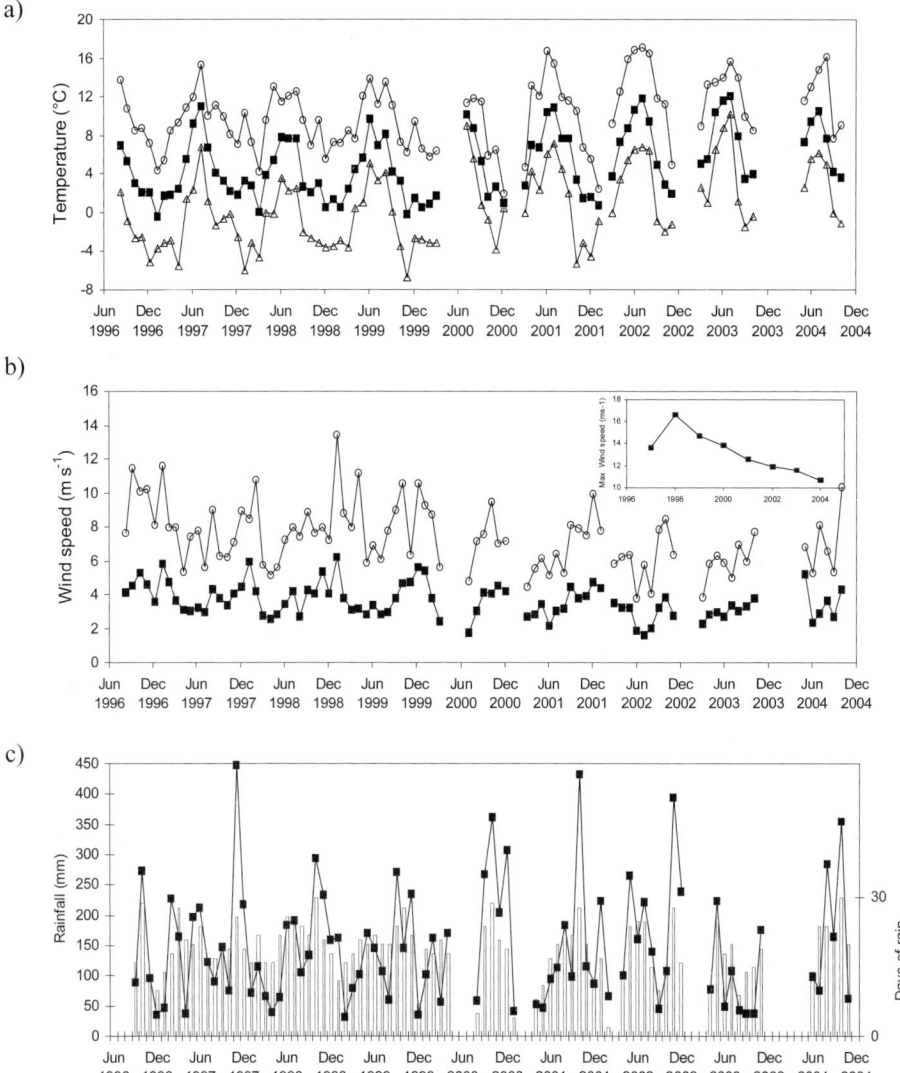

Figure 3. Lochnagar AWS data. (a) Monthly mean (solid square), minimum (open triangle) and maximum (open circle) air temperatures. (b) Monthly mean (solid squares) and monthly maximum recorded wind speeds (open circles). Insert shows annual maximum recorded wind speed. (c) Monthly rainfall data. Bars show number of days with rain, squares monthly rainfall totals.

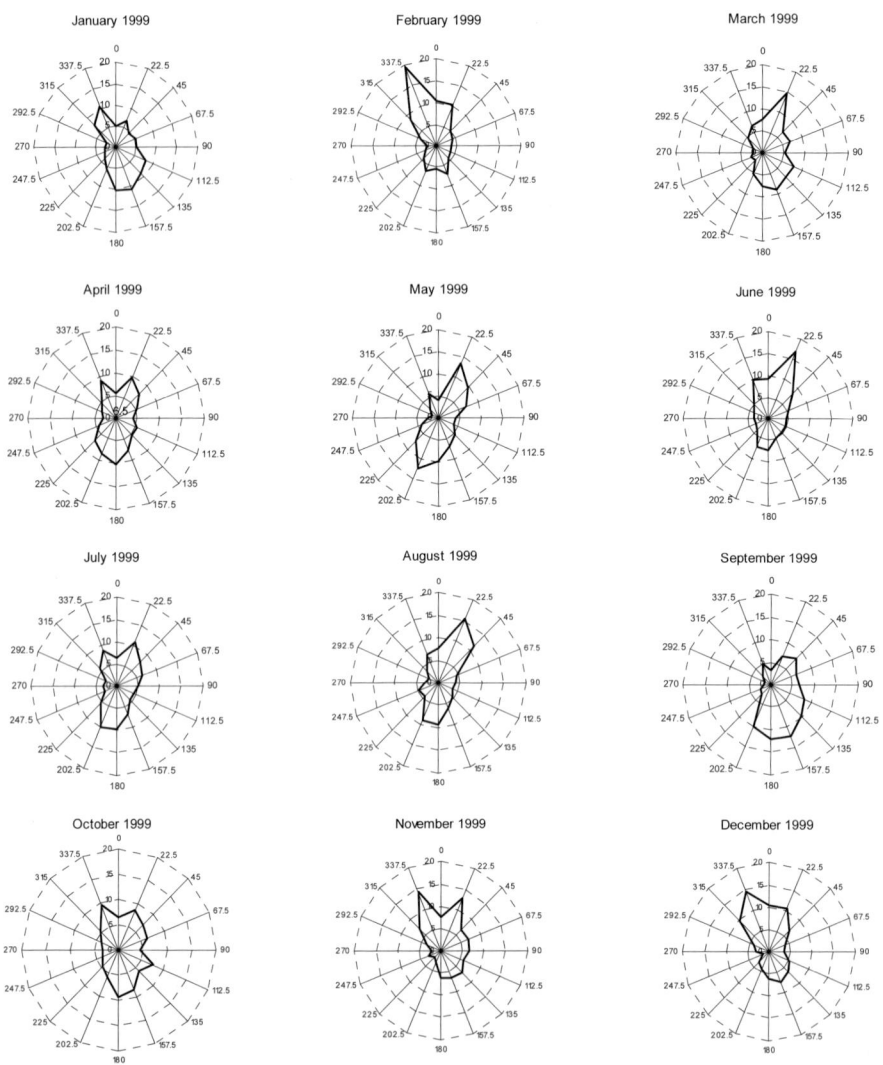

Figure 4. Lochnagar AWS data. Wind direction roses. January to December 1999. Wind direction changes little from month to month being controlled by the local topography.

irregular temperature profiles taken for the full water column since 1995 (Monteith unpublished data).

Water temperature is a fundamental parameter in the functioning of the loch and the annual variation can be seen to be driving the temporal patterns of other processes. One example is shown in Figure 6 where a good inverse agreement can be observed between the temporal trends in lake water temperature (as exemplified by the 1.5 m thermistor) and water silicon (Si) concentration data over a five year period. Si concentrations can be seen to increase as temperature decreases in the autumn and decrease in the spring as temperatures increase. This is undoubtedly due to the Si uptake by diatoms during the main growth period thereby reducing water concentrations. Diatoms have been found to be the main component of the coarse silt fraction in the sediment record of Lochnagar (Dalton et al. 2005; Chapter 8: Rose this volume) and therefore the period of diatom growth, driven by water temperatures, can be seen to be affecting Si concentrations. Lochnagar water chemistry is discussed further in Jenkins et al. (Chapter 9 this volume).

Ice-cover

There are no historical monitored data for ice cover of Lochnagar or snow cover of its catchment and the only published records that exist are the occasional observations reported in climbing journals such as the Cairngorm Club Journal and the Scottish Mountaineering Club Journal. So, for example, we know that in January 1896 "the loch reposed under a covering of ice" and was also completely ice-covered on the 1st April of that year (McKenzie 1897) whilst on the 3rd February 1899 "the corrie was in its grandest winter costume all the gullies being full of snow and the loch frozen" (Anon 1900). Similarly, we know the loch was frozen on 1st April 1899 (Anon 1900) and "early April" 1905 (Levack 1906) whilst the loch was ice-free in late-May 1894 (Rose 1895). Thus, prior to the regular recording of observations only anecdotal evidence exists and this suggests that complete ice cover occurred on the loch between January and April and partial ice cover often between November and June (Patrick et al. 1991).

Regular observations of snow and ice-cover at Lochnagar began in August 1996 and continue through to the present. These observations are weekly or fortnightly and consist of an estimate of the percentage snow and ice cover on the catchment and loch respectively, the depth of snow (mean of five measurements in an undrifted area of level ground) and ice thickness (measured if possible). These data are presented in Figure 7 and show the highly variable nature of these parameters in recent years. The winter of 2000/01 appears to be most severe of recent times with the longest period of continuous ice cover and the thickest snow and ice measurements. By contrast, the snow and ice cover data for other years show far more variability with ice cover coming and going several times through the winter and snow cover rarely attaining 100%. However, patches of snow continue to remain in sheltered cracks and gullies on the corrie back-wall through to June or July. The coarse resolution of these data requires that trends should be treated with caution, but it is interesting to note that only in the winter of 2000/01 did the loch have any significant ice cover during April whereas full ice cover in this month was mentioned in at least two years of the early climbing reports mentioned above. Anecdotal evidence for more prolonged snow cover in the past include Backhouse (1849, in Birks 1996) where four feet (1.2 m) of snow was recorded whilst climbing the Black Spout in August 1848.

Since August 2002 an automatic digital camera has taken a photograph of the loch daily at 1200 GMT thereby greatly increasing the resolution of catchment snow-cover and loch ice-cover data. Figure 8 shows examples of the daily images and illustrates the

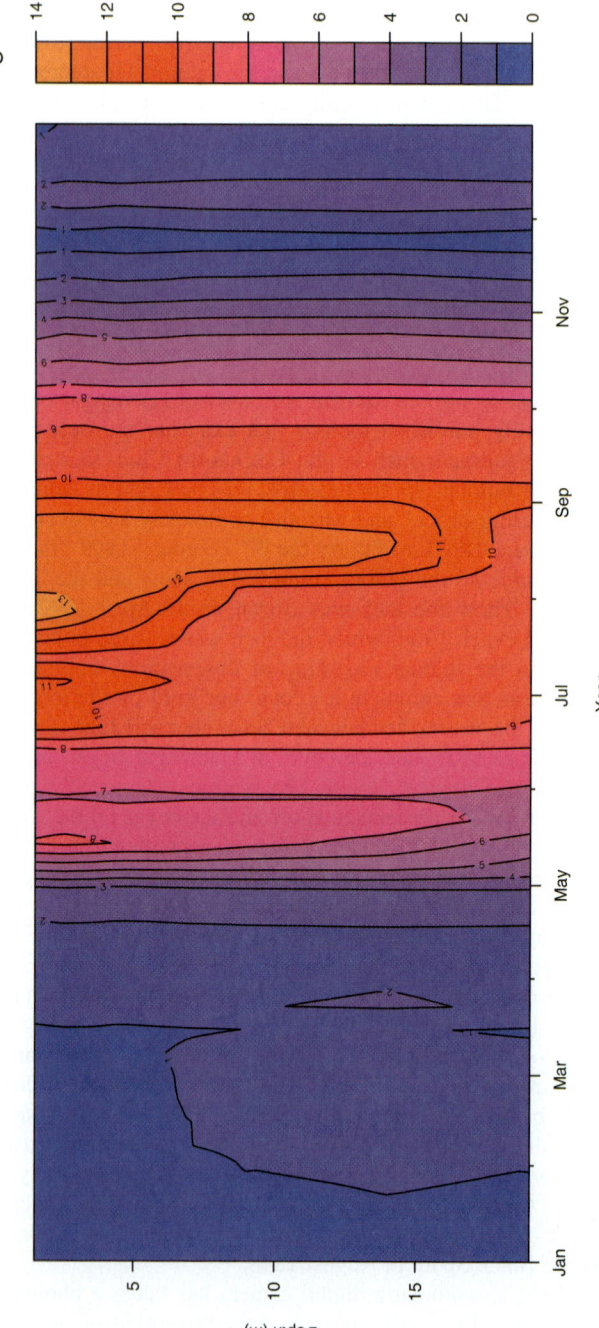

Figure 5. Summary isotherm diagram for the water-temperature data for 2000 for the 1.5 to 19m section covered by the Lochnagar thermistor chain.

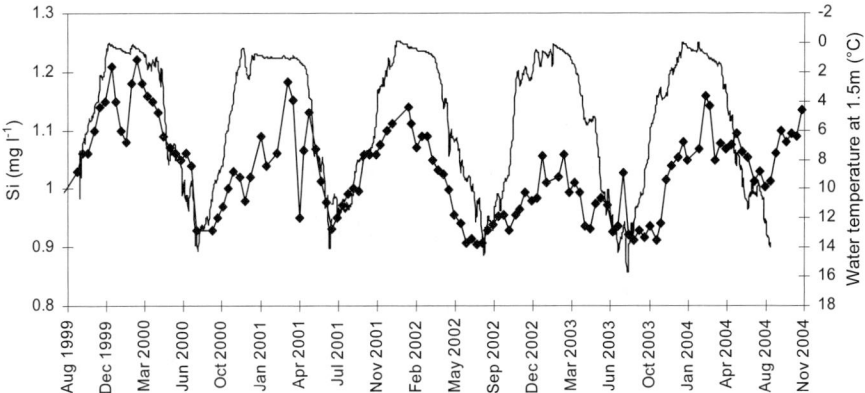

Figure 6. Silicon concentration (♦) and water temperature (inverted scale). Note how the main decrease in silicon concentration each year occurs when water temperatures are rising most rapidly. This is decrease in silicon is probably due to diatom production.

rapidity of ice and snow loss. These daily photographs allow the dates of ice-on and ice-off to be determined accurately and so help tune ice-cover modelling algorithms (see Chapter 18: Kettle and Thompson this volume). The ice-cover duration on Scottish lochs turns out to be very sensitive to climate change. As Kettle and Thompson discuss in Chapter 18, Lochnagar at 788 m happens to lie within a particularly sensitive elevation range. At present Lochnagar is frozen for about three months in the year. In general snow-covered ice reflects most (>70%) of the incident sunlight and so the loch remains frozen for a considerable time. However in a maritime climate just a modest warming will mean that ice will have great difficulties in becoming established, as convective exchange will keep the loch waters warm in early winter and inhibit the initial development of ice. Thus ice-cover duration at Lochnagar is very responsive to temperature changes. It is likely that colder winters, as encountered in the Little Ice Age, would have caused ice-cover duration to have been extended by over 6 weeks (author's unpublished calculations).

Reanalysis assimilation data

Since the advent of satellites, the quantity and variety of weather information collected has multiplied dramatically. The World Wide Web now gives access to vast quantities of data. Reanalysis assimilation provides an invaluable synthesis of these huge, multifaceted datasets. In reanalysis a frozen NWP (numerical weather prediction) analysis forecast system is used to statistically reintegrate the short-term forecasts and observations. The advantages of reanalysis data assimilation are that it offers several years of fields from an unchanging, state-of-the-art, system. Satellite temperature soundings, satellite winds, radiosondes, aircraft and surface measurements are all incorporated. All these observations, after quality control, are processed to create an

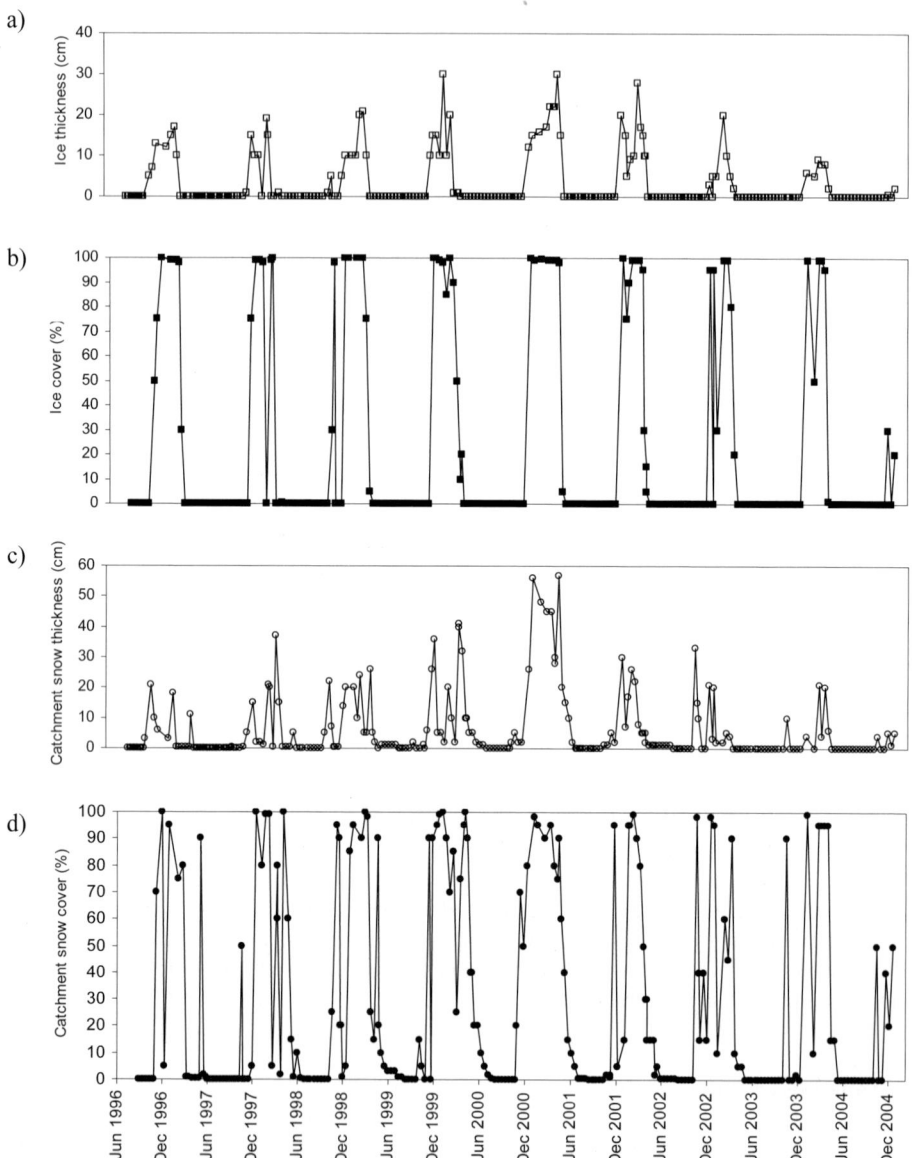

Figure 7. Snow and ice. (a) Loch-ice thickness (cm), (b) Ice cover (%), (c) Catchment snow thickness (cm), (d) Catchment snow cover (%).

optimum analysis by combining the information from the forecast and observations and by taking into account their error characteristics. The resulting assimilation describes the state of the atmosphere, at 6-hourly intervals, back to 1948 at numerous pressure levels. When downscaled, reanalysis products can provide useful weather information even in remote mountain areas normally lacking instrumental records (Kettle and Thompson 2004) at sub-daily intervals as far back as 1948 (Kalnay et el 1996). It is also invaluable for filling in missing data values.

Water temperature

Surface water-temperatures of oceans, seas and lakes are controlled by energy balances involving radiation, sensible heat exchange, evaporation and transfer of latent heat, as well as wind driven mixing, advection and circulation patterns. Deterministic models (Stefan et al. 1998; Chapter 18: Kettle and Thompson this volume) allow the underlying physical processes and dynamics to be explored. Empirical models (e.g., Kettle et al. 2004, Thompson et al. 2005) provide an alternative approach. Here the aim is to establish statistical associations between the atmosphere and hydrosphere (e.g., between air- and water-temperatures; or between ice-cover duration and air temperatures). If simple, parsimonious relationships can be established, and demonstrated to possess skill under the present-day climatic regime, then they are likely to be of value in assessments of the impacts of climate change, or reconstructions of past conditions such as those of the Little Ice Age.

Unlike water which is slow to heat up, air is very sensitive to variation in the components of the energy budget. Air temperature thus shows high frequency variability which needs to be smoothed in order to relate it to water temperature. The empirical model we suggest uses an exponential smoothing following the method described in Kettle et al. (2004). When water heats up its density changes which encourages the lake to stratify on hot days and to mix on cool days when the surface temperatures fall below the temperature of the deeper water. However, if the water temperature drops below 4 °C the temperature-density relationship reverses so that warmer water (<4 °C) is more dense than cooler water. This reversal is difficult to include in a simple empirical model, so we have chosen to run our model only for water temperatures > 4 °C. Similarly a separate model is needed to give information about ice cover on the lake. The heart of our empirical model is the exponential smoothing function, f

$$f(air, \alpha) = \alpha * air(t) + (1 - \alpha) * f(air(t - \Delta t)) \tag{1}$$

where air(t) is the air temperature at time t and Δt is a time interval (in this case, one day); α is the exponential smoothing coefficient which is in the interval 0 to 1. When α = 1 there is no smoothing, and as α approaches zero the smoothing increases so that the day by day fluctuations in air temperature have very little correlation with lake surface temperatures. This means that α is related to the speed at which the water temperatures react to air temperatures and is thus related to the volume of water above the mixed layer depth or to the total volume of the lake if the lake is fully mixed (see Kettle et al. 2004). This type of exponential smoothing model was fitted to a number of lochs in

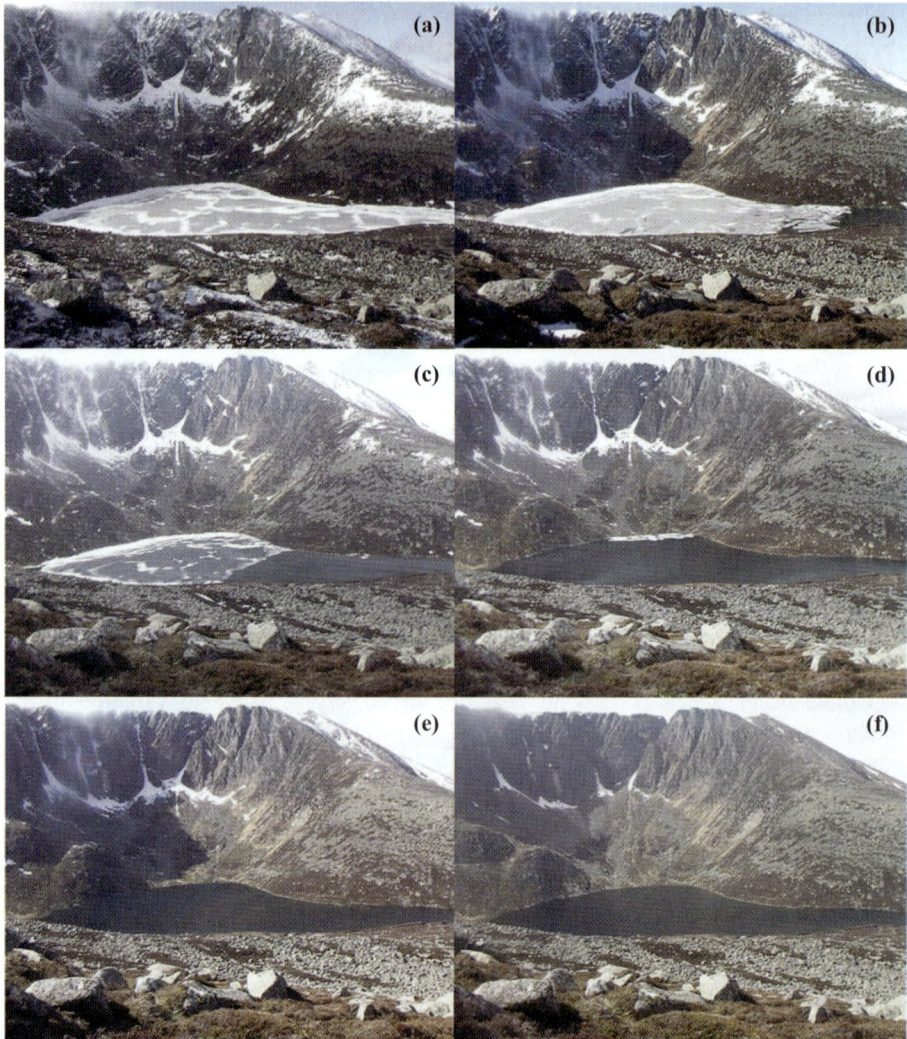

Figure 8. Montage of six digicam photographs taken during March and April 2003. (a) – (f) are 6th, 18th, 22nd, 29th and 30th March and 18th April respectively. (a) – (c) serve to illustrate how snow cover in recent years has been rather ephemeral (cf. Figure 7d) with rapid clearing of the catchment. (a) – (e) chart the demise of ice-cover from 100%, on 6th March, to 0% on 30th March. On 22nd March 2003 (c) the loch is half way through ice-melt following an initial ice-break on 11th March, eleven days previously. At this time most of the Lochnagar catchment is free of snow, although a light dusting from two days previously coats the corrie back-wall. Small snow patches remain at the base of the back-wall, and on the distant ridge tops. The thermistor chain probably became free of the wider ice sheet by the 24th, and definitely by the 25th. The ice disappears from the lake on March 29th (e). By 18th April only minor snow patches linger in back-wall gullies, while flowering of *Vaccinium myrtillus*, which gives a purple tinge to the low mounds in the foreground, is approaching full bloom (anthesis).

Scotland monitored during the EU funded EMERGE project (Patrick and Wathne 2000) and the smoothing coefficient was found to be inversely related to the area of the lake (see Figure 9). Examples of typical thermistor data for Lochnagar are shown in Figures 10 and 11.

We find that lake surface temperature (LST) at Lochnagar can be modelled as

$$LST = 0.88*f(aws,0.125) - 1.30*\sin(2\Pi\ doy/365.25) - 3.08*\cos(2\Pi\ doy/365.25) \quad (2)$$

where aws is the mean daily air temperature at the Lochnagar automatic weather station; the last two terms simply describe an annual cycle; doy is the day-of-the-year starting from 1 January and f is the exponential smoothing with a smoothing coefficient = 0.125. The fit is shown in Figure 12. The mean absolute error is found to be 0.81 °C. Using the daily water temperature model (Equations 1 and 2) it is possible to use daily air temperature data from Balmoral (9 km to the north) to reconstruct water temperatures (>4 °C) at Lochnagar back to 1960. The mean and maximum modelled water temperatures are shown in Figure 13. There is an obvious increasing trend in both mean and maximum temperatures. The regression lines indicate that during the 40-year period, from 1960 to 1999, maximum temperatures increased by 1.3 °C and mean temperatures by 0.6 °C.

In assessing the likely fidelity of applying the exponential smoothing model of Figure 9 to earlier, or future, time periods it is worth considering the temperature structure of the whole lake. Figure 11 plots the variation of Lochnagar water temperatures as measured through two typical annual cycles at depths down to 19 m. The thermistor

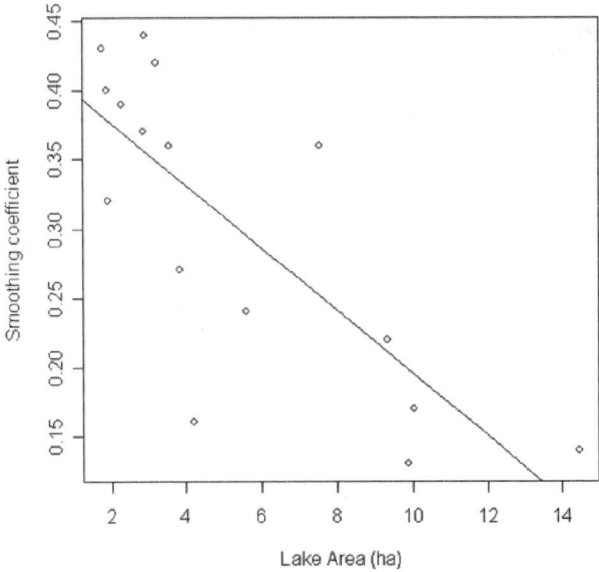

Figure 9. Relationship between smoothing coefficient α and lake area ($R^2 = 0.60$, α = 0.42-0.023*area)

(a)

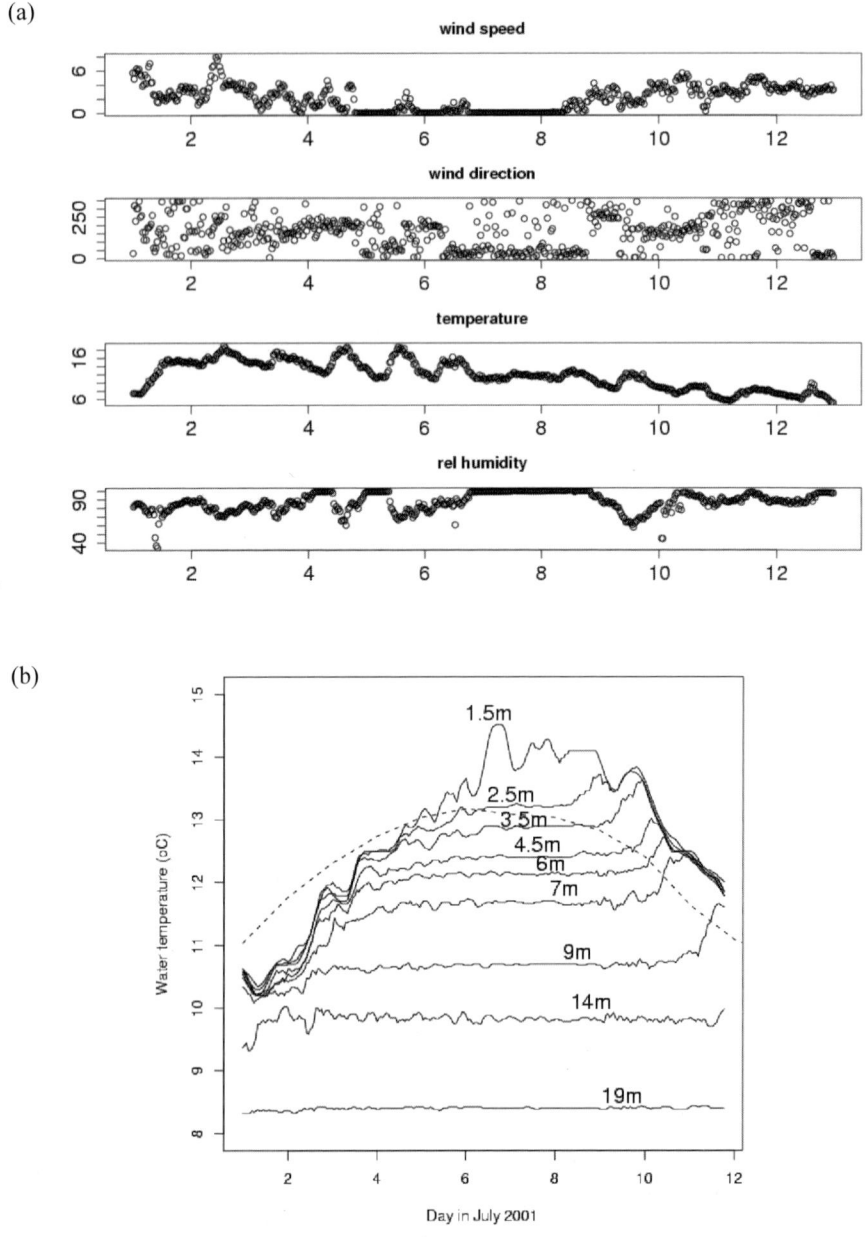

(b)

Figure 10. Example of Lochnagar AWS and water-temperature measurements. The logs span the first 12 days of July 2001, and record a water temperature stratification and mixing event. (a) AWS: Half-hourly wind speed (m s^{-1}), Wind direction (degrees), Temperature ($^{\circ}$C), Relative humidity (%). (b) Thermistor chain: Hourly water temperatures ($^{\circ}$C) at 1.5, 2.5, 3.5, 4.5, 6.0, 7.0, 9.0, 14.0 and 19.0 m depth (solid lines). Modelled average-daily epilimnion temperatures ($^{\circ}$C) (dashed line).

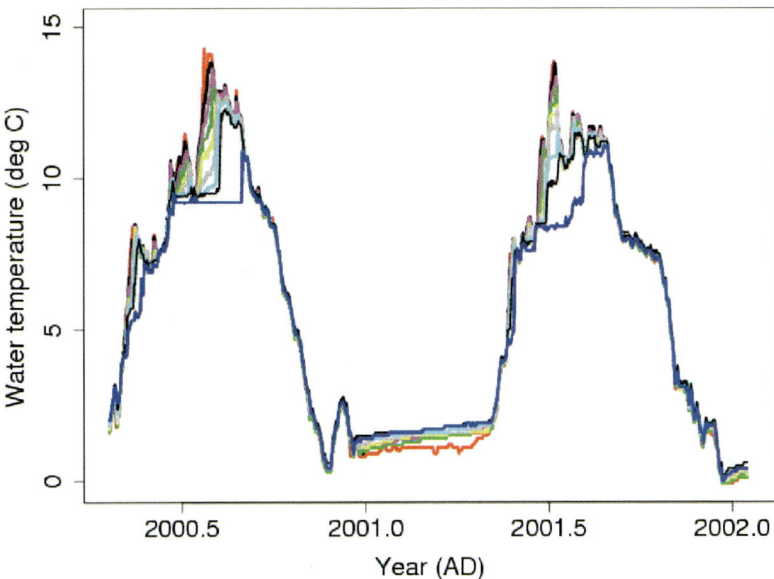

Figure 11. Lochnagar water-temperatures (°C) during two annual cycles 2001 and 2002. Colours denote different thermistor depths. Red shows shallowest and blue deepest deployment.

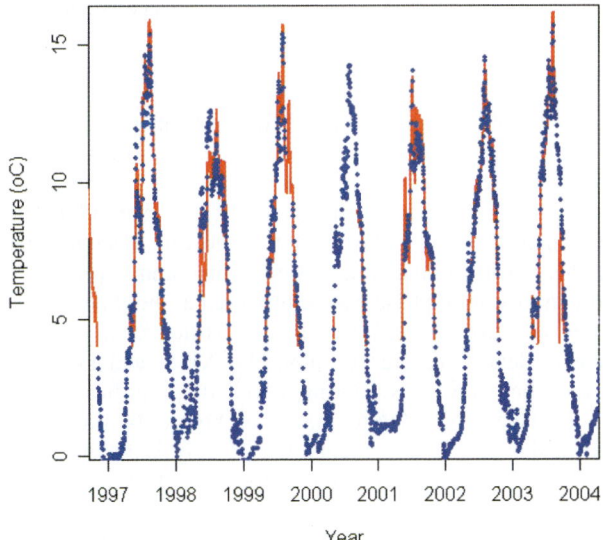

Figure 12. Modelled (red lines) and measured (blue circles) lake temperatures at 1.5 m depth between 1997 and 2004. Water temperatures above 4°C have been modelled. The fit is good, with a mean absolute error of 0.81 °C, except for very occasional, short periods (e.g., early summer in 1998, 2001 and 2003) when errors of over a degree occur (see Figure 8b).

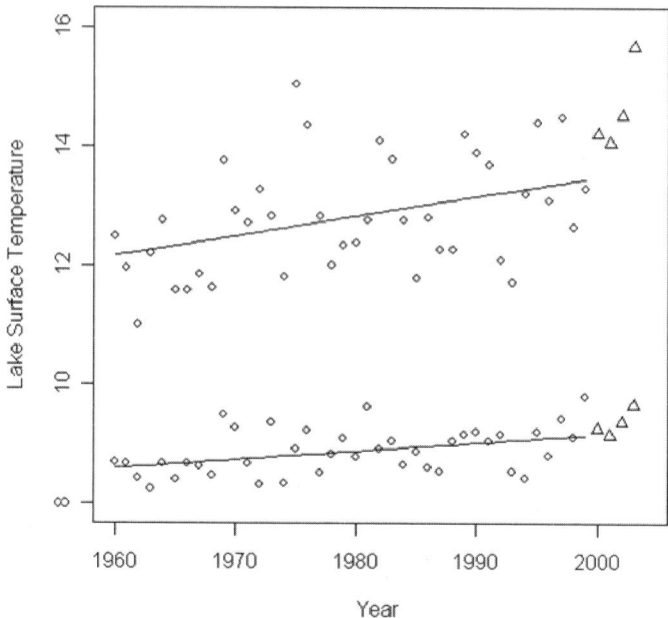

Figure 13. Reconstructed mean and maximum water temperatures at Lochnagar. Computed from daily mean air temperatures at Balmoral using Equations 1 and 2. Lines show linear regression fits. Triangles are values taken from the thermistor data. Mean values are the mean of the water temperatures > 4 °C (i.e., for the open-water period).

traces begin in April 2000 soon after ice-off. Water temperatures rise rapidly; short periods of stratification form as the surface temperature rises to 14.7 °C. By the end of summer 2000 the lake is fully mixed and can be seen to remain so throughout the period of autumnal cooling. Loch-ice forms on 20th December 2000. The rapid rise of water temperatures in the first week of May 2001, following a late ice-break, completes the first annual cycle. A very similar annual water-temperature cycle follows in 2001/2. Once again a short stratification phase can be observed at the time of the peak surface-water temperatures in early July 2001. It is instructive to consider this 2001 stratification event in detail. Figure 10a plots the AWS, and Figure 10b the water temperature records, for twelve days during which the stratification and subsequent mixing events took place. The NOAA images of Figure 14 illustrate the changing synoptic situation at the time of the events. On 1st July isothermal temperatures are found in the mixed layer down to 9m depth (Figure 10b). Over the next five days rising air-temperatures engender a stratification event as the mixed layer gradually disappears. Despite the falling air-temperatures between 5th and 8th July (Figure 10a) the stratification persists due to the exceptionally low wind-speeds. However the synoptic situation changed on July 8th. A transition from the blocked synoptic situation (Figure 14a) takes place on the morning of July 8th as the winds pick up. Immediately thereafter the water at 2.5 m depth can be seen to warm (Figure 10b), so heralding the end of the stratification. Over the next few days the wind-driven mixing steadily penetrates down into the loch, and by 11th July isothermal conditions have returned once again to the uppermost 9m. As such events are relatively scarce, these short summer periods are one of the few times when a Kettle et al. (2004) type of model somewhat misfits the Lochnagar thermistor data. Consequently even a simple (three parameter) exponential smoothing model should be capable of hindcasting epilimnion-temperature fluctuations at Lochnagar.

Past climate

Documentary records and tree-rings

Documentary records and tree-ring sequences (Hughes et al. 1984) provide the best means of extending instrumental climate series further back into the past. While such records from Scotland are not very extensive, excellent records have been established in other parts of Europe, notably Pfister's compilations for central Europe (Pfister 1992) and from pine trees at the timberline in N. Scandinavia (Briffa et al. 1990, Grudd et al. 2002). In Scotland unusually cold conditions in the 1700s resulted in the curious incident of an Inuit Eskimo paddling his kayak into the estuary of the Don River, near Aberdeen (Lamb 1982).

The Little Ice Age in Scotland

The best guide to the magnitude of the climate changes (on decadal to centennial timescales) that have taken place at Lochnagar during the last few thousand years is probably that provided by the Scandinavian tree-ring reconstructions of Grudd et al. (2002). 880 living, dead and subfossil trees were used in their detailed study. Summer temperatures 1 to 1.5 °C cooler than normal dominated the 'Little Ice Age' from around AD 1000 to the 17th century, and a generally cold 'Dark Ages' climate from about AD 500 to about AD 900. Conversely the 'Medieval Warm Period', around AD 1000, and a warm 'Roman' period in the first century AD saw temperatures around 0.5 °C warmer than normal (1971-2000). Even more pronounced fluctuations in summer temperatures are reconstructed for the period between 600-1 BC (Grudd et al. 2002). As advances of Norwegian glaciers partly reflect local precipitation changes (Nesje and Dahl 2003), they along with the advanced positions of Alpine glaciers by AD 1350 (Grove and Switsur 1994), proxy information on frost, freezing of water bodies, duration of snow cover and untimely (unseasonable) activity of vegetation (Pfister 1992, Luterbacher et al. 2004), and reconstructions of winter temperatures in the Low Countries (Shabalova and van Engelen 2003), point to enhanced wintertime precipitation along with winter temperatures over 1 °C cooler than normal during the Little Ice Age.

The Holocene

Until about 15000 years ago, the whole of the Cairngorms region was covered by part of an ice-sheet that extended across all of the northern British Isles (Chapter 4: Hall this volume). Following the retreat of the ice the initial plant communities were open, but

(a)   (b)

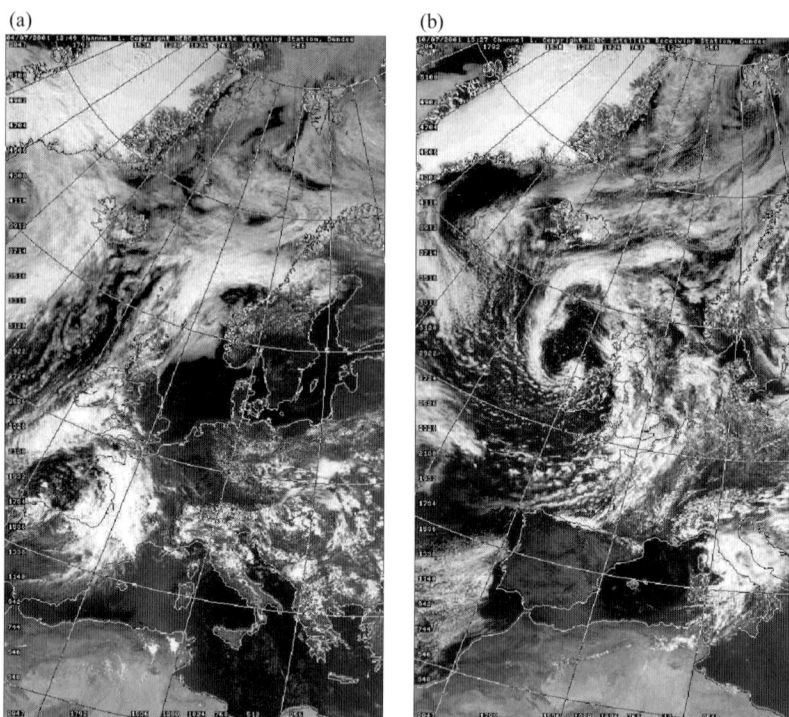

Figure 14. NOAA satellite images (channel 1: visible) obtained during the time period covered by Figure 8. (a) 12:49 on 4 July 2001: A persistent high pressure area, marked by the clear skies over Scandinavia and the North Sea, has caused a warm southerly air-flow (Figure 8a) over Scotland for much of the previous four days. Low-level sea fog lies between Scotland and Norway. Higher level clouds, in the low pressure area centred in the Bay of Biscay and over Ireland and western Scotland are blocked by the North Sea high. (b) 13:27 on 10 July 2001: Southwesterlies have returned to Lochnagar. A depression off Northern Ireland is moving in, a swirl of frontal cloud covers Scotland and England.

woodland of birch spread rapidly across the landscape in the early Holocene (Chapter 7: Birks this volume). Woodland became well established, with pine as the dominant tree, by about 8000 years ago, probably at all altitudes up to about 700 or 800 m. At higher altitudes, plant communities are likely to have remained open. After about 3000 years ago the forest declined regionally as a result of increasing anthropogenic pressures, and peatland ecosystems increased in areal extent (Bennett 1996).

At Lochnagar, more than 1.5 m of sediment was laid down during the last 9000 years (Dalton et al. 2005). Cycles in the per cent loss-on-ignition and per cent dry weight occur in a sediment core covering this period. The more organic phases may reflect warmer climate phases and increased productivity (Chapter 8: Rose this volume). While diatom assemblage changes in recent sediments reflect a response to acid deposition (Chapter 14: Monteith et al. this volume), earlier variations in species composition may correspond to alternations between cold and warm phases (Dalton et al 2005) and a chironomid response to a change from warm to cool temperatures is indicated in the early part of the core. Some of the variations in diatom species composition may correspond to alternations between cold and warm phases (Dalton et al. 2005). Changes in a century long diatom oxygen isotope curve (1892 to 1995, dated using accumulation frequencies of spheroidal carbonaceous particles) reflect changing amount of winter precipitation (snow fall) since periods of greater snowfall are isotopically more depleted and the greater volume takes longer to be flushed through the system (Leng et al. 2001). Periods of depleted oxygen isotope ratios occur around 1980 and in the 1920s.

Based on the tree-ring data the summer water temperature and ice-cover duration models discussed above, and in Chapter 18, suggest that decadal-long fluctuations of +0.5 to −1.5 °C in summer water temperature, and of over 60 days in ice-cover duration would not have been unusual at Lochnagar during the late Holocene. As training sets of chrysophyte cyst, chironomid and diatom assemblages from lake sediments have climate reconstruction errors of as low as 0.7 °C (Lotter et al. 1997) micro-palaeontological studies would appear to have the potential to detect centennial long fluctuations in relatively undisturbed mountain lakes such as Lochnagar. However while long term trends through the Holocene are found at Lochnagar, and at other Cairngorm lochs (Battarbee et al. 2001), any decadal to centennial duration fluctuations have remained extremely elusive.

Discussion

The high sensitivity of maritime mountain sites

Maritime localities are unusually sensitive to climate change on account of their equitable climates. In Scotland phenological events (such as budbreak or first flowering dates) that are controlled by springtime temperatures advance 11 days for every 1 °C rise in temperature (Clark and Thompson 2004). This compares with a change of only four days for each degree in a continental interior. The growing season in maritime situations is similarly also extremely sensitive to temperature changes (Manley 1945). In Scotland this sensitivity and it's potential for high societal impact is well illustrated by the work of Parry (1978) on the abandonment of medieval farmsteads in the Lammermuirs. The expansion of settlement and cultivation, during the medieval climatic optimum followed by a subsequent retreat in the 14th century, when reduced temperatures and increased wetness caused cultivation limits to retreat downslope by as much as 200 m, reflects the strong climatic responses of the Scottish uplands in the Little Ice Age.

Spatial coherence and up-scaling

The tightness of the synoptic setting (illustrated by the satellite images of Figure 14) with mountain weather (as recorded by the AWS, Figure 10a) and underwater temperatures (as recorded by the thermistor chain data in Figure 10b) shows how regional circulation patterns in marine west coast climates have a direct effect on the temperature structure of mountain lakes across the UK.

Good coherence is found between climatic change in mountain sites and in their neighbouring lowlands. Interannual variations in air temperature, for example, show strong correlations over distances of several hundred kilometres (Agusti-Panareda and Thompson 2002; Jones and Thompson 2003). Water temperature fluctuations similarly show good coherence over many hundreds of metres elevation range and spatially over many hundreds of square kilometres. The detailed monitoring of climate and weather being carried out at Lochnagar (by the AWS); the lake waters (by the thermistor chain) and the catchment (by the Digicam), and their relationships with the monitoring of chemical deposition (as part of the United Kingdom Acid Deposition Network and pollutants (metals, fly-ash particles and persistent organics), are of much more than just local value being of relevance to marine west coast climatic settings in general.

Summary

- Lochnagar has the typical climate of a marine west-coast situation, with mild summers, overcast winters and rain at any time.
- Instrumental records from the nearby valley stations of Braemar (since 1855) and Balmoral provide extensive meteorological data.
- Mountain weather records are provided by automatic weather-stations on Caingorm summit (since 1977) and at Lochnagar lake (since 1996).
- Water-temperature measurements began at Lochnagar in 1999.
- Ice-cover and snow-cover duration (since 2002) can be estimated from daily photographs obtained from an automatic digital-camera overlooking the loch.
- Open-water season water-temperatures and ice-cover duration are both found to be closely coupled to air temperatures. An exponential-smoothing model exploits the tight air-water temperature relationship to allow water temperatures to be reconstructed throughout the instrumental period.
- Regional circulation changes, at the synoptic scale, are tightly related to water temperatures and to stratification profiles through advection of heat and through the physical action of the wind. Water temperature changes at Lochnagar are thus representative of changes across the UK.
- Reconstructions of maximum summer water-temperatures, mean open-season water-temperatures and ice-free period all show increasing trends over recent decades.
- Maritime locations, such as Lochnagar, are unusually sensitive to climate change on account of their equitable climate.

Acknowledgements

Michael Hughes is responsible for managing the automatic camera and weather station at Lochnagar. Jo Porter undertook the recording of snow and ice cover. Anna Agusti-Panareda generated the air-temperature reconstructions shown in Figure 2 as part of the MOLAR project (contract number ENV4-CT95-0007). Gavin Simpson produced Figure 5. We are very grateful to the Dundee Satellite Receiving Station, Dundee University, Scotland for the NOAA satellite images.

References

Agusti-Panareda A. and Thompson R. 2002. Reconstructing air temperature at eleven remote alpine and arctic lakes in Europe from 1781 to 1997 AD. J. Paleolim. 28: 7–23.

Agusti-Panareda A., Thompson R., and Livingstone D.M. 2000. Reconstructing climatic variations at high elevation lake sites in Europe during the instrumental period. Verh. Internat. Verein. Limnol. 27: 479–483.

Anon. 1900. Lochnagar in February and April. Cairngorm Club J. 3: 193–195.

Baker B., Curran J.C., Harrison F., Peckham G.E. and Smith S.D. 1979. An automatic weather station for operation in severe icing climates. J. Phys. E: Sci. Instrum. 12: 734–738.

Battarbee R.W., Cameron N.G., Golding P., Brooks S.J., Switsur R., Harkness D., Appleby P.G., Oldfield F., Thompson R., Monteith D.T. and McGovern A. 2001. Evidence for Holocene climate variability from the sediments of a Scottish remote mountain lake. J. Quat. Sci. 16: 339–346.

Bennett K.D. 1996. Late-Quaternary vegetation dynamics of the Cairngorms. Bot. J. Scot. 48: 51–63.

Birks H.J.B. 1996. Palaeoecological studies in the Cairngorms: Summary and future research needs. Bot. J. Scot. 48: 117–126

Birks H.J.B. This volume. Chapter 7. Flora and vegetation of Lochnagar – past, present and future. In: Rose N.L. (ed.) 2007. Lochnagar: The natural history of a mountain lake. Springer. Dordrecht.

Briffa K.R., Bartholin T.S., Eckstein D., Jones P.D., Karlén W., Schweingruber F.H. and Zetterberg P. 1990. A 1,400-year tree-ring record of summer temperatures in Fennoscandia. Nature 346: 434–439.

Broadmeadow M.S.J., Ray D. and Samuel C.J.A. 2005. Climate change and the future for broadleaved tree species in Britain. Forestry 78: 145–161.

Clark R.M. and Thompson R. 2004. Botanical records reveal changing seasons in a warming world. Australasian Science (October) 37–39.

Dalton C., Birks H.J.B., Brooks S.J., Cameron N.G., Evershed R.P., Peglar S.M., Scott J.A. and Thompson R. 2005. A multi-proxy study of lake-development in response to catchment changes during the Holocene at Lochnagar, north-east Scotland. Palaeogeog. Palaeoclim. Palaeoecol. 221: 175–201.

Grove J.M. and Switsur R. 1994. Glacial geological evidence for the medieval warm period. Climatic Change 26: 143–169.

Grudd H., Briffa K.R., Karlen W., Bartholin T.S., Jones P.D. and Kromer B. 2002. A 7400-year tree-ring chronology in northern Swedish Lapland: Natural climatic variability expressed on annual to millennial timescales. Holocene 12: 657–665.

Hall A.M. This volume. Chapter 4. The shaping of Lochnagar: Pre-glacial, glacial and post-glacial processes. In: Rose N.L. (ed.) 2007. Lochnagar: The natural history of a mountain lake. Springer. Dordrecht.

Harrison J. 1997. Central and Southern Scotland. In: Wheeler D. and Mayes J. (eds.) Regional climates of the British Isles. Routledge, London. pp 228–253.

Hughes M. This volume. Chapter 2. Physical characteristics of Lochnagar. In: Rose N.L. (ed.) 2007. Lochnagar: The natural history of a mountain lake. Springer. Dordrecht.

Hughes M.K., Schweingruber F.H., Cartwright D. and Kelly P.M. 1984. July-August temperature at Edinburgh reconstructed from tree-ring density and width measurements. Nature 308: 341–344.

Hurrell J.W. 1995. Decadal trends in the North Atlantic Oscillation: Regional temperatures and precipitation. Science 269: 676–679.

Jenkins A., Reynard N., Hutchins M., Bonjean M. and Lees M. This volume. Chapter 9. Hydrology and hydrochemistry of Lochnagar. In: Rose N.L. (ed.) 2007. Lochnagar: The natural history of a mountain lake. Springer. Dordrecht.

Jones P.D. and Thompson R. 2003. Instrumental records. In: Mackay A.W., Battarbee R.W., Birks H.J.B. and Oldfield, F. (eds.) Global Change in the Holocene. Arnold, London. pp. 140–158.

Kalnay E., Kanamitsu M., Kistler R., Collins W., Deaven D., Gandin L., Iredell M., Saha S., White G., Woolen J., Zhu Y., Chelliah M., Ebisuzaki W., Higgins W., Janowiak J., Mo K.C., Ropelewski C., Wang J., Leetma A., Reynolds R., Jenne R. and Joseph D. 1996. The NCEP/NCAR 40-year reanalysis project. Bull. Am. Met. Soc. 77: 437–471.

Kettle H. and Thompson R. 2004. Statistical downscaling in European mountains: Verification of air temperature reconstructions. Climate Res. 26: 97–112.

Kettle H. and Thompson R. This volume. Chapter 18. Future climate predictions for Lochnagar. In: Rose, N.L. (ed.) 2007. Lochnagar: The natural history of a mountain lake. Springer. Dordrecht.

Kettle H., Thompson R., Anderson N.J. and Livingstone D.M. 2004. Empirical modeling of summer lake surface temperatures in southwest Greenland. Limnol. Oceanog. 49: 271–282.

Lamb H.H. 1982. Climate, History and the Modern World, New York: Methuen. 387 pp.

Leng M., Monteith D., Patrick S., Greenwood P., Jones V., Rose N., Battarbee R. 2001. A 100 year-long oxygen isotope record from diatom silica in Lochnagar, Scotland Abstract (PAGES - PEPIII: Past climate variability through Europe and Africa) Aix-en-Provence, France.

Levack J.R. 1906. Lochnagar and the Cairngorms in snow. Cairngorm Club J. 5: 164–165.

Lotter A.F., Birks H.J.B., Hofmann W. and Marchetto A. 1997. Modern diatom, cladocera, chironomid and chrysophyte cyst assemblages as quantitative indicators for the reconstruction of past environmental conditions in the Alps. I. Climate. J. Paleolim. 18: 395–420.

Luterbacher J., Dietrich D., Xoplaki E., Grosjean M. and Wanner H. 2004. European seasonal and annual temperature variability, trends and extremes since 1500. Science. 303: 1499–1503.

Manley G. 1945. The effective rate of altitudinal change in temperate Atlantic climates. Geog. Rev. 35: 408–417.

Manley G. 1953. The mean temperature of Central England, 1698 to 1952. Quart. J. Roy. Met. Soc. 79: 242–261.

Manley G. 1974. Central England Temperatures: Monthly means 1659 to 1973. Quart. J. Roy. Met. Soc. 100: 389–405.

McKenzie A. 1897. The White Mounth in winter. Cairngorm Club J. 2: 113–116.

Monteith D.T., Evans C.D. and Dalton C. This volume. Chapter 14. Acidification of Lochnagar and prospects for recovery. In: Rose N.L. (ed.) 2007. Lochnagar: The natural history of a mountain lake. Springer. Dordrecht.

Nesje A. and Dahl S.O. 2003. The ''Little Ice Age' – only temperature? Holocene 13: 139–145.

Parry M.L. 1978. Climate change, agriculture and settlement. Folkestone: Dawson. 214 pp.

Patrick S. and Wathne B.M. 2000. The message from the mountain lakes. RTD Info 28: 34–35.

Patrick S.T., Juggins S., Waters D. and Jenkins A. 1991. The United Kingdom Acid Waters Monitoring Network: Site descriptions and methodology report. ENSIS Publishing, London. 63 pp.

Pfister C. 1992. Monthly temperature and precipitation in central Europe 1525–1979: Quantifying documentary evidence on weather and its effects. In: Bradley R.S. and Jones P.D. (eds.) Climate since A.D. 1500. Routledge, New York. pp. 118–142.

Rose J. 1895. The Lochnagar corrie. Cairngorm Club J. 1: 297–300.

Rose N.L. 2001. Meteorological data for Lochnagar 1996–2000. A summary report. Environmental Change Research Centre, University College London, Research Report No. 78, 92 pp.

Rose N.L. This volume. Chapter 8. The sediments of Lochnagar: Distribution, accumulation and composition. In: Rose N.L. (ed.) 2007. Lochnagar: The natural history of a mountain lake. Springer. Dordrecht.

Rose N.L. and Yang H. This volume. Chapter 17. Temporal and spatial patterns of spheroidal carbonaceous particles (SCPs) in sediments, soils and deposition at Lochnagar. In: Rose N.L. (ed.) 2007. Lochnagar: The natural history of a mountain lake. Springer. Dordrecht.

Roy M. 1997. The Highlands and Islands of Scotland. In: Wheeler D. and Mayes J. (eds.) Regional climates of the British Isles. Routledge, London. pp 228–253.

Schär C., Vidale P.L. and Luthi, D. 2004. The role of increasing temperature variability in European summer heat waves. Nature 427: 332–336.

Shabalova M.V. and van Engelen A.F.V. 2003. Evaluation of a reconstruction of winter and summer temperatures in the Low Countries, AD 764–1998. Climatic Change 58: 219–242.

Stefan H.G., Fang X. and Hondzo M. 1998. Simulated climate changes effects on year-round water temperatures in temperate zone lakes. Climatic Change 40: 547–576.

Thompson R., Kamenik C. and Schmidt R. 2005. Ultra-sensitive Alpine lakes and climate change. J. Limnol. 64: 139–152.

Walker G.T. and Bliss E.M. 1932. World Weather V. Mem. Roy. Met. Soc. 4: 53–84.

Weston K.J. 1992. Objectively analyzed cloud immersion frequencies for the United Kingdom. Met. Mag. 121: 108–111.

Weston K.J. and Fowler D. 1991. The importance of orography in spatial patterns of rainfall acidity in Scotland. Atmos. Environ. 25: 1517–1522.

6. THE DEVELOPMENT, DISTRIBUTION AND PROPERTIES OF SOILS IN THE LOCHNAGAR CATCHMENT AND THEIR INFLUENCE ON SOIL WATER CHEMISTRY

RACHEL C. HELLIWELL (r.helliwell@macaulay.ac.uk)
ALLAN LILLY and JOHN S. BELL
Macaulay Land Use Research Institute
Craigiebuckler
Aberdeen AB15 8QH
Scotland
United Kingdom

Key words: acidification, Lochnagar, modelling, montane soils, soil chemistry, soil processes, Scotland

Introduction

Despite being only 108.5 ha in extent (Chapter 2: Hughes this volume), the Lochnagar catchment has a number of contrasting soil types due largely to the variability in both topography and climate. This Chapter sets out to describe the main soil forming factors that determine the extent and distribution of soils within the catchment, the consequent effects on the physico-chemical properties of the soil and how soil processes affect soil and surface water chemistry. Finally, this information is used to predict the impact of sulphur and nitrogen deposition on these sensitive soils over the next 200 years.

The soils in the Lochnagar catchment

The Lochnagar catchment is located in the southern Cairngorms lying between the Great Glen Fault to the northwest and the Highland Boundary fault to the south. Older Dalradian Age sediments dating back 700 million years were intruded by granites (Stephenson and Gould 1995; Chapter 3: Goodman this volume). These metamorphosed sediments were subsequently eroded leaving behind the more resistant granite which now forms the Cairngorm and Lochnagar massif. The catchment is an 'arm-chair' shaped hollow known as a corrie, with a precipitous backwall and steep sides (Figure 1a) and a small loch lies at the bottom. The catchment altitude ranges from 788m at the outflow to 1155m at the highest point (Figure 1b). The summit of the catchment lies at a latitude of around 57° North but the decrease in temperatures

N.L. Rose (ed.), Lochnagar: The Natural History of a Mountain Lake
Developments in Paleoenvironmental Research, 93–120.
© 2007 *Springer*.

associated with increasing altitude, mean that the plateau has a climate similar to that at sea level on the Arctic Circle (latitude 66° North). This has a dramatic effect on soil development with periglacial activity such as freezing and thawing, solifluction and stone sorting resulting in distinctive montane soils. At these higher altitudes, vegetation growth is slow and the mountains are dominated by montane communities, for example, on the exposed summits the vegetation is sparse and dominated by the three-leaved rush (*Juncus trifidus*). On the flanks of the corrie where the exposure is slightly less severe, vegetation communities are dominated by wind clipped dwarf shrubs such as heather (*Calluna vulgaris*) and the trailing azalea (*Loiseleuria procumbens*) (Chapter 7: Birks this volume). Heathers dominate the soils at lower altitudes and in late-lie snow beds, grasses such as *Nardus stricta* and sedges (*Carex* spp.) are common. The exposed rock and screes often have a cover of lichens and mosses while *Sphagnum* dominates the peat deposits occurring in low lying areas

The soils that develop in this cold mountain climate are nutrient deficient and acidic due to the very slow release of nutrients from weathering of the coarse sandy, quartz-rich parent materials such as morainic drifts and frost-shattered debris. Leaching also removes minerals and nutrients from the soils, although on waterlogged and flushed areas, the removal of nutrients from the soil is not as great. In general, the surface horizons of many soils found in the Lochnagar catchment are extremely acid, with a pH of 4 or less. Nearly all available nutrients are concentrated in these upper horizons. High carbon and nitrogen ratios indicate that decomposition of organic matter is slow with little activity by decomposing organisms (Heslop 1981).

Due to the acid nature of the granite rock and the combination of low evapotranspiration with relatively high rainfall (>1600mm per annum), many of the high altitude soils often show evidence of weak podzolisation and are classified by the Soil Survey of Scotland (1984) as alpine podzols or oroarctic podzols, reflecting both the altitudinal effect on the climate (Birse 1980) and the consequent effect on soil development. However, these soils do not have permafrost and therefore do not qualify as cryosols according to the World Reference Base although they are seasonally cryoturbated and are probably closer to skeletic podzols (FAO/ISSS/ISRIC 1998).

On gentler slopes or in topographic depressions, the soils generally develop a thick organic surface layer due to the slow decomposition of organic material, a result of the climate, runoff accumulation and acidic parent materials. Where organic surface horizons, up to 0.5 m thick, overlay a mottled and gleyed mineral subsoil, the soils are classified as peaty gleys by the Soil Survey of Scotland (1984) or histic gleysols by the World Reference Base classification (FAO/ISSS/ISRIC 1998) and characteristically are waterlogged for part of the year. Where rock is close to the surface, peaty rankers occur when the peaty layer develops on coherent rock (histosols according to the World Reference Base classification). In the base of the corrie around the loch, the organic layer has developed to a depth of over one metre in places. Soils with organic horizons > 0.5 m are classified as peat (histosols). Peat is inherently acid and provides little nourishment for plants. As a consequence peat can only support acid tolerant vegetation communities that can withstand wet conditions (Chapter 7: Birks this volume).

Immature shallow soils such as rankers and lithosols, (described as leptosols in the World Reference Base classification) have formed where rock is close to the surface

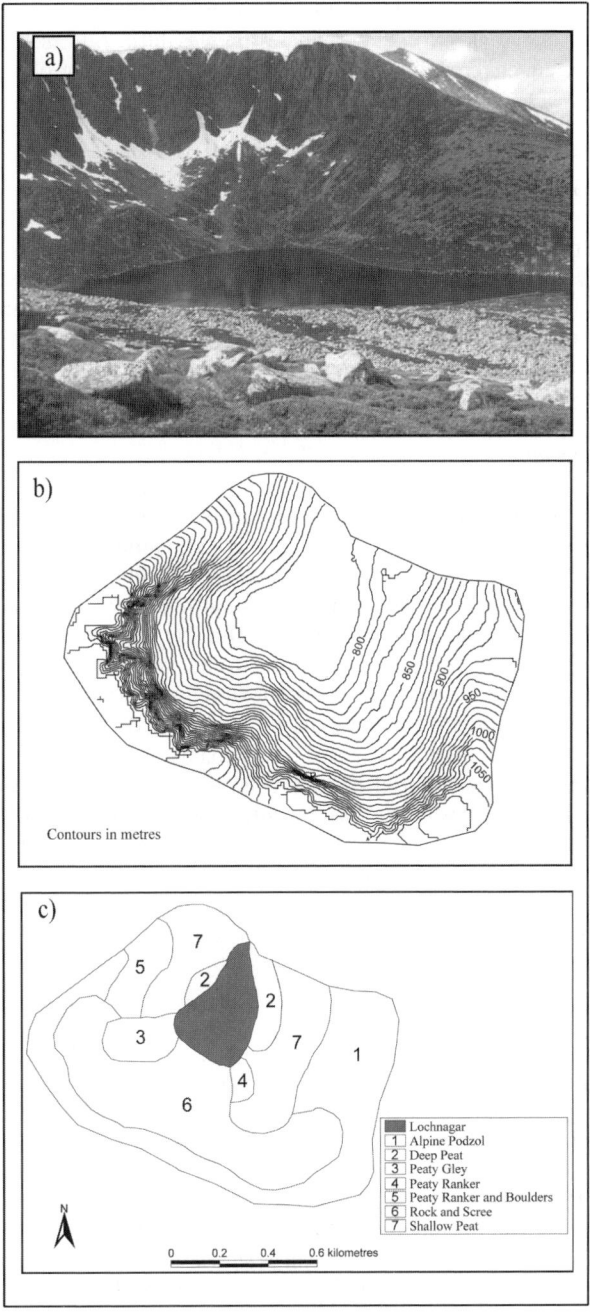

Figure 1. The Lochnagar catchment (a) with contour (b) and (c) soil maps. (Photograph courtesy of Environmental Change Research Centre). Figure first published in J. Limnol. 63(1) 2004. Reproduced with permission.

and the drifts are thin or non-existent. Screes and boulder slopes (particularly those that are still active) and rock pavements, where there is little fine material for plants to colonise, are often covered in lichens, bacteria and other micro-organisms (Heslop 1981).

Factors affecting soil development in the Lochnagar catchment

There are a number of different environmental factors that act together to determine the rate of soil development and the dominant soil forming processes. These are the nature of the parent material in particular, the inherent geochemistry and particle size, topography especially slope angle and aspect, climate, living organisms including plants and animals. The length of time that pedogenesis has been taking place is also crucial (Jenny 1941).

Soil parent material

The Lochnagar granite generally comprises coarse crystals of quartz, feldspars and biotite (Stephenson and Gould 1995; Chapter 3: Goodman this volume) and is inherently low in base cations. Much of the weathered drifts of the Tertiary period were removed during successive glaciations exposing relatively hard, coherent bedrock which is resistant to weathering under present climatic conditions. Frost shattering has given rise to coarse stony drifts on the summit and flanks of the corrie, while corrie glaciers are responsible for the stony coarse textured moraines that occur on the corrie floor near to the outflow (Stephenson and Gould 1995; Chapter 4: Hall this volume). The moraines are generally poorly sorted sandy loams or loamy sands with subangular stones indicating that the material has not been transported over large distances. The precipitous backwall of the corrie has considerable amounts of bare rock with some screes developed at the base of stone chutes. Many of the screes are still active and cryoturbation of the frost shattered debris still occurs at altitudes greater than about 850 m (depending on exposure). Solifluction lobes can still be seen on some slopes in the catchment. Organic deposits are found in low lying areas where water accumulates, for example, in the areas around the loch.

Topographic factors

The topography of the Lochnagar catchment has a major influence on soil development. The slopes range from precipitous (>45 degrees) on the backwall of the corrie to gentle (0-5 degrees) at the loch outlet and on the southerly slopes of the high plateau. The very steep slopes have little drift cover and little soil development whilst the active screes that have developed below the crags of the corrie backwall result in an environment that is unfavourable for soil development. On the easterly and westerly flanks of the corrie, the drifts are often subjected to solifluction processes, the movement of material down slope due to gravity and freeze/thaw cycles creating microtopographic features such as boulder lobes and terracettes. The morainic drifts occur as mounds with a wide range of slopes.

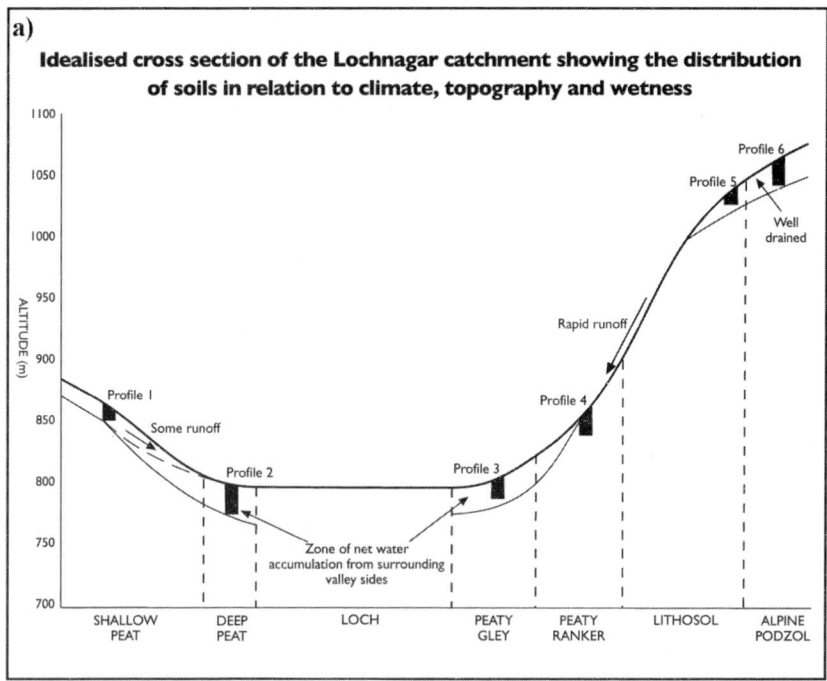

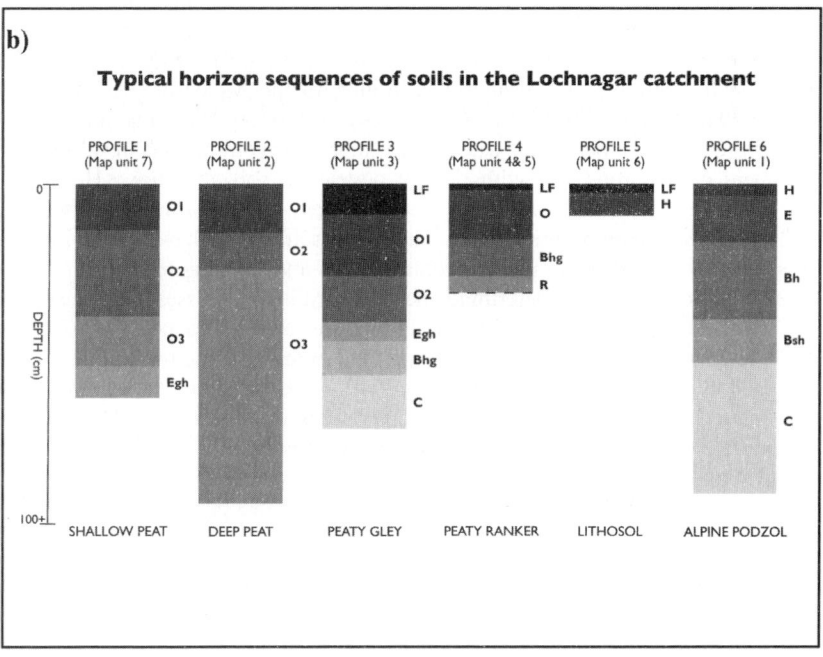

Figure 2. (a) Cross section of Lochnagar catchment and (b) typical horizon sequences.

The topography not only influences soil formation and the nature of the soil parent material in the catchment but also the soil water regime. The steeper slopes encourage greater amounts of runoff, this water accumulates in the flat areas around the loch (Figure 1b, c), resulting in, waterlogging and gleying (oxidation and reduction cycles under anaerobic and aerobic conditions) of the soils giving rise to a toposequence where soil formation is greatly influenced by the topographic position and the water regime. In the base of the corrie, close to the loch, the gentle slopes and the confined nature of the landform has contributed to the development of organic deposits and soils with organic surface layers

Climatic links

Climate determines the nature of weathering and the rate of chemical and physical processes influencing profile development. The key components of climate in soil formation in Scotland are precipitation and temperature. Like many of the corries in the Cairngorm massif (Gordon 2001) the Lochnagar catchment has a predominantly north-northeast aspect but with relatively large areas facing to the west and east. The northerly facing slopes receive less solar radiation (Chapter 2: Hughes this volume) making them colder than east or west facing slopes, allowing snow patches to persist into early summer (Chapter 5: Thompson et al. this volume), and reducing evaporation.

The Lochnagar catchment has moderate to high precipitation (more than 1600 mm per annum), frequent snowfall, and low temperatures throughout the year (Chapter 5: Thompson et al. this volume). Birse and Dry (1970) have estimated that the summit of the catchment has less than 275 day degrees of accumulated temperature above 5.6 ° C per annum. The combination of altitude and a maritime climate exposes the Lochnagar catchment to mean wind speeds in excess of 8 m s^{-1} (with gusts of up to 77 m s^{-1} being recorded on the summit of Cairngorm according to the Meteorological Office) and freezing conditions (accumulated frost in excess of 470 day °C; Birse and Robertson 1970). These low temperatures cause frost heave (cryoturbation) resulting in the development of a porous soil with weakly developed soil structure. However, the Scottish climate is not stable and throughout the year low pressure systems sweep in from the southwest often bringing warmer air that results in the thawing of lying snow.

The relatively high rainfall, low evaporation rates and freezing temperatures interact with soil parent material to determine the soil forming processes in the catchment. Where the drifts are permeable and free draining, the soils are leached, but where water accumulates, for example at the base of slopes and in topographic hollows, the soils are waterlogged exhibiting gleying and mottling within the profile or the accumulation of organic matter. The low annual average temperatures mean that the rates of chemical weathering are slow often resulting in weakly expressed pedological features and the breakdown of organic matter is also slow. Low evapotranspiration rates and high rainfall leads to leaching of base cations in the freely drained permeable soils.

Biotic factors

In the Scottish uplands there is a clear association between soil type and the vegetation communities it supports (Birks: Chapter 7 this volume), but it can be difficult to separate the influence of the vegetation on the soil from the dependence of the vegetation on certain soil processes. The cool climate experienced by the soils in the catchment combined with the acidity of the parent rock suppresses biological activity such that the mixing of soil layers, nutrient cycling, and structural stability are limited. Plants growing in the study area produce an acid litter, giving rise to an unfavourable environment for earthworms which breakdown and incorporate vegetative matter into the soil. This allows the development of distinctive soil horizons. On the warmer lower slopes and in wet areas, the breakdown of dead plant material is still sufficiently slow to allow the development of organic rich soils and ultimately peat (Pulford 1998). On the colder upper slopes, few plants can survive and the organic matter content of the soils is less.

Time factor: period of soil formation

It is generally accepted that glaciers finally disappeared from the corries of Scotland around 10000 years ago (Stephenson and Gould 1995; Gordon 2001; Chapter 4: Hall this volume). As the climate warmed, the new glacial deposits such as moraines and tills as well as glaciofluvial deposits were exposed and soils began to develop. Initially, much of Scotland would have resembled the arctic tundra of today with periglacial activity such as freezing and thawing, solifluction, stone sorting and the presence of permafrost. Although the climate has fluctuated over the past 10000 years, it has gradually warmed and Scotland now has a temperate maritime climate (Chapter 5: Thompson et al. this volume). However, the climate has not remained the same and the climate has been both warmer and wetter as well as colder (for example during the 'Little Ice Age') that may well have resulted in corrie glaciers reforming in the Scottish mountains. These climatic changes have also affected the vegetation communities that have existed in the catchment which, have in turn, influenced soil development and the composition of organic layers in particular.

Distribution of soils and their physico-chemical properties

The soils of the Lochnagar catchment were mapped at a scale of 1:10000 (Figure 1c) using free survey techniques. A total of seven individual soils types were identified according to the classification system used by the Soil Survey of Scotland (Soil Survey of Scotland 1984). The spatial distributions of these soils are particularly influenced by two soil forming factors: climate and topography. Peats and poorly draining soils with peaty surface horizons have formed in hydrological receiving sites in the valley bottom while freely draining soils, occupy the, steeper slopes (where they are often shallow), valley sides and the summit plateau (Figure 2a and 2b).

Organic soils

The Soil Survey of Scotland (Soil Survey of Scotland 1984) defines organic or peat soils as soil having a surface organic layer greater than 0.5 m thick and with more than 35% organic carbon content. These soils equate with the Histosol group of the World Reference Base (FAO/ISSS/ISRIC 1998). In general, the organic soils in Scotland have >50% organic carbon content, are usually acidic with average pH <4, possess low bulk density, high water holding capacity, low load-bearing strength and are often characterised by mire and blanket bog plant communities. Within the catchment, thick peat deposits are found in areas surrounding the loch where water from the surrounding slopes accumulates and where the slopes are <10 degrees (Chapter 2: Hughes this volume). As the slope angle increases or the slope form becomes convex, these peat deposits become thinner. The peat in the catchment commonly supports vegetation communities with *Sphagnum* spp. and *Eriophorum* spp. and has C:N ratios greater than 30 reflecting the extremely slow rate of decomposition. The peat deposits have been categorised into two classes according to thickness; shallow peat which has a thickness of 0.5 to 1 m and deep peat which is more than 1m thick.

Shallow peat

Shallow peat (Map unit 7, Figure 1c) 0.5-1 m thick, extends over 20% of the catchment and occurs on the undulating, gently and moderately sloping lower catchment floor and slopes, ranging in altitude from 788 m to 903 m, and overlying a stony, sandy loam till (Table 1). The peat is generally fibrous and semi-fibrous near to the surface but the degree of humification increases and becomes more amorphous towards the base of the deposits, the composition of which is determined by the surface vegetation present at the time of development. Where the slopes increase to the northwest and east of the loch (average slope is 13 degrees), the peat is relatively drier. The present vegetation communities are dominated by *Calluna vulgaris* and *Vaccinium myrtillus*.

Table 1. Example description of a shallow peat profile.

Horizon	Depth (m)	Description	Carbon (%)	pH (water)	Base saturation (%)
O1	0-15	Dark brown, fibrous and wet, organic with no structure	52	4.33	13.0
O2	15-43	Black, semi-fibrous and moist, organic with coarse blocky structure	56	3.73	10.8
O3	43-59	Black, amorphous and moist, organic with coarse blocky structure	57	3.92	5.6
Egh	59-69	Dark greyish brown, gleyed, wet sandy loam. Weak fine blocky structure with many subangular stones	1.5	4.4	2.1

Deep peat

A small area of the catchment (5%) has peat deposits >1m thick (Map unit 2; Figure 1c) which are classified as deep peat by the Soil Survey of Scotland (1984). This deeper peat, occurs adjacent to the east and west shores of the loch outlet and carries a

a) Peat

b) Peaty Gley

c) Peaty Ranker

d) Alpine Podzol

Figure 3. Photographs of soil profiles characteristic of those found in the Lochnagar catchment.

vegetation cover mainly of *Sphagnum spp* and *Eriophorum spp*. The peat has distinct stratification with changes in colour, structure and nature of the organic matter, resulting from different degrees of humification and development (Figure 3a and Table 2). This stratification possibly reflects the long term change in vegetation in response to the changes and fluctuations in climate since the retreat of the ice sheet. From pollen records of samples taken at Lochnagar, Dalton et al. (2005), identified a loss of sub-alpine birch scrub around 9000-8000 years before present (BP) which was replaced by the growth of pine trees between 8000-5500 yr BP and later, an expansion of grasslands (5500-4000 yr BP). These vegetation types were replaced by moorland heath and blanket bog around 1500 years BP and are still present today.

The occurrence of gullies and strongly eroded peat along the eastern shore of Lochnagar accounts for a large extent of Map unit 2 (Figure 1c) (Jones et al. 1993). The peat is eroded into hags and gullies by the action of water combined with wind, and freeze-thaw activity. The gullies and channels are generally 1-2 m in depth, often exposing the stony mineral material beneath. These channels are often 2 m and sometimes up to 8 m wide and generally occur approximately 5-20 m apart, exposing a considerable area of bare peat to further erosion. The erosion of peat has lead to a complex topography with considerable variability in the depth of peat between the channels and banks of un-eroded peat.

Table 2. Example description of a deep peat profile.

Horizon	Depth (m)	Description	Carbon (%)	pH (water)	Base saturation (%)
O1	0-13	Reddish brown, fibrous and wet, organic with no structure	51	4.30	19.0
O2	13-25	Black, fibrous and wet, organic with no structure	53	3.9	17.0
O3	25-100+	Dark brown, semi-fibrous and wet, organic layer with no structure	53	3.7	17.2

Peaty gley

Peaty gleys (histic gleysols) formed on the granitic parent materials occur to the west of the loch in hydrological receiving sites at the foot of the corrie (Map unit 3; Figure 1c). These soils have developed under conditions of intermittent or permanent waterlogging and occupy approximately 4% of the catchment at altitudes of around 788 m to over 932 m and occur adjacent to the peat soils (Figure 1c). The parent material is a loamy coarse sand to coarse sandy loam textured morainic drift. The peaty surface layer typically has organic carbon contents in excess of 35%, is less than 50 cm thick and overlies a gleyed albic horizon (Egh) that forms due to the reduction and removal of iron often with high levels of organic carbon (Table 3). This process is not the same as podzolisation but is due to the reduction of ferric iron compounds to (mobile) ferrous compounds when oxygen levels are very low or absent during periods when the soil is waterlogged. Below the albic horizon lie gleyed Bg and Cg coarse textured (loamy sand) and often stony horizons. These horizons have greyish colours with a few bright orange (or ochreous) mottles. These mottles are iron deposits formed under intermittent periods of reducing and oxidising conditions and can sometimes be associated with root channels (Figure 3b).

Peaty ranker

Peaty rankers are soils that have surface horizons of organic material between 0.1 m and 0.5 m thick, resting directly on hard coherent rock (Figure 3c). Occasionally there is a thin mineral layer below the organic surface horizon (Table 4). These soils are recognised as histosols within the World Reference Base Classification but are classified separately from peats within Scotland. Peaty rankers cover only 1% of the catchment and lie between 788 m and 831 m altitude (Figure 1c). The organic surface layer has organic carbon contents of around 40%. The peaty rankers have developed on the moderately steep north western slopes (Map unit 5; Figure 1c) where there are a large number of rock outcrops and large boulders but they are also found near to the loch (Map unit 4; Figure 1c) where there are fewer boulders.

Table 3. Example description of a peaty gley profile.

Horizon	Depth (m)	Description	Carbon (%)	pH (water)	Base saturation (%)
LF	0-10	Reddish brown, fibrous and moist, organic with no structure	-	-	-
O1	10-30	Black, semi-fibrous, moist, organic with a very coarse blocky structure. Few small granite stones	38	3.8	9.6
O2	30-45	Black, amorphous and moist with no structure. Few small granite stones	23	4.2	5.6
Egh	45-51	Dark greyish brown, moist loamy coarse sand with fine blocky structure and few small granite stones	8.5	4.3	2.8
Bhg	51-62	Dark grey, wet humose loamy sand with no structure and few small angular granite stones	13.6	4.3	2.5
C	62-80	Dark brown, wet loamy sand with single grain structure. Many small and many very large angular granite stones.	2.6	4.4	2.6

Rock and scree (Lithosol)

One of the most extensive map units in the catchment (31%) is Map unit 6 (Figure 1c). This map unit comprises bare rock and scree with a few areas of lithosols (Table 5) where the rock is within 10 cm of the soil surface and where a thin A horizon has developed on unconsolidated screes (hyperskeletic leptosols in the World Reference Base classification). The lithosols have only a thin covering of organic material overlying hard, coherent rock (lithic leptosols according to World Reference Base). The vegetation communities associated with these soils generally comprises mosses and lichens with a few dwarf scrubs.

Table 4. Example description of a peaty ranker profile.

Horizon	Depth (m)	Description	Carbon (%)	pH (water)	Base saturation (%)
LF	0-2	Dark grey, semi-fibrous and moist organic with no structure	-	-	-
O	2-28	Black, amorphous and wet, organic with a coarse blocky structure.	41	4.2	4.1
Bhg	28-36	Dark greyish brown, very wet sandy loam with weak blocky structure	2.4	5.0	2
R		Hard coherent granite rock			

Table 5. Example description of a lithosol.

Horizon	Depth (m)	Description	Carbon (%)	pH (water)	Base saturation (%)
LF	0-3	Dark reddish brown, semi-fibrous and moist organic with no structure	-	-	-
H	3-10	Black, amorphous and moist, organic with coarse blocky structure	44	3.5	3

Alpine podzol

Alpine podzols or oroarctic podzols (Map unit 1; Figure 1c) occupy an area of 33% of the catchment and occur at the highest altitudes, extending along the top of the corrie at 1150 m and down to the eastern perimeter of the catchment at approximately 850 m (Figure 1c). The soils would be classed as skeletic podzols (FAO/ISSS/ISRIC 1998) due to the large volume of angular stone fragments found throughout the profile. Stone sorting, stone orientation, cryoturbation and general downslope movement of material are part of the pedological process in this environment. This, together with the depressed chemical and biological activity due to low temperatures, generally results in weakly developed soil horizons. Above 850-1000 m there is often insufficient vegetation to stabilise the surface of the unconsolidated superficial deposit. As a result, a layer of granite grit, low in organic matter, accumulates at the surface. This loose mountain detritus with underlying finer material is a result of frost shattering of larger granite stones and exposed bedrock and is often transported by solifluction or wind. Although the depth of soil is variable, it has been recorded to depths of over 1 m in the Cairngorms (Nolan and Lilly 1985). Often finer silty textured material is deposited on the tops of stones in the subsoil indicative of the down-washing of this material (Romans et al. 1966; Heslop 1981). The drift is generally a loamy coarse sand or course sandy loam often with large angular rock fragments throughout the profile that increase in both size and volume with depth. One of the main features of these soils, however, is their open, porous structure. The bulk density of the fine earth fraction of similar soils in other parts of the Cairngorms range from 0.58 to 1.1 g cm^3 with an average value of 0.91 g cm^3 and the total porosity of these mineral soils ranged from 60 to 70 percent throughout the profile.

The alpine podzols often have a thin humose surface horizon (H) that may contain bleached sand grains and is loose, which shows some of the effects of cryoturbation (Table 6; Figure 3d). In many of these soils, there is a thin (<5cm) layer of grit that overlies this H horizon. The H horizon overlies a bleached, thin eluvial E horizon. In some case, the H/E sequence is replaced by a surface Ah horizon with an organic carbon content of around 5.5 percent. Below this layer is a dark reddish brown or dark brown humus-enriched mineral Bh or, where sesquioxides are present, a Bhs. The presence of the Bh horizon is quite distinctive and many of the soil profiles may qualify as humic podzols under the World Reference Base classification system (FAO/ISSS/ISRIC 1998). However, Romans et al (1966) suggested that the highly

Table 6. Example description of an alpine podzol profile.

Horizon	Depth (m)	Description	Carbon (%)	pH (water)	Base saturation (%)
H	0-4	Black, amorphous and moist, organic with fine blocky structure	23	4.1	12.4
E	4-19	Dark grey, humose loamy sand with medium blocky structure	1	4.5	15.4
Bh	19-44	Black, humose loamy sand with very fine blocky structure and few large angular granite stones.	5.8	4.9	1.3
Bsh	44-58	Dark brown, loamy coarse sand with fine blocky structure and common medium angular granite stones	1.6	5.1	2.7
C	58-100	Dark brown, loamy sand with medium and fine blocky structure and many very large angular granite stones.	0.5	5.1	7.1

organic Bh horizon may in fact be a buried Ah and that any E horizon found overlying these Bh layers was, in fact, a buried grit layer. Below the Bh horizon there often is a weakly developed Bs (sesquoxide enriched) that grades gradually into a relatively unaltered C horizon. Where the surface layer is an Ah horizon and there is no evidence of a bleached E horizon, the soil may be classed as a skeletic cambisol (FAO/ISSS/ISRIC 1998).

Impacts of soil processes on surface water chemistry

One of the vital roles played by soil is its function as a buffer regulating the quality of surface waters. Precipitation interacts to a substantial degree with soil before entering surface or ground waters. The precise fate of incoming precipitation depends upon the physico-chemical characteristics of the soil and the quality, duration and intensity of the precipitation, in additional to other physical characteristics of the catchment.

Soil acidification

Soil acidification involves the removal of base cations and the consequent accumulation of acid in the soil. It is a natural process which, in natural ecosystems, operates over many thousands of years. The buffering capacity of a soil determines its ability to prevent pH decline. For example, the greater the volume of organic matter (or the clay content) the greater will be the soil's pH buffering capacity. In the absence of pollution, divalent cations such as Ca^{2+} and Mg^{2+}, with their high charge, are strongly attracted to the organic matter and clay surfaces.

Deposition of pollutants such as sulphate (SO_4), nitrate (NO_3), and ammonium (NH_4) however, can accelerate the rate of acidification over decades. Inputs of nitric and sulphuric acids to soil result in a shift in ion exchange equilibrium in the short term, and thus enhance leaching of base cations. If the base cations leached from cation exchange sites during heavy rain are replenished subsequently by geochemical weathering, the pH of the soil will not decline. If, on the other hand, the soil is naturally low in readily weatherable minerals, as is the case at Lochnagar, the base saturation and pH of the soil will decline rapidly in response to acid deposition until a new equilibrium is reached. The extent to which this happens varies within the soil profile depending on the nature and properties of the individual soil horizons.

The soils in the Lochnagar catchment are formed from granitic parent material, are naturally acidic, and have a low buffering capacity to changes in pH. In addition, enhanced leaching of nutrients due to high rainfall can exacerbate soil and water acidification.

Buffering capacity

Chemical weathering of primary silicate minerals and the release of base cations is the principal mechanism by which inputs of acids to soils are buffered. Weathering rates are of particular importance in the context of soil acidification as the rate of release of base cations from the soil and uptake by vegetation will generally determine whether a soil will become acidified by internal soil processes and/or acidic inputs from the atmosphere. Long term weathering rates have been calculated for some upland catchments in Scotland and the related weathering processes have been established. Weathering rates for a freely drained alpine podzol similar to those soils found in the Lochnagar catchment were determined using three conventional methods (i) the soil type and bedrock geology approximation of de Vries (1993), (ii) assignment according to the Skokloster workshop (Werner and Spranger 1996) and (iii) calculation with the PROFILE model (Hodson et al. 1996). Table 7 shows whole profile weathering rates for the base cations calcium (Ca), magnesium (Mg), sodium (Na) and potassium (K). Whilst the soil/geology approximation predicts the greatest weathering rates, the Skokloster assignment consistently predicts the lowest weathering rate for all soil horizons. Although the weathering rates determined from the three methodologies were found to show a general lack of agreement due to the inherent difficulties in quantifying mineral weathering, the calculated weathering rates fall within the lower, most sensitive critical load classes giving these methods some credibility (Stutter 2001).

Table 7. Weathering rate values at Lochnagar for three soil horizons generated by three methods (Stutter 2001).

Soil Horizon	Horizon depth (m)	Total cation weathering rates/m depth (kmol$_c$ha^{-1}yr^{-1})		
		Soil/geology approximation	Skokloster assignment	PROFILE model
E	0.08	1.54	0.116	0.735
Bh	0.1	2.16	0.232	1.78
Bs	0.42	0.927	0.273	0.832

Erosion

In some areas of the UK peat can be eroded at rates of up to 5 cm yr^{-1} (Tallis et al. 1998) either from overland flow at the surface or by preferential flow paths within the peat which eventually form gully systems characteristic of the eroded peat typical of Map unit 2 (Figure 1c). Examination of the land cover of Scotland dataset (MLURI 1993) showed that 24% of the land classed as peatlands had significant signs of erosion. This represents around 15% of the land area of Scotland (approximately 12000 km^2). The stability of these huge carbon stores under changing climate conditions has been subject to recent concern due to expected climate warming, which has been predicted to result in significant losses of carbon from upland ecosystems (Ineson et al. 1998). The release of dissolved organic carbon (DOC) into soil water and ultimately to surface water represents a key process for the loss of carbon from organic rich ecosystems, since DOC is a significant energy source for microbial populations. To ensure effective management of these fragile mountain systems it is important to understand the main process and mechanisms responsible for the release of organic matter into solution, however this continues to be an area of scientific debate. This is covered in more detail below.

In the Lochnagar catchment, agents of peat erosion have been identified by observation, from documentary evidence and by inference. Possible causes of peat erosion can be attributed to a number of factors including the removal of peat during exceptional, high intensity rain events; a cause of increasing concern given projections for climate change in the Cairngorm region (Hulme et al. 2002; Chapter 18: Kettle and Thompson this volume) and vegetation change in the catchment (Dalton et al. 2005). Gully erosion causes a drying of the peat within 1-2 m of the gully, this change in soil wetness can lead to reduced species diversity and poor bank stability (Tallis 1997) and discontinuous vegetation cover (Anderson and Yalden 1981). In circumstances where the peat surface becomes exposed to wind, peat degradation may accelerate. In Britain, the principal components of bog vegetation and peat are species of the moss *Sphagnum*. *Sphagnum* is rootless and so cannot withstand extreme erosive surface forces. Although *Sphagnum* thrives under naturally acidic, nutrient poor conditions, they are highly sensitive to climate change and changes in the deposition of airborne pollutants. Observations of plant remains preserved in peat in other parts of the UK indicate that *Sphagnum* is particularly sensitive to anthropogenic acid deposition and its occurrence has significantly declined over the past 100-150 years, thus potentially exposing peat to erosion (Tallis 1997).

At Lochnagar, most erosion has taken place in the deep peat (on the eastern flank of the loch) but despite predictions of high intensity rain storms in the future (Hulme et al. 2002), there is considerable debate about the possibility of climate warming where generally drier climatic conditions, leading to lowered water tables, increase the susceptibility of peat to erosion (Yeo 1997). There is no evidence, past or present of over-grazing and/or burning in the catchment which can lead to a reduced plant cover and exposure of bare peat to erosion (Bragg and Tallis 2001).

Human trampling can affect soil erosion as the protective vegetation cover is damaged and soil is exposed to erosion by wind and water. Although the impact of human trampling is minimal throughout Scotland, in areas that are popular with hill walkers and rock climbers such as Lochnagar, the impacts have increased in recent decades (Lance et al. 1991). Damage due to trampling is concentrated along pathways to the main climbing routes to the corrie summits of Lochnagar. Grieve (2001) found that the mean organic matter content of alpine podzols under vegetation disturbed by trampling was less than half that under complete vegetation cover at Beinn Avon, Braeriach and Cairn Gorm in the Cairngorm mountains. This result was indicative of footpath erosion. In the study by Grieve (2001), the effects of trampling intensified where natural cryoturbation processes were predominant. The soil types investigated in this study were similar to those in the Lochnagar catchment.

Increased production of organic acids in soil and surface water

It is clear from Figure 4 that concentrations of dissolved organic carbon (DOC) in Lochnagar have significantly increased from 1988 to 2002. Peat erosion may contribute to this rising trend in DOC, however, other soil processes, driven primarily by a change in climate and/or deposition, have been proposed. The increase in DOC concentration could result from changes in hydrology, i.e., a decrease in discharge could result in a changes in concentrations (Tranvik and Jansson 2002), however there is little evidence from the short hydrological record presented in Chapter 9 (Jenkins et al. this volume) that discharge has decreased at Lochnagar. Reductions in sulphur deposition have also been linked to the increasing DOC concentrations; a trend that has been observed across many of the UK Acid Waters Monitoring Network sites (Chapter 14: Monteith et al. this volume). Increasing DOC production has been associated with decreasing mineral acidity (Krug and Frink 1983), but evidence from field observations is ambiguous. Decomposition, and therefore DOC production, is restricted by repression of the major biodegrading hydrolase enzymes. The activity of these enzymes is constrained by phenolic compounds which accumulate in peat because the activity of phenol oxidase is inhibited in the oxygen deficient peat. A loss of phenolic compounds means that decomposition can continue even after a return to saturated conditions. This has been referred to as an enzyme latch mechanism (Freeman et al. 2001), i.e., enzymes are switched on when the soil water content decreases from saturation but are not switched off immediately on return to saturated conditions. This mechanism could cause increased peat decomposition and therefore increased DOC release following periods of drought. In another study, Worrall et al. (2003) associated the observed increases in DOC to rising temperatures over the past 30 years that enhance microbial activity, decomposition of peat and production of DOC. Enchytraeid worms (*Oligochaeta*), which are common in upland soils, increase in abundance at higher temperatures, and by influencing microbial activity, litter fragmentation and soil aeration, may also increase surface water DOC (Cole et al. 2002).

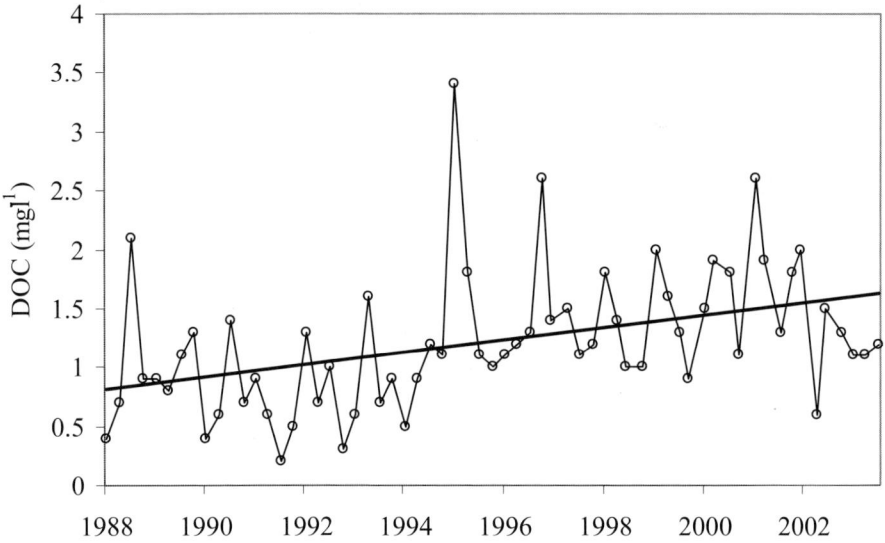

Figure 4. Temporal changes in DOC in Lochnagar (1988 to 2004). These data are from the UK Acid Waters Monitoring Network (Shilland et al. 2004). The equation of the regression line is y = 0.0044x – 3.8285 (R^2 = 0.1654; p<0.001)

The importance of soil in determining surface water nitrate concentrations

Controls on nitrate leaching from montane catchments are not fully understood. Despite a relatively constant observed flux of deposited nitrogen to the Lochnagar catchment increased surface water nitrate concentrations have been observed in the loch during the past 12 years (Figure 5) (Davies et al. 2005). A continued upward trend, could significantly affect the acid status of the loch in the future as well as potentially damaging the semi-natural terrestrial ecosystem. The process of nitrogen saturation describes the decreased immobilisation of deposited nitrogen in catchment soils as a result of increasing nitrogen richness relative to carbon (decreased C:N ratio) and subsequent nitrate leaching. Recently it has been demonstrated that the loss of nitrogen to surface waters is dependent on soil carbon storage (Evans et al. 2006; Jenkins et al. 2001). The total carbon pool and nutrient content of soils in the Lochnagar catchment were quantified for use in dynamic modelling assessments. This quantification was achieved by collecting 69 samples from the surface organic horizons of major soil types in the Lochnagar catchment. These samples were analysed for carbon and nitrogen and loss-on-ignition, and the dominant plant communities were also identified at the sampling sites. A measurement of the total depth of the organic horizons was made at the time of sampling, and bulk density was determined from pedotransfer functions. Individual C:N ratios of the organic horizons ranged from 20-50. As previously described, the organic status of soils is affected by the slope, altitude, drainage and vegetation. Eight topographically distinct areas were identified within a

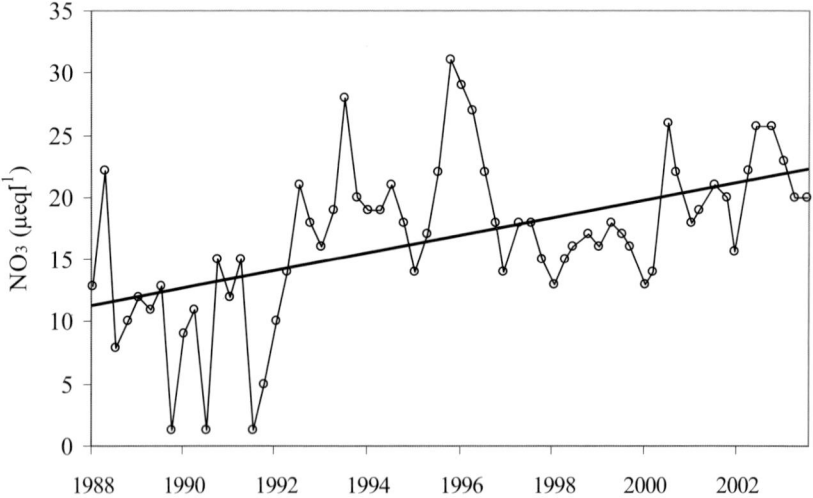

Figure 5. Long term trend in NO_3 at Lochnagar (1988 to 2004). This data is part of the UK Acid Waters Monitoring network (Shilland et al. 2004). The equation for the regression line is y = 0.0586x – 51.007 (R^2 = 0.2633; p<0.001)

Geographic Information System (GIS), based on these aforementioned characteristics and associated with point observations of percentage carbon and nitrogen in the organic layer. The area of boulders and scree in the topographic classes was calculated from photographs of the catchment, and carbon and nitrogen pools were determined within the GIS (Figure 6). With the greatest carbon pools determined for the deep organic soils at the base of the catchment (Classes 1, 2 and 3) and smallest for the lithosols (Class 8). Nitrogen pools are also related to the soil organic content, the largest pools are associated with organic rich soils such as the deep and shallow peats surrounding the loch (Table 8). The catchment weighted carbon and nitrogen pools were 967 and 26 mol m^{-2} respectively, and a C:N ratio of 37.2.

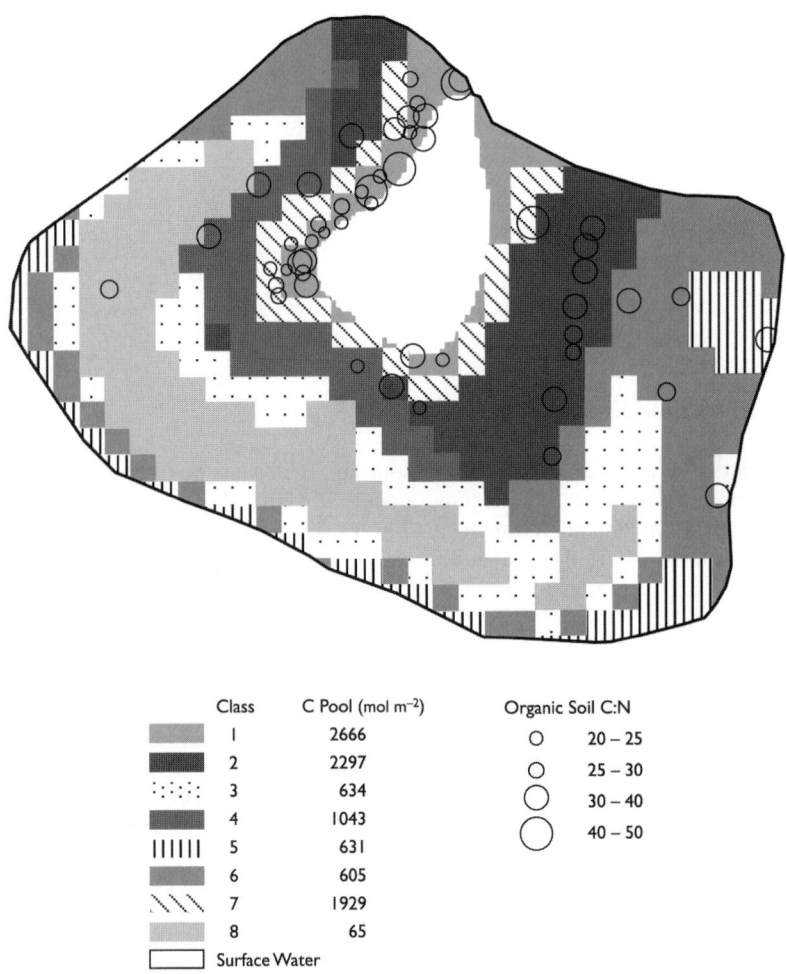

Class	C Pool (mol m^{-2})
1	2666
2	2297
3	634
4	1043
5	631
6	605
7	1929
8	65
Surface Water	

Organic Soil C:N
20 – 25
25 – 30
30 – 40
40 – 50

Figure 6. Map of soil classes at Lochnagar indicating mean C Pool (mol m^{-2}) and C:N ratio. From Jenkins et al. (2001). Reproduced with the permission of European Geosciences Union.

Although the carbon pool has been identified as an important control on surface water nitrate it is possible to hypothesise other mechanisms that might also lead to a reduction in catchment retention such as increased mineralisation of organic nitrogen or decreased plant uptake as a result of climatic conditions. Despite this, we note that 8% of the catchment area comprises the loch to which nitrogen is deposited directly (hydrological nitrogen) and another 30% of the catchment is composed of rock and scree. It is probable that the capacity of the Lochnagar catchment to immobilise nitrogen is reduced compared to other upland catchments.

Table 8. Nitrogen pools (mol m^{-2}) by soil class.

Class	N Pool (mol m^{-2})
1	78
2	57
3	18
4	27
5	16
6	18
7	44
8	2

MAGIC modelling of soil and soil solution

The impact of acidic deposition on the soils of Lochnagar has been assessed using output from the dynamic acidification model, MAGIC (Model of Acidification of Groundwater in Catchments (Cosby et al. 2001)). This model application illustrates the extent and timing of acidification and recovery in response to recent UNECE protocols to reduce levels of sulphur and nitrogen deposition.

Soil parameterisation

Organic and mineral soil data from the soil types identified within the Lochnagar catchment were vertically (by horizons within a profile) and spatially (by area of different soil types) weighted within the catchment according to Helliwell et al. (1998). The soils were segregated into upper and lower compartments based on the dominant hydrological pathways through the catchment soils, determined from the HOST (Hydrology of Soil Types) classification (Boorman et al. 1995). Thus one set of soil parameters were generated to represent the upper organic rich and mineral horizons (O, H and E) and another set for the lower mineral horizons (B and C). Each HOST class was also assigned a standard percentage runoff (SPR) value, which indicates the percentage of rainfall that contributes to the short term increase in flow. Soil physico-chemical properties were vertically and spatially weighted for soil layer 1 (upper organic and mineral horizons) and layer 2 (mineral horizons mainly from alpine podzols and peaty gleys) based on the hydrological separation of horizons (Table 9).

All precipitation was routed through soil layer 1 and the proportion of runoff entering the loch directly from this soil layer is determined by the SPR, the remaining proportion drains through soil layer 2.

Model Calibration

The MAGIC model was calibrated to surface water parameters observed in Lochnagar (Aherne et al. in press) and soil parameters (Table 9) from the surrounding catchment. A detailed account of the hydrochemistry used to calibrate the MAGIC model is discussed in Chapter 9 (Jenkins et al. this volume). Full details of the model structure, the calibration procedure and the general methodology are given in Evans et al. (2001), Helliwell et al. (2003) and Helliwell et al. (2004).

Deposition scenario

Deposition estimates from 1999-2001 were derived from regional models at a 5km^2 resolution (Helliwell et al. 2003) for the Lochnagar area. The historical trend in wet, non-marine sulphate deposition has been the main driver of change for both soil and surface water chemistry in recent years at Lochnagar (Chapter 14: Monteith et al this volume). The sequence adopted for this model application is shown in Figure 7, with observed deposition trends to 2000 and future deposition based on current emission legislation emission controls as modelled by EMEP (Co-operative Programme for Monitoring and Evaluation of the Long Range Transmission of Air Pollutants in Europe; Schöpp et al. 2003).

Table 9. Soil physico-chemical parameters for the Layer 1 (organic horizons) and Layer 2 (mineral horizons) for Lochnagar (Aherne et al. in press).

	Units	Layer 1	Layer 2
Depth	m	0.12	0.39
Bulk density	g cm^{-3}	0.39	1.06
Loss on Ignition	%	23	7
Carbon: Nitrogen ratio		27	17
Cation Exchange Capacity (CEC)	meq m^{-3}	303	105
Exch. Ca	% CEC	6.33	3.2
Exch. Mg	% CEC	5.54	2.6
Exch. Na	% CEC	0.98	0.6
Exch. K	% CEC	1.14	0.4
Soil pH	pH unit	4.17	4.7
Soil water percolation	% of PPT	100	58

Exch. = Exchangeable cation; PPT = Precipitation

Response of the soils at Lochnagar to changes in acid deposition

The segregation of soil parameters between the organic and mineral layers result in a significantly different base status throughout the simulation period (1850-2050). The organic layer was more responsive to changes in atmospheric deposition and soils were predicted to acidify rapidly from the mid 1850s to the 1980s, a time of peak sulphur deposition. The recovery phase was initiated in the 1990s in response to a decline in sulphur emissions, and this resulted in an increased recovery rate compared to the base status of the mineral layer (Figure 8). Although the base saturation of the mineral layer declined very gradually from 1850s, emission reduction protocols were shown to have little effect on the recovery of the base status, indeed a continued decline of 0.7% is simulated from 1990 to 2050.

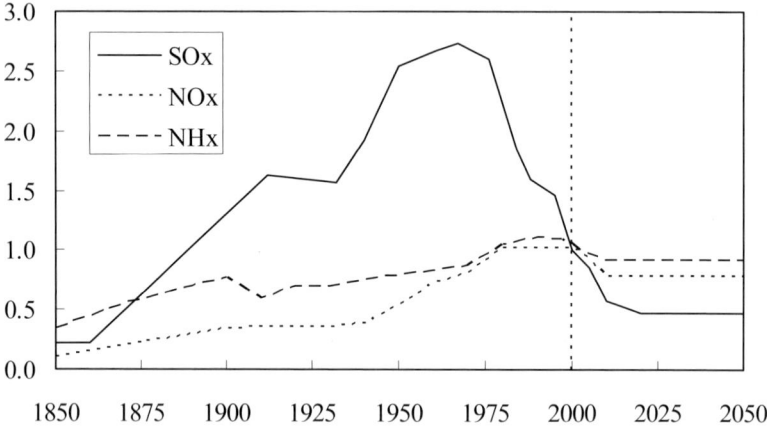

Figure 7. Historic and future sulphate and nitrogen deposition sequences (y-axis scale factor: 2000 deposition is equivalent to a factor of 1) for the period 1850-2050.

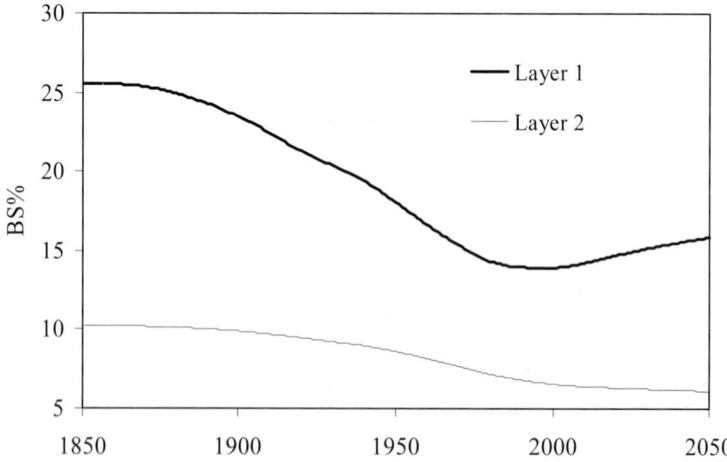

Figure 8. Simulated time-series of base saturation (%) at Lochnagar (Aherne et al. in press).

Large differences are evident between the calibrated soil parameters from the organic (Layer 1) and mineral (Layer 2) soil compartments (Table 10). In 1850, when acid inputs were negligible, base cations derived from atmospheric deposition interacted with exchange sites in the organic soil complex. With time, increasing acid deposition depleted the base cations from the exchange complex resulting in rapid loss of base cations predominantly from the organic compartment. The low selectivity co-efficients of the organic soil (Layer 1) indicate that the remaining base cations are held tightly on the exchange complex. The base cation flux from weathering was greater in the mineral layer than the organic layer, yet the importance of base cations derived from atmospheric deposition, far outweighs the contribution to the soil systems from weathering.

Table 10. Calibrated soil parameters for Lochnagar (Aherne et al. in press).

	Units	Layer 1	Layer 2
Selectivity co-efficient Al-Ca	Log10	-1.72	1.74
Selectivity co-efficient Al-Mg	Log10	-1.24	1.92
Selectivity co-efficient Al-Na	Log10	-1.26	0.28
Selectivity co-efficient Al-K	Log10	-6.06	-2.11
Ca weathering	meq m^{-2} yr^{-1}	0.9	8.4
Mg weathering	meq m^{-2} yr^{-1}	0.5	5.0
Na weathering	meq m^{-2} yr^{-1}	1.8	16.7
K weathering	meq m^{-2} yr^{-1}	0.4	3.5
Base saturation 1850	%	25.5	10.2
Al solubility coefficient	Log10	7.3	8.5

Potential impacts of climate change on the soils of Lochnagar

Climate change, as predicted by the UKCIP02, is expected to produce higher temperatures, largely unpredictable changes in rainfall and an increase in extreme weather in the Scottish highlands (Hulme et al. 2002; Chapter 18: Kettle and Thompson this volume). The consequence for soils and their associated vegetation will therefore be complex and difficult to predict.

Whilst climate change predictions are generally made on a decadal timescale, the soils in the catchment have developed over thousands of years in which the climate has changed considerably. A soil develops in response to many factors and, in time, the soil itself will have an influence on its response to changes in climate, for example, the high degree of porosity of the alpine podzols mean that the soil will still remain unsaturated throughout the year even if the rainfall increases dramatically. Some soil processes however, may be affected by shorter term changes in the climate. Within a decade, changes in temperature and precipitation will have modified soil temperature and moisture regimes, pH, base saturation, fertility status, surface litter and biological activity. Due to the range of microclimates found in the Scottish Highlands, it is difficult to make predictions about the impact of climate change at any specific location, however, climate warming is likely to cause a decrease in the extent of the more vulnerable plant species found at these high altitudes (Coll et al. 2005; Chapter 7:

Birks this volume) and this will likely cause changes in the type and quantity of litter input.

It is likely then, that one of the greatest impacts of climate change on montane soils will be on nutrient cycling if the climate was to become warmer. Undisturbed mountain ecosystems conserve nutrients; in other words the losses are small relative to the turnover within the ecosystem. Changes in weathering rates and litter input may well perturb the current nutrient balance. Another impact could be changes in carbon content. The loss of carbon from soil following warming may be partially or completely offset by increases in the carbon input to soils as a result of enhanced vegetation productivity under elevated CO_2 and higher temperatures but little is known definitively about the effect of high CO_2 on soil organic matter quality and how this might affect the soil carbon pool. The size of the soil carbon store also depends on the depth of the soil profile, its water regime (wet soils tend to accumulate more carbon than dry soils in this mountain environment), temperature (cold soils store more carbon than warm soils) and the type and quantity of clays present (Scholes et al. 1999). However, increased precipitation will lead to increased runoff and is likely to exacerbate current gullying in the eroded peat soils leading to a loss of carbon from the catchment.

Increasingly, it is recognised that a consequence of climate change may be the collapse of the Thermohaline Circulation that is responsible for bringing warmer waters from lower latitudes to the UK. This would lead to a cooling of the climate rather than current predictions of warming and it is, therefore, difficult to predict the future climate of the Lochnagar catchment.

Conclusion

The soils of the Lochnagar catchment are typical of the eastern Cairngorm granite plateau as are the processes that are currently affecting the chemistry of the surface waters. The climate is harsh, suppressing both soil processes and biological activity which means that the terrestrial and aquatic ecosystems are sensitive to environmental change such as acidification. Sulphur and nitrogen deposition are two of the main drivers of acidification in these ecosystems. Current work in this environment is focussed on the interaction of acid deposition and climate change (Wright et al. 2006). Although not unique in a UK context, the extent of the soils, landform and vegetation communities found in the Lochnagar catchment and the Cairngorms in general means that the area is of national significance in terms of its wildlife habitat and ecosystems and has recently been assigned National Park status.

Summary

- The climate within the catchment has changed dramatically over the last 10000 years.
- The catchment has a relatively wide range of soils that are typical of much of the Cairngorms.

- The soils at higher altitudes or on exposed ridges are subject to persistent freeze/thaw and are strongly leached while those at lower altitudes and hydrological receiving sites are dominated by an accumulation of organic carbon within surface layers.
- Soil and surface water chemistry are heavily influenced by both chemical and physical processes.
- The main issue is acidification of the soil and surface water. The main agents of acidification are acids deposited through precipitation (sulphate, oxidised and reduced nitrogen) and organic acids from erosion of the peat soils.
- The extent of soil and soil water acidification is dependant on the soil buffering capacity and weathering rate but weathering rate is difficult to quantify.
- From the mid-19th century until the 1980s, the soils and surface waters of Lochnagar slowly acidified as a result of sulphur deposition. Internationally agreed protocols to reduce acid emissions have led to a recovery in the soil base saturation and a concomitant increase in the base status of soil water at Lochnagar.
- More recently the increased production of organic acids and release of nitrate to Lochnagar are of more concern to scientists than sulphur deposition

References

Aherne J., Helliwell R.C., Lilly A., Ferrier R.C. and Jenkins A. in press. Simulation of soil and surface water acidification: Divergence between one-box and two-box models? Applied. Geochem..

Anderson P. and Yalden D.A. 1981. Increased sheep numbers and loss of heather moorland in the Peak District, England. Biol Cons. 20: 195–213.

Birks H.J.B. This volume. Chapter 7. Flora and vegetation of Lochnagar – past, present and future. In: Rose N.L. (ed.) 2007. Lochnagar: The natural history of a mountain lake. Springer. Dordrecht.

Birse E.L. 1980. Suggested changes to the world soil classification to accommodate Scottish mountain and aeolian soils. J. Soil Sci. 31: 117–124.

Birse E.L.and Dry F.T. 1970 Assessment of climatic conditions in Scotland. 1. Based on accumulated temperature and potential water deficit. The Macaulay Institute for Soil Research, Aberdeen. 25 pp.

Birse E.L. and Robertson L. 1970. Assessment of climatic conditions in Scotland. 2. Based on exposure and accumulated frost. The Macaulay Institute for Soil Research, Aberdeen 41 pp.

Boorman D.B., Hollis J.M. and Lilly A. 1995. Hydrology of soil types: A hydrologically based classification of the soils of the United Kingdom. Report no. 126. Institute of Hydrology, UK 137 pp.

Bragg O.M and Tallis J.H. 2001. The sensitivity of peat-covered upland landscapes. 2001. Catena 42: 345–360.

Cole L., Bardgett R.D., Ineson P. and Adamson J.K. 2002. Relationships between enchytraeid worms (*Oligochaeta*), climate change, and the release of dissolve organic carbon from blanket peats in North England. Soil Biol Biochem 34: 599–607.

Coll J., Gibb S.W. and Harrison S.J. 2005. Modelling future climates in the Scottish Highlands – an approach integrating local climatic variables and regional climate model outputs. In Thompson D.B.A., Price M.A. and Galbraith C.A. (eds.) Mountains of northern Europe: Conservation, management, people and nature. TSO Scotland. Edinburgh. pp. 103–120.

Cosby B.J., Ferrier R.C., Jenkins A., and Wright R.F. 2001. Modelling the effects of acid deposition: refinements, adjustments and inclusion of nitrogen dynamics in the MAGIC model. Hydrol. Earth Sys. Sci. 5: 499–517.

Dalton C., Birks H.J.B., Brooks S.J., Cameron N.G., Evershed R.P., Peglar S.M., Scott J.A. and Thompson, R. 2005. A multi-proxy study of lake development in response to catchment changes during the Holocene at Lochnagar, north-east Scotland. Palaeogeog. Palaeoclim. Palaeoecol. 221: 175–201.

Davies J.J.L., Jenkins A., Monteith D.T., Evans C.D. and Cooper, D.M. 2005. Trends in surface water chemistry of acidified UK Freshwaters, 1988–2. Environ. Pollut. 137: 27–39.

De Vries W., Posch M., Reinds G.J. and Kamari, J. 1993. Critical loads and their exceedance on forest soils in Europe (Report no.58; revised version). Wageningen, Netherlands 27 pp.

Evans C., Jenkins A., Helliwell R.C., Ferrier R. and Collins R. 2001. Freshwater acidification and recovery in the United Kingdom. Centre for Ecology and Hydrology, Wallingford 80 pp.

Evans C.D., Reynolds B., Jenkins A., Helliwell R.C., Curtis C.J., Goodale C.L, Ferrier R.C., Emmett B.E., Pilkington M.G., Caporn S.J.M, Carroll J.A., Norris D., Davies J. and Coull M.C. 2006. Evidence that soil carbon pool determines susceptibility of semi-natural ecosystems to elevated nitrogen leaching, Ecosystems 9: 453–462.

FAO/ISSS/ISRIC. 1998. World reference base for soil resources. Food and Agriculture Organisation/International Society of Soil Science/International Soil Reference and Information Centre publication. World Soil Resources Reports, No. 84. Food and Agriculture Organisation Rome 91pp.

Freeman C., Ostle N. and Kang, H. 2001. An enzyme 'latch' on a global carbon store - a shortage of oxygen locks up carbon in peatlands by restraining a single enzyme. Nature, 409: 149.

Goodman S. This volume. Chapter 3. Geology of Lochnagar and surrounding region. In: Rose N.L. (ed.) 2007. Lochnagar: The natural history of a mountain lake. Springer. Dordrecht.

Gordon J.E. 2001. The corries of the Cairngorm Mountains. Scot. Geog. J 117: 49–62.

Grieve I.C. 2001. Human impacts on soil properties and their implications for the sensitivity of soil systems in Scotland. Catena 42: 361–374.

Hall A.M. This volume. Chapter 4. The shaping of Lochnagar: Pre-glacial, glacial and post-glacial processes. In: Rose N.L. (ed.) 2007. Lochnagar: The natural history of a mountain lake. Springer. Dordrecht.

Helliwell R.C., Ferrier R.C. and Jenkins A. 1998. Comparison of methods for estimating soil characteristics in regional acidification models; an application of the MAGIC model to Scotland. Hydrol. Earth Sys. Sci. 2: 509–520.

Helliwell R.C., Jenkins A., Ferrier R.C., and Cosby B.J. 2003. Modelling the recovery of surface water chemistry and the ecological implication in the British uplands. Hydrol. Earth Sys. Sci. 7: 456–466.

Helliwell R.C. and Kernan M. 2004. Modelling hydrochemical and ecological trends in acid sensitive surface waters in the Scottish Highlands. J. Limnol. 63: 111–122.

Heslop R.E.F. 1981. Soils. In Nethersole-Thompson D. and Watson A. (eds.) The Cairngorms, their natural history and scenery. The Melvin Press, Perth. 222–227.

Hodson M.E., Langan S.J. and Wilson J.F. 1996. A sensitivity analysis of the PROFILE model in relation to the calculation of soil weathering rates. App. Geochem. 11: 835–844.

Hughes M. This volume. Chapter 2. Physical characteristics of Lochnagar. In: Rose N.L. (ed.) 2007. Lochnagar: The natural history of a mountain lake. Springer. Dordrecht.

Hulme M., Jenkins G.J., Lu X., Turnpenny J.R., Mitchell T.D., Jones R.G., Lowe J., Murphy J.M., Hassell D., Boorman P., McDonald R. and Hill S. 2002. Climate change scenarios for the United Kingdom: UKCIP02 scientific report. Tyndall Centre for Climate Change Research, School of Environmental Sciences, University of East Anglia, Norwich, UK 124 pp.

Ineson P., Taylor K., Harrison A.F., Poskitt J., Benham D.G., Tipping E. and Woof C. 1998. Effects of climate change on nitrogen dynamics in upland soils. I. A transplant approach. Global Change Biol. 4: 143–152.

Jenkins A., Ferrier R.C. and Helliwell R.C. 2001. Modelling nitrogen dynamics at Lochnagar, N.E. Scotland. Hydrol. Earth Sys. Sci. 5: 519 – 527.

Jenkins A., Reynard N., Hutchins M., Bonjean M. and Lees M. This volume. Chapter 9. Hydrology and hydrochemistry of Lochnagar. In: Rose N.L. (ed.) 2007. Lochnagar: The natural history of a mountain lake. Springer. Dordrecht.

Jenny H. 1941. Factors of soil formation. McGraw-Hill. New York

Jones V.J., Flower R.J., Appleby P.G., Natkanski J., Richardson N., Rippey B., Stevenson A.C. and Battarbee R.W. 1993. Palaeolimnological evidence for the acidification and atmospheric contamination of lochs in the Cairngorm and Lochnagar areas of Scotland. J Ecol 81: 3–24.

Kettle H. and Thompson R. This volume. Chapter 18. Future climate predictions for Lochnagar. In: Rose, N.L. (ed.) 2007. Lochnagar: The natural history of a mountain lake. Springer. Dordrecht.

Krug E.C. and Frink C.R. 1983. Acid rain on acid soil: A new perspective. Science 221: 520–525

Lance A., Thaxton R. and Watson A. 1991. Recent changes in footpath width in the Cairngorms. Scot. Geog. Mag. 107: 106–109.

MLURI 1993. The Land Cover of Scotland 1988. Macaulay Land Use Research Institute, Aberdeen 81 pp.

Monteith D.T., Evans C.D. and Dalton C. This volume. Chapter 14. Acidification of Lochnagar and prospects for recovery. In: Rose N.L. (ed.) 2007. Lochnagar: The natural history of a mountain lake. Springer. Dordrecht.

Nolan A. and Lilly A. 1985. The soils of the Allt a' Mharcaidh catchment. Report produced as part of the Surface Water Acidification Programme. Macaulay Land Use Research Institute 45 pp.

Pulford I.D. 1998. Factors influencing the formation of Scottish Soils. Scot. Geog. Mag. 114: 10–12.

Romans J.C.C., Stevens J.H. and Robertson L. 1966. Alpine soils of north-east Scotland. J. Soil Sci. 17: 184–199.

Tallis J.H. 1997. The southern Pennine experience: An overview of blanket mire degradation. In: Tallis J.H., Meade R. and Hulme P.D. (eds.) Blanket mire degradation causes, consequences and challenges. University of Manchester 9-11 April. British Ecological Society. pp 7–15.

Tallis J.H. 1998. Growth and degradation of British and Irish blanket mires. Environ. Rev. 6: 81–122.

Tranvik L.J. and Jansson M. 2002. Terrestrial export of organic carbon. Nature 415: 861–862.

Scholes R.J., Schlze E.D., Pitelka L.F. and Hall, D.O. 1999. Biogeochemistry of terrestrial ecosystems. In: Walker B., Steffen W., Canadell J. and Ingram J. (eds.) The terrestrial biosphere and global change: Implications for natural and managed ecosystems. Cambridge University Press. pp 271–303.

Schöpp W., Posch M., Mylona S. and Johansson M., 2003. Long-term development of acid deposition (1880–2030) in sensitive freshwater regions of Europe. Hydrol. Earth Sys. Sci. 7: 436–446.

Shilland E.M., Monteith D.T., Bonjean M. and Beaumont W.R.C. (eds.) 2004. The United Kingdom Acid Waters Monitoring Network. Data Report for 2003–2004 (Year 16). Accessed from: http://www.ukawmn.ucl.ac.uk/report/AWMNAnnualReport2003–04.pdf

Soil Survey of Scotland 1984. Organisation and methods of the 1:250 000 soil survey of Scotland. The Macaulay Institute for Soil Research. Aberdeen 81 pp.

Stephenson D. and Gould D. 1995. British regional geology: The Grampian Highlands (4th Edition). British Geological Survey. HMSO. London 261 pp.

Stutter M. 2001. The use of atmospheric sea salt deposition effects in the quantification of weathering rates for United Kingdom upland catchments. PhD Thesis, Aberdeen.

Thompson R., Kettle H., Monteith D.T. and Rose N.L. This volume. Chapter 5. Lochnagar water-temperatures, climate and weather. In: Rose N.L. (ed.) 2007. Lochnagar: The natural history of a mountain lake. Springer. Dordrecht.

Werner B. and Spranger T. 1996. Mapping critical loads / levels and geographical areas where they are exceeded in 1996 (Federal Environment Agency Texte 71/96) Federal Environment Agency, Germany.

Wright R.F., Aherne J., Bishop K., Camerero L., Cosby B.J., Evans C.D., Hardekopf D., Helliwell R.C., Forsius M., Holberg M., Hruska J., Jenkins A., Moldan F., Posch M. and Rogora M. 2006. Modelling the effect of climate change on recovery of acidified freshwaters: Relative sensitivity of individual processes in the MAGIC and SMART models. Sci. Tot. Environ. 365: 154–166.

Worrall F., Reed M., Warburton J. and Burt T. 2003. Carbon budget for a British upland peat catchment. Sci. Total Environ. 312: 133–146

Yeo M. 1997. Blanket mire degradation in Wales. In: Tallis J.H., Meade R. and Hulme P.D. (eds.) Blanket mire degradation causes, consequences and challenges. University of Manchester 9-11 April. British Ecological Society. pp 101–115.

7. FLORA AND VEGETATION OF LOCHNAGAR – PAST, PRESENT, AND FUTURE

H. JOHN B. BIRKS (John.Birks@bio.uib.no)
Department of Biology
University of Bergen
Allégaten 41
N-5007 Bergen, Norway

and

Environmental Change Research Centre
University College London
Pearson Building, Gower Street
London, WC1E 6BT
United Kingdom

Key words: alpine flora, conservation, mountain vegetation, pollen analysis, recent climate change, vegetational history

Introduction

The granite massifs of Lochnagar and the Cairngorms in the Eastern Highlands of Scotland are areas of considerable scientific, ecological, and conservation importance nationally and internationally (Curry-Lindahl 1974; Nethersole-Thompson and Watson 1974; Ratcliffe 1977; World Wilderness Congress 1983; Gimingham 2002a; Shaw and Thompson 2006a). The large granite mass of the Cairngorms rising to 1309 m on Ben Macdui lies on the junction of the former counties of Inverness-shire, Banffshire, and Aberdeenshire and forms the watershed between the River Spey and the River Dee. The smaller granite massif of Lochnagar (1155 m altitude) lies south of the River Dee and is the only area in the British Isles that is ecologically comparable to the Cairngorms (Ratcliffe 1977). In a recent conservation plan, Gimingham (2002b) defines the 'Cairngorms area' for National Park designation to include the Cairngorms, Lochnagar, and the upper parts of the glens of Angus and Glenshee. This emphasises the close ecological links between the Cairngorm massif and the Lochnagar massif. As Gimingham (2002b) notes, the Cairngorms and its surroundings "merit description as the jewel in the crown of Scotland's natural heritage". In 2003, almost 60% of the

N.L. Rose (ed.), Lochnagar: The Natural History of a Mountain Lake
Developments in Paleoenvironmental Research, 121–151.
© 2007 *Springer*.

Cairngorm area was designated as Scotland's second National Park, making it the largest National Park in the United Kingdom (D.B.A. Thompson et al. 2006). Substantial parts of the area have also been designated as Special Protection Areas, Special Areas of Conservation, and Sites of Special Scientific Interest (see D.B.A. Thompson et al. 2006).

The extensive high plateaux of the Cairngorms and Lochnagar, dissected by steep-sided valleys such as the Lairig Ghru, Glen Avon, Gleann Einich, and Glen Muick, support some of the finest areas of near-natural or semi-natural vegetation in Britain (Ratcliffe 1977; Horsfield 2006). These high mountain areas, with their many rivers, streams, and lochs, gradually merge towards the surrounding valleys where remnants of native semi-natural woodland occur within a cultural landscape of moorland and heath (Ratcliffe 1977; Hall 2006). The lower ground is a diverse cultural landscape of farmland, meadows, alluvial grasslands, plantation forests, small areas of semi-natural woodland, and thriving human settlements (Gimingham 2002b; Rowse 2006). The Cairngorm area as a whole is, without doubt, a magnificent region, with its outstanding mountain scenery, its extensive plateaux, its rocky corries and boulder fields, its sense of 'remoteness', 'wilderness', and 'naturalness', and its tundra-like or arctic-alpine plant and animal communities occupying a larger continuous area than anywhere else in Britain (Gimingham 2002b; Campbell and Anderson 2006; D.B.A. Thompson et al. 2006). It provides important refuges for many rare and scarce species of plant and animal (Shaw and Thompson 2006a, 2006b). At lower altitudes, the foothills and the straths contain some of the finest and most important remnants of native 'Caledonian' Scots pine forest in Scotland and also stands of native deciduous woodlands of birch, alder, oak, aspen, and rowan (Ratcliffe 1974), as well as extensive areas of heathland, both dry and wet, and a range of mires and other wetlands, forming a distinctive and attractive mosaic of habitats (Ratcliffe 1974; Gimingham 2002b; Hall 2006; Bean 2006; Rowse 2006). The Cairngorms is one of the few places in Britain where a natural tree-line is still present at about 640 m (McConnell and Legg 1995; Gimingham 2002c).

The Cairngorm area including Lochnagar is very important from a conservation viewpoint not only nationally and as part of Scotland's 'natural heritage' (Ratcliffe 1977; Gimingham 2002b) but also internationally. It is unique as a very large area of high acid ground above 1000 m in a transitional oceanic-continental climate (Green 1974; McClatchey 1996) at a latitude of about 57°N and formed largely of granite (Johnstone 1974) with small outcrops of base-rich rocks in some corries and around the edge of the granite massifs (Gimingham 2002b). The spectacular landscape of deeply dissected plateaux with rocky summit tors, massive boulder fields, cliffs, deeply cut steep corries, and high-altitude lochs is a result of the interaction between bedrock geology and glaciation (Sugden 1974; Brazier et al. 1996; Brown and Clapperton 2002; Gordon and Wignall 2006; Hall: Chapter 4 this volume). Although glaciers have long gone, there is clear evidence of their erosional and depositional activities everywhere in the area (Sissons and Grant 1972; Sugden 1977; Sissons 1979; Rapson 1985; Hall: Chapter 4 this volume). In late summer the high areas are largely free from snow, apart from a few more-or-less permanent snow-patches on north-facing slopes, in sheltered hollows on the plateau, and in shaded gullies in the Cairngorms (Watt and Jones 1948). Snow only occasionally survives through the summer on Lochnagar itself in shaded gullies (Watson 2005). Since 1996 the catchment of the Lochnagar loch has been snow-free by the end of June (Neil Rose, UCL pers. commun.) in contrast to the situation in the late 1960s and early 1970s when snow persisted in gullies and shaded areas below cliffs until mid- or late-August (author's personal observations). Late-snow-influenced vegetation is more extensively developed in the Cairngorm area as a whole than anywhere else in Britain (McVean and Ratcliffe 1962; Averis et al. 2004; Horsfield 2006). In winter the snow-cover is variable due to intermittent thaw or, on exposed summits and ridges, removal by strong winds. A diverse range of montane dwarf-shrub heaths is widespread on well-drained acid soils and in contrast to almost all other Scottish mountains, dry montane dwarf-shrub heaths are more extensive here than in any other mountain area in Britain or Ireland (Ratcliffe 1977; Horsfield 2006). The environment at high altitudes is very harsh (McClatchley 1996) and gives the Cairngorms and Lochnagar a strong 'arctic' flavour climatically and ecologically (D.B.A. Thompson et al. 2006).

Using a wide range of conservation-evaluation criteria, Ratcliffe (1977) concluded that the Cairngorm massif was the most important mountain system in Britain and accorded it a rating of 1* as being an internationally important site in the context of nature conservation as a whole within Great Britain. The Cairngorms have, in the last 40 years, become a strong attraction for visitors in search of the inspiration to be derived from the 'wilderness' quality of the area or the stimulus of participating in a range of open-air pursuits such as skiing, rock or ice climbing, walking, or exploring the natural history of the area (Gimingham 2002b). There has been a corresponding increase in tourist facilities and this development has reduced, to some extent, the conservation value and feeling of 'wilderness' of parts of the Cairngorm massif (Ratcliffe 1977; Mackay 2002; Neville et al. 2006). The Lochnagar massif receives less tourist pressure than the Cairngorms. It was originally regarded by Ratcliffe (1977) as a top-grade site but because it duplicates the range of ecological variation found in the Cairngorms and is less extensive than the Cairngorms, Ratcliffe (1977) gave it a grade 2 status in *A Nature Conservation Review*. (A grade 2 site is a site of equivalent or only slightly inferior merit to those in grade 1. Grade 2 sites often duplicate some essential features of related grade 1 sites which have priority in nature conservation (Ratcliffe 1977). However, Ratcliffe (pers. commun. 2004) considered that Lochnagar now warranted grade 1 status because of the lower tourist impact there and the absence of any tourist development compared to the Cairngorms.

The Cairngorms and the Lochnagar area can conveniently be divided into three broad ecological and altitudinal zones (Gimingham 2002b):

1. The mountain and plateau zone, consisting of 'alpine' areas above the former tree-line (about 600-650 m) such as low-alpine moorlands, dwarf-shrub heaths, and grass-heaths, corries, summit plateaux, and high wind-exposed tops;

2. The forest and moorland zone, consisting of native semi-natural woodlands and planted forests, moorlands, heaths, and bogs between the 'alpine' zone and the lowland valley zone;

3. The valley zone, consisting of a diverse cultural landscape of farmland, meadows, plantation forests, and human settlements, as well as rivers, lochs, moors, mires, and semi-natural woodland fragments. The zone lies below about 300 m in the north around the Cairngorm massif and below about 425 m in the south and east near Lochnagar.

This chapter describes the flora and vegetation primarily of the mountain and plateau zone of Lochnagar. Brief mention is made of the flora and vegetation of the surrounding forest and moorland zone and the valley zone. It also outlines current knowledge on the post-glacial vegetational history of Lochnagar and nearby areas, and discusses the possible effects of future climate warming and other environmental changes on the nationally rare and scarce plants of Lochnagar. It concludes by summarising the botanical interest of Lochnagar and its flora and vegetation. Plant nomenclature follows Stace (1997) for vascular plants, Blockeel and Long (1998) for bryophytes (mosses and liverworts), and Purvis et al. (1992) for lichens. Common English names for the vascular plants are listed in Table 1 along with their Latin names. Spelling of place names follows the Ordnance Survey 1:50000 second series.

Throughout this chapter, the Cairngorms refers to the Cairngorm massif, the Cairngorm area refers to the Cairngorm massif, the Lochnagar massif, and the upper parts of the glens of Angus and Glenshee (Gimingham 2002b), and Lochnagar refers to the Lochnagar massif. When the loch also called Lochnagar is mentioned, some reference is given to indicate that the loch rather than the massif is being discussed.

Table 1: Scientific Latin names of vascular plants mentioned in the text and their common English names according to Stace (1997).

Agrostis canina	velvet bent	*Hordeum*	barley
A. capillaries	common bent	*Huperzia selago*	fir clubmoss
Alchemilla alpina	alpine lady's-mantle	*Isoetes lacustris*	quillwort
A. glabra	lady's-mantle	*Juncus trifidus*	three-leaved rush
Alnus glutinosa	alder	*Juniperus communis*	juniper
Alopecurus borealis	alpine foxtail	*Linnaea borealis*	twinflower
Angelica sylvestris	wild angelica	*Listera cordata*	lesser twayblade
Antennaria dioica	mountain everlasting	*Loiseleuria*	trailing azalea
Arctostaphylos	bearberry	*procumbens*	
uva-ursi		*Luzula arcuata*	curved wood-rush
Athyrium distentifolium	alpine lady-fern	*L. spicata*	spiked wood-rush
Betula nana	dwarf birch	*L .sylvatica*	great wood-rush
B. pendula	silver birch	*Lycopodium annotinum*	interrupted clubmoss
B. pubescens	downy birch	*Melampyrum pratense*	common cow-wheat
Blechnum spicant	hard-fern	*M. sylvaticum*	small cow-wheat
Brachypodium	false brome	*Mercurialis perennis*	dog's mercury
sylvaticum		*Molinia caerulea*	purple moor-grass
Calluna vulgaris	heather	*Moneses uniflora*	one-flowered
Caltha palustris	marsh marigold		wintergreen
Campanula	harebell	*Myrica gale*	bog-myrtle
rotundifolia		*Nardus stricta*	mat-grass
Carex	sedge	*Narthecium ossifragum*	bog asphodel
C. bigelowii	stiff sedge	*Oreopteris*	lemon-scented fern
C. curta	white sedge	*limbosperma*	
C. lachenalii	hare's-foot sedge	*Orthilia secunda*	serrated wintergreen
C. pauciflora	few-flowered sedge	*Oxalis acetosella*	wood-sorrel
C. rariflora	mountain bog-sedge	*Oxyria digyna*	mountain sorrel
C. rostrata	bottle sedge	*Persicaria vivipara*	alpine bistort
Cerastium alpinum	alpine mouse-ear	*Phegopteris*	beech fern
C. cerastioides	starwort mouse-ear	*connectilis*	
Chamerion	rosebay willowherb	*Phleum alpinum*	alpine cat's-tail
angustifolium		*Pinus sylvestris*	Scots pine

Table 1 (cont.)

Cicerbita alpina	alpine blue sow-thistle	*Plantago lanceolata*	ribwort plantain
Cirsium heterophyllum	melancholy thistle	Poaceae	grasses
Cochlearia officinalis	common scurvygrass	*Poa flexuosa*	wavy meadow-grass
C. pyrenaica ssp.	Pyrenean scurvygrass	*Poa* x *jemtlandica*	Swedish meadow-grass
alpina		*Populus tremula*	aspen
Cornus suecica	dwarf cornel	*Pyrola media*	intermediate
Corylus avellana	hazel		wintergreen
Cryptogramma crispa	parsley fern	*P. minor*	common wintergreen
Deschampsia cespitosa	tufted hair-grass	*Quercus petraea*	sessile oak
D. cespitosa ssp.	alpine tufted hair-	*Q. robur*	pendunculate oak
alpina	grass	*Ranunculus acris*	meadow buttercup
D. flexuosa	wavy hair-grass	*Rubus chamaemorus*	cloudberry
Diphasiastrum alpinum	alpine clubmoss	*Rumex acetosa*	common sorrel
Dryopteris	buckler- & male-fern	*R. acetosella*	sheep's sorrel
D. dilatata	broad buckler-fern	*Salix*	willow
D. expansa	northern buckler-fern	*S. herbacea*	dwarf willow
D. oreades	mountain male-fern	*S. lapponum*	downy willow
Empetrum nigrum	crowberry	*Saussurea alpina*	alpine saw-wort
E. nigrum ssp.	alpine crowberry	*Saxifraga cernua*	drooping saxifrage
hermaphroditum		*S. oppositifolia*	purple saxifrage
Epilobium alsinifolium	chickweed willowherb	*S. rivularis*	highland saxifrage
E. anagallidifolium	alpine willowherb	*S. stellaris*	starry saxifrage
Erica cinerea	bell heather	*Sedum rosea*	roseroot
E. tetralix	cross-leaved heath	*Selaginella*	lesser clubmoss
Eriophorum	common cottongrass	*selaginoides*	
angustifolium		*Sibbaldia procumbens*	sibbaldia
E. vaginatum	hare's-tail cottongrass	*Silene dioica*	red campion
Festuca vivipara	viviparous sheep's-	*Solidago virgaurea*	goldenrod
	fescue	*Sorbus aucuparia*	rowan
Filipendula ulmaria	meadowsweet	*Trichophorum*	deergrass
Fraxinus excelsior	ash	*cespitosum*	
Galium saxatile	heath bedstraw	*Trientalis europaea*	chickweed wintergreen
Geranium sylvaticum	wood crane's-bill	*Trollius europaeus*	globeflower
Geum rivale	water avens	*Ulmus*	elm
G. urbanum	wood avens	*Urtica*	nettle
Gnaphalium	highland cudweed	*Vaccinium myrtillus*	bilberry
norvegicum		*V. uliginosum*	bog bilberry
G. supinum	dwarf cudweed	*V. vitis-idaea*	cowberry
Goodyera repens	creeping lady's-tresses	*Veronica alpina*	alpine speedwell
Gymnocarpium	oak fern	*V. serpyllifolia* ssp.	thyme-leaved
dryopteris		*humifusa*	speedwell
Hieracium	hawkweed	*Viola riviniana*	common dog-violet
holosericeum			

The vegetation types present on Lochnagar are, wherever possible, equated with the vegetation types delimited and described for the British Uplands by Averis et al. (2004). The name and National Vegetation Classification code letter and number for each vegetation type mentioned follow Averis et al. (2004) and are listed in Table 2 (see also Horsfield 2006 for the EC Habitats Directive Annex 1 Codes and the UK Biodiversity Action Plan priority habitats for these vegetation types).

Table 2: Vegetation types defined by Averis et al. (2004) that occur on the Lochnagar massif or surroundings, along with their National Vegetation Classification code letter and number.

H-10	*Calluna vulgaris-Erica cinerea* heath
H-12	*Calluna vulgaris-Vaccinium myrtillus* heath
H-13	*Calluna vulgaris-Cladonia arbuscula* heath
H-16	*Calluna vulgaris-Arctostaphylos uva-ursi* heath
H-18	*Vaccinium myrtillus-Deschampsia flexuosa* heath
H-20d	*Vaccinium myrtillus-Racomitrium languinosum* heath, *Rhytidiadelphus loreus-Hylocomium splendens* sub-community
H-22	*Vaccinium myrtillus-Rubus chamaemorus* heath
M-2	*Sphagnum cuspidatum/fallax* bog pool community
M-7	*Carex curta-Sphagnum russowii* soligenous mire
M-15	*Trichophorum cespitosum-Erica tetralix* wet heath
M-19	*Calluna vulgaris-Eriophorum vaginatum* blanket mire
M-19b	*Calluna vulgaris-Eriophorum vaginatum* blanket mire, *Empetrum nigrum* ssp. *nigrum* sub-community
M-19c	*Calluna vulgaris-Eriophorum vaginatum* blanket mire, *Vaccinium vitis-idaea-Hylocomium splendens* sub-community
M-33	*Pohlia wahlenbergii* var. *glacialis* spring
S-9	*Carex rostrata* swamp
U-7	*Nardus stricta-Carex bigelowii* grass-heath
U-9	*Juncus trifidus-Racomitrium lanuginosum* rush-heath
U-10	*Carex bigelowii-Racomitrium lanuginosum* moss-heath
U-11	*Polytrichum sexangulare-Kiaeria starkei* snow-bed
U-12	*Salix herbacea-Racomitrium heterostichum* snow-bed
U-13	*Deschampsia cespitosa-Galium saxatile* grassland
U-14	*Alchemilla alpina-Sibbaldia procumbens* dwarf-herb community
U-16	*Vaccinium myrtillus-Luzula sylvatica* tall-herb community
U-16a	*Vaccinium myrtillus-Luzula sylvatica* tall-herb community, *Dryopteris dilatata-Dicranum majus* tall-herb sub-community
U-17	*Luzula sylvatica-Geum rivale* tall-herb community
U-18	*Cryptogramma crispa-Athyrium distentifolium* snow-bed
W-18	*Pinus sylvestris-Hylocomium splendens* woodland

Present-day flora and vegetation

Lochnagar is formed from a granite intrusion surrounded by a 'ring complex' of metamorphosed rocks which show progressively less alteration away from the intrusion (Brown and Clapperton 2002; Goodman: Chapter 3 this volume). The lower slopes on the south side of Lochnagar rise gradually from about 400 m at Loch Muick to the summit of Lochnagar (1155 m). The northern side of the mountain is more rugged with four corries. The two western corries are virtually one, for the arête between them has been almost destroyed. The eastern corries comprise the spectacular main northeastern corrie more than 1 km across and containing the deep (26 m), large (10.4 ha) corrie lake of Lochnagar at 788 m altitude, and a poorly developed corrie to the northwest below Cac Càrn Beag. These four corries all supported small glaciers in the Younger Dryas stadial about 11550-12650 years ago (Sissons and Grant 1972).

The present-day mean July air temperature near the loch of Lochnagar is 9.8°C (seven years, Neil Rose, UCL pers. commun.), whereas mean January temperature is

1.3°C (six years, Neil Rose, UCL pers. commun.). Soils are well-drained podsolic soils, shallow peats, ranker soils, or bare rock depending on topography and location (see Bruneau 2006; Helliwell et al.: Chapter 6 this volume).

The flora and vegetation of the Lochnagar massif as a whole are discussed first in terms of the major vegetation types and ecologically prominent species. More detailed attention is then given to the flora and vegetation of the northeastern corrie. In conclusion, brief mention is made of the flora and vegetation of the forest and moorland zone and the valley zone.

The Lochnagar massif as a whole

The lower drift-covered slopes of the forest and moorland zone support large areas of dry or moist *Calluna vulgaris* heather moor consisting of a mosaic of *Calluna vulgaris-Erica cinerea* heath (vegetation unit H-10 in Averis et al. 2004 and Table 2), *Calluna vulgaris-Vaccinium myrtillus* heath (H-12), and *Calluna vulgaris-Arctostaphylos uva-ursi* heath (H-16). These grade through open pine heath into pine woods (*Pinus sylvestris-Hylocomium splendens* woodland, W-18), for example in Ballochbuie Forest and in parts of Glen Muick (see below). Fine stands of *Calluna vulgaris-Vaccinium myrtillus* heath (H-12) with abundant *Empetrum nigrum* ssp. *hermaphroditum, Lycopodium annotinum,* and a range of lichens occur on the outlying spurs of White Mounth and Cairn Bannoch (Ratcliffe 1977). Wetter ground is dominated by *Trichophorum cespitosum-Erica tetralix* wet heath (M-15) occurring on shallow, wet, or intermittently waterlogged acid peat or peaty mineral soils. *Calluna vulgaris* heath extends to a higher elevation here than anywhere else in Britain, but changes with increasing altitude to a more prostrate montane *Calluna* heath rich in *Arctostaphylos uva-ursi* and *Loiseleuria procumbens* (*Calluna vulgaris-Arctostaphylos uva-ursi* heath, H-16). This heath type finally peters out at about 800 m (Ratcliffe 1977). Where snow cover is greater than average, the *Rhytidiadelphus-Hylocomium* sub-community of *Vaccinium-Racomitrium* heath (H-20d), *Vaccinium myrtillus-Rubus chamaemorus* heath (H-22), and *Nardus stricta-Carex bigelowii* grass-heath (U-7) communities are well developed. 'Peat alpines' such as *Cornus suecica, Lycopodium annotinum, Rubus chamaemorus, Trientalis europaea,* and *Vaccinium uliginosum* occur frequently and *Betula nana* occurs locally in blanket mires, with *Sphagnum fuscum.*

In the Lochnagar area there are three main types of blanket-mire and related wet-heath vegetation (Ratcliffe 1977). On the lower slopes and valley floors (e.g. Glen Muick, Glen Allt, Glen Gelder) there is a prevalence of *Trichophorum cespitosum-Erica tetralix* wet heath (M-15), varying from a dry type in well-drained or much disturbed situations to a *Sphagnum*-rich type where the water table is high. On the high watersheds (e.g. below Cac Càrn Beag) this community grades into *Calluna vulgaris-Eriophorum vaginatum* blanket mire (M-19). At altitudes above about 850 m the disappearance of *Calluna vulgaris* results in the montane sub-communities of M-19 with dominant *Empetrum nigrum* ssp. *nigrum* and ssp. *hermaphroditum* (M-19b, M-19c) becoming locally prominent. High-level blanket mires are rather local in the Lochnagar area, probably because of the porosity of the granite soils. Blanket mire occurs here (e.g. parts of the White Mounth and between Cuidhe Cròm and Cac Càrn Mòr) and in the Cairngorms at a higher elevation (at least 1000 m) than in any other mountain range in Britain (Ratcliffe 1977). Alpine *Carex curta-Sphagnum russowii* soligenous mires (M-7) with *C. lachenalii* and *C. rariflora,* and the bryophytes *Scapania paludosa, S. uliginosa,* and *Sphagnum lindbergii* are well represented in waterlogged hollows below springs and snow-beds on or near the Lochnagar plateau. Areas of wet *Sphagnum cuspidatum/fallax* bog pools (M-2) occur locally near Loch Muick and in Glen Allt and support *Carex pauciflora,* the elegant *Sphagnum riparium,* and the national rarity *S. balticum.*

On well-drained soils above about 800 m, prostrate *Calluna vulgaris* heath (*Calluna vulgaris-Cladonia arbuscula* heath, H-13), often of the 'wave' variety, is locally frequent between 800-900 m, usually with an abundant lichen cover. With increasing altitude, the heath becomes more patchy, and the most characteristic vegetation types of the high summit plateaux are *Carex bigelowii-Racomitrium lanuginosum* moss-heath (U-10) and *Juncus trifidus-Racomitrium lanuginosum* rush-heath (U-9). Various facies of this *Juncus trifidus* heath can be found, ranging from stands where *J. trifidus* is co-dominant with the moss *R. lanuginosum* to an open tussocky growth in which lichens (e.g. *Cetraria islandica, Cladonia arbuscula, C. uncialis* ssp. *biuncialis, Ochrolechia frigida*) are the main associates. There are also substantial areas on the high plateaux of dominant *Carex bigelowii,* with common associates of the mosses *Dicranum fuscescens* and *Polytrichum alpinum,* or a range of lichens (U-8). The porous gravelly soils at over 1100 m mostly have a very discontinuous vegetation cover, but where snow lies a little deeper, there are stands of dense, short *Nardus stricta* with a few lichens and bryophytes (U-7) (Ratcliffe 1974, 1977).

Snow cover is an ecological factor of great importance in the mountain and plateau zone of Lochnagar and in the Cairngorms as a whole (Ratcliffe 1974, 1977). There is a good range of late-snow-influenced vegetation and a greater extent of such vegetation than in any other mountain area in Britain. The strictly chionophilous vegetation types are *Cryptogramma crispa-Athyrium distentifolium* snow-beds (U-18) on block screes below corrie back-walls; *Polytrichum sexangulare-Kiaeria starkei* snow-beds (U-11) and *Salix herbacea-Racomitrium heterostichum* snow-beds (U-12) in shaded hollows below high cliffs, in sheltered hollows on the plateau, and in steep gullies of north- and east-facing cliffs; *Alchemilla alpina-Sibbaldia procumbens* dwarf-herb community (U-14) in areas irrigated by water from melting snow; and bryophyte-dominated spring communities (e.g. *Pohlia wahlenbergii* var. *glacialis* spring (M-33)) often associated with snow melt-water. Some of these snow-bed types, especially U-11 and U-12, support a range of nationally rare bryophytes including *Conostomum tetragonum, Kiaeria falcata, K. starkei, Polytrichum sexangulare, Anthelia juratzkana, Marsupella adusta, M. brevissima, M. sprucei, M. boeckii* var. *stableri, Moerckia blyttii, Nardia breidleri,* and *Pleurocladula albescens,* particularly on Coire Bhoidheach on White Mounth. Irrigated rocks near snow-beds on Coire Bhoidheach support small colonies of the nationally rare bryophytes *Andreaea mutabilis* and *Marsupella condensata.*

The northeast corrie of Lochnagar

The most famous feature of Lochnagar is the magnificent, large northeast corrie with its 240 m gully-riven granite cliffs. The flora and vegetation of this corrie are markedly sub-alpine or alpine in character, with strong botanical and ecological affinities to the sub-alpine and low-alpine zones of western Norway today (Moen 1999). The corrie presently supports a mosaic of vegetation types with a total plant cover of about 75%, alternating with block scree, boulders, and bare rock. The dominant vegetation consists of dwarf, often rather prostrate *Calluna vulgaris* and stunted *Vaccinium myrtillus* heath, growing with the dwarf-shrubs *Arctostaphylos uva-ursi, Empetrum nigrum* ssp. *hermaphroditum, V. uliginosum,* and *V. vitis-idaea* (H-12 and H-16 in Averis et al. 2004 and Table 2). Common associates include *Antennaria dioica, Carex bigelowii, Cornus suecica, Deschampsia flexuosa, Diphasiastrum alpinum, Huperzia selago, Loiseleuria procumbens, Rubus chamaemorus, Salix herbacea,* mosses such as *Hypnum jutlandicum, Pleurozium schreberi, Polytrichum alpinum,* and *Rhytidiadelphus loreus,* and lichens such as *Cetraria islandica, Cladonia floerkeana, C. portentosa, C. uncialis* ssp. *biuncialis,* and *Coelocaulon aculeatum.* This mixed-heath complex occurs on well-drained skeletal, ranker or thin acid podsolic soils. It alternates with areas dominated by *Calluna vulgaris, Eriophorum vaginatum, Molinia caerulea, Nardus stricta,* and *Trichophorum cespitosum,* and with some moss cover of *Polytrichum strictum, Sphagnum capillifolium, S. fallax, S. fuscum,* and *S. subnitens.* These stands of the *Calluna vulgaris-Eriophorum vaginatum* blanket mire (M-19) occur on shallow (50-80 cm) peat in waterlogged hollows.

In wind-exposed areas around the edges of the corrie, open stands of *Carex bigelowii, Loiseleuria procumbens, Luzula spicata,* and *Juncus trifidus* are locally frequent with the moss *Racomitrium lanuginosum* and a range of lichens including *Cetraria islandica, Cladonia arbuscula, C. gracilis, C. uncialis* ssp. *biuncialis, Coelocaulon aculeatum, Ochrolechia frigida, Stereocaulon saxatile,* and *Thamnolia vermicularis.* These wind-blasted heaths belong most closely to the *Juncus trifidus-Racomitrium lanuginosum* rush-heath vegetation type (U-9). In less exposed areas, a more closed vegetation occurs dominated by *Carex bigelowii* and *Racomitrium lanuginosum* (U-10), with *Deschampsia flexuosa, Festuca vivipara, Galium saxatile,* and *Salix herbacea.* In the most sheltered areas around the corrie edges and at slightly lower altitudes between Cuidhe Cròm and Cac Càrn Mor, prostrate *Calluna vulgaris* heath rich in lichens (H-13) occurs locally with a patchy abundance of *Agrostis canina, Carex bigelowii, Diphasiastrum alpinum, Empetrum nigrum* ssp. *hermaphroditum, Gnaphalium supinum, Huperzia selago, Juncus trifidus,* and *Nardus stricta,* growing within a near-continuous carpet of lichens such as *Alectoria nigricans, Cetraria nivalis, Cladonia arbuscula, C. portentosa, C. rangiferina, C. uncialis* ssp. *biuncialis, Coelocaulon aculeatum,* and *Thamnolia vermicularis.* This lichen heath is more extensive on Lochnagar than in the Cairngorms, probably because of the lower rainfall in the east (Ratcliffe 1974).

Extensive areas of stable block-scree below the north- and northeast-facing cliffs of the corrie support a rich assemblage of tall ferns such as *Athyrium distentifolium, Blechnum spicant, Cryptogramma crispa, Dryopteris expansa, D. oreades, Gymnocarpium dryopteris, Oreopteris limbosperma,* and *Phegopteris connectilis.* Amongst the boulders and between the ferns, there is a range of small herbs such as *Alchemilla alpina, Gnaphalium supinum, Oxalis acetosella,* and *Sibbaldia procumbens,* tall grasses such as *Deschampsia cespitosa,* and tall wood-rushes such as *Luzula sylvatica.* Bryophytes are locally abundant on and between the boulders, with common species such as *Barbilophozia floerkei, Dicranum scoparium, Hylocomium splendens, Polytrichum alpinum,* and *Rhytidiadelphus loreus,* and more local species such as *Arctoa fulvella, Bazzania tricrenata, Conostomum tetragonum, Dicranum glaciale, Gymnomitrion concinnatum, Kiaeria blyttii, K. falcata, K. starkei, Lophozia opacifolia,*

L. sudetica, Mylia taylorii, Pseudoleskea patens, Rhizomnium magnifolium, and *Tetralophozia setiformis.* These fern-dominated stands belong to the *Cryptogramma crispa-Athyrium distentifolium* snow-bed vegetation type (U-18) of Averis et al. (2004). The prolonged snow cover provides protection against spring frosts for the young, emerging fern fronds (McVean and Ratcliffe 1962) and provides a humid atmosphere in which western liverworts such as *Bazzania tricrenata, Mylia taylorii,* and *Scapania ornithopodioides* can flourish in the Eastern Highlands (Ratcliffe 1968). Lochnagar is currently the easternmost known locality in the British Isles for *S. ornithopodioides* (Gordon Rothero, pers. commun.).

In areas of late snow-lie and in hollows within the block screes, small stands of *Vaccinium myrtillus-Deschampsia flexuosa* heath (H-18) occur locally, with a strong dominance of *V. myrtillus* and locally *V. uliginosum,* growing with *Alchemilla alpina, Nardus stricta,* and an abundance of large mosses such as *Dicranum scoparium, Hypnum jutlandicum,* and *Pleurozium schreberi.*

On the high summit ground above the massive northeast corrie and above the western corries below The Stuic and near Loch nan Eùn, there are fine stands of the rare *Carex curta-Sphagnum russowii* mire (M-7) in hollows flushed by snow melt-water and by springs. These soligenous mires support the nationally rare *Alopecurus borealis, Carex lachenalii,* and *C. rariflora,* and the rare bryophytes *Scapania paludosa, S. uliginosa,* and *Sphagnum lindbergii,* growing with *Anthelia julacea, C. bigelowii, Eriophorum angustifolium, Nardus stricta, Rubus chamaemorus, Salix herbacea,* and *Saxifraga stellaris.*

Most of the ledges on the main cliffs in the northeast corrie support species-poor *Vaccinium myrtillus-Luzula sylvatica* tall-herb vegetation (U-16) with few associates except common large mosses such as *Hylocomium splendens, Hypnum jutlandicum,* and *Rhytiadelphus loreus.* Damper, slightly flushed ledges are distinguished by a striking growth of tall ferns including *Athyrium distentifolium, Dryopteris dilatata, D. expansa, Gymnocarpium dryopteris, Oreopteris limbosperma,* and *Phegopteris connectilis,* associated with *Alchemilla alpina, Galium saxatile, Oxalis acetosella, Ranunculus acris, Silene dioica,* and the nationally rare moss *Plagiomnium medium.* These stands are close to the *Dryopteris dilatata-Dicranum majus* sub-community U-16a described by Averis et al. (2004).

To the northwest of the large Black Spout gully in the corrie back-wall, there are luxuriant ledges of 'tall herbs' growing on ungrazed, near-inaccessible ledges that have some moderately base-rich rocks above them such as lime-bearing pockets in thin veins within the granite. Water from these veins seeps down the cliffs to produce damp, base-rich soils on the ledges. These ledges support fine stands of the *Luzula sylvatica-Geum rivale* tall-herb community (U-17). Common associates include *Alchemilla alpina, A. glabra, Angelica sylvestris, Caltha palustris, Campanula rotundifolia, Chamerion angustifolium, Cirsium heterophyllum* (Figure 1a), *Cochlearia officinalis, C. pyrenaica* ssp. *alpina, Deschampsia cespitosa, Filipendula ulmaria, Geranium sylvaticum, Geum rivale, Luzula sylvatica, Oxalis acetosella, Oxyria digyna, Persicaria vivipara, Rumex acetosa, Salix lapponum, Saussurea alpina, Sedum rosea, Solidago virgaurea, Trollius europaeus* (Figure 1b), *Vaccinium myrtillus,* and *Viola riviniana,* and the ferns *Athyrium distentifolium, Oreopteris limbosperma,* and *Phegopteris connectilis.* On smaller ledges and in crevices in this area of the cliff, *Antennaria dioica, Cerastium alpinum* (Figure 2a), and the Scottish endemic *Hieracium holosericeum,* and other endemic or rare species of *Hieracium* sect. *alpina* and *Hieracium* sect. *subalpina* occur locally (Peter Sell, pers. commun.).

The national rarity, *Cicerbita alpina* (Figure 1c), occurs in one of its four known British localities on the upper ledges of Black Spout. It is locally abundant and forms almost pure patches amidst the fern-rich (U-16a) and tall-herb-rich (U-17) ledges (Marren et al. 1986). The deep blue-violet heads of flowers usually appear in late July and persist until early September. Seed-set appears to be very irregular and seed is often not viable (Wigginton 1999). Another national rarity, *Gnaphalium norvegicum* (Figure 1d), grows nearby in closed turf on the stony sides of steep gullies, growing with *Alchemilla alpina, Deschampsia cespitosa, Kiaeria blyttii, K. falcata,* and *Marsupella emarginata*. In areas below the cliffs there is a species-poor *Deschampsia cespitosa* grassland (*Deschampsia cespitosa-Galium saxatile* grassland U-13) with *Agrostis capillaris, Galium saxatile, Phleum alpinum,* and *Rhytidiadelphus loreus*, evidently derived by animal grazing from tall-herb vegetation (U-17) that, by analogy with sub-alpine areas in western Norway today, would have once been more common on the slopes in the Lochnagar corrie. Today such tall-herb vegetation (U-17) is confined to ungrazed, near inaccessible ledges on the high cliffs.

Moist areas within the Black Spout gully are characterised by a local abundance of the conspicuous pale-green moss *Pohlia wahlenbergii* var. *glacialis*, forming small stands of the *Pohlia wahlenbergii* var. *glacialis* spring vegetation type (M-33). These springs support *Cochlearia officinalis, Deschampsia cespitosa* ssp. *alpina, Epilobium alsinifolium, E. anagallidifolium, Saxifraga stellaris, Veronica alpina, V. serpyllifolia* ssp. *humifusa,* the bryophytes *Calliergon sarmentosum, Philonotis fontana, P. seriata, Pohlia ludwigii,* and *Scapania paludosa,* and the national rarity *Saxifraga rivularis* (Figure 2b). These springs are fed by snow melt-water from long-lasting snow-beds at the head of the gully. The populations on Lochnagar of *S. rivularis* all appear to be small and are easily damaged by erosion caused by increased use of the Black Spout by climbers and walkers. Three additional nationally rare flowering plants grow in small quantities at or near the top of the Black Spout. These are *Luzula arcuata, Poa flexuosa,* and *P.* x *jemtlandica. P. flexuosa* and *P.* x *jemtlandica* grow in loose scree and on stone ledges with *Alchemilla alpina, Juncus trifidus,* and *Saxifraga stellaris,* whereas *L. arcuata* occurs in open stony fell-field and rock debris near the top of the Black Spout, growing with *Carex bigelowii, Juncus trifidus,* and *Racomitrium lanuginosum.*

The forest and moorland zone and the valley zone

Turning to the forest and moorland (425-600 m) and valley (below 425 m) zones, the most striking vegetation type in these zones in the Lochnagar area is Scots pine woodland (e.g. Ballochbuie Forest along Deeside). These pine woods occur between 300 and 600 m on coarse, sandy, and gravelly drift soils derived from granite, giving markedly acid and base-deficient soils. Although some pine woods occur on rocky, steep craggy slopes and some Deeside woods are steep gorge woodlands, the most extensive woods are on rather gentle slopes or on the mildly undulating morainic foothills below about 400 m (Ratcliffe 1974, 1977; Hall 2006). Although there is a general appearance of naturalness in the woodlands (*Pinus sylvestris-Hylocomium splendens* woodland, W-18 in Averis et al. 2004), almost all the pine woods in the Lochnagar area appear to have been managed for timber production at some time and can thus be regarded as semi-natural woodland (Ratcliffe 1974, 1977). These pine woods have a distinctive group of locally rare boreal-forest plants, including two small orchids, *Goodyera repens* and *Listera cordata,* a range of wintergreens (*Pyrola media, P. minor,* the local *Orthilia secunda,* and the very rare *Moneses uniflora*), *Melampyrum pratense, M. sylvaticum, Trientalis europaea,* the beautiful but very local twin-flower *Linnaea borealis,* the elegant moss *Ptilium crista-castrensis,* and the rare liverwort *Lophozia longidens.* Whilst some areas are left to regenerate naturally after felling, there has been a significant amount of replanting. Some of the areas of open heather moor between or amongst blocks of pine wood remain relatively treeless because natural regeneration can be poor locally. In other areas, young trees grow up rapidly and abundantly in cleared areas. Natural pine regeneration appears to depend on many factors such as the intensity of deer grazing and the coincidence of good seed years with heather burning of a clearing. In general, regeneration is poorer in the Deeside pine woods than in the Speyside woods, probably because of heavier deer grazing on Deeside. Successful regeneration has, however, been achieved by fencing against deer in parts of Deeside (Ratcliffe 1974, 1977). For example, 350 ha (of a 3000-ha wood) has been fenced at Ballochbuie to allow regeneration and there are at present no plans to use the wood for timber production. Deer numbers have been reduced from 5000 to 3500 animals in the Balmoral Estate since 1990. The nearby Invercauld and Mar Estates have schemes for similarly reducing deer numbers and for encouraging regeneration by short-term fencing (Alison Averis, pers. commun.).

Besides pine woods, birch (*Betula pendula, B. pubescens*) and juniper (*Juniperus communis*) are widespread and locally abundant and can occur in pure stands as well as mixed with pine. Their presence, especially birch, is probably related to better than average soil conditions, and the interesting mixed pine, birch, and juniper stands on a south-facing slope at Crathie on Deeside occur, at least partly, on soils derived from basic schist. There are also stands of hazel (*Corylus avellana*) there at low altitude, growing with *Brachypodium sylvaticum, Geum urbanum,* and *Mercurialis perennis* on fertile, brown-earth soils. In the area, there is also much rowan (*Sorbus aucuparia*), some aspen (*Populus tremula*), and an abundance of alder (*Alnus glutinosa*) on damp richer soils, especially river and stream alluvium (Ratcliffe 1974, 1977). The status of oak (*Quercus petraea, Q. robur*) in the lowlands of the Lochnagar area is obscure. There are several small oak woods on Deeside, for example at Craigendarroch near Ballater at 250-300 m and at Dinnet at 200 m. These woods have tall, well-grown trees over 100 years old and although they have been coppiced, they have the appearance of being semi-natural, with a 'typical' associated flora of shrubs, dwarf-shrubs, grasses, herbs, bryophytes, and lichens (Ratcliffe 1974, 1977).

Calluna vulgaris moorland, managed by rotational burning for red grouse (*Lagopus lagopus scoticus*) (Hudson 2002), dominates the non-forested areas of the forest and moorland zone. These moorlands are part of the *Calluna vulgaris-Vaccinium myrtillus* heath vegetation type (H-12), but some of these moorlands include stands of the *Calluna vulgaris-Arctostaphylos uva-ursi* heath (H-16). Other vegetation types in the forest and moorland zone include blanket mires, wet bogs, and sedge swamps around small lochans (e.g. *Carex rostrata* swamp, S-9; Bean 2006).

The valley zone below 425 m consists primarily of plantation forests, and small areas of semi-natural woodland, particularly on steep rocky crags, some sheep and cattle farmland, meadows, and alluvial grasslands (see Rowse 2006).

(a)

(b)

(c)

(d)

Figure 1. (a) *Cirsium heterophyllum* and (b) *Trollius europaeus,* characteristic 'tall herbs' of the mildly basic, ungrazed ledges on the cliffs in the northeast corrie of Lochnagar. (c) *Cicerbita alpina* and (d) *Gnaphalium norvegicum* - two of Lochnagar's nationally rare species, growing on ledges near the Black Spout. Photographs by John Birks.

(a)

(b)

Figure 2. (a) *Cerastium alpinum,* an 'arctic-alpine' plant of mildly basic crevices and ledges that grows locally on the cliffs of the northeast corrie of Lochnagar. (b) *Saxifraga rivularis,* one of Lochnagar's nationally rare species that grows in the Black Spout gully of the northeast corrie. Photographs by John Birks.

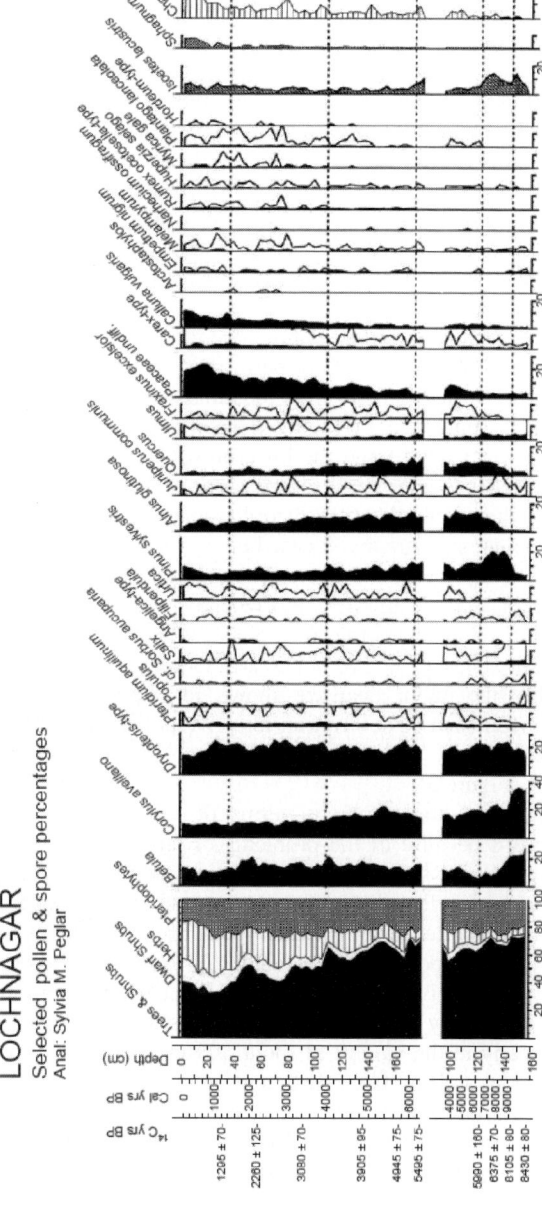

LOCHNAGAR
Selected pollen & spore percentages
Anal: Sylvia M. Peglar

Figure 3 Summary pollen diagram for Lochnagar based on cores NAG 27 (upper part) and NAG 30 (lower part). Only the major pollen and spore taxa are shown. All pollen and spores are expressed as percentages of total terrestrial pollen and spores (ΣP), except for *Isoetes lacustris*, *Sphagnum*, and charcoal particles. These are expressed as shown on the diagram. The pollen and spore values are plotted on a depth basis, with 0 cm being the core-top (sediment-water interface). An estimated calibrated age scale (calibrated years BP (BP = before present 1950)) is shown, based on the 10 radiocarbon dates. The pollen stratigraphy is divided into five pollen assemblage zones, prefixed PNAG. From Dalton et al. (2005) with permission from Elsevier.

Past flora and vegetation

Much of the Lochnagar massif lies above the presumed natural tree-line in the Eastern
Highlands of Scotland today (ca. 640 m) (Birks 1988; McConnell and Legg 1995).
However, seedlings of pine and birch can be found very rarely as high as 850 m on
crags and sheltered rock outcrops. Fossil pine stumps about 7000 years old occur in
eroding peat at 790 m on the eastern shore of the loch (Rapson 1985) and along its
outflow stream at 700 m (Dalton et al. 2005). In a reconstruction of the potential forest
vegetation of the Scottish Highlands under present climatic conditions by McVean and
Ratcliffe (1962), the loch at 788 m lies near or just above the tree-line, whereas the
valleys to the south and east are proposed to be dominated by *Betula* spp. to about 650
m, by *Pinus sylvestris* locally on poor soils to about 550 m, and by *Quercus* spp. to
about 400 m. The occurrence of fossil pine stumps and the potential forest
reconstruction suggest that the vegetation of Lochnagar, at least at the altitude of the
loch, has changed markedly over the last 7000 years.

   Detailed evidence for such vegetational change is provided by pollen analyses of two
sediment cores from the loch (NAG27, NAG30) performed by Sylvia M. Peglar
(Dalton et al. 2005) (see Figure 2 in Chapter 8 for locations). The available pollen
stratigraphy from Lochnagar (summarised in Figure 3) consists of the entire core
NAG27 (collected from 21 m water depth) and the lowermost 60 cm of core NAG30
(collected from 4.2 m water depth). The NAG27 record represents the last 6000 years
of sedimentation in the loch, whereas the NAG30 record extends to about 9500 years
ago. The chronology for the pollen stratigraphy is based on six radiocarbon dates from
the NAG27 core and four dates from the NAG30 core. The chronology shown in Figure
3 is based on these radiocarbon dates calibrated into calendar years and subsequent
statistical regression modelling to estimate the age of every sediment sample (full
details are given in Dalton et al. 2005). Over 110 pollen and spore types were identified
from the two cores, but only the major taxa are shown in Figure 3. The pollen
stratigraphy is divided for convenience of description into five pollen zones (PNAG-1
to PNAG-5). The zones and the core chronology show that there is stratigraphical
overlap between the two cores (Figure 3).

   The interpretation of the pollen stratigraphy in terms of vegetational history of
Lochnagar is not straightforward because of the problems of far-distance transported
pollen blown from the lowlands. Even though there are no trees or shrubs today in the
catchment of the loch, the modern pollen deposition (uppermost sample in Figure 3)
contains about 40% tree and shrub pollen (mainly *Betula* pollen but with high values of
Corylus avellana and *Alnus glutinosa* pollen and low values of *Pinus sylvestris* pollen).
The pine values are surprisingly low, given the modern abundance of pine trees in a 10-
15 km radius of Lochnagar. It is many kilometres to the nearest *Corylus* populations,
and *Alnus glutinosa* grows locally in the valleys at about 300 m elevation about 8 km
away, for example on Deeside and in Glen Esk and Glen Clova. The modern
assemblage suggests that the relevant pollen-source area of the loch today is very large
and much greater than the theoretical expectation based on Sugita's (1994) pollen-
source area model, which would predict a relevant source area with a radius of about
800-1500 m from the loch's margin. It seems that today (and one presumes in the past
too), the relevant pollen-source area is likely to have had a radius of several kilometres.

If this is correct, interpretation of the Lochnagar pollen stratigraphy in terms of local floristic and vegetational history is difficult and inevitably rather conjectural.

Inferred floristic and vegetation history

The lowest pollen zone (PNAG -1, Figure 3) with a basal radiocarbon date of 8430 ± 80 ^{14}C years BP (before present = AD 1950) and a modelled calibrated age of about 9500 years BP may reflect the local occurrence of fern-rich *Betula* and *Corylus* scrub with some *Populus tremula, Salix,* and *Sorbus aucuparia* growing on relatively fertile brown-earth soils in or near the corrie. Tall herbs such as *Angelica*-type, *Filipendula, Rumex acetosa,* and *Urtica* and tall ferns such as *Dryopteris*-type, *D. filix-mas*-type, and *Gymnocarpium dryopteris* may have grown in this scrub, as they do today in high-altitude birch woods in Glen Clova and on Deeside (Ratcliffe 1974; Huntley and Birks 1979a; b).

Pinus sylvestris and, a little later, *Alnus glutinosa* pollen values increase in zone PNAG-2 (about 8350 calibrated years BP) (Figure 3). These rises reflect the expansion of these trees in the Eastern Highlands of Scotland from about 8500 years ago (Birks and Mathewes 1978; Bennett 1994; 1996; Birks 1996a; Huntley et al. 1997). Pine certainly grew near the loch at the time of zone PNAG-2 as fossil pine stumps occur near the outflow and along Lochnagar Burn at about 700 m. Rapson (1985) obtained a date of 6080 ± 50 radiocarbon years BP (ca. 6800-7000 calibrated years BP) for a pine stump near the eastern shore of the loch. Fossil pine stumps or macrofossils (e.g. bark, needles, seeds) are recorded up to 700-800 m elevation and dating from 7500 to 2500 radiocarbon years BP (ca. 8200-2800 calibrated years BP) in the Cairngorms and neighbouring areas (e.g. Birks 1975; Dubois and Ferguson 1985; 1988; Bennett 1996; Allen and Huntley 1999). It is likely that during zone PNAG-2 (8800-6000 calibrated years BP), the Lochnagar corrie was wooded with sub-alpine birch scrub with *Juniperus communis, Populus tremula, Salix* spp., and *Sorbus aucuparia*, abundant tall ferns, and some tall herbs on the slopes and block-screes below the cliffs. Small stands of stunted pine may have occurred locally in marginal habitats, such as wet, peaty areas near the loch and along the outflow. It is unclear if *Alnus glutinosa, Fraxinus excelsior, Quercus* spp., or *Ulmus* grew near the loch at this time. Overall it seems unlikely ecologically. These trees may have extended their altitudinal ranges by about 200 m if early- and mid-Holocene (post-glacial) mean summer temperatures were about 1-2°C warmer than today (Dalton et al. 2005). Given such an elevational rise, their upper altitudinal limit in the area may have been 500-600 m; about 200 m lower than the loch and its hydrological catchment. The vegetation of the catchment of the loch about 8000-7000 years ago was probably open birch scrub, with small stands of pine on wet soils. The vegetation may have resembled vegetation that occurs on steep rocky slopes and in corries at about 700-900 m elevation today on acid or mildly basic soils in western Norway (e.g. Odland 1981; 1991a; b; Fremstad 1997; Moen 1999).

Pollen zones PNAG-3 (6000-4000 calibrated years BP), PNAG-4 (4000-1200 calibrated years BP), and PNAG-5 (1200-0 calibrated years BP) (Figure 3) show progressively increasing values of *Calluna vulgaris, Carex*-type, Poaceae undiff., and *Vaccinium*-type pollen and correspondingly decreasing percentages of tree and shrub pollen, especially *Corylus avellana, Pinus,* and *Quercus*. These pollen-analytical changes are interpreted as reflecting the progressive loss of woodland and scrub in the catchment of the loch and in the relevant pollen-source area of the core and the expansion of grassland, moorland, heath, and blanket mire. Pollen and spores of plants such as *Arctostaphylos, Empetrum nigrum, Huperzia selago, Melampyrum, Narthecium ossifragum, Ranunculus acris, Rubus chamaemorus, Vaccinium, Rumex acetosa, R. acetosella, Sphagnum,* and *Urtica* all increase in these zones. The increase in microscopic charcoal particles in the last 1000 years may reflect 'muir-burning' and the subsequent management of grouse-moor, as the charcoal values are closely paralleled by changes in *Calluna vulgaris* pollen percentages.

The scattered occurrences of pollen and spores of alpine and montane taxa, all of which grow today in the Lochnagar corrie, on the cliff ledges, or in the gullies of the back-wall cliffs, indicate the persistence of these plants locally over the last 9500 years. These include *Cerastium cerastioides*-type, *Cryptogramma crispa, Diphasiastrum alpinum, Huperzia selago, Lycopodium annotinum, Rubus chamaemorus, Salix herbacea*-type, *Saxifraga cernua/S. rivularis, S. oppositifolia*-type, *S. stellaris*-type, *Sedum (S. rosea?), Selaginella selaginoides,* and *Trollius europaeus.*

There are two major long-term trends in the Lochnagar pollen stratigraphy and inferred vegetational history: (1) decreasing tree and shrub pollen, suggesting decreasing tree- and shrub-cover and associated changes in light and shade conditions from open sub-alpine woodland 9000-6000 years ago to the treeless vegetation of the corrie today; and (2) changing soil fertility from brown earths to less fertile podsols, acid podsols, and shallow peat, and to more open peaty and mineral soils in the last 1400 years due to the long-term processes of leaching, podsolisation, and paludification (Birks 2005), and more recently of the short-term processes of erosion, burning, animal grazing, and trampling. Dalton et al. (2005) show that this trend in declining catchment soil fertility is reflected by long-term changes in the pH of the loch's water.

Comparisons with nearby areas

The pollen stratigraphy from Lochnagar (Figure 3) is almost unique amongst the available pollen stratigraphies from the Eastern Highlands of Scotland in that it is one of the few pollen stratigraphies prepared from a high altitude (788 m) site that has been a lake throughout its history. Its pollen stratigraphy is thus not strongly influenced by local pollen deposition (*sensu* Janssen 1973). Such local deposition in small mires or ponds can obscure the regional pollen stratigraphy and can create some problems in interpretation (e.g. Huntley 1981; 1994). However, being a high altitude site the interpretation of the Lochnagar pollen stratigraphy is complicated by the high amount of extra-regional pollen (*sensu* Janssen 1973) blown in from far distances. If allowances are made for these various palynological and site biases, the most striking feature of the Lochnagar vegetational history is that pine was never a major component. This contrasts with the over-riding dominance of pine in sites up to about 600 m altitude in the Cairngorms to the north and west (e.g. Birks 1970; 1975; O'Sullivan 1974, 1975, 1976; Birks and Mathewes 1978; Bennett 1996). Pollen stratigraphies from Caenlochan Glen (550 m altitude) and Corrie Fee, Glen Clova (450 m altitude) about 15 km to the south of Lochnagar similarly show that pine was not a major component of the post-glacial vegetation (Huntley 1981). Together, these data suggest that Lochnagar was close to the southern boundary of the zone of dominant pine as a forest tree in the mid post-glacial about 6000-7000 years ago. The available pollen stratigraphical data provide some support for the reconstructed distribution by McVean and Ratcliffe (1962) of pine in the Eastern Highlands. They suggested that pine was relatively local in this area and confined to poor soils between 400-550 m. Their reconstruction was based on a combination of ecological, historical, and palaeoecological data available in the early 1960s. Despite the abundance of pine on Deeside today and the occurrence of fossil pine stumps along Lochnagar Burn, in Glen Muick, and in Glen Callater, pine appears to have been a rare or local tree in the Lochnagar area, being confined to wet or waterlogged sites (cf. Bennett 1984).

In contrast to the surrounding lowlands, substantial areas of woodland remained in many parts of the Eastern Highlands until relatively recently (Huntley 1981; Birks 1988; Huntley et al. 1997). Some decline in woodland cover may have occurred in the late post-glacial as a result of presumed climatic changes that resulted in the lowering of the altitudinal tree-line and as a result of an increase in the extent of blanket mires (Huntley et al. 1997). Progressive leaching of soils during the post-glacial may also have been responsible for the gradual shift from woodland towards peat development at higher altitudes (Bennett 1996). Human activities, the introduction of domesticated grazing animals, especially sheep, and red deer (*Cervus elaphus*) grazing may have greatly reduced woodland regeneration and thus enabled or accelerated the shift to moorland and blanket mire in the last few hundred years (Birks 1988; Bennett 1996; Huntley et al. 1997). This shift appears to have been more pronounced in the south and east of the Cairngorm area (e.g. Lochnagar, Caenlochan, Clova) than in the north and west (e.g. Loch Einich). Widespread sheep grazing following the Highland Clearances of the 1860s-1880s contributed greatly to the extensive loss of woodland. Sheep and deer grazing is the principal factor resulting in regeneration failure of trees. The regular managed burning of moorlands and heathlands that occupy areas once supporting woodland is associated in part with sheep husbandry and in part with the management of grouse moor. High populations of red deer are common in the Lochnagar area, adding further to the grazing pressure that prevents tree regeneration in so much of the Eastern Highlands today (Nethersole-Thompson and Watson 1974; J. Thompson et al. 2006).

Future flora and vegetation

Lochnagar with its extensive plateaux, corries, summit and gully snow-beds, and precipices supports a high number of rare, so-called 'Red Data Book', and 'scarce' species of vascular plants (Table 3; Nagy et al. 2006) and bryophytes (Rothero 2006). There are nine rare species of vascular plant (Wigginton 1999; Cheffings and Farrell 2005) and eight rare species of bryophyte (Church et al. 2001), and 15 'scarce' species of vascular plant (Stewart et al. 1994) and 50 'scarce' species of bryophyte (data from Hill et al. 1991, 1992, 1994). A 'scarce' species is defined as having presumed native occurrences since 1970 in 16-100 10-km squares of the Ordnance Survey National Grid. Data on rare and scarce lichens are not yet available for Lochnagar (cf. Gilbert and Fox 1985; Fryday 2006). The occurrence of 24 rare or scarce vascular plants and 58 rare or scarce bryophytes is a remarkable concentration of plant biodiversity for a single mountain that is almost entirely granite and is very acid almost everywhere (Shaw and Thompson 2006b).

Table 3: 'Red Data Book' species (critically endangered, endangered, vulnerable, and 'scarce' species (recorded as a native species from 16-100 10-km squares of the Ordnance Survey National Grid since 1970) of vascular plants that occur on or near Lochnagar.

'Red Data' species (Wigginton 1999, Cheffings and Farrell 2005)	
Carex lachenalii	Not threatened
Carex rariflora	Lower risk, near threatened
Cicerbita alpina	Vulnerable
Gnaphalium norvegicum	Lower risk, nationally scarce
Luzula arcuata	Lower risk, near threatened
Moneses uniflora	Vulnerable
Poa flexuosa	Vulnerable
Poa x *jemtlandica*	Vulnerable
Saxifraga rivularis	Lower risk, near threatened

Scarce species (Stewart et al. 1994)	
Alopecurus borealis	*Melampyrum sylvaticum*
Athyrium distentifolium	*Phleum alpinum*
Betula nana	*Pinus sylvestris* ssp. *scotica*
Cerastium alpinum	*Pyrola media*
Epilobium alsinifolium	*Salix lapponum*
Goodyera repens	*Sibbaldia procumbens*
Linnaea borealis	*Veronica alpina*
Lycopodium annotinum	

The rare and scarce species of vascular plants and bryophytes primarily occur in seven major types of habitat on Lochnagar:

1. Pine and birch woodlands (W-18) on the lower slopes along Deeside and in Glen Muick (*Goodyera repens, Linnaea borealis, Lycopodium annotinum, Melampyrum sylvaticum, Moneses uniflora, Pinus sylvestris* (ssp. *scotica*), *Pyrola media*, and the bryophytes *Lophozia longidens* and *Ptilidium pulcherrimum*).
2. Blanket mires (M-2, M-19, M-19b, M-19c) at low-high altitudes (to about 850 m) (*Betula nana*, and the bryophytes *Calypogeia azurea, Cephalozia pleniceps, Sphagnum balticum, S. riparium*, and *Tetraplodon angustatus* (on decaying bones amidst eroding peat and by peaty tracks)).
3. Wet rocks by streams and cascades at medium to high altitudes (600-1000 m) (*Andreaea mutabilis, A. nivalis, Bryum riparium, Hygrohypnum smithii*, and *Marsupella sphacelata*, all bryophytes).
4. Long-lasting snow-beds (U-11, U-12, U-18) amidst boulders and screes in the corries, in cliff gullies, and in sheltered hollows on the plateaux, and associated open, wet or flushed rocks and soil and boulders and open soil within screes (*Athyrium distentifolium* and *Saxifraga rivularis*, and the bryophytes *Andreaea alpestris, A. mutabilis, A. nivalis, Anthelia juratzkana, Arctoa fulvella, Brachythecium glaciale, Conostomum tetragonum, Dicranum glaciale, Dicranoweisia crispula, Diplophyllum taxifolium, Haplomitrium hookeri, Kiaeria blyttii, K. falcata, K. starkei, Lophozia opacifolia, Marsupella adusta, M. alpina, M. boeckii* var. *stableri, M. brevissima, M. condensata, M. sparsifolia,*

M. sphacelata, M. sprucei, Moerckia blyttii, Nardia breidleri, N. geoscyphus, Pleurocladula albescens, Polytrichum sexangulare, Pseudoleskea patens, Pterigynandrum filiforme, Rhizomnium magnifolium, Scapania ornithopodioides, and *Tetralophozia setiformis).*

5. Open soil on high-altitude plateaux and gully heads (*Luzula arcuata, Poa flexuosa,* and *P.* x *jemtlandica,* and the bryophytes *Ditrichum lineare* and *D. zonatum* var. *zonatum*).
6. Ungrazed cliff ledges (U-16, U-16a), crevices, gullies, and cliff faces (*Cerastium alpinum, Cicerbita alpina, Gnaphalium norvegicum, Salix lapponum, Sibbaldia procumbens,* and *Veronica alpina,* and the bryophytes *Amphidium lapponicum, Arctoa fulvella, Barbilophozia lycopodioides, Oedipodium griffithianum,* and *Plagiomnium medium*).
7. High-altitude soligenous mires (M-7), springs and flushes (M-33), and flushed grasslands (U-14), often associated with snow-melt water (*Alopecurus borealis, Carex lachenalii, C. rariflora, Epilobium alsinifolium, Phleum alpinum,* and *Veronica alpina,* and the bryophytes *Bryum weigelii, Harpanthus flotovianus, Philonotis seriata, Pohlia ludwigii, Scapania paludosa, S. uligonosa, Sphagnum lindbergii,* and *Splachnum vasculosum*).

Possible changes in the future

There is much concern amongst conservationists about the possible impact of future 'global change' on the flora and fauna of Scottish mountains (e.g. Watt et al. 1997; Mackey et al. 2001; Kerr and Ellis 2002; Ellis and Good 2005; Ellis and McGowan 2006), in particular the effects of increased winter and summer temperatures, changes in precipitation, and increased atmospheric nitrogen deposition (Fowler et al. 2002). There is a recorded increase of 0.8°C for annual combined land and sea-surface temperature for Scotland for the period of 1970-2000 (Mackey et al. 2001). This is likely to continue, with predictions varying from a 1.5 to 5.8°C increase during the current century. Precipitation has increased by 5-10% over the last century, associated with an estimated 2-4% increase in the frequency of heavy rain events and a tendency for summers to become drier (Ellis and Good 2005). Predictions suggest that annual precipitation may increase by 5-20% by 2100, but almost entirely in the winter months. The seasonal balance of precipitation may thus change markedly (Ellis and Good 2005). Changes in land-use, increased recreational use, and increased atmospheric nitrogen deposition may also have important impacts on Scottish mountains (Fowler and Battarbee 2005).

Although many climatic predictions suggest increasing temperatures in the future (Ellis and Good 2005), it is more difficult to predict what the effects of such temperature increases will be on the ecologically important factors of snow-lie and wind in Scottish mountains (McClatchey 1996; Brown and Clapperton 2002; Kettle and Thompson: Chapter 18 this volume). As March and April weather appears to be crucial in influencing the extent of summer snow-patches in the Cairngorms (Spink 1980), changes in temperature and precipitation in these months may have a major effect on snow-lie (McClatchey 1996). The relatively mild winters of the late 1980s and throughout the 1990s and 2000s probably resulted from a dominance of warm maritime westerly air-streams, leading to low snow-fall in the Cairngorms (Harrison 1993). In an analysis of the extent of summer snow-patches in the period 1979-1989, Watson et al. (1994) found that snow-patch survival into the summer showed a surprising absence of any statistical relationship with summer climate. They proposed that irregular build-up of snow and occasionally melting during winter can influence the structure of the snowpack which then affects melting in spring and summer. They conclude that the extent of summer snow-beds is a useful index of increasing variability in climate, rather than a reflection of winter or summer temperature *per se*. Increased variability in climate appears to be a feature of recent decades and is predicted by climatologists for the near future. If this is correct, the extent of long-lasting snow on Lochnagar and the Cairngorms may show considerable variability in the near future although actual snow-fall may decline dramatically (Chapter 18: Kettle and Thompson this volume).

All the rare or scarce alpine and montane plants that grow on Lochnagar today also occur in the mountains of western or central Norway. Some species are absent from the most oceanic mountains in western Norway (e.g. *Carex lachenalii, C. rarifolia, Cicerbita alpina, Epilobium alsinifolium, Poa flexuosa, Saxifraga rivularis*) (Moe 1995) or are extremely rare in these areas (e.g. *Betula nana, Gnaphalium norvegicum*). The populations of some of these species on Lochnagar or in the Cairngorms are growing at or near to their southern and/or western limits of range in Europe (Birks 1996b). It is clearly very important to monitor the performance and status of these species and also of snow-bed bryophytes and lichens in relation to climatic and other environmental changes, because as they are growing at their range limits they are likely to be particularly sensitive to small shifts in climate (Woolgrove and Woodin 1994; Birks 1996b).

Future changes in the atmospheric deposition of nitrogen are also important for snow-bed plants, as snow is a very efficient scavenger of atmospheric pollutants, thereby concentrating pollutants into the first stages of snowmelt. At snowmelt, 'acid flushes' can then occur that can cause damage to snow-bed bryophytes such as *Kiaeria starkei* that are physiologically active even when growing under snow (Woolgrove and Woodin 1996a, 1996b, 1996c). Careful monitoring of snow-bed vegetation is thus an important priority for future research on Lochnagar (Rothero 2003).

In the absence of such long-term monitoring, only tentative predictions about future floristic and vegetational change on Lochnagar can be made. These predictions are based on the results of a botanical survey of the Jotunheimen Mountains in central Norway where the flora was studied in 265 localities on 23 mountains in 1998 and compared with the flora recorded at the same 265 localities in 1930-31 (Klanderud and Birks 2003). In addition, results from a statistical modelling study on the possible impacts of future climate change on Norwegian mountain plants (Sætersdal and Birks 1997) are relevant in predicting possible future floristic changes on Lochnagar.

First, if, as seems likely, the extent of long-lasting snow-beds on Lochnagar decreases, the abundance of many chionophilous plants restricted to this extreme habitat will decline, including the national rarity *Saxifraga rivularis* (Figure 2b). Lochnagar and the Cairngorms are ideal terrain for collecting and holding snow and their high plateaux and corries currently carry the most extensive and long-lasting snow-beds and associated vegetation in Britain (Ratcliffe 1977). Changes in the extent of snow-cover could have important impacts on much of the flora and vegetation of Lochnagar, especially for the many rare and scarce ferns, bryophytes, and lichens that are confined to late snow-beds in the corries and on the plateaux.

Second, a decrease in snow-cover and hence of snow-melt water will have impacts on the specialised high-altitude soligenous mires, and springs and flushes that are closely associated with snow-melt. Species such as *Alopecurus borealis, Carex lachenalii, C. rariflora,* and *Sphagnum lindbergii* are likely to decline (cf. Sandvik and Heegaard 2003).

Third, with decreased snow-cover, the growth and abundances of dwarf-shrubs such as *Calluna vulgaris, Empetrum nigrum* ssp. *hermaphroditum, Vaccinium myrtillus,* and *V. uliginosum* are likely to increase in the corries, resulting in the spread of dwarf-shrub heath at the expense of fern-rich snow-bed vegetation.

Fourth, it is possible that under conditions of warmer summers and increased atmospheric nitrogen deposition, the overall growth of sedges, grasses, and rushes on the high plateaux and gully tops may increase and the growth of bryophytes such as *Racomitrium lanuginosum* may decrease (e.g. Pearce and van der Wal 2002; Pearce et al. 2003; van der Wal et al. 2003). As a result, the extent of open soil and *Racomitrium* heath may decrease, thereby reducing the available habitat for extreme rarities such as *Luzula arcuata, Poa flexuosa,* and *P. x jemtlandica.*

Unless there are major changes in the recreational use of the northeast corries of Lochnagar, the remarkable flora and vegetation of the ungrazed and near inaccessible cliff ledges near the Black Spout are unlikely to change appreciably. However, one population of *Gnaphalium norvegicum* (Figure 1d) on Lochnagar has varied greatly in size from year to year (author's personal observations) and another population has markedly declined as a result of an avalanche sweeping most of the population away (Wigginton 1999). The possibility of such small populations becoming extinct by natural processes and extreme events can never be excluded, especially in a steep, mountainous area such as Lochnagar (Marren et al. 1986). It is possible that warm summers may benefit the flowering and fruiting of *Cicerbita alpina* (Figure 1c). At present, seed-set appears to be very irregular and seed produced is often not viable (Marren et al. 1986; Wigginton 1999). As *C. alpina* flowers late, warmer summers may be advantageous to its seed production.

The extensive moorlands and associated blanket mires at low to medium altitudes around Lochnagar are likely to remain unchanged if the present land management continues. Similarly, the existing woodland areas are unlikely to change unless there are major changes in forestry practice. Some of the boreal and northern plants such as *Goodyera repens, Linnaea borealis, Melampyrum sylvaticum,* and *Moneses uniflora* may decrease if summer and winter temperatures increase. In Scandinavia today they are mainly plants of areas with moderately mild summers but cold winters (Dahl 1998).

Discussion and conclusions

To botanists and plant ecologists, Lochnagar is an area of very considerable interest and importance for several reasons.

First, its present-day flora and vegetation are outstanding in the context of British uplands and mountains (Ratcliffe 1977). Lochnagar supports one of the richest and least disturbed ranges of alpine vegetation types on acid soils in a transitional oceanic/continental climate at its latitude anywhere in Europe. It has extensive dry dwarf-shrub heaths, lichen-rich summit heaths, and high-altitude mires, springs, and flushes.

Late-snow-influenced vegetation is more extensively developed here and in the Cairngorms than anywhere else in Britain. Its magnificent northeast corrie, with its 240 m vertical cliffs with broad, sloping ledges inaccessible to deer provide a glimpse of what Scottish mountain vegetation may have once been like before centuries of animal grazing. The major ecological gradients that are clearly illustrated on Lochnagar are (1) altitude, from the lowland valley zone, through the forest and moorland zone, to the mountain and alpine plateau zone; (2) exposure, from wind-exposed ridges and cols to sheltered snow-hollows and gullies; (3) moisture, that parallels in part the exposure gradient, from very dry, well-drained sites to waterlogged hollows; and (4) grazing from ungrazed inaccessible ledges rich in tall herbs and tall ferns to grazed grasslands in unprotected areas. Lochnagar has a high number of notable nationally rare or scarce species of vascular plant and bryophytes. Such a concentration is unusual for an acidic mountain consisting of granite, as high diversity and many rare species are often associated with base-rich sites. The reasons for the richness on Lochnagar are the varied topography of the mountain and hence the high diversity of habitats there. Within a relatively small area, there are many habitats along the altitudinal, exposure, moisture, and grazing gradients. Historical factors are also important, as it appears that the northeast corrie of Lochnagar was never densely forested, even during the so-called post-glacial 'thermal maximum'. Open birch scrub with some scattered pine stands may have been present up to the base of the cliffs in the corrie during the mid-post-glacial (9000-6000 years ago). Permanently treeless habitats have therefore always been present in the corrie during the post-glacial and the cliffs, ledges, and gullies would have provided ideal habitats for arctic-alpine plants intolerant of shade and high temperatures to grow (Pigott and Walters 1954).

Second, Lochnagar has a detailed pollen-stratigraphical record from the loch that covers the last 9500 years. This is one of the few pollen stratigraphies available from a high-altitude lake (788 m) in Britain. It provides unique insights into the vegetational history not only of the northeast corrie of Lochnagar but also of the sub-alpine zone in eastern Scotland. It shows that tree and shrub cover had a maximum density to about 6000 years ago. It progressively declined in the last 6000 years as a result of natural soil deterioration and paludification and, more recently, of animal grazing and possibly human activity. There is remarkably little evidence for major human impact on the vegetation until recently (Dalton et al. 2005).

Third, Lochnagar currently supports a large number of nationally rare or scarce species of vascular plant and bryophyte. Many are present as small populations and are closely associated with areas of prolonged snow-lie. The future of these populations is uncertain as the impacts of 'global change' become more and more pronounced. In light of recent evidence from Norway, it is very likely that many of these populations will decline or even go extinct as long-lasting snow-beds decrease in response to milder winters and warmer summers. Lochnagar is an ideal site for long-term monitoring of these populations, as many of them are growing at or near their southern or western limits of range in Europe and are thus likely to be sensitive to environmental changes.

Lochnagar is a major botanical and plant ecological resource in both a British and a European context. It has unique ecological links with, on one hand, the mountains of western Norway, and, on the other hand, the Cairngorms and other mountain areas in Scotland. It has enormous potential for detailed botanical monitoring and plant ecological research. It is indeed a botanical "jewel in the crown of Scotland's natural heritage".

Summary

- The present flora and vegetation of the Lochnagar massif are described in terms of the major vegetation types that occur from the lowland valley zone, through the forest and moorland zone, to the alpine mountain and plateau zone. The flora and vegetation of the famous northeast corrie of Lochnagar are discussed in greater detail, as this corrie is the focus of much modern ecological and limnological research.
- The past floristic and vegetational history of Lochnagar over the last 9500 years is outlined, based on a detailed pollen diagram prepared from two sediment cores collected from the loch in the northeast corrie. The pollen analytical results suggest that up to about 6000 years ago, the corrie may have supported open birch scrub with willow, rowan, and aspen. Pine may have occurred locally. The extent of the scrub progressively declined in the last 6000 years, probably in response to climate change, natural soil deterioration, animal grazing, and possibly human activity.
- Lochnagar supports a surprisingly high number of nationally rare and scarce species of flowering plants and bryophytes. Many of these species are associated with long-lasting snow-beds. If the patterns and extent of snow-cover change as a result of future 'global change' some of the rare species on Lochnagar will decline or even go extinct locally.
- Lochnagar is a major botanical resource in Scotland as it provides a strong link between the mountains of the Scottish Highlands and of western Norway. Much ecological research needs to be done on the flora and vegetation of Lochnagar, particularly the long-term monitoring of populations of rare and endangered species and of snow-bed vegetation.

Acknowledgements

I am greatly indebted to the late Derek A. Ratcliffe and the late J. Grant Roger for freely sharing their considerable knowledge of Lochnagar and its ecology and botany with me, to Hilary Birks, Brian Huntley, and the late Derek Ratcliffe for their companionship in the field on Lochnagar, to Des Thompson and colleagues at Scottish Natural Heritage for many helpful discussions, to Don Monteith and Neil Rose for maintaining my interest in Lochnagar, to Sylvia Peglar for her detailed pollen-analytical studies at Lochnagar, to Gordon Rothero for generously giving me access to his unpublished bryophyte records from the Lochnagar area, to Ben and Alison Averis, Neil Rose, and Des Thompson for their very helpful and insightful comments on the manuscript, and to Cathy Jenks for her invaluable help in preparing this chapter.

References

Allen J.R.M. and Huntley B. 1999. Estimating past floristic diversity in montane regions from microfossil assemblages. J. Biogeog. 26: 55–73.

Averis A.M., Averis A.B.G., Birks H.J.B., Horsfield D., Thompson D.B.A. and Yeo M.J.M. 2004. An illustrated guide to British upland vegetation. Joint Nature Conservation Committee, Peterborough. 454 pp.

Bean C. 2006. Freshwaters. In: Shaw, P. and Thompson, D.B.A. (eds.) The Nature of the Cairngorms: Diversity in a changing environment. The Stationery Office, Edinburgh. pp. 119–131.

Bennett K.D. 1984. Post-glacial history of *Pinus sylvestris* in the British Isles. Quat. Sci. Rev. 3: 133–155.

Bennett K.D. 1994. Post-glacial dynamics of pines (*Pinus sylvestris* L.) and pinewoods in Scotland. In: Aldhous J.R. (ed.), Our pinewood heritage. Forestry Commission, Farnham. pp. 23–39.

Bennett K.D. 1996. Late-Quaternary vegetation dynamics of the Cairngorms. Bot. J. Scot. 48: 51–63.

Birks H.H. 1970. Studies in the vegetational history of Scotland I. A pollen diagram from Abernethy Forest, Inverness-shire. J. Ecol. 58: 827–846.

Birks H.H. 1975. Studies in the vegetational history of Scotland IV. Pine stumps in Scottish blanket peats. Phil. Trans. Roy. Soc. Lond. B 270: 181–226.

Birks H.H. and Mathewes R.W. 1978. Studies in the vegetational history of Scotland V. Late Devensian and early Flandrian pollen and macrofossil stratigraphy at Abernethy Forest, Inverness-shire. New Phytol. 80: 455–484.

Birks H.J.B. 1988. Long-term ecological change in the British Uplands. In: Usher M.B. and Thompson D.B.A. (eds.). Ecological change in the uplands. Blackwell Scientific Publications, Oxford. pp. 37–56.

Birks H.J.B. 1996a. Great Britain – Scotland. In: Berglund B.E., Birks H.J.B., Ralska-Jasiewiczowa M. and Wright H.E. (eds.), Palaeoecological events during the last 15000 years. Regional syntheses of palaeoecological studies for lakes and mires. J. Wiley and Sons, Chichester. pp. 95–143.

Birks H.J.B. 1996b. Palaeoecological studies in the Cairngorms – summary and future research needs. Bot. J. Scot. 48: 117–126.

Birks H.J.B. 2005. Fifty years of Quaternary pollen analysis in Fennoscandia 1954–2004. Grana 44: 1–22.

Blockeel T.L. and Long D.G. 1988. A check-list and census catalogue of British and Irish bryophytes. British Bryological Society, Cardiff. 208 pp.

Brazier V., Gordon J.E., Hubbard A. and Sugden D.E. 1996. The geomorphological evolution of a dynamic landscape: the Cairngorm Mountains, Scotland. Bot. J. Scot. 48: 13–30.

Brown I.M. and Clapperton C.M. 2002. The Physical Geography. In: Gimingham C.H. (ed.), The ecology, land use and conservation of the Cairngorms. Packard Publishers, Colchester. pp. 8–22.

Bruneau P.M.C. 2006. Geodiversity: Soils. In: Shaw P. and Thompson D.B.A. (eds.) The nature of the Cairngorms: Diversity in a changing environment. The Stationery Office, Edinburgh. pp. 43–51.

Campbell L. and Anderson C. 2006. Landscape character. In: Shaw P. and Thompson D.B.A. (eds.) The nature of the Cairngorms: Diversity in a changing environment. The Stationery Office, Edinburgh. pp. 53–59.

Cheffings C.M. and Farrell L. (eds.) 2005. The vascular plant Red Data List for Great Britain. Joint Nature Conservation Committee, Peterborough. 116 pp.

Church J.M., Hodgetts N.G., Preston C.D. and Stewart N.F. (eds.) 2001. British Red Data Books – mosses and liverworts. Joint Nature Conservation Committee, Peterborough. 168 pp.

Curry-Lindahl K. 1974. IUCN Survey of northern and western Europe National Parks and equivalent reserves. Report of Great Britain. International Union of Conservation Nature, London, 187 pp.

Dahl E. 1998. The phytogeography of northern Europe (British Isles, Fennoscandia and adjacent areas). Cambridge University Press, Cambridge. 297 pp.

Dalton C., Birks, H.J.B., Brooks S.J., Cameron N.G., Evershed R.P., Peglar S.M., Scott J.A, and Thompson R. 2005. A multi-proxy study of lake-development in response to catchment changes during the Holocene at Lochnagar, north-east Scotland. Palaeogeog. Palaeoclim. Palaeoecol. 221: 175–201.

Dubois A.D. and Ferguson D.K. 1985. The climatic history of pine in the Cairngorms based on radiocarbon dates and stable isotope analysis, with an account of the events leading up to its colonisation. Rev. Palaeobot. Palynol. 46: 55–80.

Dubois A.D. and Ferguson D.K. 1988. Additional evidence for the climatic history of pine in the Cairngorms, Scotland, based on radiocarbon dates and tree-ring D/H ratios. Rev. Palaeobot. Palynol. 54: 181–185.

Ellis N.E. and Good J.E.G. 2005. Climate change and effects on Scottish and Welsh montane ecosystems: A conservation prospective. In: Thompson D.B.A., Price M.F. and Galbraith C.A. (eds.). Mountains of Northern Europe: Conservation, management, people, and nature. The Stationery Office, Edinburgh. pp. 99–102.

Ellis N. and McGowan G. 2006. Climate Change. In: Shaw P. and Thompson D.B.A. (eds.) The nature of the Cairngorms: Diversity in a changing environment. The Stationery Office, Edinburgh. pp. 353–365.

Fowler D., McFadyen G.G., Ellis N.E. and MacGregor W.G. 2002. Air pollution and atmospheric deposition in Scotland: Environmental and natural heritage trends. In: Usher M.B., Mackey E.C. and Curran J.C. (eds.). The State of Scotland's environment and natural heritage. The Stationery Office, Edinburgh. pp.89–110.

Fowler D. and Battarbee R.W. 2005 Climate change and pollution in the mountains: the nature of change. In: Thompson D.B.A., Price M.F. and Galbraith C.A. (eds.) Mountains of Northern Europe: Conservation, management, people, and nature. The Stationery Office, Edinburgh. pp. 71–88.

Fremstad E. 1997. Vegetasjonstyper I Norge. NINA Temahefte 12: 1–279.

Fryday A. 2006. Lichens. In: Shaw, P. and Thompson, D.B.A. (eds.) The Nature of the Cairngorms: Diversity in a changing environment. The Stationery Office, Edinburgh, pp. 177–193.

Gilbert O.L. and Fox B.W. 1985. Lichens of high ground in the Cairngorm Mountains, Scotland. The Lichenologist 17: 51–66.

Gimingham C.H. (ed.) 2002a. The ecology, land use and conservation of the Cairngorms. Packard Publishers, Colchester, 224 pp.

Gimingham C.H. 2002b. Introduction. In: Gimingham C.H. (ed.). The ecology, land use and conservation of the Cairngorms. Packard Publishers, Colchester. pp. 1–6.

Gimingham C.H. 2002c. Vegetation. In: Gimingham C.H. (ed.). The ecology, land use and conservation of the Cairngorms. Packard Publishers, Colchester. pp. 23–42.

Goodman S. This volume. Chapter 3. Geology of Lochnagar and surrounding region. In: Rose N.L. (ed.) 2007. Lochnagar: The natural history of a mountain lake. Springer, Dordrecht.

Gordon J. and Wignall R. 2006. Geodiversity: geology and landforms. In: Shaw, P. and Thompson, D.B.A. (eds.) The nature of the Cairngorms: Diversity in a changing environment. The Stationery Office, Edinburgh. pp. 13–41.

Green F.H.W. 1974. Climate and weather. In: Nethersole-Thompson D. and Watson A. (eds.). The Cairngorms - Their natural history and scenery. Collins, London, pp. 228–236.

Hall A.M. This volume. Chapter 4. The shaping of Lochnagar: Pre-glacial, glacial and post-glacial processes. In: Rose N.L. (ed.) 2007. Lochnagar: The natural history of a mountain lake. Springer, Dordrecht.

Hall J. 2006. Forests and Woodlands. In: Shaw P. and Thompson D.B.A. (eds.) The nature of the Cairngorms: Diversity in a changing environment. The Stationery Office, Edinburgh. pp. 91–107.

Harrison S.J. 1993. Differences in duration of snow cover on Scottish ski-slopes between mild and cold winters. Geog. Mag. 109: 34–44.

Helliwell R.C., Lilly A. and Bell J.S. This volume. Chapter 6. The development, distribution and properties of soils in the Lochnagar catchment and their influence on soil water chemistry. In: Rose N.L. (ed.) 2007. Lochnagar: The natural history of a mountain lake. Springer, Dordrecht.

Hill M.O., Preston C.D. and Smith A.J.E. 1991. Atlas of the Bryophytes of Britain and Ireland. Volume 1 Liverworts (Hepaticae and Anthocerotae). Harley Books, Colchester. 347 pp.

Hill M.O., Preston C.D. and Smith A.J.E. 1992. Atlas of the Bryophytes of Britain and Ireland. Volume 2 Mosses (except Diplolepideae). Harley Books, Colchester. 400 pp.

Hill M.O., Preston C.D. and Smith A.J.E. 1992. Atlas of the Bryophytes of Britain and Ireland. Volume 3 Mosses (Diplolepideae). Harley Books, Colchester. 419 pp.

Horsfield D. 2006. Upland plant communities. In: Shaw P. and Thompson D.B.A. (eds.) The nature of the Cairngorms: Diversity in a changing environment. The Stationery Office, Edinburgh. pp. 71–89.

Hudson P.J. 2002. Grouse and moorland management. In: Gimingham C.H. (ed.). The ecology, land use and conservation of the Cairngorms. Packard Publishers, Colchester. pp. 139–147.

Huntley B. 1981. The past and present vegetation of the Caenlochan National Nature Reserve, Scotland II. Palaeoecological investigations. New Phytol. 87: 189–222.

Huntley B. 1994. Late Devensian and Holocene palaeoecology and palaeoenvironments of the Morrone Birkwoods, Aberdeenshire, Scotland. J. Quat. Sci. 9: 311–336.

Huntley B. and Birks H.J.B. 1979a. The past and present vegetation of the Morrone Birkwoods National Nature Reserve, Scotland I. A primary phytosociological survey. J. Ecol. 67: 417–446.

Huntley B. and Birks H.J.B. 1979b. The past and present vegetation of the Morrone Birkwoods National Nature Reserve, Scotland II. Woodland vegetation and soils. J. Ecol. 67: 447–467.

Huntley B., Daniell J.R.G. and Allen J.R.M. 1997. Scottish vegetation history: The Highlands. Bot. J. Scot. 49: 163–175.

Janssen C.R. 1973. Local and regional pollen deposition. In: Birks H.J.B. and West R.G. (eds.) Quaternary Plant Ecology. Blackwell Scientific Publications, Oxford. pp. 31–42.

Johnstone G.S. 1974. Geology. In: Nethersole-Thompson D. and Watson A. (eds.) The Cairngorms - Their natural history and scenery. Collins, London. pp. 200–209.

Kerr A. and Ellis N. 2002. Managing the impacts of climate change on the natural environment. In: Usher M.B., Mackey E.C. and Curran J.C. (eds.). The State of Scotland's Environment and Natural Heritage. The Stationery Office, Edinburgh. pp. 239–250.

Kettle H. and Thompson R. This volume. Chapter 18. Future climate predictions for Lochnagar. In: Rose, N.L. (ed.) 2007. Lochnagar: The natural history of a mountain lake. Springer, Dordrecht.

Klanderud K. and Birks H.J.B. 2003. Recent increases in species richness and shifts in altitudinal distributions of Norwegian mountain plants. Holocene 13: 1–6.

Mackay J.W. 2002. Open air recreation in the Cairngorms. In: Gimingham C.H. (ed.). The ecology, land-use and conservation of the Cairngorms. Packard Publishers, Colchester. pp. 160–168.

Mackey E.C., Shaw P., Holbrook J., Shewry M.C., Saunders, G., Hall J. and Ellis N.E. (eds.). 2001. Natural Heritage Trends Scotland 2001. Scottish Natural Heritage, Battleby. 200 pp.

Marren P.R., Payne A.G. and Randall R.E. 1986. The past and present status of *Cicerbita alpina* (L) Wallr. in Britain. Watsonia 16: 131–142.

McClatchey J. 1996. Spatial and altitudinal gradients of climate in the Cairngorms – observations from climatological and automatic weather stations. Bot. J. Scot. 48: 31–49.

McConnell J. and Legg C. 1995. Are the upland heaths in the Cairngorms pining for climate change? In: Thompson D.B.A., Hester A.J. and Usher M.B. (eds.). Heaths and moorland: Cultural landscapes. HMSO, Edinburgh. pp. 154–161.

McVean, D.N. and Ratcliffe D.A. 1962. Plant communities of the Scottish Highlands. HMSO, London. 445 pp.

Moe B. 1995. Studies of the alpine flora along an east-west gradient in western Norway. Nordic J. Bot. 15: 77–89.

Moen A. 1999. National Atlas of Norway: Vegetation. Norwegian Mapping Authority, Hønefoss. 200 pp.

Nagy L., Sydes C., McKinnell J. and Amphlett A. 2006. Vascular Plants. In: Shaw P. and Thompson D.B.A. (eds.) The nature of the Cairngorms: Diversity in a changing environment. The Stationery Office, Edinburgh. pp. 215–241.

Nethersole-Thompson D. and Watson A. 1974. The Cairngorms – Their natural history and scenery. Collins, London. 286 pp.

Neville G., Duncan K. and Mackay A. 2006. Recreation. In: Shaw P. and Thompson D.B.A. (eds.) The nature of the Cairngorms: Diversity in a changing environment. The Stationery Office, Edinburgh. pp. 381–393.

Odland A. 1981. Pre- and subalpine tall herb and fern vegetation in Røldal, W. Norway. Nordic J. Bot. 1: 671–690.

Odland A. 1991a. On the ecology of *Thelypteris limbosperma* – a synecological investigation of *T. limbosperma*-dominated stands in W Norway. Nordic J. Bot. 10: 637–659.

Odland A. 1991b. A synecological investigation of *Athyrium distentifolium*-dominated stands in western Norway. Nordic J. Bot. 11: 651–673.

O'Sullivan P.E. 1974. Two Flandrian pollen diagrams from the east-central Highlands of Scotland. Pollen et Spores 16: 33–57.

O'Sullivan P.E. 1975. Early and middle-Flandrian pollen zonation in the eastern Highlands of Scotland. Boreas 4: 197–207.

O'Sullivan P.E. 1976. Pollen analysis and radiocarbon-dating of a core from Loch Pityoulish, eastern Highlands of Scotland. J. Biogeog. 3: 293–302.

Pearce I.S.K. and van der Wal R. 2002. Effects of nitrogen deposition on growth and survival of montane *Racomitrium lanuginosum* heath. Biol. Conserv. 104: 83–89.

Pearce I.S.K., Woodin S.J. and van der Wal R. 2003. Physiological and growth responses of the montane bryophyte *Racomitrium lanuginosum* to atmospheric nitrogen deposition. New Phytol. 160: 145–155.

Pigott C.D. and Walters S.M. 1954. On the interpretation of the discontinuous distribution shown by certain British species of open habitats. J. Ecol. 42: 95–116.

Purvis O.W., Coppins B.J., Hawksworth D.L., James P.W. and Moore D.M. 1992. The lichen flora of Great Britain and Ireland. The British Museum (Natural History), London. 710 pp.

Rapson S.C. 1985. Minimum age of corrie moraine ridges in the Cairngorm Mountains, Scotland. Boreas 14: 155–159.

Ratcliffe D.A. 1968. An ecological account of Atlantic bryophytes in the British Isles. New Phytol. 67: 365–439.

Ratcliffe D.A. 1974. The Vegetation. In: Nethersole-Thompson D. and Watson A. (eds.). The Cairngorms - Their natural history and scenery. Collins, London. pp. 42–76.

Ratcliffe D.A. (Ed.) 1977. A Nature Conservation Review Volumes 1 and 2. Cambridge University Press, Cambridge. 401 pp, 320 pp.

Rothero G.P. 2003. Bryophyte conservation in Scotland. Bot. J. Scot. 55: 17–26.

Rothero G.P. 2006. Bryophytes. In: Shaw P. and Thompson D.B.A. (eds.). The nature of the Cairngorms: Diversity in a changing environment. The Stationery Office, Edinburgh. pp. 195–213.

Rowse C. 2006. Farmland. In: Shaw P. and Thompson D.B.A. (eds.). The nature of the Cairngorms: Diversity in a changing environment. The Stationery Office, Edinburgh. pp. 109–117.

Sandvik S.M. and Heegaard E. 2003. Effects of simulated environmental changes on growth and growth-form in a late snow-bed population of *Pohlia wahlenbergii* (Web. et Mohr) Andr. Arct. Antarct. Alp. Res. 35: 341–348.

Shaw P. and Thompson D.B.A. (eds.) 2006a. The Nature of the Cairngorms: Diversity in a changing environment. The Stationery Office, Edinburgh, 444 pp.

Shaw P. and Thompson D.B.A. 2006b. Patterns of species diversity in the Cairngorms. In: Shaw P. and Thompson D.B.A. (eds.). The nature of the Cairngorms: Diversity in a changing environment. The Stationery Office, Edinburgh. pp. 395–411.

Sisson J.B. 1979. The Loch Lomond Advance in the Cairngorm Mountains. Scot. Geog. Mag. 95: 66–82.

Sissons J.B. and Grant A.J.H. 1972. The last glaciers in the Lochnagar area, Aberdeenshire. Scot. J. Geol. 8: 85–93.

Spink P.C. 1980. Scottish snow-beds in summer: 1979 survey and some comments on the last 30 years. Weather 35: 256–259.

Stace C.A. 1997. New Flora of the British Isles (second edition). Cambridge University Press, Cambridge. 1130 pp.

Stewart A., Pearman D.A. and Preston C.D. 1994. Scarce Plants in Britain. Joint Nature Conservation Committee, Peterborough, 515 pp.

Sugden D.E. 1974. Landforms: In: Nethersole-Thompson D. and Watson A. (eds.). The Cairngorms - Their natural history and scenery. Collins, London. pp. 210–221.

Sugden D.E. 1977. Did glaciers form in the Cairngorms in the 17–19th centuries? Cairngorm Club J. 97: 189–201.

Sugita S. 1994. Pollen representation of vegetation in Quaternary sediments: theory and method in patchy vegetation. J. Ecol. 82: 881–897.

Sætersdal M and Birks H.J.B. 1997. A comparative ecological study of Norwegian mountain plants in relation to possible climate change. J. Biogeog. 24: 127–152.

Thompson D.B.A., Shaw P. and Gordon J. 2006. Introduction: A sense of being in the Cairngorms. In: Shaw P. and Thompson D.B.A. (eds.). The nature of the Cairngorms: Diversity in a changing environment. The Stationery Office, Edinburgh. pp. 1–11.

Thompson J., Bryce J., Scott R. and Horsfield D. 2006. Deer management. In: Shaw P. and Thompson D.B.A. (eds.). The nature of the Cairngorms: Diversity in a changing environment. The Stationery Office, Edinburgh. pp. 367–379.

van der Wal R., Pearce I., Brooker R., Scott D., Welch D. and Woodin S. 2003. Interplay between nitrogen deposition and grazing causes habitat degradation. Ecol. Letters 6: 141–146.

Watson A. 2005. Warmer climate and Scottish snow. Scottish Mountaineering Club J. 39: 51–59.

Watson A., Davison E.W. and French D.D. 1994. Summer snow patches and climate in northeast Scotland, UK. Arctic Alpine Res. 26: 141–151.

Watt A.S. and Jones E.W. 1948. The ecology of the Cairngorms I. The environment and the altitudinal zonation of the vegetation. J. Ecol. 36: 283–304.

Watt A.D., Carey P.D. and Eversham B.C. 1997. Implications of climate change for biodiversity. In: Fleming L.V., Newton A.C., Vickery J.A. and Usher M.B. (eds.). Biodiversity in Scotland: Status, trends and initiatives. The Stationery Office, Edinburgh. pp. 147–159.

Wigginton M.J. (ed.). 1999. British Red Data Books 1. Vascular Plants (third edition). Joint Nature Conservation Committee, Peterborough. 465 pp.

Woolgrove C.E. and Woodin S.J. 1994. Relationships between the duration of snow-lie and the distribution of bryophyte communities within snowbeds in Scotland. J. Bryol. 18: 253–260.

Woolgrove C.E. and Woodin S.J. 1996a. Current and historical relationships between the tissue nitrogen content of a snow-bed bryophyte and nitrogenous air pollution. Envir. Poll. 91: 283–288.

Woolgrove C.E. and Woodin S.J. 1996b. Effects of pollutants in snowmelt on *Kiaeria starkei*, a characteristic species of late snow-bed bryophyte-dominated vegetation. New Phytol. 133: 519–529.

Woolgrove C.E. and Woodin S.J. 1996c. Ecophysiology of a snow-bed bryophyte *Kiaeria starkei*
    during snowmelt and uptake of nitrate from meltwater. Can. J. Bot. 74: 1095–1103.
World Wilderness Congress 1983. Congress resolutions. World Wilderness Congress, 3.

PART II: THE CONTEMPORARY PHYSICAL
         AND BIOLOGICAL STATUS OF LOCHNAGAR

8. THE SEDIMENTS OF LOCHNAGAR: DISTRIBUTION, ACCUMULATION AND COMPOSITION

NEIL L. ROSE (nrose@geog.ucl.ac.uk)
Environmental Change Research Centre
University College London
Pearson Building, Gower Street
London WC1E 6BT
United Kingdom

Key words: Lochnagar, organic matter, sediment accumulation rates; sediment distribution

Introduction

The c. 10000 year sediment record accumulated within the loch basin (Chapter 4: Hall this volume; Chapter 7: Birks this volume) has been central to much of the research undertaken at Lochnagar. The first major studies of the loch were palaeolimnological and assessed the acidification history of the site in relation to its contemporary water quality status (Chapter 14: Monteith et al this volume). This rapidly expanded into determining the sediment archived histories of long-term vegetation changes via the pollen record (Chapter 7: Birks this volume) as well as changes in the atmospheric deposition of trace metals (Chapter 15: Tipping et al. this volume), persistent organic pollutants (POPs) (Chapter 16: Muir et al. this volume) and fly-ash particles (Chapter 17: Rose and Yang this volume). These in turn have developed into monitoring programmes, the data from which form the basis for other chapters in this book. More recently the palaeolimnological focus has shifted to include the evidence for climatic change and its impacts (Dalton et al 2005; and below).

In the course of these studies information on the sediment itself has been gained in terms of its distribution (i.e., knowing the 'best' place to core in order to obtain appropriate records), its accumulation in various parts of the sediment basin (from ^{210}Pb-dating and other chronological markers) and its composition. The aim of this chapter is to provide this information as background for the other palaeolimnological studies described in this book as well as showing how temporal changes in the sediment matrix itself have been used to obtain useful information on environmental change within the loch and in its catchment.

N.L. Rose (ed.), Lochnagar: The Natural History of a Mountain Lake
Developments in Paleoenvironmental Research, 155–175.
© *2007 Springer.*

Estimating the boundary of sediment accumulation

With the exception of the southeastern 'corner' of Lochnagar, the southern end of the loch, and stretches of the northeast and northwest shorelines, the littoral area is dominated by a boulder and stone substrate (Chapter 10: Flower et al. this volume). The substrate of the southeastern corner has a sandy nature being the result of coarse debris in-washed from the slopes of the Red Spout above (see Figure 6 in Chapter 4: Hall this volume), while the southern end continues the steep slope of the backwall (Chapter 2: Hughes this volume) and is composed of bed-rock below which rocks and gravels can be found. Along some parts of the northwestern and in particular the northeastern shores, the shoreline is affected by eroding peats.

Contiguous sediment accumulation within Lochnagar, estimated from both the UK Acid Waters Monitoring Network macrophyte surveys (Chapter 10: Flower et al., this volume) and attempts at coring along transects (Chapter 17: Rose and Yang this volume; Chapter 15: Tipping et al. this volume), therefore only starts beyond this rocky littoral area. This empirical estimate of the limit to sediment accumulation is shown in Figure 1a (pale yellow) although sediment deposits undoubtedly occur between the boulders and stones, especially in areas affected by the eroded catchment peats. This assessment of the boundary of sediment accumulation is, of course, only speculative and such estimates can lead to significant uncertainties in full basin sediment accumulation.

Evans and Rigler (1980a;b) discuss the distribution of sediments above and within apparently well-defined non-accumulating zones and suggest that the most likely situation is a rapidly decreasing sediment accumulation above this boundary rather than a sharp 'cut-off' or a gradual decrease from the accumulation zone to no accumulation at zero water depth. This would also be the most likely situation for Lochnagar were it not for the boulders within the littoral. Evans and Rigler (1980a) also suggest that the "depth above which no permanent sediments accumulate" can be defined by the bottom of the summer epilimnion. At Lochnagar, the mid-summer epilimnion occurs at about 6 – 8 m depth, well below the depth of permanent sediment accumulation (see below). Furthermore, the epilimnion in Lochnagar appears to get deeper as the summer progresses (see Figure 5 in Chapter 5: Thompson et al this volume for an example; Don Monteith UCL: unpublished data, for other years) and the loch is also well-mixed for 9 – 10 months of the year. Thus, this model does not work for Lochnagar. Hilton et al. (1986) suggest that there is negligible sediment accumulation on slopes greater than 14% (orange and red shading in Figure 3b in Chapter 2) and these slopes in Lochnagar account for 18% of the area of the loch basin mainly in the south and west. The distribution of shallow slopes within the basin agrees well with that of sediment accumulation in Lochnagar (see below).

Rowan et al. (1992) use estimates of wave energy as a means to define a mud deposition boundary depth ('Mud DBD') and show how this can be estimated from the fetch of a lake. They also show how this depth is altered by the effects of slope (Rowan et al. 1995 a; b). For Lochnagar, in the areas of least slope (< 4%) Rowan et al.'s mud DBD occurs at a depth of about 1 m while for slopes of 14% the mud DBD lies at c. 3m and for slopes approaching 30% (i.e., steepest parts of the loch basin in the south and west) the mud DBD is greater than the maximum depth of the loch implying no accumulation in these zones. Furthermore, Rowan et al. (1995a) show how the mud

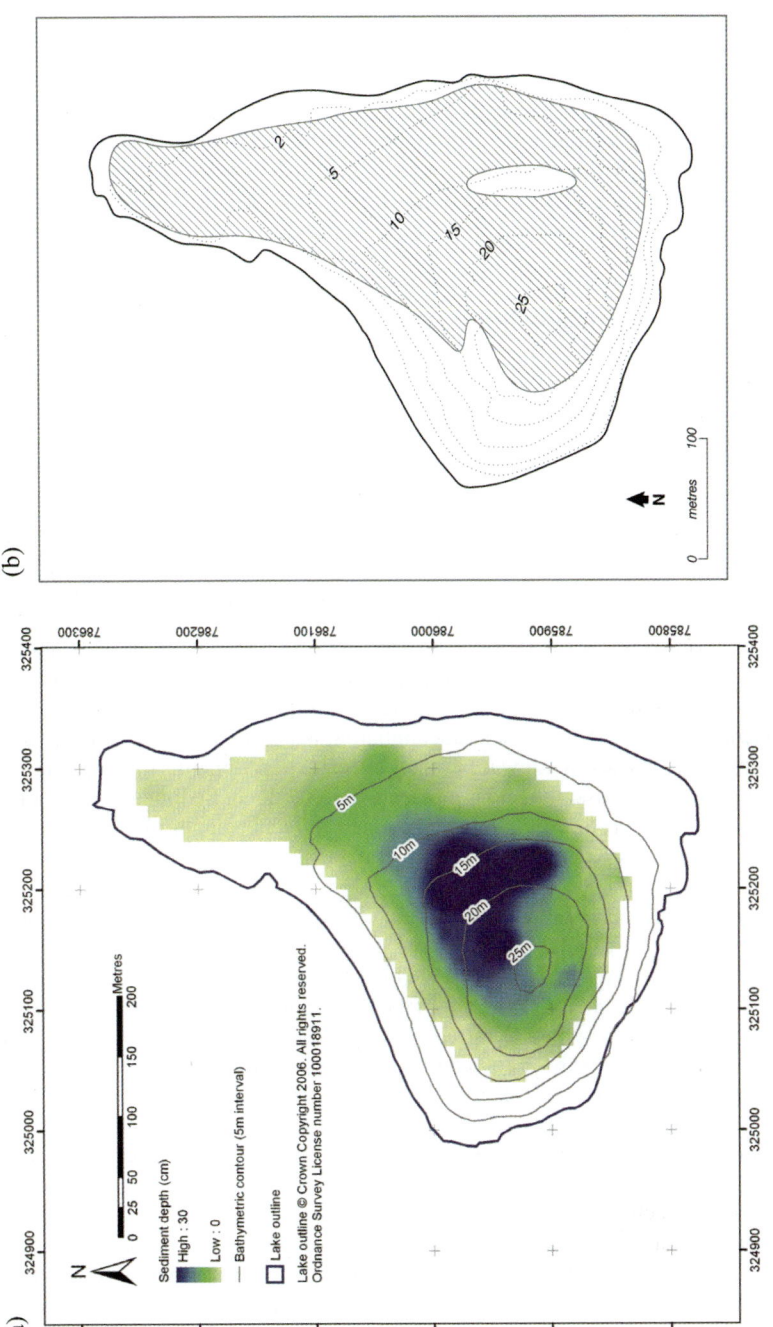

Figure 1. (a) Distribution of post-1850 sediment accumulation as determined by ^{210}Pb and spheroidal carbonaceous particle dated sediment cores. Pale yellow shading in littoral area denotes empirically estimated region of little or no sediment accumulation. (b) Region of sediment accumulation in Lochnagar estimated using the 'Mud DBD' approach of Rowan et al. (1992).

DBD at erosional sites impacted by the largest waves should be increased by a factor of 1.34. This increases the shallowest mud DBD estimate by only 20 cm, the 14% estimate by c. 1m (to 4 m) and makes little difference to the steepest slopes as they are already below the maximum depth of the loch. Plotting these areas onto a bathymetric map of Lochnagar shows the areas in which there should be accumulating sediment based on wave theory and slope. This is shown in Figure 1b and agrees reasonably well with the empirical estimate of this boundary. The implications of this mud DBD are that below these threshold depths (within the area of accumulation shown in Figure 1b) sediments do not become resuspended (Blais and Kalff 1993) and that any re-worked sediment from shallower areas is derived from above this threshold.

Sediment distribution and movement

A plot of the depth of sediment accumulated since 1850 (corrected for differences in date of coring to a uniform date of 2000) determined from radiometric and spheroidal carbonaceous particle (SCP) chronologies is presented in Figure 1a. It can be seen that, in agreement with other studies on sediment core variability (e.g., Anderson 1990a; b) there is considerable variation across the loch basin, but reasonable consistency in the central area of the basin. The greatest accumulation is in the centre of Lochnagar where c. 30 cm has accumulated in the 150 year period, but this area of maximum accumulation is not in the deepest part of the loch. In the deepest area (see bathymetry in Hughes: Chapter 2 this volume) sediment accumulation seems to be more heterogeneous, with some cores showing a moderate post-1850 depth of accumulation (e.g., NAG6 – 17cm; NAG8 – 16 cm; NAG23 – 17cm) while other cores show little (e.g., NAG3 – 8cm; NAG17 – 4cm) (see Figure 2 for core locations). Elsewhere in the basin, the depth of post-1850 sediment accumulation decreases from the central 'high' accumulation area outwards towards the estimated sediment limit suggesting considerable sediment focussing is taking place within the loch. Sediment accumulation in Lochnagar cannot therefore be simply explained by water depth alone as found in other studies (Evans and Rigler 1980a; b) in agreement with the work of Rowan et al (1995b).

A great many factors have been cited as being the cause of, or at least influential in, sediment accumulation patterns. Lehman (1975) states that the lake morphology is the critical criterion especially with respect to the changing nature of accumulation patterns as the lake in-fills, Håkanson (1977) suggests a large number of influencing factors including frequency, velocity and duration of winds, fluctuations in water level, water circulation, fetch, water depth, lake bottom roughness and sedimentological factors such as compaction and, of course, rate and source of input. Håkanson (1977) also states that water content of the surficial sediments provides an indication as to the dynamic process dominating in that part of the lake, thus for sediment with a water content of < 50%, 50 – 75% and > 75%, the dominant processes are erosion, transportation and accumulation respectively. Rowan et al (1995b) suggest that 60% water content is the critical threshold for erosional and depositional zones.

Applying Håkanson's water content thresholds to Lochnagar sediments suggests that the only non-accumulating area is in the southwest of the loch (Figure 3a) and that the majority of the loch basin is an accumulation zone. However, Håkanson goes on to state that for slopes within accumulation zones transport also occurs where the slope is greater than a value generated by:

$$W_{0-1} = -8.55s + 114.7 \tag{1}$$

where W_{0-1} is the water content of the surface sediment and 's' is the slope. Thus for Lochnagar where a typical water content across the accumulating area is 87%, slopes >3.2 % correspond to transportation zones. As 79% of the Lochnagar sediment basin has a slope >3.2% this would imply significant transport within the accumulation area which may explain the level of focussing observed from the post-1850 sediment accumulation data (Figure 1a). In contradiction to this, Blais and Kalff (1995) suggest that sediment water content does not denote the boundary of the transportation zone and that this is better defined by using the mean slope of the lake basin. They suggest that this is the "single best determinand" for this definition as the slope is not only critical to the water turbulence which moves sediment downslope but also, in steeper areas, to

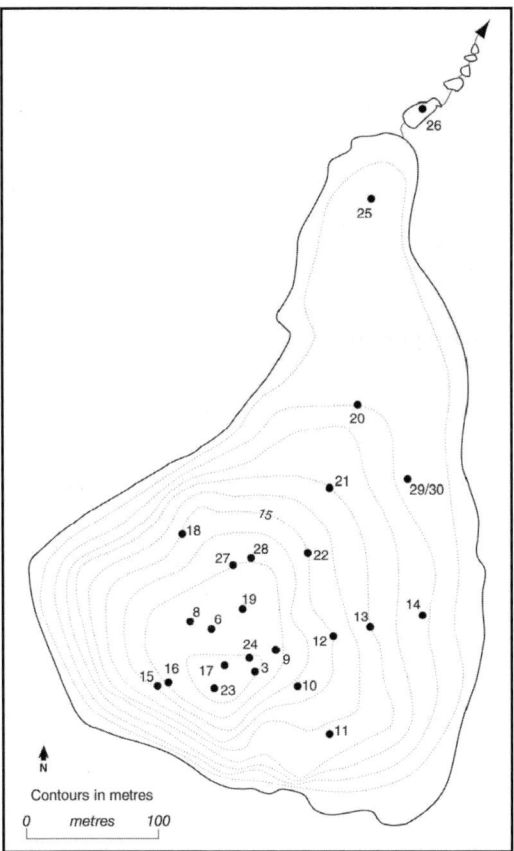

Figure 2. Sediment coring locations. (i.e., 9 denotes the location of core NAG9 etc.)

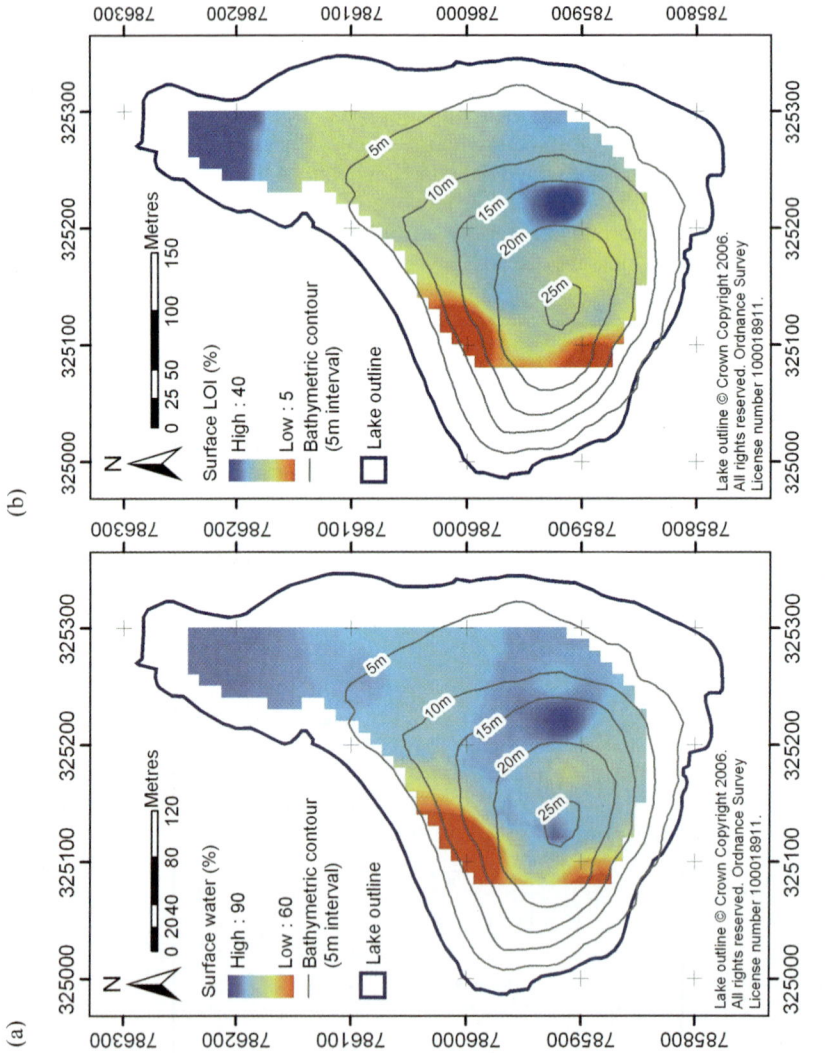

Figure 3. (a) Water content and (b) organic matter content (as estimated by loss-on-ignition at 550 °C) for Lochnagar surface sediments.

sediment slumping. Focusing is therefore a function of lake form. The percentage of the basin in the accumulation zone (%ZA) using this approach is defined as:

$$\%ZA = 49.92(\pm 3.73) - (2.5 (\pm 0.31)\acute{a}_p) \qquad (2)$$

where $\acute{a}_p$ is the mean basin slope (Blais and Kalff 1995). For Lochnagar, the mean basin slope is 8.4% (range 0 - 28.45%, SD 5.97%; Chapter 2: Hughes this volume) and hence, using Equation 2, 22.6 – 35.2% of the loch basin is in the accumulation zone.

An alternative approach to estimating these dynamic process areas is proposed in Shteinman and Parparov (1997) where the percentage of the lake bed subject to erosion and transportation (a_{E+T}) (i.e., with a high probability of resuspension) and the area of sediment accumulation (a_A) are given by:

$$a_{E+T} = 100 - a_A = 25*(DR)*41^{0.061/(DR)} \qquad (3)$$

where DR, the dynamic ratio = $\sqrt{(a/D)}$; a is the lake surface area in km^2 and D is the average depth of the lake in metres. Substituting values of 'a' and 'D' for Lochnagar into Equation 3 results in the estimate that 22% of the loch basin is subject to erosion and transportation and therefore 78% is accumulating (a_A).

Estimates of the area of the Lochnagar basin that form the sediment accumulation area therefore vary considerably from 30% using the Blais and Kalff (1995) approach to 87% employing that of Håkanson (1977). In reality it is likely that the accumulation area lies between these two extreme estimates and the approaches of Shteinman and Parparov (1997), Rowan et al (1992) and an estimate based on empirical evidence provide values of 78%, 62% and 60% respectively. Without more detailed surveying it is not possible to determine which of these is the more accurate.

In terms of the uneven distribution of sediments within any defined accumulating zone, Hilton et al. (1986) suggest ten distribution mechanisms that could result in sediment focussing. Of these, 'riverine delta formation' and 'river plume sedimentation' are obviously not applicable at Lochnagar due to the lack of major inflows. 'Continuous complete mixing', where sedimenting material is continuously mixed over the entire volume of the lake, can also be discounted as this results in shallow and deep water sediment traps recording similar fluxes and this is not the case here (see Figure 4). Similarly, 'organic degradation', where deep water traps record lower fluxes than shallow traps due to organic decomposition in the water column, is also discounted by these data and further, is usually only applicable to sites with high algal productivity. Neither 'intermittent complete mixing' whereby sediment is resuspended from all over the lake-bed and completely mixed such that sediment accumulation is proportional to depth, nor 'intermittent epilimnetic complete mixing', where sediment cores below the epilimnion show constant accumulation rate, explain the sediment distribution in Lochnagar as the deepest areas show both variable accumulation and accumulation rates lower than other, shallower areas of the loch.

Of Hilton et al.'s four remaining distribution mechanisms, 'peripheral wave attack' would seem to be a likely means by which sediment deposited amongst the boulders and stones in the littoral area could be resuspended and moved to deeper water. However, it is suggested that if this is the dominant process then sediment accumulation rates should increase linearly with water depth and that the regression line of this linear

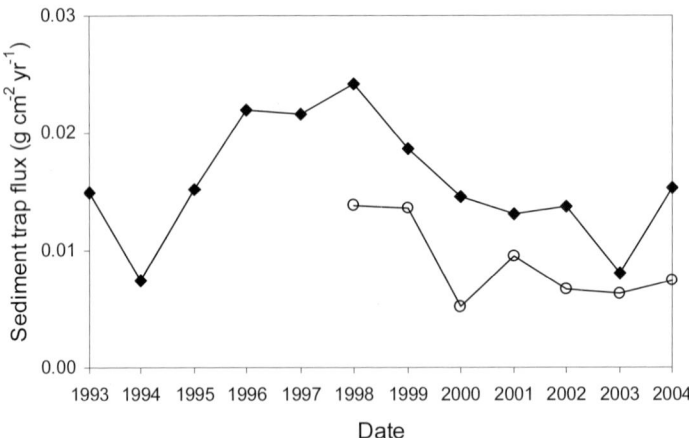

Figure 4. Annual dry matter fluxes for deepwater (filled symbols) and shallow water (open symbols) sediment traps in Lochnagar 1993 – 2004.

relationship intercepts the depth axis at the depth where the areas of transport and accumulation meet. Such a linear relationship does not exist at Lochnagar, where a plot of accumulation versus depth is far more random (Figure 5) and thus more indicative of 'random redistribution'. However, 'random redistribution' also leads to sediment accumulation in traps being far higher than cores (Hilton et al. 1986) which is clearly not the case at Lochnagar (compare data in Figure 4 with Figure 6).

Movement down-slope is also undoubtedly a factor in the redistribution of sediments in Lochnagar especially in the steeper southern and western parts of the loch. Hilton et al. (1986) suggest that there should be negligible accumulation on slopes greater than 14% (also Blais and Kalff 1995) whilst for slopes between 4% and 14% accumulation is reduced with respect to accumulation on more horizontal areas. For Lochnagar this means that 18% of the loch basin area should not accumulate sediment. Conversely, 82% of the basin has a slope of < 14% and should accumulate sediment to a greater or lesser degree, with 29% of the basin area having a slope of less than 4% and 53% having a slope of 4 – 14%. From this it would be expected that the flattest 29% of the basin would be the area in which the greatest accumulation occurs, but this is not the case and the area of maximum post-1850 accumulation (Figure 1a) lies predominantly within the 4 – 14% slope region.

Even in areas of lake basins unaffected by any of the sediment redistribution mechanisms discussed above, and it is debateable whether there are any areas of the Lochnagar basin where this is the case, significant spatial heterogeneity is known to occur, even at quite small spatial scales. Downing and Rath (1988) state that sediments can exhibit this spatial heterogeneity in ways unrelated to focussing or other large-scale redistribution processes. Therefore, it should be expected that spatial patchiness will occur in the sediment record of Lochnagar and data from multiple cores used for interpretation wherever possible. A number of processes leading to this heterogeneity are proposed by Downing and Rath (1988). These include: accumulation in the shadow

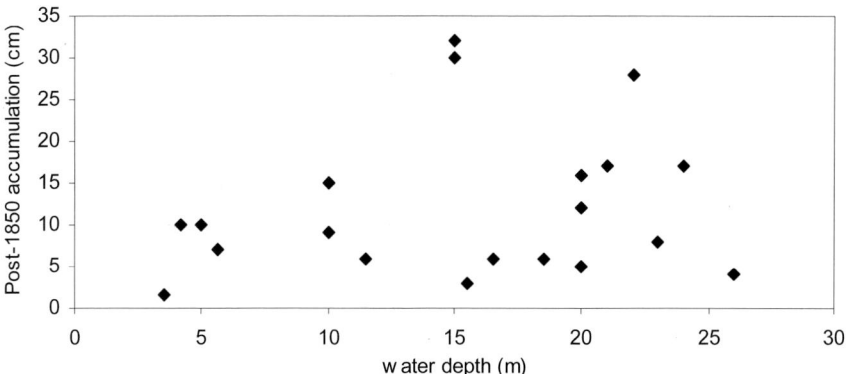

Figure 5. Depth of post-1850 sediment accumulation versus water depth for all sediment cores taken from Lochnagar.

of macrophytes and outcrops; local temperature differences resulting from groundwater leading to differential decomposition of organic matter; patchily distributed benthic invertebrates accumulating material around them; redistribution by slow, deep currents; small scale variations in the bottom profile and ice-rafting of debris to deepwater areas. Of these, the latter two would seem to be of most relevance to Lochnagar. Rockfalls from the back-wall could easily have produced variations in the bottom profile and may explain the irregularity of the sediment record in the deepest area. Ice-rafting is also known to occur and debris falling from the backwall could be transported to any point in the loch by this route. However, dominant wind directions (Chapter 5: Thompson et al. this volume) are such that the last remaining ice tends to move towards the back-wall and hence most debris is likely to remain in this area of the basin.

In summary, it seems likely that no single dominant process explains the distribution of sediment and focussing within Lochnagar. The movement of sediment from littoral areas to deeper waters by wave resuspension, movement downslope and ice-rafting are all potential and likely factors. Rock-falls in the past may have influenced the bottom topography and could explain the heterogeneous sediment record in the deepest areas. However, none of these processes explain why the main area of accumulation is not in the deepest water, but in an area central to the loch.

One hypothesis for this distribution is that suspended sediment is moved towards the back-wall in surface waters driven by prevailing wind directions (Chapter 5: Thompson et al. this volume). The return flow from this movement must be either around the edges of the loch or below the surface. The steep back-wall is the most shaded area of the catchment and has snow patches lingering until June (and much longer in the past). The cooling of the surface water in this shaded area and input of cold meltwater from the catchment could therefore cause waters in this area to descend, making the return flow via deeper waters. Suspended sediment could then be deposited as the water moves back slowly at depth and this could result in sediment deposition in the central area. This process is described in Hilton et al (1986) as 'current redeposition' and is thought to occur where cooling surface waters move down adjacent steep slopes eroding sediment on those slopes and redepositing them in central areas. This would seem to fit the situation for Lochnagar, where the northern end of the loch is both the coolest, as a result of shading and meltwater inputs, and also the steepest (Figure 3b in Chapter 2).

Scale of sediment focussing in Lochnagar

Sediment focussing describes the non-uniform distribution of the sediment over the lake bed due to erosional processes in some parts of the lake (see above) transporting material and preferentially depositing it in other areas. In Lochnagar, the main area of focussing appears to be the central, rather than the deepest, part of the basin. The scale of this process can be estimated using ^{210}Pb data (Peter Appleby, University of Liverpool pers. commun.).

If the total input to the loch is Q Bq m^{-2}, and the inventory at a particular point is A Bq m^{-2} then the focussing factor (f) would be defined as:

$$f = A/Q \qquad (4)$$

The term A can be determined by measurements on a sediment core from that location while Q will be the sum of direct deposition from the atmosphere onto the surface of the loch itself plus erosional inputs from the catchment. Using the data from three dated sediment cores (NAG3, 6 and 8) taken from the deep area of the loch (Figure 2) and assuming that these cores are equally representative of this deep area, then Q is the mean of the inventories for these cores:

$$Q = (6823 + 9144 + 7422)/3 = 7796 \text{ Bq m}^{-2} \qquad (5)$$

The focussing factors for each core would then be 6923/7796 = 0.88, 9144/7796 = 1.17, 7422/7796 = 0.95 respectively. However, as these cores are all from the deeper parts of the loch they will overestimate the value of Q, and the focussing factors will all be larger than these. If erosional inputs from the catchment are negligible then:

$$Q = 113/0.03114 = 3628 \text{ Bq m}^{-2} \qquad (6)$$

where 113 Bq m^{-2} is an estimate of the direct fallout of ^{210}Pb at Lochnagar from soil core (106 Bq m^{-2} yr^{-1}) and rainfall (120 Bq m^{-2} yr^{-1}) measurements. The focussing factors for the three cores would then be 1.88, 2.52, and 2.05, with a mean value of 2.15. Although the three cores are from deep area of the loch it is probably mis-leading to assume that the influence of the catchment is zero at these points. The focussing factor for the deep area of the loch is therefore likely to be slightly below this latter estimate, in the region of 2.0.

Sediment accumulation rates

Four sediment cores taken from the deep area of Lochnagar (NAG3 – 1986; NAG6 – 1991; NAG8 – 1996; NAG23 - 1998) (See Figure 2 for locations) have been radiometrically dated. Although the records contained within these cores show good agreement with respect to ecological changes and the record in atmospherically deposited contaminants (e.g. Chapter 15: Tipping et al. this volume; Chapter 16: Rose and Yang this volume) the post-1850 sediment accumulation is found to vary considerably (see above). This is in agreement with multi-core studies from other lakes (e.g., Anderson 1989; 1990a; Rose et al 1999). The cores from Lochnagar reflect this variability both in total accumulation since 1850 (8cm, 16cm, 15cm and 17 cm respectively) and in temporal variability in accumulation rates over this period.

Earliest sediment accumulation rates in NAG3, 6 and 23 show reasonable agreement and range between 0.0068 and 0.012 g cm^{-2} yr^{-1} (Figure 6) whilst early sediment accumulation rates in NAG8 vary considerably from rates at the same sort of magnitude as these other cores (0.05 – 0.08 g cm^{-2} yr^{-1} for 1868 ± 30 to 1887 ± 25) to a period of much higher sediment accumulation between 1890 ± 23 and 1908 ± 13 of 0.11 – 0.17 g cm^{-2} yr^{-1}. This period of rapid sedimentation coincides with an increase in sediment density and therefore could be due to slumping of catchment derived inorganic material (Peter Appleby, University of Liverpool pers. commun.). An earlier and less distinct episode of a similar nature may have occurred in the mid-19th century and may explain the elevated accumulation rates at the very base of the core (Figure 6). The accumulation rate of NAG8 between these events is similar to that of the other cores at this time suggesting, that under normal conditions, there is a reasonable agreement in sediment accumulation rate in this part of the loch. However, it also highlights the vulnerability of this deep area to these events. Analysis of mineral magnetism of the NAG8 core provides supporting evidence for this interpretation of the sediment accumulation data. The upper part of the core shows a signal typical of increasing atmospheric pollution since the 1890s and the suggestion of a recovery in the few surface samples. However, prior to this late-19th century increase the magnetic record suggests an episode of erosion that transported haematite-rich minerals to the sediment (John Dearing, University of Liverpool pers. commun.). Such an event agrees with the input of catchment derived inorganic material at this time indicated by the sediment accumulation rate data.

Variability in sediment accumulation rate also occurs between the temporal records of these cores. In NAG3 and NAG8 accumulation rate increases to the surface of the core, with maxima at the surface of 0.026 and 0.041 g cm^{-2} yr^{-1} respectively (0.16 and 0.15 cm yr^{-1}) corresponding to an increase by a factor 2 and 3.8 over the uppermost 150 years (Figure 6). Alternatively, NAG6 shows an increase in accumulation rate from 0.0095 g cm^{-2} yr^{-1} (0.098 cm yr^{-1}) before 1930 to over 0.025 g cm^{-2} yr^{-1} (0.22 cm yr^{-1}) in the mid-1980s before declining again to lower values at the surface of 0.02 g cm^{-2} yr^{-1} (0.18 cm yr^{-1}). NAG23 differs again by showing no particular trend (Figure 6) but varying between 0.01 and 0.015 g cm^{-2} yr^{-1} throughout.

Over the longer time-scale Dalton et al. (2001; 2005) reported ^{14}C AMS dates for the centrally located core NAG27 taken from 20.6 m and NAG30 taken from 4.2 m nearer the eastern shore (Figure 2). Although the dating interval is rather coarse, the depth / date profile for NAG27 (Figure 7) appears to be remarkably consistent over the 5000 years between the oldest and most recent ^{14}C dates, with accumulation rates estimated to range between 0.026 and 0.054 cm yr^{-1} throughout. The depth / date profile for NAG30 covers an older time period (5100 – 8400 BP) and shows even slower accumulation rates of 0.007 – 0.033 cm yr^{-1} (Dalton et al. 2005). These data are thus in agreement with the spatial distribution shown in Figure 1a where slower accumulation rates are observed closer to the limits of the accumulating zone within Lochnagar, but

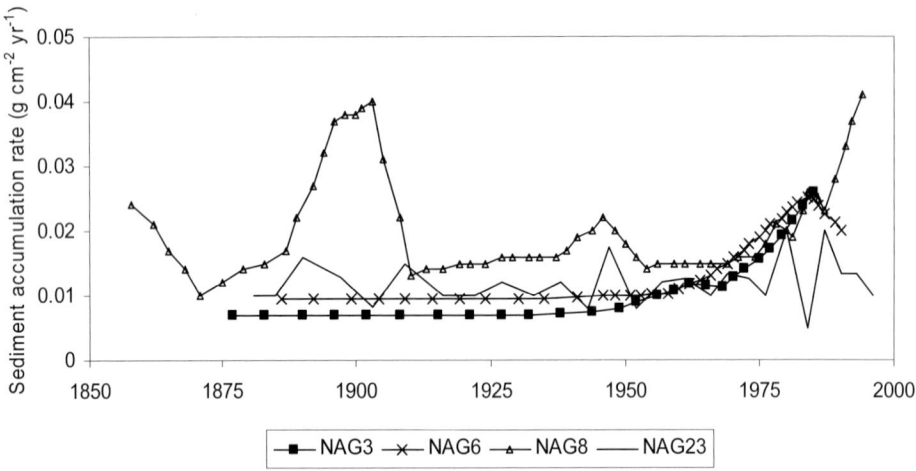

Figure 6. Sediment accumulation rates for sediment cores NAG3, 6, 8 and 23 determined by radiometric dating.

they also show sediment accumulation rates up to an order of magnitude slower than those observed at the bases of the four ^{210}Pb-dated cores.

Earlier ^{14}C dates in NAG27 show reasonable agreement with expected pollen changes (e.g., elm decline c. 5900 years BP; Dalton et al. 2005) while dating of the uppermost section of NAG27 using spheroidal carbonaceous particles (Rose unpublished data) shows that the mid-19th century is at c.20 cm depth, in good agreement with the post-1850 sediment accumulation from other cores in this area of the loch (Figure 1a). Therefore the lowest presence of SCPs (20cm) is very close to the uppermost ^{14}C date (29.6 – 30cm; 1094 – 1302 years BP). If all these data are correct, and there is no hiatus in sediment accumulation, then there is a dramatic increase in sediment accumulation at this depth although there is no other reported sediment data to suggest this has occurred in these (i.e., Dalton et al. 2001; 2005) or any other core from Lochnagar. Combining these dating approaches suggests that sediment accumulation rates in Lochnagar have increased in at least two phases. First, an increase over the estimated basal rate 5 – 6000 years BP and second, a further doubling of accumulation rate over the last 150 – 200 years.

The influence of climate change on future sediment accumulation rates

Anecdotal and modelled evidence indicate that Lochnagar now has a shorter and more ephemeral period of winter ice cover than it did in the past (Chapter 5: Thompson et al. this volume) and future climate predictions for the site suggest that the loch could be completely ice-free by the end of the 21st century (Chapter 18: Kettle and Thompson this volume). The loch water is therefore predicted to increase in temperature and this will result in a potentially longer growing season for algae and aquatic macrophytes.

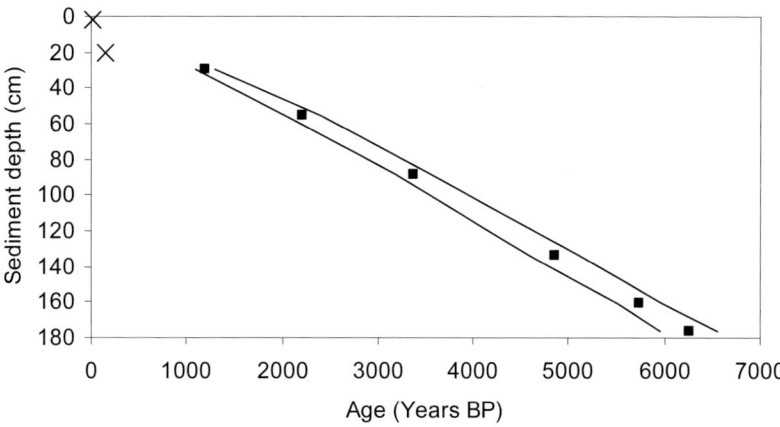

Figure 7. [14]C AMS (filled symbols) and SCP (×) dates determined for the core NAG27. Errors for ^{14}C dates are shown by the lines. (Errors for SCP dates are too small to be shown on this scale, but are ± 25 years for the lower date and ± 5 years for the upper date.)

Further, warmer temperatures would elevate the rate of mineralisation of organic matter in catchment soils releasing both carbon and nitrogen. While the lengthened growing season would increase biological demand and the elevated nitrogen may lead to phosphorus limitation in the loch, increased soil temperatures could enhance soil microbial activity leading to higher phosphorus as well as nitrogen loading from catchment soils. Particulate phosphorus input may also be elevated from increased erosion of catchment soils while recovery from acidification may reduce aluminium complexation of phosphorus (Kopáček et al. 2001) thereby also making it more biologically available. In short, the effects of climate could elevate both nitrogen and phosphorus inputs to the loch as water temperatures increase resulting in more productivity, reduced light penetration and greater autochthonous inputs to the sediment Chapter 19: Rose and Battarbee this volume).

Predicted changes in climate also include increased summer drought and higher winter rainfall and these could elevate catchment peat erosion which is already quite severe in some areas of the Lochnagar catchment (Chapter 6: Helliwell et al. this volume). This would also increase the amount of allochthonous material reaching the sediment record.

An increase in both autochthonous and allochthonous inputs to the sediment will increase the sediment accumulation rate at Lochnagar. The sediment accumulation rate at Lochnagar is known to have doubled in the last 200 years and these climate induced increases in sediment accumulation rate will increase this further to an unknown extent. As the lake infills, there will be changes in sediment distribution as the mean basin slope decreases and the sediment accumulation zone widens (Blais and Kallf 1995). If sediment deposition in shallow waters exceeds that which can be transported downslope by the available energy of the basin then sediments will accumulate in shallower areas and be permanently deposited in littoral areas (Engstrom and Swain 1986). Sediment resuspension will increase, affecting the light regime and influence the habitats and growth patterns of aquatic plants in the loch and the fauna dependent upon them.

Sediment description

Except in the southwestern area of the loch, where the sediments can be paler and coarse as a result of in-wash from the backwall, the sediment of Lochnagar is largely homogeneous in texture and colour. A description of the 1986 core, NAG3, describes the uppermost 50 cm of sediment as dark brown (Munsell Colour 10YR/2/2) and fine-grained (Patrick et al. 1989). Between 50 – 60 cm the sediment became greyish-brown and more sandy (Munsell Colour 10YR/5/2) becoming darker again lower down (Munsell Colour 10YR/3/3). Gritty, and slightly lighter coloured layers were observed at 9 – 13 cm and 29 – 30 cm and a gritty fibrous layer at 40 – 43cm. It was suggested that these changes were due to major fluxes of terrestrially-derived debris and terrestrial plant remains as a result of accelerated phases of catchment erosion (Patrick et al. 1989).

Sediment composition and sources

The sediment of Lochnagar shows considerable variation in the composition of its major fractions. The percentage water content of recent sediments varies from more than 80% (and frequently up to 90%) over the majority of the area of accumulating sediment to less than 60% in the southwest of the loch in the area dominated by coarse inputs from the back-wall (Figure 3a). Similarly, organic matter content (estimated by loss-on-ignition (LOI) at 550 °C) ranges between 20 – 30% over the majority of the basin, falling to less than 10% in the southwest corner but also with occasional areas of elevated organic content (35 - 40%) in central and northern areas (Figure 3b). These higher values are almost certainly due to the in-wash of eroded peat material from the northwest and, in particular, the northeastern shores. This hypothesis is supported by the organic content of the sediment from the outflow pool which reaches almost 80% due to the eroding peat almost entirely surrounding the area.

The water and organic matter content also vary considerably through time at sites within the loch. Figure 8 shows the temporal trends in percentage dry weight (Figure 8a) and LOI (Figure 8b) for cores NAG27 and 28 taken from the central area of the loch and, from ^{14}C dating, estimated to cover c. 6000 years (Dalton et al. 2005). The cores, taken close together, show remarkable temporal agreement in trends over their full length. Dry weights are typically 10 – 20% but there are three prominent departures from this to higher values. These occur at 155 – 170 cm, 125 – 140cm and 70 – 100cm in NAG27 dated to 5500 – 6000 BP, 4500 – 5000 BP and 2500 – 3500 BP respectively. The latter feature comprises three very distinct dry weight peaks within it, the largest reaching almost 60%. All these features are replicated in NAG28. The LOI profiles similarly show very good replication between the cores and considerable variability through time. In the lower 100 cm of the cores (80 – 180 cm in NAG27 – Figure 8b) there appears to be a cyclicity to the data with LOI peaks of c. 30% occurring at 175cm, 145 – 155 cm and 105 – 115 cm (dated to 6200 BP, 5000 – 5500

a)

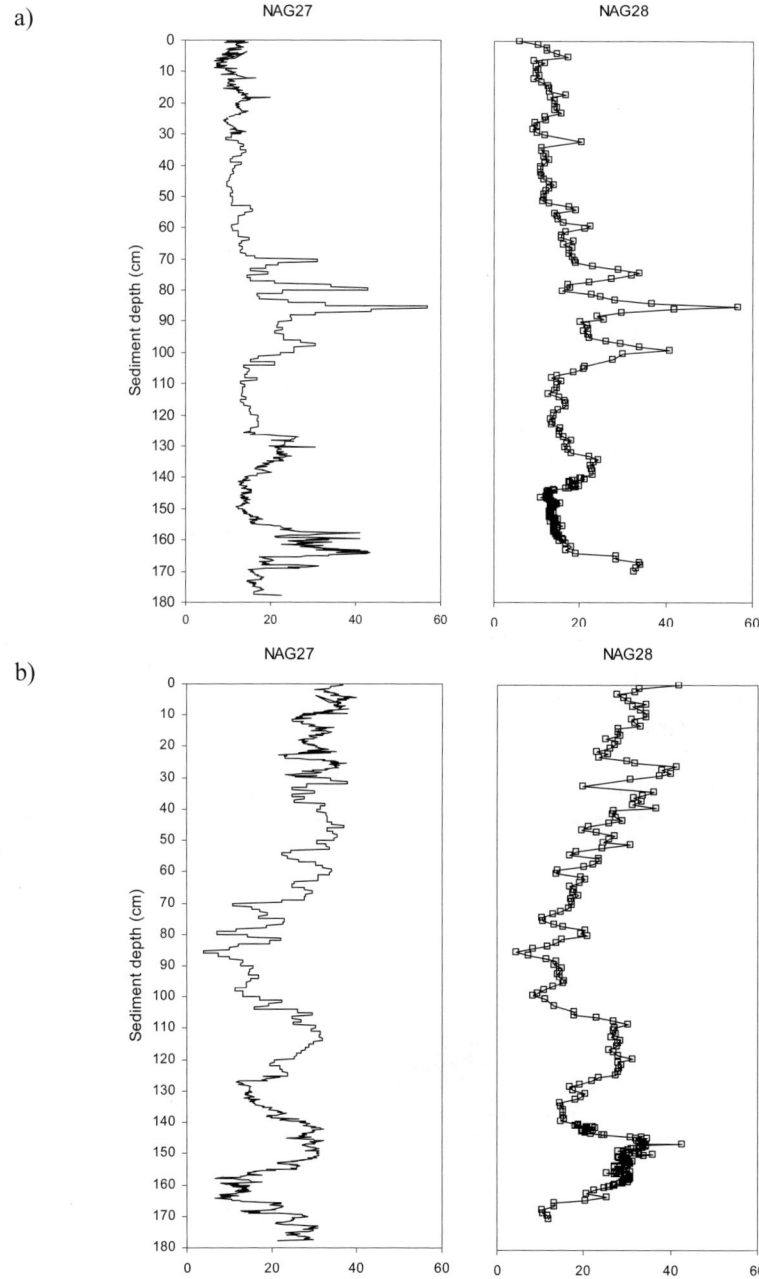

b)

Figure 8. (a) Dry weight and (b) loss-on-ignition (at 550 ºC) profiles for the two long cores NAG27 and NAG28, showing the LOI cycles in the early part of the records. Data from Dalton et al. (2001).

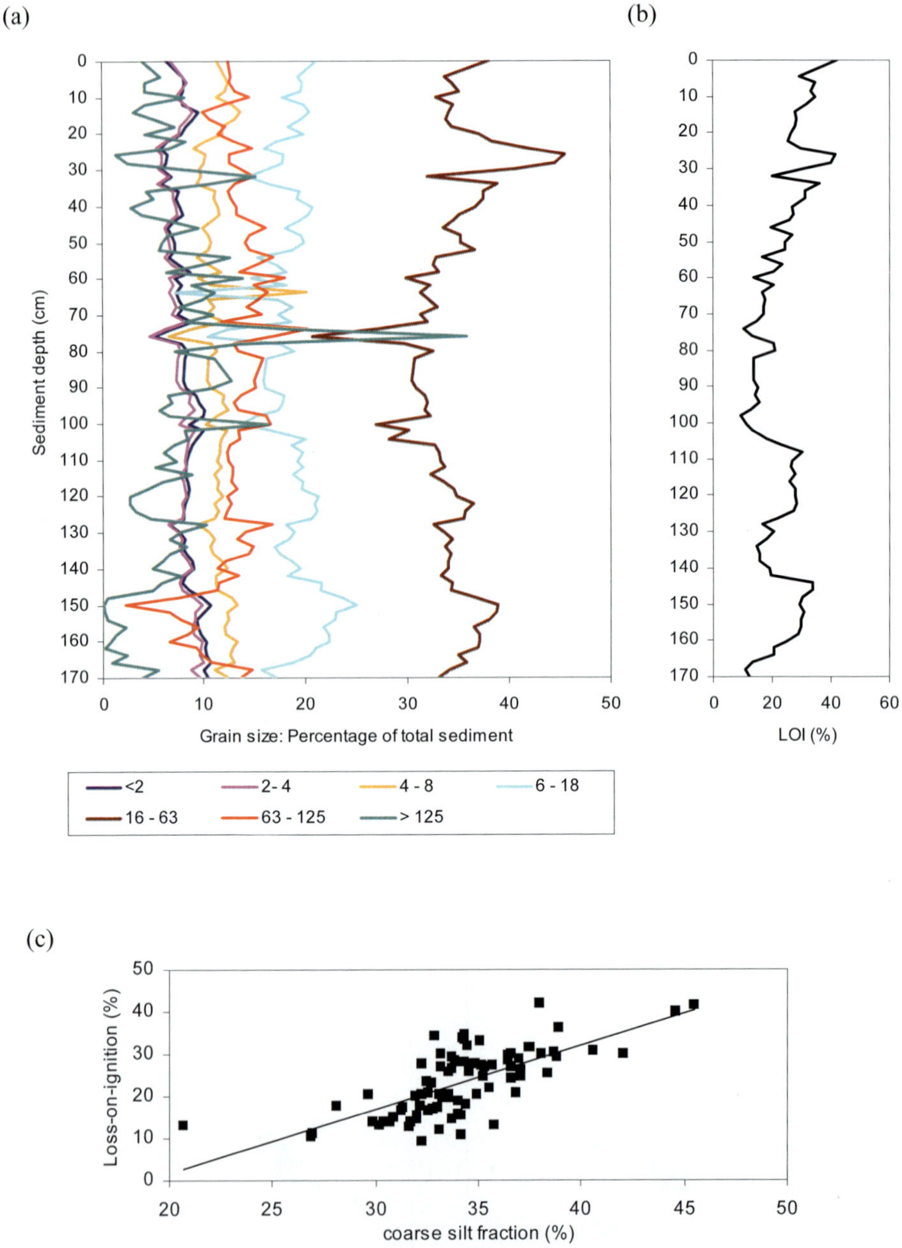

(a)   (b)

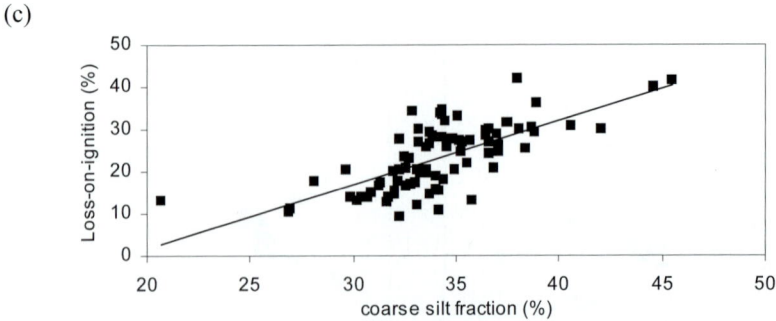

(c)

Figure 9. Particle size data for NAG28 as percentage of the total sediment showing the dominance of the coarse silt fraction; (b) the profile of organic matter content as estimated by loss-on-ignition; and (c) the relationship between loss-on-ignition (at 550°C) and the coarse silt fraction ($R^2 = 0.486$).

BP and 3750 – 4250 BP respectively; Dalton et al. 2005) and LOI 'troughs' of 10 – 15% in-between. Above 80cm, however, this cyclicity breaks down as LOI values increase to c. 30% at 60cm depth and remain at 25 – 35% for the remainder of the more recent period. This early cyclicity is intriguing as it covers c. 3000 years and Dalton et al. (2001; 2005) used a range of sediment proxies to determine that the cause was predominantly driven by catchment-derived sources. Their evidence is briefly outlined below.

C:N ratios

Carbon: nitrogen ratios vary between 12 – 20 and average around 15. These levels are consistent with those of higher plants, but not algae. Autochthonous organic matter generally has lower C:N of below 10. A change to higher C:N (c. 20) is observed at about 20cm in NAG28 (Dalton et al. 2001) which is dated to c. 1000 years BP using the Dalton et al. [14]C chronology, but possibly only occurs 150 – 200 years BP if the SCP dating for NAG27 is used and cross-correlated to NAG28. If the SCP dating is used, then the change in C:N could be due to post-Little Ice Age peat erosion from the catchment (Rick Battarbee, University College London pers. commun.).

Lipid analysis

The lipid fraction, in particular the n-alkanes and n-alkanoic acids indicate a dominantly terrestrial input. Total n-alkanes show two major periods of elevated concentration in NAG28, at 100 – 120cm and above 20 cm (4000 – 5000 yrs BP and after 500 years BP respectively (using Dalton et al. (2005) ^{14}C chronology). It is suggested that these reflect periods of peat in-wash from the catchment. The more recent n-alkane concentration increase is coincident with the step-change in C:N ratio and therefore using the SCP chronology, could also support the idea of a post-Little Ice Age enhancement in peat erosion. The alcohol and sterol fractions are dominated by dinostanol (4α,23,24-trimethyl-5α-cholestan-3β-ol) (Dalton et al 2005)   an unambiguous marker of dinoflagellate derived organic matter. The concentration of chlorins, an early diagenetic product of chlorophyll, can be used as a proxy for lake productivity and shows a peak at 40 – 60cm depth in NAG28. The ^{14}C dates would put this peak at 3 - 4000 years BP.

Particle size analysis

Particle size data covering c. 6000 years from the ^{14}C dated NAG28 core (Dalton et al. 2005) are shown in Figure 9. The dominant fraction of Lochnagar sediment throughout this period is coarse silt-sized particles (16 – 63 μm), with minor fractions of clay and sand (Figure 9a and 10a). The coarse-silt fraction includes a high component of diatom frustules and is positively correlated with LOI (Figure 9b, 9c and 10b) indicating that biogenic sources are a major component. Percentage dry weight is positively correlated to the sand fraction and both sand and clay fractions seem to be catchment derived, the clays from erosion and the sands from coarse inputs probably via winter snow falls from the corrie backwall and ice-rafting. The balance between catchment inputs and lake productivity therefore probably drive the fluctuations in dry weight and LOI as

a)

b)

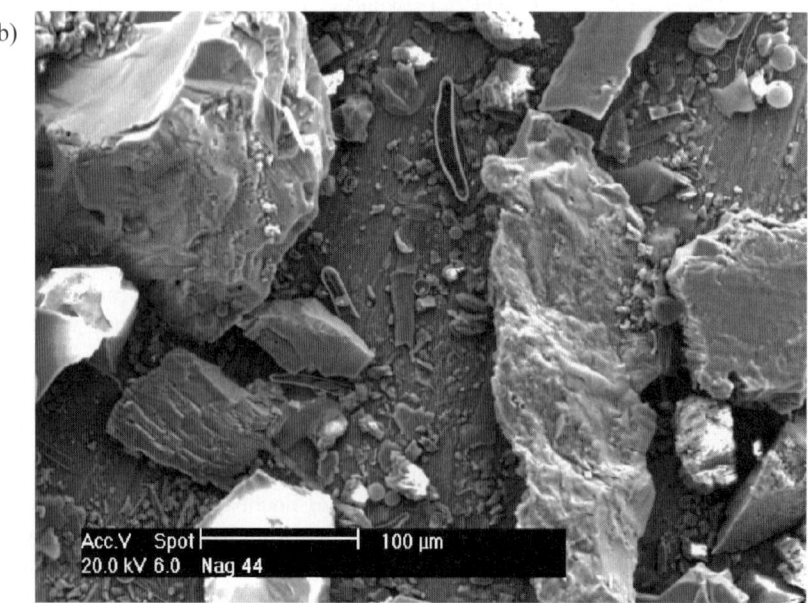

Figure 10. Scanning electron micrographs of sediment samples from (a) 16.2 – 16.4 cm and (b) 88 – 88.2 cm from NAG28. The latter is taken from a dry weight peak (see Figure 8a).

well as temporal changes in particle size spectra. These particle fractions lead to multi-peaked spectra and there is considerable variability through time from a mean particle size of c. 35 μm to 110 μm in more sandy layers (Dalton et al. 2005) for example at 74 – 76 cm (c. 3000 years BP from ^{14}C dates).

Summary

- There is considerable variability both spatially and temporally in sediment composition and distribution within Lochnagar. While there is a great deal of evidence for the catchment being a major source of sediment, the composition is driven by the balance between mineral inputs from the back-wall, phases of peat erosion and lake productivity (evidenced by the dominance of diatom frustules in the coarse silt fraction).
- Distribution of sediment within the loch is focussed on the central area of the basin, rather than the deepest point, and accumulation decreases away from this area to the limit of sediment accumulation at the base of the steep slopes or beyond the rocky littoral (although sediment temporarily settles between the boulders).
- While the reasons for this distribution pattern remain uncertain, movement of sediment towards the centre of the loch is likely to be a combination of resuspension of littoral sediment by wave action, slumping and sliding of material on steep slopes, ice rafting of snow-fall material, and water movements within the loch affected by cold melt-water input from the back-wall.
- Rates of sediment accumulation in this central area also appear to have increased both over a millennial time-scale and by a factor of 2-4 within the last 200 years. Given the composition of the sediment it seems likely that this recent increase is due to increasing catchment inputs and / or lake productivity.
- Catchment inputs are probably derived from increased peat erosion while elevated lake productivity could be due to warmer waters and a longer algal growing season.
- Future climate scenarios for Lochnagar over the 21st century (Chapter 18: Kettle and Thompson this volume) predict that the criteria for further exacerbation of both these processes will continue and hence it would appear that sediment accumulation rates at Lochnagar will also continue to accelerate over this period.

Acknowledgements

I would like to thank Mike Hughes for generating the sediment distribution figures; Peter Appleby for the radiometric dating of NAG3, 6 and 8 and for his helpful comments on sediment focussing; John Dearing and John Boyle (University of Liverpool) for the use and interpretation of their magnetics data; Roy Thompson (University of Edinburgh) for the use of his grain size data and the scanning electron micrographs and Catherine Dalton (University of Limerick) for the use of her sediment data. I am grateful to John Dearing and N. John Anderson for their comments on an earlier version of this chapter.

References

Anderson N.J. 1989. A whole-basin diatom accumulation rate for a small eutrophic lake in Northern Ireland and its palaeoecological implications. J. Ecol. 77: 926–946.

Anderson N.J. 1990a. Variability of diatom concentrations and accumulation rates in sediments of a small lake basin. Limnol. Oceanogr. 35: 497–508.

Anderson N.J. 1990b. Variability of sediment diatom assemblages in an upland, wind-stressed lake (Loch Fleet, Galloway, S.W. Scotland). J. Paleolim. 4: 43–59.

Birks H.J.B. This volume. Chapter 7. Flora and vegetation of Lochnagar – past, present and future. In: Rose N.L. (ed.) 2007. Lochnagar: The natural history of a mountain lake. Springer. Dordrecht.

Blais J.M. and Kalff J. 1993. Atmospheric loading of Zn, Cu, Ni, Cr and Pb to lake sediments: The role of catchment, lake morphometry and physico-chemical properties of the elements. Biogeochem. 23: 1–23.

Blais J.M. and Kalff J. 1995. The influence of lake morphometry on sediment focusing. Limnol. Oceanogr. 40: 582–588.

Dalton C., Battarbee R.W., Birks H.J.B., Brooks S.J. Cameron N.G., Derrick S., Evershed R.P., Peglar S.M., Scott J.A. and Thompson R. 2001. Holocene lake sediment core sequences from Lochnagar, Cairngorm Mts., Scotland – UK Final report for CHILL-10,000. Environmental Change Research Centre, University College London. Research Report No. 77. 94 pp.

Dalton C., Birks H.J.B., Brooks S.J. Cameron N.G., Evershed R.P., Peglar S.M., Scott J.A. and Thompson R. 2005. A multi-proxy study of lake-development in response to catchment changes during the Holocene at Lochnagar, north-east Scotland. Palaeogeog. Palaeoclimat. Palaeoecol. 221: 175–201.

Downing J.A. and Rath L.C. 1988. Spatial patchiness in the lacustrine sedimentary environment. Limnol. Oceanog. 33: 447–458.

Engstrom D.R. and Swain E.B. 1986. The chemistry of lake sediments in space and time. Hydrobiol. 143. 37–44.

Evans R.D. and Rigler F.H. 1980a. Measurement of whole lake sediment accumulation and phosphorus retention using lead-210 dating. Can. J. Fish. Aquat. Sci. 37: 817–822

Evans R.D. and Rigler F.H. 1980b. Calculation of the total anthropogenic lead in the sediments of a rural Ontario lake. Environ. Sci. Technol. 14: 216–218.

Flower R.J., Monteith D.T., Tyler J. Shilland E. and Pla S. This volume. Chapter 10. The aquatic flora of Lochnagar. In: Rose N.L. (ed.) 2007. Lochnagar: The natural history of a mountain lake. Springer. Dordrecht.

Håkanson J. 1977. The influence of wind, fetch and water depth on the distribution of sediments in Lake Vänern, Sweden. Can. J. Earth Sci. 14: 397–412.

Hall A.M. This volume. Chapter 4. The shaping of Lochnagar: Pre-glacial, glacial and post-glacial processes. In: Rose N.L. (ed.) 2007. Lochnagar: The natural history of a mountain lake. Springer. Dordrecht.

Helliwell R.C., Lilly A. and Bell J. This volume. Chapter 6. The development, distribution and properties of soils in the Lochnagar catchment and influence on surface water chemistry. In: Rose N.L. (ed.) Lochnagar: The natural history of a mountain lake. Springer.

Hilton J., Lishman J.P. and Allen P.V. 1986. The dominant processes of sediment distribution and focusing in a small, eutrophic, monomictic lake. Limnol. Oceanog. 31: 125–133.

Hughes M. This volume. Chapter 2. Physical characteristics of Lochnagar. In: Rose N.L. (ed.) Lochnagar: The natural history of a mountain lake. Springer.

Kettle H. and Thompson R. This volume. Chapter 17: Future climate predictions. In: Rose N.L. (ed.). Lochnagar: The natural history of a mountain lake. Springer. Dordrecht.

Kopáček J., Ulrich K-U., Hejzlar J., Borovec J. and Stuchlik E. 2001. Natural inactivation of phosphorus by aluminium in atmospherically acidified water bodies. Wat. Res. 35: 3783–3790.

Lehman J.T. 1975. Reconstructing the rate of accumulation of lake sediment: The effect of sediment focussing. Quat. Res. 5: 541–550.

Monteith D.T., Evans C. and Dalton C. This volume. Chapter 14. The acidification of Lochnagar and prospects for recovery. In: Rose N.L. (ed.) Lochnagar: The natural history of a mountain lake. Springer.

Muir D.C.G. and Rose N.L. This volume. Chapter 16. Persistent organic pollutants in the sediments of Lochnagar. In: Rose N.L. (ed.) 2007. Lochnagar: The natural history of a mountain lake. Springer. Dordrecht.

Patrick S.T., Flower R.J., Appleby P.G., Oldfield F., Rippey B., Stevenson A.C., Darley J. and Battarbee R.W. 1989. Palaeoecological evaluation of the recent acidification of Lochnagar, Scotland. Palaeoecology Research Unit, University College London. Research Paper No. 34. 57 pp.

Rose N.L. and Battarbee R.W. This volume. Chapter 19. Past and future environmental change at Lochnagar and the impacts of a changing climate. In: Rose N.L. (ed.) 2007. Lochnagar: The natural history of a mountain lake. Springer. Dordrecht.

Rose N.L. and Yang H. This volume. Chapter 17. Temporal and spatial patterns of spheroidal carbonaceous particles (SCPs) in sediments, soils and deposition at Lochnagar. In: Rose N.L. (ed.) 2007. Lochnagar: The natural history of a mountain lake. Springer. Dordrecht.

Rose N.L., Harlock S. and Appleby P.G. 1999. Within-basin profile variability and cross-correlation of sediment cores using the spheroidal carbonaceous particle record. J. Paleolim. 21: 85–96.

Rowan D.J., Rasmussen J.B. and Kalff J. 1992. Estimating the mud deposition boundary depth in lakes from wave theory. Can. J. Fish. Aquat. Sci. 49: 2490–2497.

Rowan D.J., Cornett R.J., King K. and Risto B. 1995a. Sediment focusing and ^{210}Pb dating: A new approach. J. Paleolim. 13: 107–118.

Rowan D.J., Rasmussen J.B. and Kalff J. 1995b. Optimal allocation of sampling effort in lake sediment studies. Can. J. Fish. Aquat. Sci. 52: 2146–2158.

Shteineman B.S. and Parparov A.S. 1997. An approach to particulate matter transfer studies in littoral zones of lakes with changing morphometry. Wat. Sci. Res. 36: 199–205.

Thompson R., Kettle H., Monteith D.T. and Rose N.L. This volume. Chapter 5. Climate and weather. In: Rose N.L. (ed.) Lochnagar: The natural history of a mountain lake. Springer.

Tipping E., Yang H., Lawlor A.J., Rose N.L. and Shotbolt L. This volume. Chapter 15. Trace metals in the catchment, loch and sediments of Lochnagar: Measurements and modelling. In: Rose N.L. (ed.) 2007. Lochnagar: The natural history of a mountain lake. Springer. Dordrecht.

9. HYDROLOGY AND HYDROCHEMISTRY OF LOCHNAGAR

ALAN JENKINS (jinx@ceh.ac.uk), NICK REYNARD,
MIKE HUTCHINS, MURIEL BONJEAN and MARTIN LEES
Centre for Ecology and Hydrology
Wallingford
Oxfordshire OX10 8BB
United Kingdom

Key words: chemistry, hydrograph, rainfall, runoff, snowmelt

Introduction

Lochnagar is located to the southeast of the Cairngorms in the headwaters of the River Dee which flows eastwards to the North Sea near Aberdeen (Chapter 1: Rose this volume). The Dee is an important salmon fishery and the fish population is sustained by seasonally high spring flows usually associated with snowmelt in the surrounding mountains. The Cairngorms represent the largest area of land above 1000 m in the UK and are characterised by significant annual snowfall and accumulation. Average rainfall for the region is 1300 mm and approximately 20% of this falls as snow in an average year (National River Flow Archive).

Hydrology

The loch has a surface area of 10.4 ha, mean depth of 8.9 m and maximum depth of 26 m to give an estimated volume of 0.926 m^3 x 10^6 (Chapter 2: Hughes this volume). This gives a mean turnover time of c. 9 months. Flow monitoring was started at the outflow of Lochnagar in April 1999 when a pressure transducer was installed in a stilling pipe into the main outflow channel. The water depth (stage) is monitored at 15-minute intervals. A rating curve has been constructed to convert stream stage (cm) into discharge (m^3 sec^{-1}) using salt dilution gauging (Figure 1). The channel at the outflow comprises loose rocks and boulders and is unstable, however, and as a result the rating curve must be considered to be inaccurate, particularly at higher flows. This precludes an assessment of water balance. Nevertheless, the detailed flow record can be broadly interpreted to infer the key rainfall-runoff characteristics of the catchment.

177

N.L. Rose (ed.), Lochnagar: The Natural History of a Mountain Lake
Developments in Paleoenvironmental Research, 177–198.
© 2007 *Springer*.

Rainfall monitoring, collected via a tipping bucket rain gauge, was started in 1996 and is available for the whole period of flow record. It is not, however, possible to estimate the annual water balance for the site because a significant proportion of the precipitation falls as snow at Lochnagar and this is not measured routinely. In addition, deposition of cloud water (occult deposition) is likely to be significant at the altitude of Lochnagar. These problems combined with the potential errors in the discharge rating-curve would limit the value of a water balance assessment.

Five years of flow data represents too short a record for an assessment of hydrological regime and so this chapter puts the Lochnagar data in the regional context of long-term records (greater than 30 years) from other flow gauges on the River Dee.

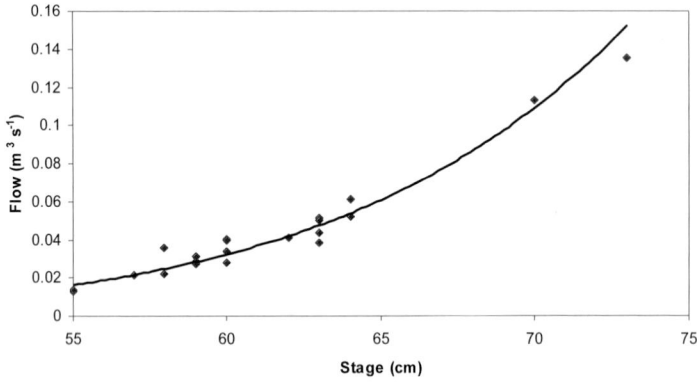

Figure 1. The rating curve for the outflow stream from Lochnagar. Equation for the line is $y = 3E\text{-}16x^{7.894}$ ($R^2 = 0.9335$). Discharge was measured using salt dilution gauging.

The National River Flow Archive (http://www.nerc-wallingford.ac.uk/ceh/nrfa) contains data from two flow gauges on the Dee downstream of Lochnagar; the Girnock Burn and the River Muick (Figure 2), both of which drain catchments which are significantly bigger than the c.1 km^2 of the Lochnagar catchment. The Girnock Burn gauged about 5 km west of Ballater has a catchment area of 30 km^2. The River Muick is gauged just south of Ballater draining a catchment of 110 km^2. The headwaters drain the southern and eastern hills around Lochnagar and there is storage in Loch Muick 6 km southeast of Lochnagar.

The flow duration curves from these two sites are steep and demonstrate the regional flow regime to be essentially rather flashy, that is, a rapid response to precipitation input (Figure 3). In this respect, the Girnock Burn is marginally steeper than the River Muick being characterised by higher peak flows and lower baseflows. This is likely to reflect the influence of Loch Muick on the river which acts to dampen the downstream flow response to rainfall input. This effectively provides storage for high flows and sustains low flows. Annual total flow per unit catchment area is considerably higher in the Muick than in the Girnock Burn reflecting higher altitude, higher rainfall and lower evapotranspiration. The flow duration curve for Lochnagar is similar to that of the

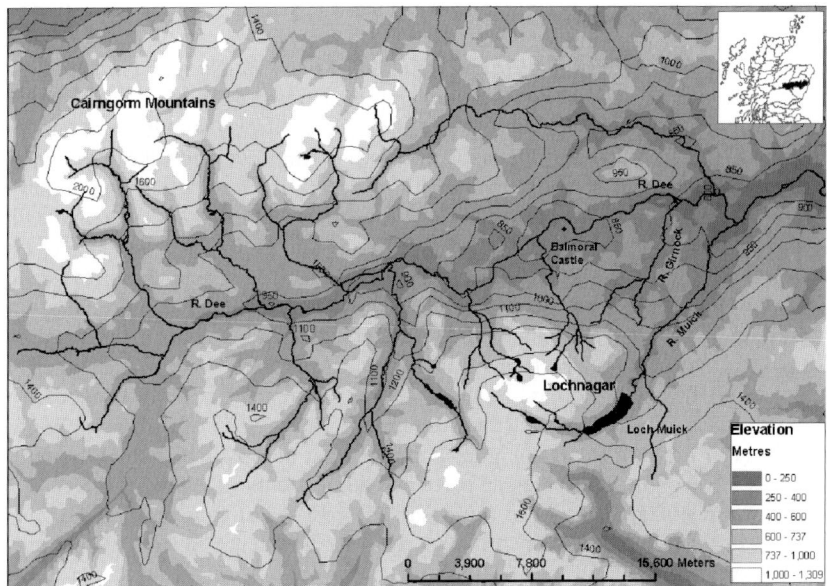

Figure 2. Map showing the location of Lochnagar, the Girnock Burn and Muick catchments in relation to the River Dee.

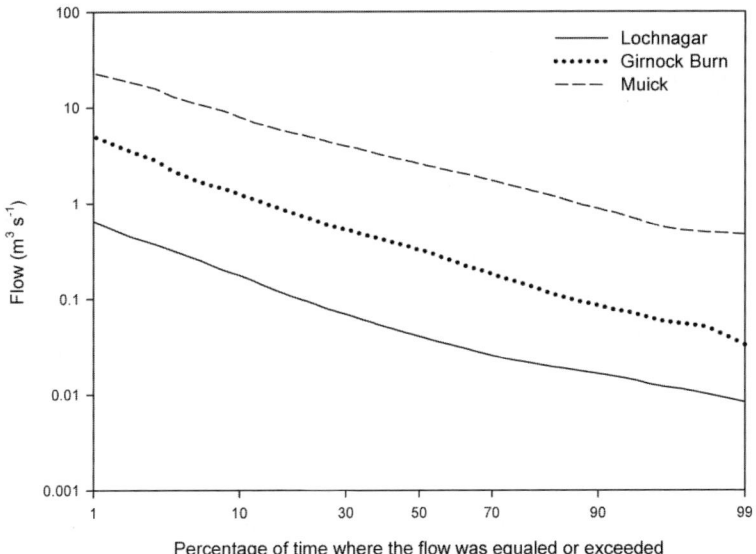

Figure 3. Flow duration curves for Lochnagar, Girnock Burn and River Muick for the period 1999 – 2004.

River Muick as the storage effect of the loch reduces the steepness of the flow duration curve.

Time series of mean daily flow data from the River Muick and Girnock Burn demonstrates the strong seasonality in the flow regime (Figure 4). Winter and spring are characterised by baseflows which are significantly higher than in summer and autumn. This reflects input from the melting snowpack. Superimposed upon these seasonally high baseflows are large events associated with rain storms. These are also common in summer and autumn and indicate the flashy response of the catchments. Note the significantly higher flows generated in the larger catchment of the River Muick.

The hydrological regime at Lochnagar reflects this regional picture of strong seasonality in flow regime but does not demonstrate the sustained winter and spring flows observed on Girnock Burn and River Muick (Figure 4). Periods of sustained and significant snowmelt can be clearly identified in late April/early May of each year but superimposed on this are short duration (1-2 days) extremely high flows occurring throughout the year. These represent the response to short-term high intensity rainfall events. Lochnagar drains a small catchment via a number of mostly perennial first order streams. Soils are shallow and scree and bare rock dominate the catchment surface. This promotes a relatively rapid response to rainfall, despite the storage in the lake, since little water is stored in the catchment although the loch itself tends to dampen the runoff response to some extent.

Storm hydrographs can be characterised into essentially two types; those generated by snowmelt and those generated by rainfall. This hydrological response has also been documented at other high altitude sites in the Cairngorms (e.g., the Allt a' Mharcaidh; Jenkins 1989). This picture is complicated, however, by the combination of rain-on-snow events during the spring melt season. Snowmelt generated hydrographs are typified by a slow rising limb compared to rainfall generated events and a slow recession (Figure 5). Snowmelt generated events also tend to have lower peak flows than rainfall generated events. The highest flows are associated with extremely intense rainfall which generates rapid flow on the relatively impervious catchment surfaces.

The event occurring on 29th January to 11th February 2000 (Figure 5a), began as a typical snowmelt event with flow increasing slowly over 2-3 days, 29th January to 1st February. Prior to this the catchment held significant snow accumulation and temperatures were low preventing thaw conditions. As a consequence, prior to 29th January the flow was very low and stable. In response to slowly increasing air temperature and rainfall of 10.2 mm on 29th January, the flow increased slowly as a melt began. Rainfall of 26 mm over the 31st January to 1st February was sufficient to prolong the melt and sustain the flow above baseflow level. On the 4th and 5th February, however, only 1.6 mm rainfall was recorded but the flow increased slowly to reach a peak of c.380 l s^{-1} in response to a sharp rise in temperature. Further rain over the next week prolonged the hydrograph but the general and slow recession is a response to the reduced meltwater contribution as the snowpack volume reduced.

In contrast, the event in August 2001, (Figure 5b) is a typical rainfall generated hydrograph. 68 mm of rainfall on 19th and 20th August caused flow to rise rapidly to reach a peak within 12 hours. The recession limb is steep and baseflow returns to its pre-storm level within five days. A further 11mm of rain on the 22nd August causes only a slight rise in flow and the recession effectively continues unabated.

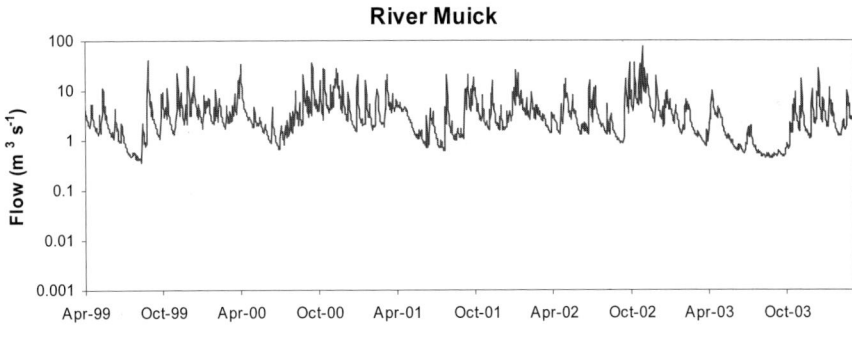

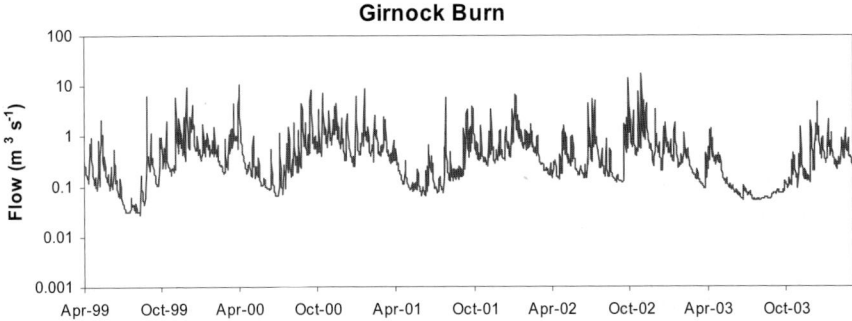

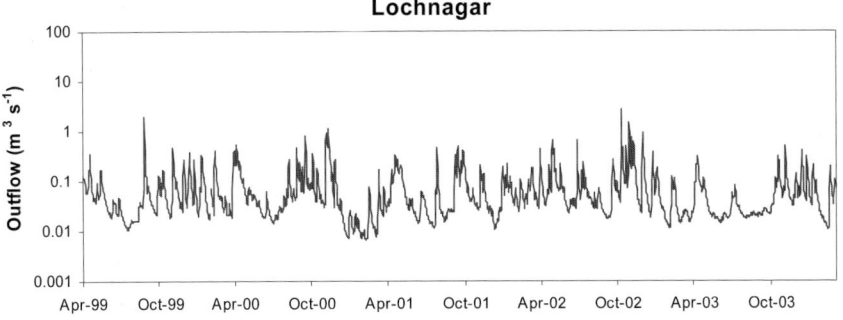

Figure 4. Mean daily runoff for Lochnagar, Girnock Burn and River Muick for the period April 1999 to April 2004.

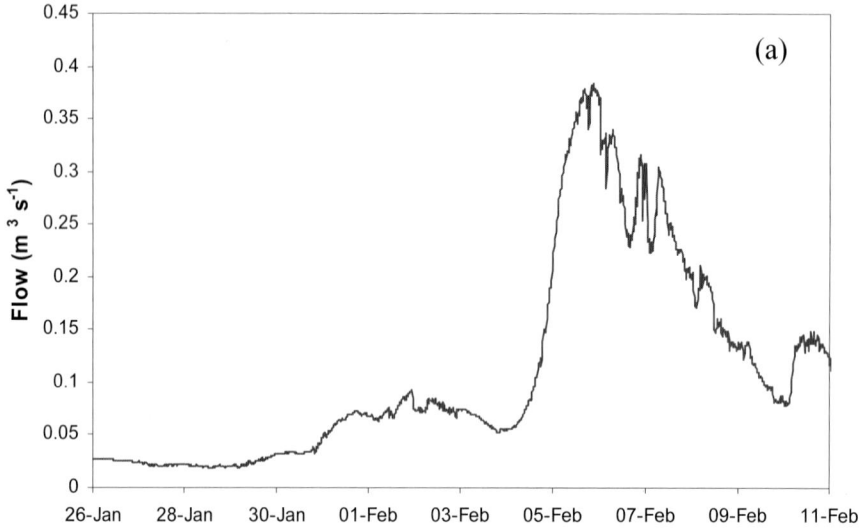

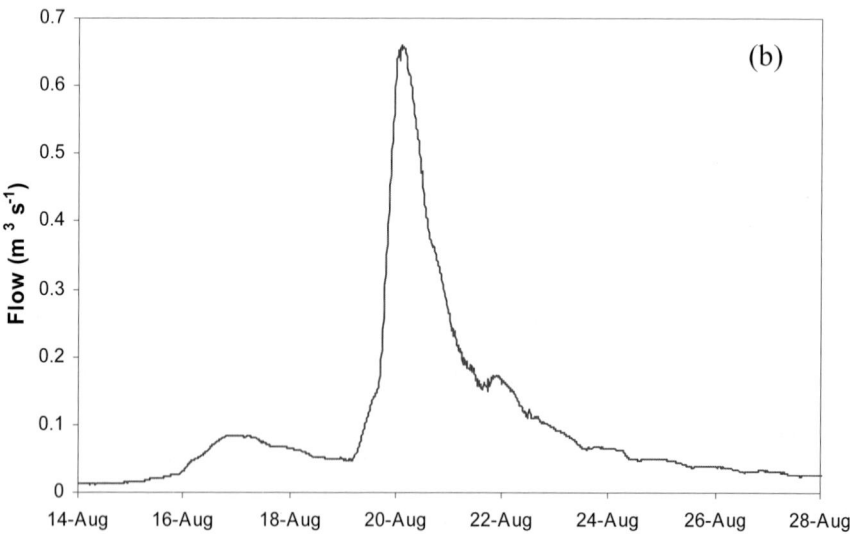

Figure 5. Characteristic storm hydrographs at Lochnagar compiled from 15 minute flow data. (a) shows a snowmelt generated event during Jan/Feb 2000 and (b) a rainfall generated event during August 2001.

Over the 5 year period 1999 – 2004, peak flows greater than the 98th-percentile flow (0.45 m^3 s^{-1}) were recorded on 25 distinct occasions. Of these 25 storm episodes, ten occurred in the autumn months (September to November), seven in the winter (December to February), six in the spring (March to May) and two in the summer (June to August).

Hydrochemistry

A high altitude loch such as Lochnagar with a pristine catchment area and no land use activity except low density deer grazing might be expected to be oligotrophic with chemical characteristics that reflect the major ion content of the local bedrock. The distance from the west coast precludes a significant sea-salt influence on sodium and chloride concentrations although some impact is inevitable and has been observed in continuous monitoring of a 290 km^2 sub-catchment of the River Dee (Jarvie et al. 2001). The water chemistry of the loch, however, has been severely impacted by historically high levels of deposition of oxides of sulphur and oxides and reduced forms of nitrogen from the atmosphere (Jenkins et al. 2001; Chapter 14: Monteith et al this volume). These acidic compounds are emitted mainly in northern England and the Midlands but with contribution from southern England and Europe (Chapter 13: Rose et al. this volume) and are transported north on the prevailing winds. Significant industrial emissions began during the industrial revolution in the mid-1800s and for sulphur, reached a peak during the early 1970s whereafter they have decreased sharply almost to background level (RGAR 1990). For nitrogen, the peak was reached in the 1990s and there has been no appreciable decline to present day. These 'acidic' compounds are deposited onto the catchment surfaces. Since the soils at Lochnagar are shallow and the granitic bedrock weathers very slowly, there is only a small pool of base cations available to buffer (neutralise) the acid anions (SO_4^{2-} and NO_3^-) from the atmosphere. As a result the deposited acid anions are transported to the loch in conjunction with 'acidic' cations (hydrogen and aluminium). The net result is to acidify the surface water. This process can take many decades depending on the deposition flux of the acid anions and the buffering capacity of the catchment with respect to base cation availability.

As a result of many decades of relatively high atmospheric deposition of sulphur and nitrogen and catchment soils that are no longer capable of providing base cations to buffer the anions, the mean chemistry of the loch today (Table 1) shows a relatively low pH and acid neutralising capacity (ANC). This would be marginally capable of supporting a brown trout population according to empirical studies (Lien et al. 1996). pH and ANC reached their lowest observed levels in the early 1990s, coinciding with a short-lived period of elevated aluminium concentrations. Aluminium becomes markedly more soluble as pH drops below 5.5. In many studies such elevated concentrations have been linked directly to ecological damage (Baker et al. 1990).

The current chemistry status has substantially improved, however, since the early 1980s and significant trends are apparent (Figure 6). Further data from the UK Acid Waters Monitoring Network (Shilland et al. 2004), whereby 3-monthly samples have been collected since 1988, have been formally analysed for trends (Table 2). Most notable is the decline in excess sulphate (that fraction of the total sulphate which is

Table 1. Mean lake chemistry characteristics 2003/04. Data from the AWMN (Shilland et al. 2004). Units are μeql^{-1} except for Al species and DOC (μgl^{-1}) and pH (log).

	Mean 4/2003-3/2004	Std.dev 4/2003-3/2004
pH	5.51	0.09
Acid Neutralising Capacity	5.87	1.23
Calcium	29.12	1.89
Magnesium	33.33	0.68
Sodium	84.78	4.35
Potassium	5.26	0.44
Total Soluble Aluminium	0.59	0.25
Soluble Labile Aluminium	0.31	0.09
Chloride	78.87	6.09
Sulphate	46.35	1.04
Non-seasalt Sulphate	38.07	0.51
Nitrate	22.04	2.65
Silica	74.64	11.96
Dissolved Organic Carbon	97.92	7.98

Table 2. Seasonal Kendall (SK) trend statistics slope estimate and significance level (P) for 1988 – 2004. Units for SK are $\mu eql^{-1}\ yr^{-1}$, except soluble aluminium, soluble labile aluminium and dissolved organic carbon ($\mu mol\ l^{-1}\ yr^{-1}$).

	SK	P
pH	0.01	0.17
Acid Neutralising Capacity	0.31	0.03
Calcium	-0.01	0.03
Magnesium	0.00	0.06
Sodium	-0.02	0.03
Potassium	-0.01	0.00
Total Soluble Aluminium	-0.84	0.16
Soluble Labile Aluminium	-0.29	0.42
Chloride	-0.04	0.10
Sulphate	-0.05	0.00
Non-seasalt Sulphate	-0.05	0.00
Nitrate	0.01	0.03
Silica	0.00	0.79
Dissolved Organic Carbon	0.05	0.00

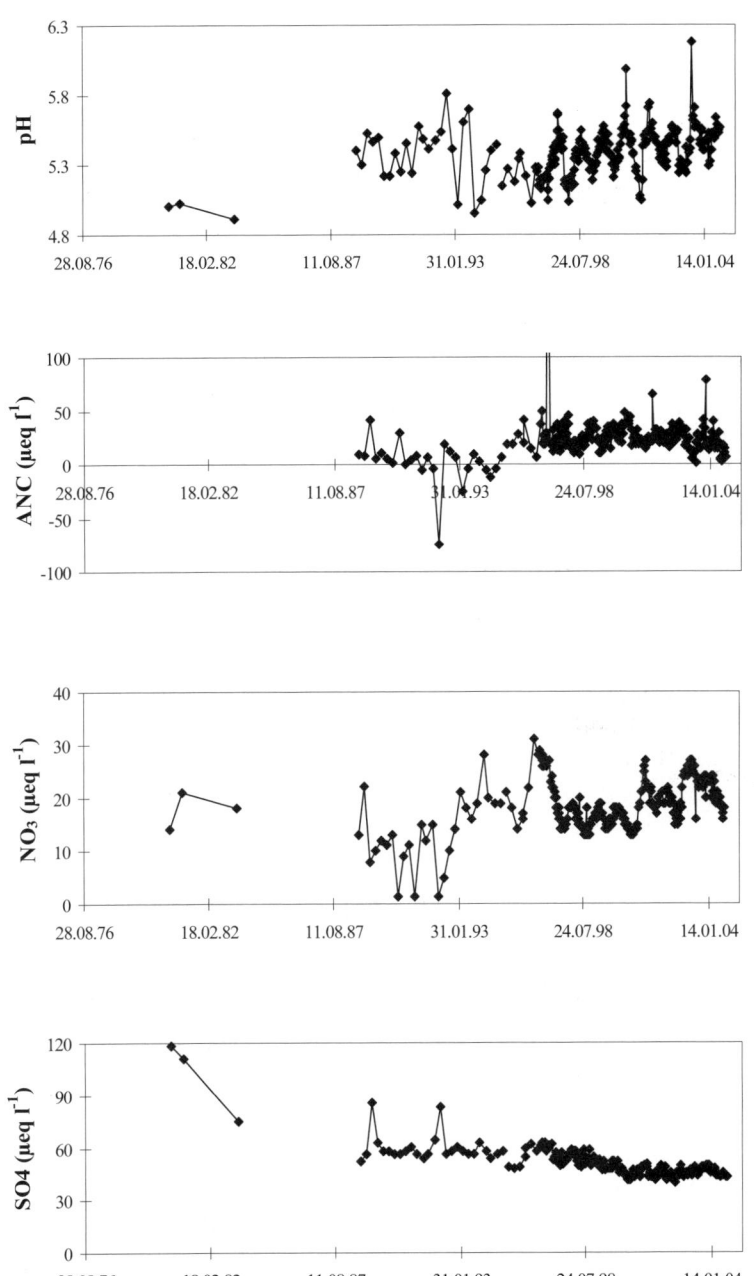

Figure 6. Water chemistry data for Lochnagar. Samples are taken at 3-monthly interval since 1988 (as part of the UKAWMN) and at 2-weekly interval since 1996.

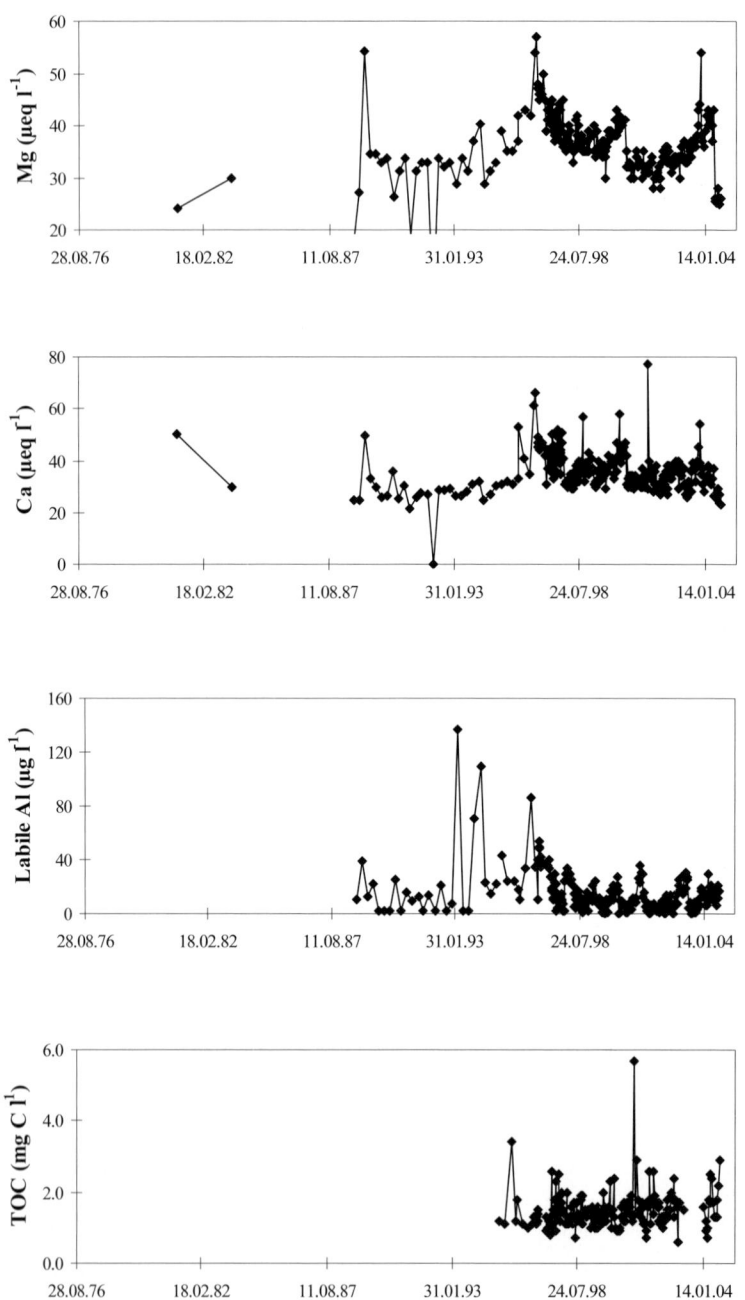

Figure 6 (continued). Water chemistry data for Lochnagar. Samples are taken at a 3-monthly interval since 1988 (as part of the UKAWMN) and at a 2-weekly interval since 1996.

anthropogenic and not derived from natural sea-salts). This undoubtedly results from the decline in sulphur deposition resulting from reduced sulphur emission as international agreements have taken effect. This has prompted an increase in pH and ANC since the early 1990s and the loch may be considered as showing signs of chemical recovery from the impact of acidification (Chapter 14: Monteith et al. this volume).

The concentration of major ions in the lake show different relationships with flow (Figure 7). Sulphate is essentially constant at all flow levels and probably reflects the 'damping' influence of the loch. Calcium, however, tends towards its minimum concentration at high flows indicating a possible dilution effect. On the other hand, at very low flow, calcium concentration ranges from the levels associated with high flows to significantly higher levels which might be expected to be associated with baseflow conditions (Neal et al. 1997). This would seem to indicate a complex set of hydrological pathways and storages operating. Acid Neutralising Capacity and pH also demonstrate a similar behaviour to calcium but with yet greater variability at very low flows.

Much of the scatter observed in these flow and ion relationships, however, is removed when the summer (May – October) samples are split from the winter (November – April) samples (Figure 8). In summer, low flows tend to be less acidic than high flows. In winter, however, pH is generally always below 5.5 irrespective of the flow level and this may well indicate the influence of meltwater contribution from the snowpack which in most years is significantly large and persistent into the early summer. This further indicates that the meltwater moves to the loch without significant interaction with catchment soils, perhaps as a result of frozen ground conditions.

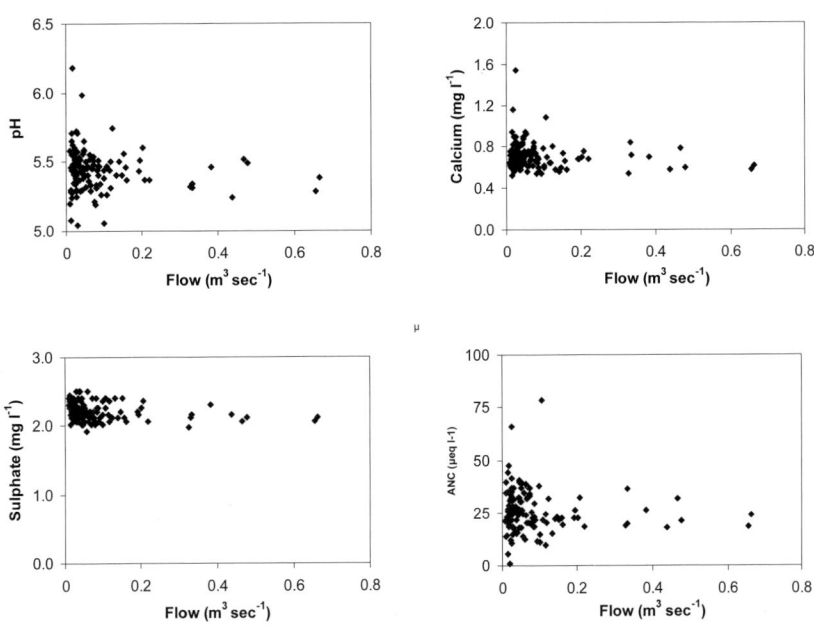

Figure 7. Relationship between water chemistry and mean daily flow 1999 – 2004.

Of some current concern is the apparent increase in nitrate concentration since the early 1990s (Curtis et al. 2005) despite an observed decline in nitrogen deposition in the region (Fowler et al. 2005). Indeed, data suggest that this trend has suppressed and delayed the pH recovery expected during the 1990s as a consequence of reduced sulphur deposition and lower levels of excess sulphate in surface water. A number of explanations have been considered for this since the understanding of the processes that govern nitrogen leakage to surface waters and how these change over time in response to nitrogen deposition and climate is not known. The observed response may be considered to represent an indication of nitrogen saturation whereby the processes which immobilise nitrogen in the terrestrial catchment (primarily plant and microbial uptake) reduce in intensity as the available nitrogen in the system builds up to a level where increased nitrogen is not required or cannot be utilised by the biomass (Stoddard 1994). If nitrogen deposition continues, therefore, a greater proportion will be released to the surface water. The lack of seasonality in the NO_3 concentrations further indicates that nitrogen saturation has occurred although the relatively sparse soil cover and lack of significant vegetation would be unlikely to immobilise a significant fraction of the nitrogen deposition even during the growing season. It is possible, therefore, that the nitrogen concentration simply reflects changes in input from year to year.

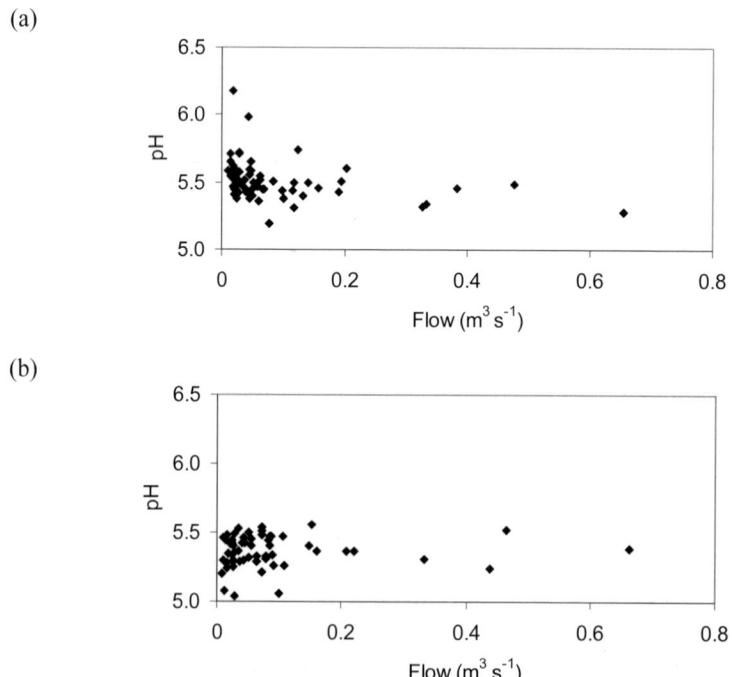

Figure 8. Seasonal flow and pH relationship for (a) summer represented by May to October and (b) winter represented by November to April.

A further possibility is that the increased nitrogen is a response to changing climatic conditions (Wright and Jenkins 2001; Evans and Monteith 2001). Evidence for generally warmer conditions on average throughout the year is widespread from many locations in the UK. At Lochnagar, this may well lead to an increase in the mineralisation rate of organic matter in catchment soils releasing both carbon and nitrogen. The observed increase in surface water DOC since the mid-1990s tends to lend weight to the climatic explanation for the increase in nitrogen. Experimental manipulation of temperature to a headwater catchment in southern Norway, the CLIMEX project (Wright and Jenkins 2001), led to increased carbon and nitrogen concentrations in run-off and increased mineralisation was directly measured as a concomitant response. Some of the increase in DOC, however, is likely to be due to decreasing sulphate levels, to which there is a negative correlation across many sites in the AWMN (Evans et al. 2005). The solubility of DOC in stream water increases in response to changes brought about by reductions in acid deposition.

The future?

The climate change scenarios in this section are based on the UKCIP data published in 2002 (Hulme et al 2002). These data represent a set of scenarios generated from the Hadley Centre global and regional climate models. Scenarios are available for 30-year periods centred on the 2020s, 2050s or the 2080s. Assumptions about future greenhouse gas emissions are encapsulated within the scenarios and presented as low, medium-low, medium-high or high emissions. These four emissions scenarios correspond to four of the Intergovernmental Panel on Climate Change (IPCC) SRES (Special Report on Emissions Scenarios) futures: B1, B2, A2, A1F1 (IPCC 2000), although they were actually generated by re-scaling the A2 scenario making the spatial patterns of change the same for each of the four UKCIP02 scenarios (Chapter 18: Kettle and Thompson this volume). For the purposes of this chapter, an illustration of possible changes to the average climate of Lochnagar is given using the UKCIP02 medium-high scenario for the 2050s.

The precipitation of Lochnagar is currently around 1300 mm as an average annual total for the 1961-1990 period (NRFA). Despite uncertainties over future predictions of rainfall at Lochnagar (Chapter 18: Kettle and Thompson this volume), under the UKCIP02 medium-high scenario for the 2050s this annual total changes little (decreasing by just 2%) although this masks a significant change in the monthly distribution (Figure 9). The precipitation totals are predicted to increase in winter (November to March), but monthly averages decrease in the spring and summer. These decreases are particularly marked during the June to September period. The changes in temperature show a more consistent pattern through the year, with average temperature increasing in all months by between 1.2 and 2°C.

The estimates of current and future potential evaporation (PE) increase in all months (Figure 10), driven primarily by the increased temperature, but also by changes in the other climate variables used to model PE (radiation, wind speed and humidity). The increases are small, in absolute terms, during the winter, becoming quite significant during the May to September period. While these figures represent potential evaporation, a catchment as small as that of Lochnagar, with such a high ratio between

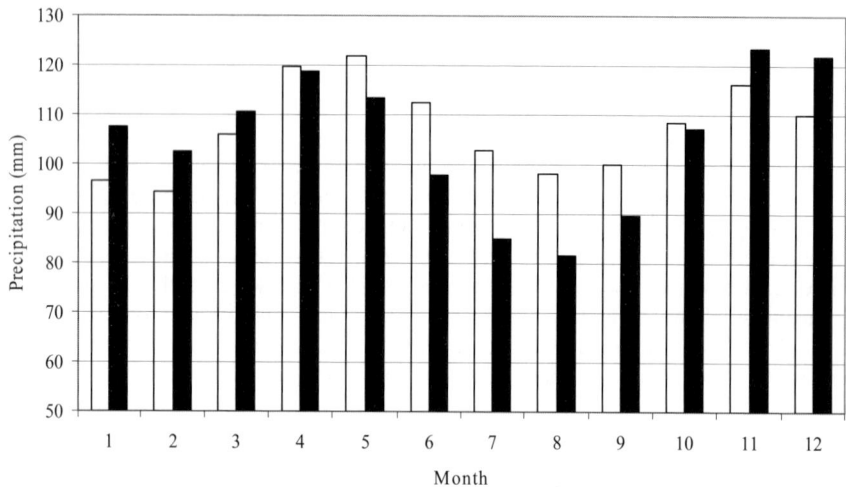

Figure 9. Estimates of 1961-1990 (open bars) and future precipitation (solid bars) derived from the UKCIP02 medium-high scenario for the 2050s.

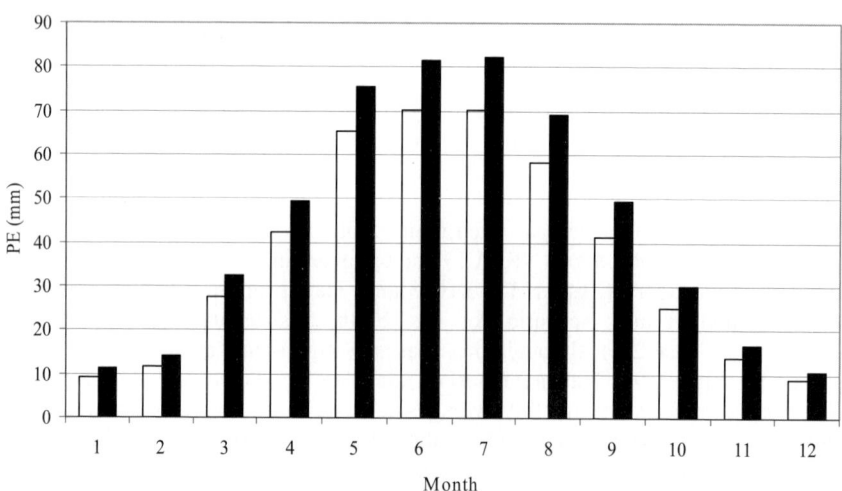

Figure 10. Estimates of MORECS 1961-1990 (open bars) and future (solid bars) potential evaporation derived from the UKCIP02 medium-high scenario for the 2050s.

catchment area and percentage coverage of open water in the loch, it is likely that these predicted increased potential losses would be close to actual increases in evaporation.

The monthly figures for changes in precipitation, temperature and PE are listed in Table 3. The UKCIP02 medium-high emissions scenario for the 2050s has been used as a guide to the type of changes in climate that might be seen in Lochnagar. These scenarios are based on the output from just one global climate model and so do not capture the full range of uncertainty in future predictions of climate (see Chapter 18: Kettle and Thompson this volume). Other models will undoubtedly give different values, sometimes in terms of the sign of change as well as the magnitude. For example, for the UK, the Hadley Centre global climate model is one of the drier models, predicting decreases in summer rainfall of over 20% by the 2080s. The Australian (CSIRO) and the Canadian model (CCCma) actually predict precipitation increases for the UK in the summer. In terms of temperature change, the Hadley Centre model falls in the middle, but all models predict increases, varying from 1.5°C to over 5°C (Hulme et al. 2002). Indeed, of all the various sources of uncertainty involved when considering the possible of impacts climate change in the future, it has been estimated that uncertainty due to the choice of global climate model might be largest single source (Jenkins and Lowe 2003; Prudhomme et al. 2005).

Table 3. Baseline (1961-1990) and future change in precipitation, temperature and potential evaporation. The changes are derived from the UKCIP02 medium-High scenarios for the 2050s.

	Jan	Feb	Mar	Apr	May	Jun	Jul	Aug	Sep	Oct	Nov	Dec
Precip (mm)	97	95	106	120	122	113	103	98	100	109	116	110
Precip Change (%)	11	9	4	-1	-7	-13	-17	-17	-10	-1	6	11
Temp (°C)	2.1	2.2	3.1	5.2	8.2	10.9	12.3	12	10	7.1	4.4	2.7
Temp Change (°C)	1.2	1.3	1.4	1.5	1.6	1.6	1.8	2.0	2.0	1.9	1.6	1.4
PE (mm)	9	12	27	42	65	70	70	58	41	25	14	9
PE Change (%)	23	21	19	17	16	16	17	19	19	19	21	24

In terms of runoff from Lochnagar, perhaps one of the profound impacts of climate change will be due to the partitioning of precipitation between rain and snow. Increased temperatures will mean less precipitation falls as snow and the average temperature increase suggests that snow is less likely to accumulate over long periods of time (Chapter 18: Kettle and Thompson this volume). This will result in lower spring high flow peaks as a result of reduced snowmelt whilst on average, flows throughout the winter will increase. This flow regime change has been simulated in many northern catchments with a spring high flow season (Reynard et al. 2004, Kay et al. 2006).

Through the summer, rainfall is predicted to decrease and PE predicted to increase. This combination will reduce flows out of Lochnagar into November. To understand more of the possible changes to the flow regime of the catchment, analogies may be drawn from a nearby catchment that has been modelled under both current and future climates (Reynard et al. 2004). This modelling was undertaken using the Probability Distributed Moisture (PDM) model (Moore 1985) under a commission for DEFRA and the Environment Agency looking at the potential impacts of climate change on flood flows in the UK. The Dee at Mar Lodge represents the upstream part of the Dee catchment, but is 289 km^2, compared with the 1 km^2 of the Lochnagar catchment.

The future predicted flow duration curve for the Dee at Mar Lodge (Figure 11) under the UKCIP02 medium-high scenario of the 2050s, shows a marked decrease against the modelled baseline at lower flows, reflecting the climate scenarios with decreased summer rainfall and elevated evaporation losses. There is remarkably little change for higher flows.

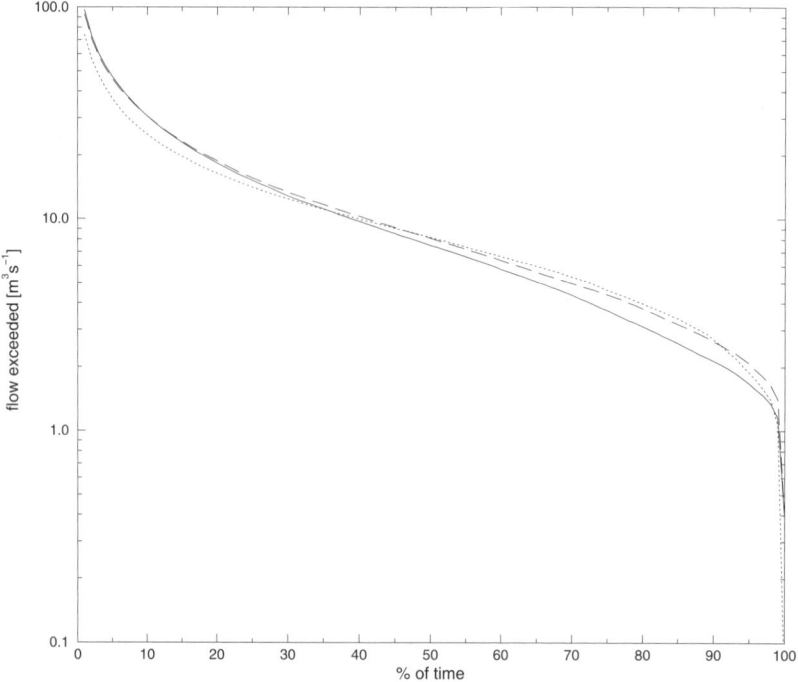

Figure 11. Flow duration curve for the Dee at Mar Lodge for the baseline and future (UKCIP02 medium-high scenario for the 2050s) climate. The black dotted line represents observed flow, the black dashed line is from modelled flow using observed rainfall and the black solid line is the future curve.

The change in the seasonal patterns of flow becomes apparent from the seasonal flow duration curves (Figure 12). There is a slight increase (comparing the black dashed baseline with the solid future curves) in higher flows during the winter and lower flows during the autumn, but a decrease across the entire flow regime during the summer.

Considering in more detail the changes in high flows, the annual flow duration curve shows an increase in the flow exceeded 3% of the time (Q3) of just 3%. However, this hides a change in seasonal peak flows. The Q3 flow increases by 10% during the winter, but decreases by nearly 30% during the summer. Increases in flow at all return periods (Figure 13) are predicted compared with the modelled baseline, with the percentage increase in flows of various return periods (Table 4) peaking at an 18.5% increase in the 50-year flow.

It should be remembered that these changes in flows are from a nearby catchment, under very similar climate change scenarios, but they are not specifically for Lochnagar. The Dee at Mar Lodge is a far larger catchment than Lochnagar, containing a greater mix of land cover types and elevation. These different catchment characteristics might suggest the effects of changes in future snowfall and accumulation rates will have a greater affect on the flow regime in Lochnagar, than those simulated

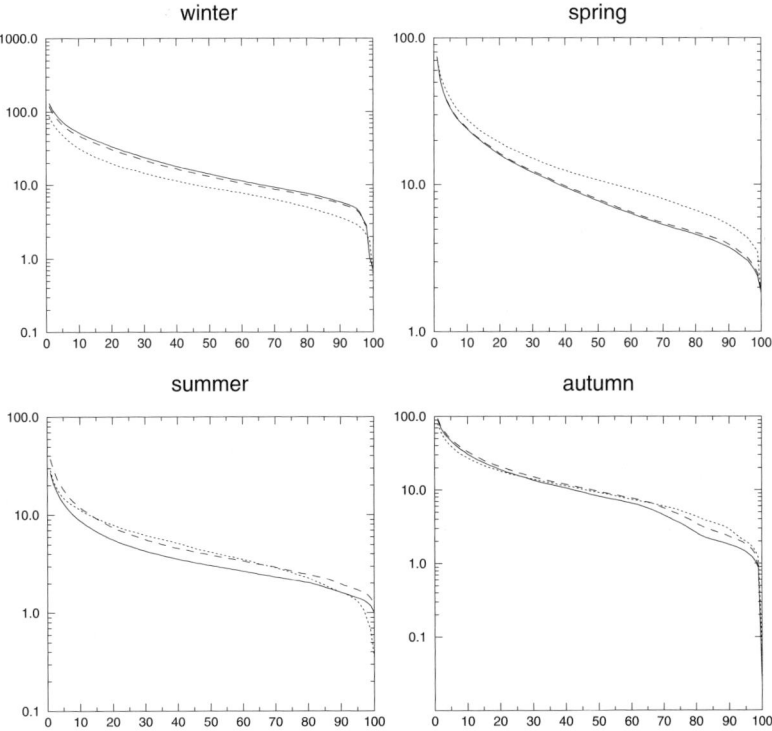

Figure 12. Seasonal flow duration curves for the Dee at Mar Lodge for the baseline and future (UKCIP02 medium-high scenario for the 2050s) climate. The black dotted line represents observed flow, the black dashed line being that from modelled flow using observed rainfall and the black solid line is the future curve.

Table 4. Percentage change in return period flows for the Dee at Mar Lodge under the UKCIP02 medium-high scenarios for the 2050s.

Return period (years)	1.0	2.0	5.0	10.0	20.0	25.0	30.0	50.0
Change in flow (%)	4.8	6.6	9.7	12.2	14.9	15.8	16.5	18.5

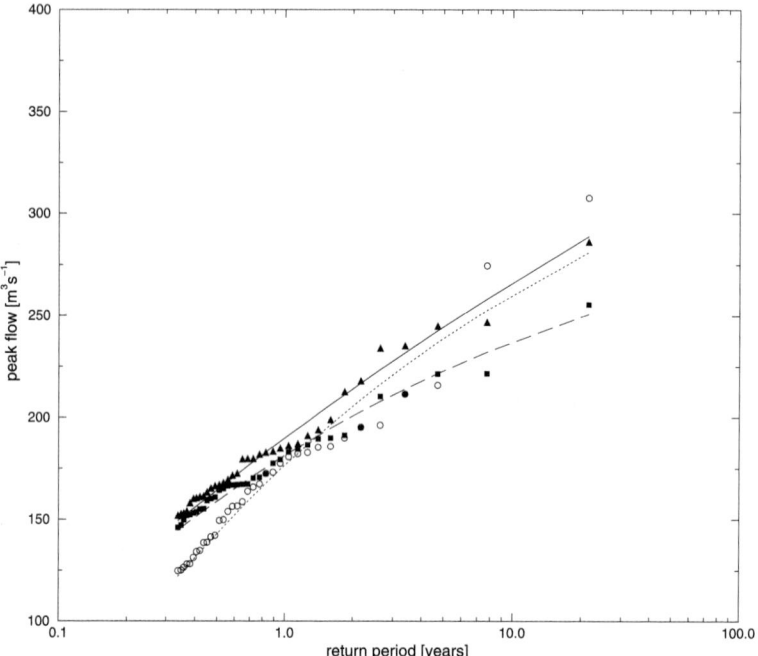

Figure 13. Flood frequency curves for the Dee at Mar Lodge for the baseline (open circles and dotted line represent observed flow; solid squares and dashed line is modelled flow with observed rainfall) and under the UKCIP02 medium-high scenario for the 2050s (solid triangles and solid line).

for the Dee. Also, only one future scenario has been presented as an illustration of some possible effects of climate change on the hydrology on this region. Scenarios sourced from alternative global or regional climate models would undoubtedly produce different changes to the flow regime.

In terms of hydrochemistry, while long-term trends in acid deposition can clearly result in gradual change in surface water chemistry (Figure 6), much of the shorter-term and sub-annual variation is a consequence of the variability and seasonality in weather conditions. It is notable that the fluctuations in outflow stream chemistry at Lochnagar are amongst the smallest in the AWMN network, in particular being considerably less than at sites closer to the west coast. This probably reflects the residence time of the loch and the shallow soils of the catchment as well as the geographic location. Mean residence time for the loch is 9 months, much greater than sites further west. Estimates for Loch Coire nan Arr, Loch Chon and Loch Tinker have approximate residence times of 1 week, 3 months and 2 months, respectively.

Rain storm events and snowmelt events are known to cause episodic acidification in streams. The dilution of well-buffered baseflow relatively rich in base cations, with rainwater low in base cations causes transient pulses of low pH. These acidic events are likely to be exacerbated by a number of chemical processes. Flushes of acidifying sulphate when rainstorms follow dry periods (e.g., at the end of the summer) have been observed in upland catchments (Huntingdon et al. 1994). Long-term monitoring data from sites in UK and Scandinavia are already showing that this mechanism of episodic acidification is decreasing in importance as sulphate deposition declines (Dick Wright, NIVA unpublished data). Large inputs of marine ions (predominantly chloride, sodium) accompanying intense depressions and frontal rainfall are known to trigger acidic pulses whereby the large input of sodium is immobilised by the soil, in turn releasing hydrogen and aluminium if the soil is poorly buffered (Wright et al. 1988; Langan 1989; Heath et al. 1992). Flushes of organic matter during storms have also been correlated with acidic episodes (Easthouse et al. 1992). Snowmelt is often accompanied by a concentrated release of accumulated dry deposits of acid anions (chloride, sulphate and nitrate) also enhancing surface water acidification (Helliwell et al. 1998).

The processes controlling mean chemical status and episodic responses are all sensitive to future changes in climate. The overall impacts of climatic trends on acidification are uncertain at Lochnagar as the many individual effects will to a greater or lesser extent work in opposition with each other. For example, it is likely that the incidence of seasalt-induced episodes will increase, yet with rising temperatures snowmelt driven episodes which can result in acute ecological stress, are likely to decrease in importance. Higher temperatures may also increase the rate of supply through bedrock and soil weathering of base cations which neutralise acidifying processes. On the other hand increasing rates of decomposition of organic matter will increase the release of nitrate as an acidifying anion. This particular control on acidity is itself likely to be confounded as continuing future increases, in part through increased decompositional release, in DOC are likely to aid surface water recovery from acidification. At Lochnagar, scenarios generated by dynamic models suggest that a 50% increase in DOC in soil solution and runoff will result in increased ANC of more than 15 μeq l^{-1} by 2030 (Wright et al. in press). Hydrochemical modelling studies are at present ongoing to ascertain the overall effect of climate change on the hydrochemistry of upland catchments impacted by the effects of acidification. Modelling studies are undergoing iterative revision and improvement as projections of climate change are themselves refined.

Summary

- The flow regime at Lochnagar is flashy with a rapid response to rainfall although the relatively large volume of the loch (turnover time of c. 9 months) moderates high and low flows.

- The hydrograph is characterised by high frequency high flow events caused by rainfall (mainly in summer), snowmelt (mainly in winter) and a combination of rain-on-snow (spring). These events cause distinct hydrograph responses.
- The loch chemistry has been acidified through time as a result of the deposition of sulphur and nitrogen compounds from the atmosphere. Observed trends are, however, towards a recovery from acidification although nitrogen concentrations remain high.
- The relationship between flow and chemistry indicates a complex array of flowpaths which change seasonally.
- At a nearby river catchment (River Dee at Mar Lodge) modelled climate change scenarios indicate that the flow regime will tend towards lower summer flows with increased winter high flows.
- Changes to future winter snowfall and accumulation at Lochnagar (relative to the larger River Dee) may have a more significant impact on the seasonal flow regime changes.
- In terms of hydrochemistry, changes in the frequency of extremes may impact biota but the longer term recovery from acidification is likely to continue.

Acknowledgements

Much of the data used to construct this chapter were collected at great personal effort by Jo Porter who has walked into the catchment every week for many years. He has our grateful thanks. Chemical analysis has been undertaken by the Freshwater Laboratory in Pitlochry, Macaulay Institute in Aberdeen and Centre for Ecology and Hydrology in Wallingford. Chemistry data were provided by Neil Rose at UCL. Susie Beresford provided assistance with the manuscript.

References

Baker J.P., Bernard D.P., Christensen S.W. and Sale M.L. 1990. Biological effects of changes in surface water acid-base chemistry. In: Acid Deposition: State of Science and Technology, NAPAP Report No. 9, National Acid Precipitation Assessment Program, Washington, D.C.

Curtis C.J., Evans C.D., Helliwell R.C. and Monteith D.T. 2005. Nitrate leaching as a confounding factor in chemical recovery from acidification in UK upland waters. Environ. Pollut. 137: 73–82.

Easthouse K.B., Mulder J., Christophersen N. and Seip H.M. 1992. Dissolved organic carbon fractions in soil and streamwater during variable hydrological conditions at Birkenes, S. Norway. Water Resour. Res. 28: 1585–1596.

Evans C.D. and Monteith D.T. 2001. Trends in surface water chemistry at AWMN sites 1988–2000. Hydrol. Earth Syst. Sci. 5: 351–366.

Evans C.D., Monteith D.T. and Cooper D.M. 2005. Long-term increases in surface water dissolved organic carbon: Observations, possible causes and environmental impacts. Environ. Pollut. 137: 55–71.

Fowler D., Smith R.I., Muller J.B.A., Hayman G. and Vincent K.J. 2005. Changes in the atmospheric deposition of acidifying compounds in the UK between 1986 and 2001. Environ. Pollut. 137: 15–25.

Heath R.H., Kahl J.S., Norton S.A. and Fernandez I.J. 1992. Episodic stream acidification caused by atmospheric deposition of seasalts at Acadia National Park, Maine, US. Water Resour. Res. 28: 1081–1088.

Helliwell R.C., Soulsby C., Ferrier R.C., Jenkins A. and Harriman R. 1998. Influence of snow on the hydrology and hydrochemistry of the Allt a Mharcaidh. Sci. Total Environ. 217: 59–70.

Hughes M. This volume. Chapter 2. Physical characteristics of Lochnagar. In: Rose N.L. (ed.) 2007. Lochnagar: The natural history of a mountain lake. Springer. Dordrecht.

Hulme M., Jenkins G.J., Lu X., Turnpenny J.R., Mitchell T.D., Jones R.G., Lowe J., Murphy J.M., Hassell D., Boorman P., McDonald R. and Hill S. 2002. Climate changes scenarios of the United Kingdom: The UKCIP02 Scientific Report. Tyndall Centre for Climate Change Research, School of Environmental Sciences, University of East Anglia, UK. 120 pp.

Huntingdon T.G., Hooper R.P. and Anlenbach B.T. 1994. Hydrologic processes controlling sulphate mobility in a small forested watershed. Water Resour. Res. 30: 283–295.

Intergovernmental Panel on Climate Change. 2000. Emissions Scenarios. In: Nakicenovic N. and Swart R. (eds.) Special report of the IPCC: A special report of Working Group III of the Intergovernmental Panel on Climate Change. Cambridge U. Press, Cambridge. 599 pp.

Jarvie H.P., Neal C., Smart R., Owen R., Fraser D., Forbes I and Wade A. 2001. Use of continuous water quality records for hydrograph separation and to assess short term variability and extremes in acidity and dissolved carbon dioxide for the River Dee, Scotland. Sci. Total Environ. 265: 85–98.

Jenkins A. 1989. Storm period hydrochemical response in an unforested Scottish catchment. Hydrol. Sci. J. 34: 393–404.

Jenkins G.J. and Lowe J. 2003. Handling uncertainties in the UKCIP02 scenarios of climate change. Hadley Centre Technical Note, 44. Met Office, UK. 15 pp.

Jenkins A., Ferrier R.C. and Helliwell R.C. 2001. Modelling nitrogen dynamics at Lochnagar, N.E. Scotland. Hydrol. Earth Syst. Sci. 5: 519–527.

Kay A.L., Jones R.G. and Reynard N.S. 2006. RCM rainfall for UK flood frequency estimation. II. Climate change results. J. Hydrol. 318: 163–172.

Kettle H. and Thompson R. This volume. Chapter 18. Future climate scenarios for Lochnagar. In: Rose, N.L. (ed.) Lochnagar: The natural history of a mountain lake. Springer.

Langan S.J. 1989. Seasalt induced streamwater acidification. Hydrol. Proc. 3: 25–42.

Lien L., Raddum G.G., Fjellheim A. and Henriksen A. 1996. A critical limit for acid neutralizing capacity in Norwegian surface waters, based on new analyses of fish and invertebrate responses. Sci. Total Environ. 177: 173–193.

National River Flow Archive. Accessed from http://www.nerc-wallingford.ac.uk/ceh/nrfa

Neal C., Wilkinson J., Neal M., Harrow M., Wickham H., Hill L. and Morfitt C. 1997. The hydrochemistry of the headwaters of the River Severn, Plynlimon. Hydrol. Earth Syst. Sci. 1: 583–617.

Monteith D.T., Evans C.D. and Dalton C. This volume. Chapter 14. Acidification of Lochnagar and prospects for recovery. In: Rose N.L. (ed.) 2007. Lochnagar: The natural history of a mountain lake. Springer. Dordrecht.

Moore R.J. 1985. The probability-distributed principle and runoff production at point and basin scales. Hydrol. Sci. J. 30: 273–297.

Prudhomme C., Piper B., Osborn T. and Davies H. 2005. Climate change uncertainty in water resource planning. Report to UKWIR, UK. 305 pp.

Reynard N.S., Crooks S.M. and Kay A.L. 2004. Impact of climate change on flood flows in river catchments. Report for Defra/EA. Project W5B-01–050. 100 pp.

Rose N.L. This volume. Chapter 1. An introduction to Lochnagar. In: Rose N.L. (ed.) 2007. Lochnagar: The natural history of a mountain lake. Springer. Dordrecht.

Rose N.L., Metcalfe S.M., Benedictow A.C., Todd M., Nicholson J. This volume. Chapter 13. National, international and global sources of contamination at Lochnagar. In: Rose N.L. (ed.) 2007. Lochnagar: The natural history of a mountain lake. Springer. Dordrecht.

RGAR 1990. Acid deposition in the UK, 1986–1988. UK Review Group on Acid Rain. Third Report. Department of the Environment, London. 124 pp.

Shilland E.M., Monteith D.T., Bonjean M. and Beaumont W.R.C. (eds.) 2004. The United Kingdom Acid Waters Monitoring Network, Data Report for 2003–2004 (Year 16). Accessed from: http://www.ukawmn.ucl.ac.uk/report/AWMNAnnualReport2003–04.pdf

Stoddard J.L. 1994. Long-term changes in watershed retention of nitrogen: Its causes and aquatic consequences. In: Baker L. (ed.) Environmental chemistry of lakes and reservoirs. Advances in Chemistry Series No. 237. American Chemical Society, Washington, D.C. pp 223–284.

Wright R.F. and Jenkins A. 2001. Climate change as a confounding factor in reversibility of acidification: RAIN and CLIMEX projects. Hydrol. Earth Syst. Sci., 5: 477–486.

Wright R.F., Norton S.A., Brakke D.F. and Frogner T. 1988. Experimental verification of episodic acidification of freshwaters by seasalts. Nature 334: 422–424.

Wright R.F., Aherne J., Bishop K., Camarero L., Cosby B.J. Erlandsson M., Evans C.D., Forsius M., Hardekopf D.W., Helliwell R., Hruska J., Jenkins A., Kopáček J., Moldan F., Posch M and Rogora M. (2006) Modelling the effect of climate change on recovery of acidified freshwaters: relative sensitivity of individual processes in the MAGIC model. Sci. Total Environ. 365: 154–166.

10. THE AQUATIC FLORA OF LOCHNAGAR

ROGER J. FLOWER (r.flower@geog.ucl.ac.uk),
DONALD T. MONTEITH, JONATHAN TYLER and EWAN SHILLAND
Environmental Change Research Centre
University College London
Pearson Building, Gower Street
London WC1E 6BT
United Kingdom

and

SERGI PLA
Department of Geography
Loughborough University
Loughborough
Leicestershire LE11 3TU
United Kingdom

Key words: aquatic plants, chrysophytes, diatoms, environmental change, Scottish highland loch

Introduction

In any attempt to evaluate the aquatic flora of a particular lake or loch, water quality is of primary consideration but there are also several other factors that influence plant occurrences and abundances in natural aquatic systems. Water quality is an intrinsic factor that includes water acidity/alkalinity, dissolved solids, nutrient concentrations and water colour. Other intrinsic factors include hydromorphy and biological attributes and interactions. For lakes in rock basin catchments, hard or softwater status is largely determined by the acidity or alkalinity of run-off water which in turn is strongly influenced by catchment geology. Water bodies on siliceous rocks are typified by softwaters with low concentrations of total dissolved solids. Nutrients also influence water quality and more nutrients generally promote more plant growth and, in most natural waters, tend to be linked with total dissolved solids and alkalinity. Low nutrient status can favour certain plant species, but productivity is limited and the aquatic system is referred to as oligotrophic. Water transparency also limits aquatic plant growth by controlling the underwater light regime and it is influenced by both dissolved

N.L. Rose (ed.), Lochnagar: The Natural History of a Mountain Lake
Developments in Paleoenvironmental Research, 199–229.
© 2007 *Springer.*

and particulate material as well as by light intensity. Irrespective of the nutrient status of an aquatic system, underwater light availability has a primary influence on aquatic plant growth but very high intensity light can suppress photosynthesis.

Extrinsic factors influencing a water body can be considered as those which occur beyond the drainage basin. Of these climate is particularly important for aquatic vegetation in upland water bodies. Temperature diminishes with altitude and, in harsh northern climates, light is often reduced by frequent low cloud cover over higher ground (Chapter 5: Thompson et al. this volume). Consequently, aquatic plant communities in upland water bodies are exposed to a variety of natural climate stressors that can either select or de-select particular species. However, it is the impact of human activities that, indirectly or directly, has promoted the most rapid environmental changes in the UK uplands (e.g., Darling and Boyd 1964; Flemming et al. 1997; Battarbee et al. 2005a). Direct impacts can include introduction of unwanted species or promotion of local landscape changes, usually by enhancing soil erosion or by causing hydrological disturbances. Probably the most pervasive form of human disturbance, however, results from the deposition of chemical pollutants from the atmosphere. In sensitive upland water bodies across the UK, acid deposition has caused sustained ecological changes since the 1800s (e.g., Mason 1990). Consequently, the composition of the aquatic flora in many upland UK waters today is the result of a multiplicity of impacting factors that have acted over a range of time-scales.

In the remoter parts of the Scottish Highlands, in the Cairngorm and Grampian Mountains, the direct effects of human disturbance are often relatively low. High altitude lochs in this region, such as Lochnagar, are however impacted by both acidification (Jones et al. 1993) and contamination from atmospheric sources (Rose et al. 2001; Bennion et al. 2002). Located at an altitude of almost 800 m, with high precipitation, high winds and relatively low temperature (see Chapter 5: Thompson et al. this volume), atmospheric pollution and a severe climate combine to restrict the growth of aquatic plants. The Lochnagar catchment is granitic with localised peat deposits; it delivers runoff to the loch that is particularly poor in nutrients. Consequently, the loch water was acidic before atmospheric pollution acidified the site (see below) and it has been strongly oligotrophic for millennia. Furthermore, sedimentary microfossil records indicate that aquatic plant productivity and diversity has always been low (Dalton et al. 2005). The oligotrophic nature of high altitude lochs is natural and reflects location and elevation. According to Mackereth (1957), the water in such sites is primary in nature having evolved relatively little in chemical composition from meteoric water. Formerly, successional changes in lake type and vegetation were related to an evolutionary series that began with oligotrophy (Pearsall e.g., 1921). However, it is now clear that lake systems primarily respond to environmental drivers that are time dependant on landscape characteristics (Engstrom et al. 2000).

Lochnagar is a good example of a water body that has become more acid with time and this is clearly shown by sediment records of the micro-algal flora (diatoms). Since about the mid-19th century, more acid tolerant species began to increase indicating a period of relatively rapid acidification of the loch water (Jones et al. 1993; Chapter 14: Monteith et al. this volume). Despite this recent acidification and uninformed opinions about upland lochs being of little botanical interest, the physiological adaptations of plants growing in these waters are notable (Rörslett and Brettum 1989; Farmer 1990) as is the botanical diversity, albeit low (Duigan 2005). Whilst it is true that the aquatic plant communities are species poor (compared with richer, lowland sites), the species that do persist at Lochnagar are sensitive to environmental change and are characteristic of rather extreme conditions. Hence, monitoring the specialised communities of sites such as Lochnagar can yield valuable ecological information about current biological status and about the nature, pace and influences of environmental change processes. In some acidified Highland lochs the aquatic plant communities are now showing signs of recovering from 20th century acidification, species abundances are changing but the processes responsible for the perceived improvements in water quality are not well understood (Monteith and Evans 2000).

This chapter summarises the aquatic flora of Lochnagar from several aspects. Following some general comments on aquatic plants in acid Scottish Highland lochs, the phytobenthos of the loch is examined and includes both the rooted, or otherwise substratum bound plants, as well as the free-floating or planktonic plant communities. The history of the aquatic vegetation is pertinent to assessment of the current status of the aquatic biota especially in relation to acidification (see Chapter 14: Monteith et al. this volume) and some information about past conditions is gleaned from the analysis of the sub-fossil plant remains in sediment cores. Finally, the aquatic plant diversity of the loch is considered in the context of future environmental change, including further threats from human activities.

Aquatic plants in acidic Scottish Highland lochs

The Highlands of Scotland are not necessarily on high ground and according to Darling and Boyd (1964), the Highland Region is that part of Scotland north of a line from the River Clyde to Stonehaven. Highland lochs therefore include a large range of water quality types and many were first investigated during the well known and extensive survey of Scottish lochs carried out in the early-20th century by Murray and Pullar (1910). Studies of the aquatic plants (Murray and Pullar 1910; Spence 1964; Palmer 1992; Palmer et al. 1994) and phytoplankton (West and West 1909; Brook 1964) of Highland lochs nearly all focused on Scottish lochs in the mid- to low-altitude regions. There was a paucity of information available for the high altitude lochs, so much so, that Burnett (1964), in his edited review of Scottish vegetation, omitted the 'mountain loch vegetation' entirely. Nevertheless, some upland (though relatively few above 500 m altitude) acidic lochs were included in former surveys for classifying aquatic vegetation in the UK for the JNCC (Joint Nature Conservation Committee) (see Palmer 1992; Rodwell 1995). Currently, the JNCC Standing Waters Database held collectively by Scottish Natural Heritage (SNH), Countryside Commission for Wales (CCW) and English Nature (EN) contains aquatic plant inventories for over 3000 sites across the UK. For classifying sites according to aquatic plants, Palmer et al. (1994) used 1124 standing water bodies, over half of which were in Scotland. The first three (out of ten) aquatic plant categories were allocated to acid water systems (pH less than c. 6.5). However, only Type 1 sites (dystrophic/very acid) accord with the current pH of Lochnagar by having a mean pH below 6. Seventeen plant taxa characterise these first three acid groups and include *Potamogeton polygonifolius*, *Lobelia dortmanna*, *Sparganium angustifolium*, *Isoetes lacustris*, *Juncus bulbosus* var. *fluitans*, *Littorella uniflora* and *Sphagnum* spp. These characteristic species are primarily restricted to low pH environments. For example, they all occurred in a low altitude acid lochan on Skye (Birks 1973) but, despite possessing similar acidity, much higher altitude Lochnagar supports only two of the preceding vascular plant species and several bryophyte species (see below).

Systematic monitoring of the aquatic vegetation in several upland acid lochs in the Scottish Highlands zone began in the late 1980s as a result of increasing concern about the impact of water acidification (Monteith and Evans 2000). The UK Acid Waters Monitoring Network (UKAWMN) was established and regular surveys of the macrophytic vegetation in Highland lochs, Loch Corrie nan Arr (in the Torridon area) and Lochs Chon and Tinker (in the Trossachs) as well as in Lochnagar were begun in 1988. Several species of aquatic plants are common to these sites and to upland water bodies elsewhere in the UK but some species show more restricted distributions. The occurrences of twelve aquatic plant taxa selected from a suite of sixteen upland UK lakes and lochs surveyed by the ECRC over the past decade are shown in Figure 1. The sites are arranged according to altitude and, although acidification impacts vary between sites, all have fairly similar water pH. The Figure indicates a trend towards lower diversity at higher altitudes that culminates with just two species of vascular aquatic plants in Lochnagar. Data in Figure 1 are not intended to imply altitudinal limits for the aquatic plants listed and for these, reference should be made to general accounts of aquatic vegetation (e.g., Preston and Croft 1997). Temperature related to altitude may exert a major influence on diversity but one outlier in the altitudinal trend displayed in Figure 1 is Llyn Clyd (53° 7.02'N; 4° 2.46'W) in Snowdonia which indicates that factors other than altitude are important. This Welsh lake is over 750 m above sea level but possesses a significant population of *Myriophyllum alterniflorum*, a plant that is typically more common in much lower altitude sites. Similarly, shallow Sandy Loch (56° 57.76'N; 3°16.23'W) lies near Lochnagar yet the JNCC Standing Waters Database records twelve aquatic plant species for this site. Sandy Loch water has higher alkalinity and calcium values than in Lochnagar (Wathne et al. 1995) and these water quality differences as well as nutrient, site aspect and bathymetric factors probably all promote aquatic plant diversity.

The aquatic plants of Lochnagar

Despite Raven and Walters (1956) devoting a chapter to the vegetation of the Cairngorms and Lochnagar, their account refers only to the terrestrial plants on the mountain of Lochnagar. Only during the 1960s was the aquatic vegetation within the loch of Lochnagar investigated and Light (1975) noted two species of aquatic macrophyte and several bryophytes. Only as a consequence of the UKAWMN was Lochnagar regularly assessed for aquatic macrophytes as well as for benthic diatoms (siliceous micro-algae) and water quality. It is now the highest altitude water body in the UK subject to regular hydro-biological sampling (for methods, see Monteith and Evans 2000).

The Lochnagar catchment lies above the present tree-line; it is a granitic boulder strewn landscape mostly above 800 m altitude and, according to one definition (see Chapter 7: Birks this volume) it lies within the alpine zone. There are difficulties with this definition but the presence of a distinctive arcto-alpine terrestrial vegetation element on Lochnagar (Raven and Walters 1954; Chapter 7: Birks this volume) supports the view that the loch lies in the biological alpine zone. Consequently, this remote and very oligotrophic loch is subjected to relatively long periods of low temperature and prolonged periods of ice cover (Chapter 5: Thompson et al. this volume) that exert a strong influence on the aquatic vegetation.

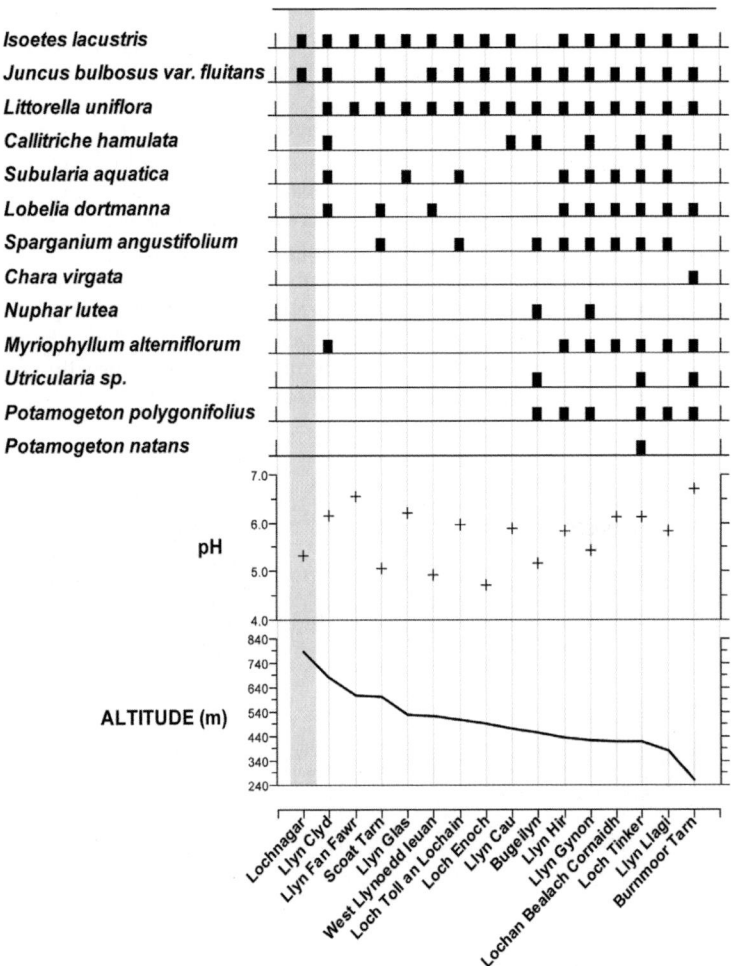

Figure 1. A diagram to show the occurrences of twelve common vascular aquatic plants in sixteen oligotrophic upland UK lochs and lakes. The sites were selected according to altitude from 788 m (Lochnagar) to 250 m (Burmoor Tarn in the Lake District) and indicate that plant diversity declines with altitude. Note that water pH shows no consistent trend over this altitudinal gradient. The vegetation records are from the Acid Waters Monitoring Network database held at the Environmental Change Research Centre (ECRC, UCL) and are not intended to indicate altitudinal limits for the plants listed (see text).

Aquatic environments

Despite its high altitude and physical characteristics, Lochnagar possesses a variety of aquatic habitats that accord with the marginal and open water zones. The loch margins are composed either of granitic boulders and cobbles or of small sand/gravel beach areas. A typical section across a rocky shore on the eastern side of Lochnagar is shown in Figure 2. Typically, a wave-cut peat bank, fringed with *Calluna* and/or moorland grasses, lies on granite boulders exposed at the waters edge. The marginal zone is divisible into distinct zones (e.g., Wetzel 1975). The upper part, known as the supralittoral, constitutes the loch margin and the wave cut bank at the waters edge; it lies above the water level of the loch but is subject to wave spray. The lower exposed rocky area (Figure 2) is referred to as the eulittoral. It is affected by water level fluctuations and bryophytes and lichens are often common here. Both the peat bank and granite boulders lie on unconsolidated glacial deposits. These boulders tend to diminish in size out from the bank as the water depth increases and give way to permanently submerged cobbles and gravels in the upper littoral zone. Further out from the bank, these deposits grade into sandy gravels mixed with organic matter (principally eroded peat debris) that constitute the lower littoral zone where rooted aquatic plants occur. Still more distant from the shore, with distance dependant on basin slope, the profundal zone begins. This zone is beyond the light limit and the sediment usually consists of fine black organic detrital mud (Chapter 8: Rose this volume).

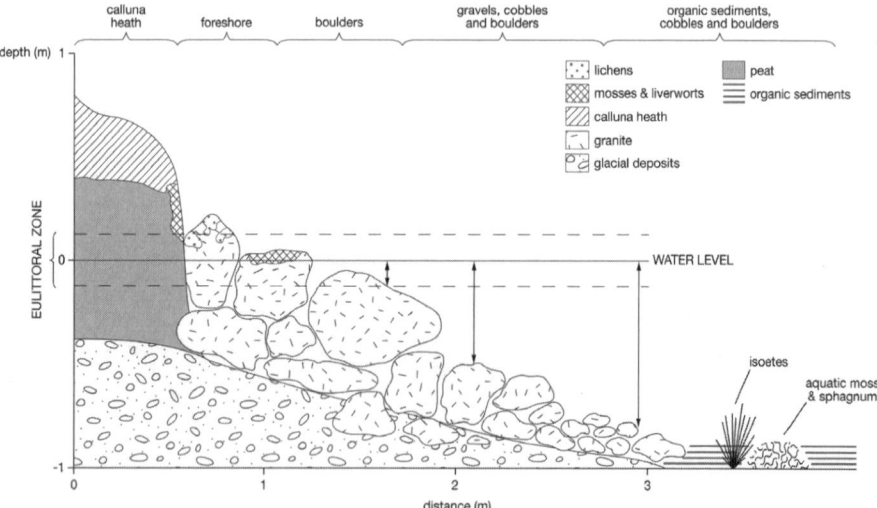

Figure 2. Depth transect for a typical section of the shore zone in Lochnagar showing the nature of the substratum from the shore bank across the littoral zone to the profundal zone (see text) and the distributions of some aquatic plants. The vertical downward arrows indicate the locations of three epilithic biofilm samples of micro-algae collected (at depths of c. 10, 45 and 80cm) in December 2004 (see text) and the shore section is taken from Transect A on Figure 3.

Elsewhere around the loch shore, granite bed-rock outcrops or sandy beaches occur (Chapter 8: Rose this volume) and, on the more sheltered western shores, organic matter in the form of redeposited fine-grained peat frequently accumulates on, and is mixed with, the coarse littoral sands. Occasionally, wave-cut thin sections of blanket peat extend to the waters edge, probably indicating periods of lower water level in the past. On the corrie backwall however the granite bed rock plunges steeply into the loch and shelving rock surfaces extend to considerable depth (c. 7 m). Also, in the region of the outflow, underwater boulder fields provide an extensive epilithic habitat and shallow water sandy silts occur within and beyond this rocky area.

The other main habitat for plant growth in Lochnagar is the illuminated open water zone where, depending on season, sparse phytoplankton communities develop. The acid, very dilute water of Lochnagar (Chapter 9: Jenkins et al. this volume), together with the extreme climate and deep mixing of the water (Chapter 5: Thompson et al. this volume), exert major influences on the distributions of aquatic plants in the open water as well as in the littoral zone. These distributions are described firstly according to spatial and then to temporal differences. Regarding the latter, the distributions of both micro-algae and aquatic macrophytes are considered over several timescales, from seasonal to decadal to centennial.

Distributions of aquatic macrophytes

There are a number of plants associated with the loch margins and damp depressions that are essentially terrestrial in habitat preference. At the edge of the supra-littoral zone and on exposed rocks of the eulittoral, bryophytes are common (Chapter 7: Birks this volume); lichens and bryophytes are usually confined to rock or eroded peat surfaces (cf. Figure 2). Within the sub-aquatic zone there are only two species of vascular aquatic macrophyte (as described by Light 1974), the pteridophyte *Isoetes lacustris* and the aquatic rush *Juncus bulbosus* var. *fluitans*. The aquatic moss *Sphagnum auriculatum* and the liverworts *Scapania undulata* and *Nardia compressa* are common in deeper water. There are no emergent aquatic plants in the loch. The vegetational structure of the littoral in Lochnagar is therefore very different from that commonly portrayed in most limnological text books (e.g., Figure 52, Spence 1964; Figure 15-1, Wetzel 1975) where the nature of aquatic vegetation is used to subdivide the littoral zone. As noted previously (Light 1974), macrophytes tend to be scarce in shallow water and so Lochnagar is similar to other alpine lochs where severe climate and ice formation prevents higher plants from establishing in the upper littoral zone. Beyond the areas most influenced by wave and ice and especially on the more sheltered sides of the loch, aquatic plants are most abundant (Figure 3). The two species of vascular aquatic plants are distributed according their life form and growth requirements. They are adapted either to a benthic (*I. lacustris*) or part floating (*J. bulbosus* var. *fluitans*) existence.

Isoetes lacustris exhibits a typical isoetid growth form with a rosette of slender leaves developed at the sediment surface. It is extremely well adapted to a harsh oligotrophic environment and has been subject to much scientific study (for a detailed review see Rörslett and Brettum 1989). As with other isoetid species, a significant proportion of total biomass is in the form of roots and rhizomes, perhaps illustrating the particular importance of the substratum as a source of carbon dioxide (CO_2) and nutrients. The

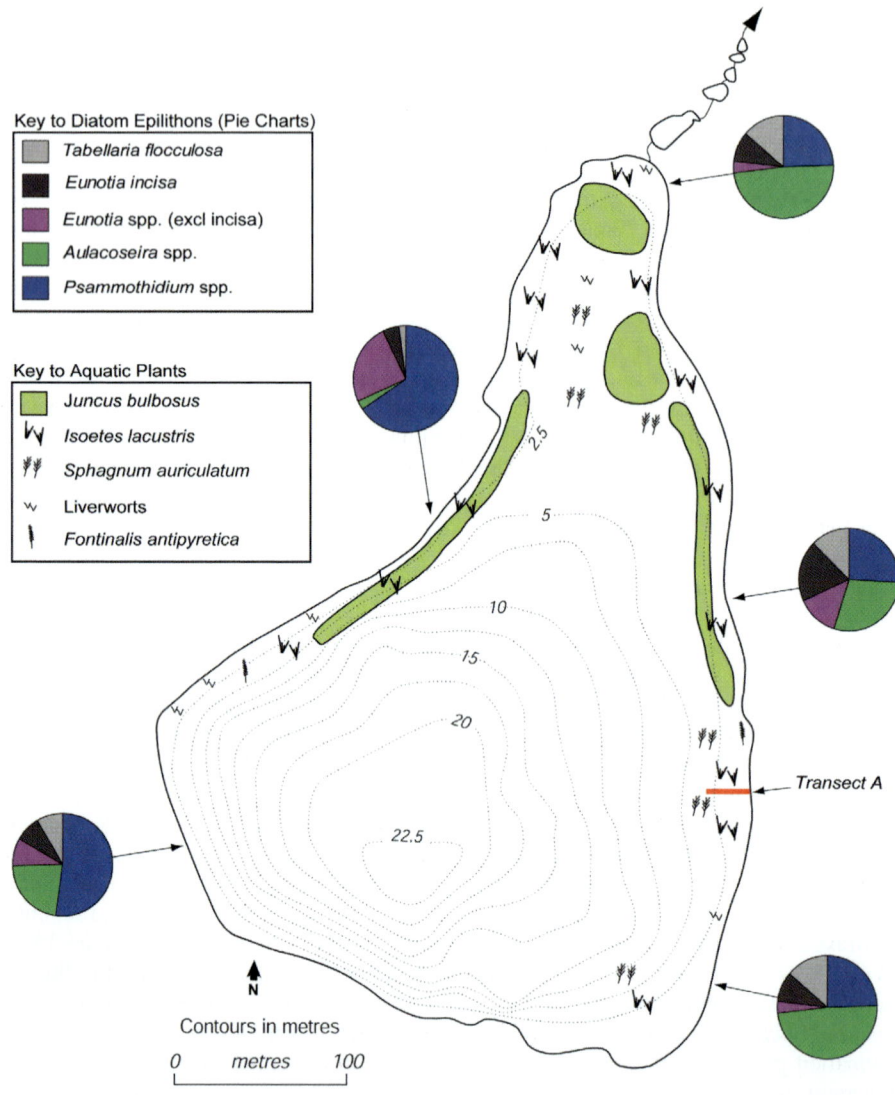

Figure 3. A map of Lochnagar to show the distribution of the two aquatic macrophytes *Isoetes lacustris* and *Juncus bulbosus* var. *fluitans* and several common bryophytes as assessed in the UKAWMN survey for 2005. Frequency abundances of diatom taxa in epilithon samples collected from submerged stones in the littoral zone at five locations around the loch are indicated as pie charts. Dotted contour lines indicate the bathymetry (in metres) of the loch and the shore section Transect A indicates the location of the littoral transect shown in Figure 2.

roots are linked to shoots by internal air channels, or lacunae, which allow efficient internal transport of carbon dioxide to shoots and oxygen to roots. Perhaps most importantly, *I. lacustris*, is able to acquire inorganic carbon for photosynthesis through crassulacean acid metabolism (CAM), which allows CO_2 to be captured at night (when it is most available) and stored intra-cellularly as malic acid, before it is decarboxylated during the following day. Boston and Adams (1987) showed that approximately half of the carbon assimilated by a related species, *I. macrospora*, was obtained in this way. Furthermore, *I. lacustris* is particularly well adapted to shade and it has been found growing at water depths of up to 8 m in clear water lakes (Hutchinson 1975). Rörslett (1988) estimated a light compensation point for this plant of just 2.5 W m^{-2} of photosynthetically active radiation (PAR). The limitations on the general distribution of *I. lacustris* distribution probably relate primarily to the plant's relatively small size (leaf length is less than 15 cm) and a slow growth rate; these render the species a poor competitor for light in more alkaline systems. In contrast to *I. lacustris*, most non-isoetid plants found in oligotrophic systems obtain their inorganic carbon directly from the water column, through their leaves. These species are often classed as "elodeid" on the basis of their elongate morphology and leaf shape which minimises boundary layer resistance to gas exchange. The absence of most elodeids from Lochnagar may in part reflect low bicarbonate availability, particularly during the summer growth season when the loch can thermally stratify for short periods. The one exception is *Juncus bulbosus* var. *fluitans*, which has an elongate growth form and forms substantial stands in Lochnagar (Figure 3). This species is ubiquitous in UK oligotrophic waters; it thrives across a range of acidities but particularly where there is an abundance of CO_2 and ammonia (Roelofs et al. 1994; Schuurkes et al. 1986). Such conditions are known to occur when lakes with calcareous sediments acidify (Arts et al. 2002). *J. bulbosus* var. *fluitans* also has a low light compensation point and is adapted to low temperatures, which gives it a competitive advantage during the early part of the year (Svedäng 1990). In Lochnagar it typically forms occasional large patches of essentially floating submerged vegetation that arise from stolons ascending from the sediment. The vivid green submerged leaf fronds of this plant can extend several metres above the sediment surface.

From below around 0.6 m water depth, the littoral substratum of the loch is generally suitable for aquatic macrophyte growth. Since 1988 vegetation survey transects have been carried out (Monteith and Evans 2000) and the depth range for *Isoetes lacustris* growth is indicated to be from about 0.6 m to about 3 m. This cut-off depth is rather shallow given the clarity of the water in Lochnagar (Secchi disc estimate of water transparency is in the order of c. 10 m). Plants growing near the depth cut-off point are larger with shoot lengths around 10 cm which is about twice that found for plants growing in shallow water. Light availability may be influencing the plant phenotype at depth but light *per se* seems not to be the factor limiting growth of *I. lacustris* in the loch below c. 3 m. Furthermore, the maximum depth for *I. lacustris* growth has changed during the UKAWMN period and the depth range has apparently increased during the first 12 years (1988-2000) of monitoring. Reasons why this plant is depth restricted in Lochnagar are not clear but since light seems not limiting, water temperature or water quality changes are suspected and these issues warrant further research.

Juncus bulbosus var. *fluitans* has a patchy distribution and generally occurs between 1m and 3 m depth and is most frequent in the northern, shallower parts of the loch (Figure 3). As for *I. lacustris*, the lower depth limit for this plant in Lochnagar does not seem to be controlled by light. The areal extents of *Juncus bulbosus* var. *fluitans* have increased significantly during the course of the UKAWMN sampling from 1988 to 2005. At the start of the monitoring period, the distribution of this plant in Lochnagar was recorded as rare (Paul J. Raven, unpublished notes). Figure 3 refers to the situation for 2005 when this plant occupied sections of the lower littoral shoreline totalling some 500 m in length or about one quarter of the entire shoreline. *J. bulbosus* var. *fluitans* is now the most common vascular plant in the loch in terms of areal cover of the littoral zone. Again, reasons why this plant has expanded so markedly in the past two decades are unclear but water quality changes resulting from trends in atmospheric deposition chemistry are thought to be the most likely cause.

For bryophytes the depth range for growth in the loch is greater; it extends from the supralittoral (splash zone) to about 7 m depth. *Sphagnum auriculatum* occupies a greater depth range on the east side of the loch compared with the west side (Figure 3) and is also abundant in the shallow water (c. 3 m depth) at the north end of the loch.

Distributions of micro-phytobenthos

The extensive areas of rocks, sand and detrital muds in the littoral zone of Lochnagar offer ample habitat opportunities for colonisation by benthic micro-algae. Although green filamentous algae (notably of the genus *Mougeotia*) occur on some surfaces, particularly during the autumn, diatom micro-algae are generally abundant throughout the littoral of the loch. Diatom micro-algae flourish in oligotrophic upland waters and are well known to be excellent indicators of water quality (Battarbee et al. 2001). Hence they are often utilised in freshwater monitoring programmes (Monteith and Evans 2000). Where light and nutrients are sufficient, diatoms form biofilms on submerged surfaces of rocks (the 'epilithon'), on aquatic plants (the 'epiphyton') and on sands and gravels (the 'epipsammon'). In the deeper littoral, the epipelon is best developed and this includes those mobile diatoms that can glide across surfaces and maintain their position on the illuminated sediment surface (see Round 1981).

Although diatoms occur in both planktonic and benthic habitats, most species in Lochnagar are benthic. An exception is *Aulacoseira distans* var. *nivalis*; this diatom appears to have both planktonic and benthic phases. Samples of epilithon collected from three depths (arrowed in Figure 2) in the autumn of 2004 were composed of several common species, including *Psammothidium marginulatum*, *Aulacoseira distans* var. *nivalis*, *Eunotia incisa* and *Tabellaria flocculosa* (Table 1). Other species of *Psammothidium* and *Eunotia*, *Frustulia rhomboides*, *Brachysira brebissonii* and *Suriella linearis* made up a large proportion of the residual diversity. A variety of *Frustulia rhomboides* also occurred in this material and a specimen is shown in Figure 4b. Also, sparsely present in the epilithon were cysts of chrysophycean algae. These siliceous flagellated micro-algae are planktonic (see below) but their siliceous resting spores or stomatocyts are deposited in the benthos. A spined or 'echinate' stomatocyst from this material is depicted in Figure 4a.

There was no discernible trend in the epilithic diatom community over the short depth section sampled in the littoral zone (c. 10-70 cm depth, see Table 1). However, a spatial

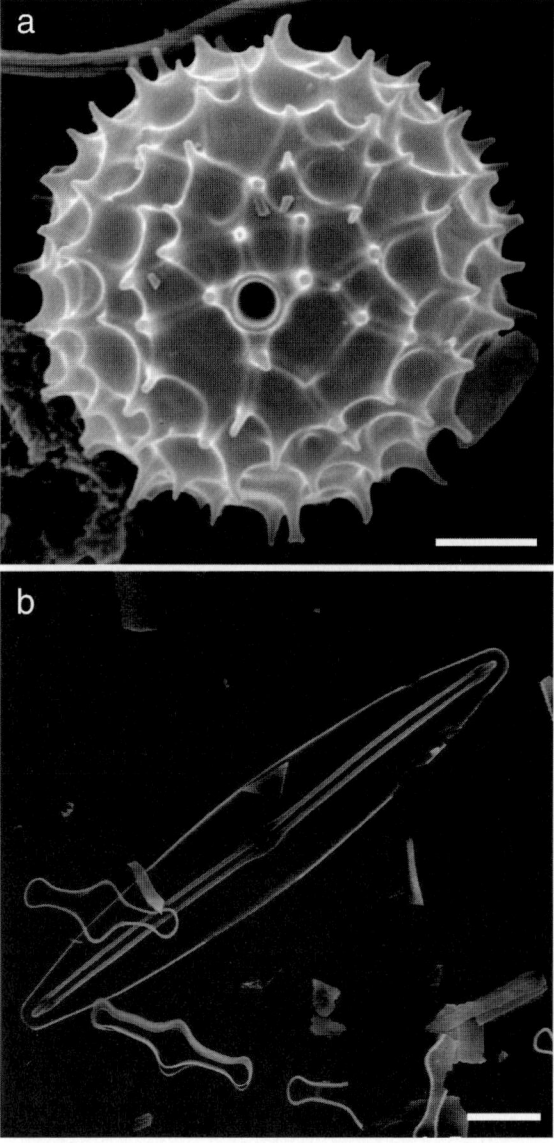

Figure 4. Two examples of siliceous algae present in epilithon biofilm samples collected from Lochnagar at 45 cm depth in December 2004 (cf. Figure 2). (a) a stomatocyst from an unidentified species of chrysophycean alga. This stomatocyst morphotype is known as S171 (see Pla 2001); it possesses a reticulate external surface with depressions and spines and a distinctive exit pore; cyst diameter is 9 μm. Scale bar is 2 μm (b) *Frustulia rhomboides* var. *viridula*, a relatively large lanceolate diatom shown as a single 105 μm long valve (two valves make one diatom cell) with a raphe (a longtitudinal slit-like structure) and central node. Scale bar is 10 μm.

Table 1. The percentage frequencies of common diatoms in samples of epilithon collected during December 2004 along the short depth transect indicated in Figure 3 (the diatom count in each sample was about 200 valves).

Water depth (cm)	Species	Frequency (%)	No. of taxa
10	*Tabellaria flocculosa*	48	12
	Fragilaria exiguiformis	20	
	Psammothidium marginulatum	12	
	Eunotia minutissima	5	
	Aulacoseira perglabra	3	
	Others	12	
45	*Fragilaria exiguiformis*	32	19
	Tabellaria flocculosa	10	
	Psammothidium marginulatum	12	
	Aulacoseira distans var. *nivalis*	13	
	Frustulia rhomboides var. *saxonica*	5	
	Eunotia denticula	5	
	E. minutissima	3.5	
	Cymbella perpusilla	5	
	Others	14.5	
70	*Tabellaria flocculosa*	20	17
	Eunotia incisa	5	
	F. exiguiformis	10	
	P. marginulatum	22	
	P. helveticum	12	
	Aulacoseira distans var. *nivalis*	7	
	Pinnularia subcapitata var. *hilseana*	5	
	Brachysira brebissonii	5	
	Others	14	

survey of the epilithon in July 2004 (Jonathan Tyler, University College London unpublished) revealed some interesting differences in the distributions of several common diatom species. The distributions of taxa recorded in these epilithon samples are indicated in Figure 3 as percentage abundance charts and show marked differences in sample composition according to location. The factors that cause a preponderance of some taxa (i.e., *Aulacoseira*) over others (i.e., *Psammothidium*) are currently unclear. Also, within the genus *Eunotia*, the abundance of the most common species, *Eunotia incisa*, appeared to be inversely related to the abundances of other *Eunotia* species. This trend is most evident when comparing the species assemblages between different shores of the loch. Along the eastern shore (north, towards the outflow stream) *E. incisa* was most abundant. However, on the western and southern shore near the steep rock wall, other *Eunotia* taxa were more frequent. In addition, these latter sites were notable for the relative lack of *Tabellaria flocculosa*. Since the morphology of the sampled substrata differed little, water quality is approximately constant around the loch and algal grazing is only slight. The different assemblages may reflect spatial effects of local peat erosion which is more prevalent in the catchment near the eastern shore of the loch. Shading effects by deposited material are known to influence stream diatom communities (e.g., Cox 1984). However, differences in exposure to light (differential shading by rock walls and aspect differences, see Chapter 2: Hughes this volume) or in physical disturbance (north and eastern facing shores receive less shelter from wind and turbulence) are thought more likely explanations for spatial differences in the distributions of shallow water benthic diatoms.

Temporal changes in diatom epilithon communities in upland acid waters are of interest from the point of view of detecting trends in water quality changes (Monteith and Evans 2000). As part of the UKAWMN programme, benthic diatom algae have been sampled in Lochnagar since 1988. Epilithon samples pooled from several sampling points were collected annually and some species changes are clearly detectable over the past 15 years (Figure 5). The percentage abundances of common epilithic diatoms shown in this Figure reveal several trends. Certain species, such as *Eunotia incisa*, *E. pectinalis* and *Brachysira brebissonii* have declined over the monitoring period while species like *Aulacoseira distans* var. *nivalis*, *Fragilaria exiguiformis*, *F. vaucheriae* and *Tabellaria flocculosa* have tended to increase. Other species such as *Psammothidium marginulatum*, *Psammothidium altiacum* var. *minor* and *P. helveticum* showed highest abundances during the middle of the sampling period and others showed an increasing trend towards the end of the period (*F. vaucheriae; Eunotia* spp.) or an irregular distribution (*E. exigua*). These floristic changes indicate biological responses to environmental change trends. Water acidity has been essentially stable during the monitoring period up to the late-1990s but in the most recent years has shown a tendency to decline (Chapter 14: Monteith et al. this volume). The post-1988 floristic trends are doubtless a manifestation of water quality changes mediated during the sampling period either by changing climate or by changing patterns in deposited pollutants. Separating these effects is the focus of current research.

Distributions of phytoplankton

The planktonic habitat in Lochnagar is characterised by micro-algae. Desmids are green algae (Chlorophyceae) that are often common in the phytoplankton in oligotrophic lochs (Brook 1964) but in Lochnagar they are scarce and chrysophytes (Chrysophyceae) usually dominate. The phytoplankton of the loch is less well monitored than the benthic micro-algae but it was sampled intermittently over three years, from 1996 to 1998 (Figure 6). The recorded changes were very similar to those found in some other mountain lakes in Europe (cf. Brettum and Halvorsen 2004; Felip et al. 1999; Fott et al. 1999; Pugnetti and Bettineti 1999). During this sampling period chrysophytes dominated the phytoplankton in terms of biomass and a common species was *Dinobryon cylindricum* Imhof. The most numerically abundant group, however, comprised spherical nanophytoplankton and dominance by this group is recognised elsewhere in high altitude oligotrophic lakes (Brettum and Halvorsen 2004). Ephemeral blooms of *Oocystis* (Chlorophyceae), an unknown green flagellate, and the dinophyte *Wolozsynskia* also made large contributions to the biomass over the studied period. Overall diversity (indicated mainly as algal genera) of the phytoplankton communities is indicated in Table 2.

Phytoplankton assemblages and biomass show a marked seasonal pattern in Lochnagar. Chrysophytes (Chrysophyceae and Synurophyceae) peaked in May during 1997 and 1998 (Figure 6) and this pattern is confirmed by subsequent observations (Jonathon Tyler, University College London unpublished). Chrysophytes appear to

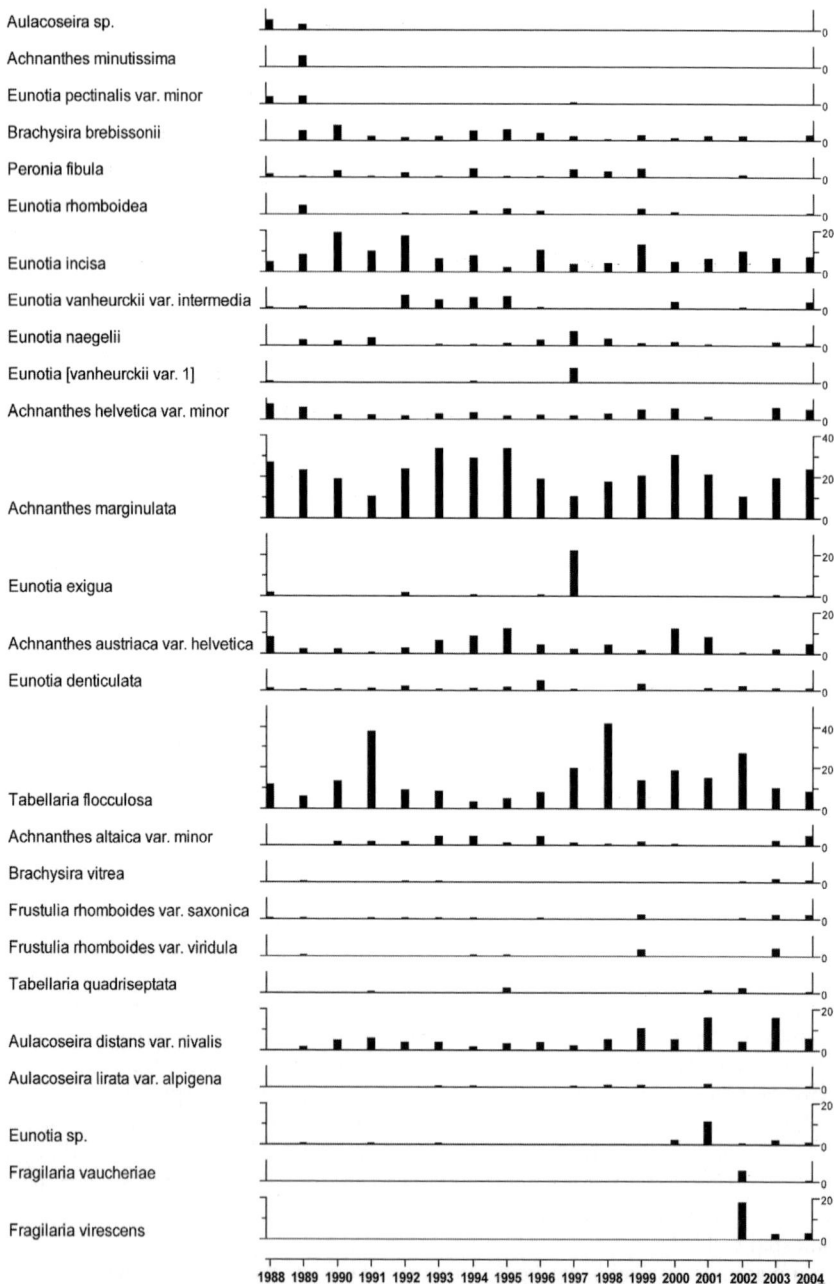

Figure 5. A time series in the percentage frequencies of diatom taxa in pooled epilithon samples collected annually from Lochnagar as part of the UK Acid Water Monitoring Network (Shilland et al. 2005) programme from 1988 to 2004.

over-winter beneath the ice, as indicated by their high abundance in November 1996 in deeper water, prior to ice formation. This may confer a competitive advantage when the spring ice-out and lake overturn occurs (usually in May). Chlorophyceans (mainly *Oocystis*, *Chlamydomonas* and an unknown flagellated species) peaked in June throughout the mixed water column. Cryptomonads (class Cryptophyceae, mainly *Cryptomonas*, *Rhodomonas*) bloomed in August followed by diatoms (class Bacillariophyceae, mainly *Aulacoseira*) growing in October. Both cryptomonads and diatoms preceded their surface water blooms by having high abundances in deeper water, suggesting incubation at depth prior to proliferation in the loch surface water. This pattern has been observed for *Aulacoseira* species in other lakes (e.g., Gibson et al. 2003) but diatom life cycles in alpine lochs are poorly researched. The abundances of spherical nanophytoplankton increased in June but generally showed a progressive rise towards an annual maximum in October. It is quite likely that the irregular sampling period missed the absolute peak abundance values for phytoplankton cells and it is also possible that any short term peaks by particular phytoplankton species were missed entirely.

Because chrysophytes, like diatoms, preserve in lake sediments where they can provide evidence of past environmental changes, they are worthy of further comment. Chrysophytes are often a key component of phytoplankton in temperate oligotrophic and mountain lakes (Hutchinson 1975; Eloranta 1995; Rott 1988) where they exhibit a distinct seasonality and are present in a wide range of environmental conditions (Siver 1995). They are motile in the water column and can migrate up or down using flagella.

Table 2. The algal taxa encountered in seven samples of phytoplankton collected from Lochnagar during 1996 to 1998 (data from J. Fott, M. Blazo and E. Stuchlík).

Bacillariophyta	**Chrysophyta**
Bacillariophyceae	Chrysophyceae
Aulacoseira	*Bitrichia*
Naviculaes g.sp.div.	*Chrysolykos*
Stephanodiscus	*Dinobryon*
Tabellaria	*Mallomonas*
	Ochromonas crenata
	Chrysophyceae g.sp.div.
Chlorophyta	
Chlorophyceae	**Cryptophyta**
Botryococcus braunii	Cryptophyceae
Chlamydomonas spp. (ovoid and spherical)	*Cryptomonas*
Cosmarium	*Rhodomonas*
Dictyosphaerium	
Koliella	**Dinophyta**
Oocystis	Dinophyceae
Tetraedron	*Kathodinium*
Tetrastrum	*Peridinium willei*
	Wolozsynskia
Cyanophyta	
Cyanophyceae	
Chroococcus	
Merismopedia	

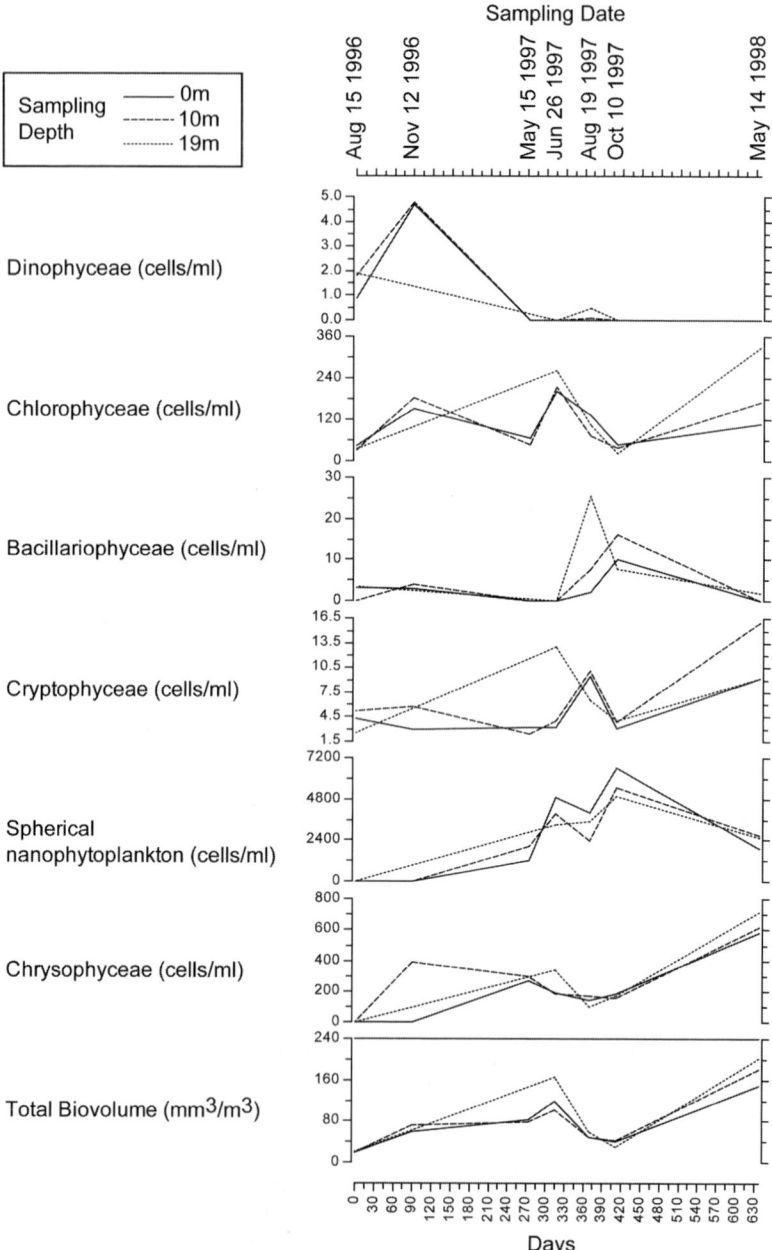

Figure 6. Phytoplankton abundance data for Lochnagar according to open water samples collected from three depths in the water column at irregular periods between August 1996 and May 1998. Algal concentrations are expressed as cells ml^{-1} and, except for the spherical nanoplankton, they are grouped according to Class. Total cell volumes are expressed as mm^3 m^{-3}. Data kindly supplied by J. Fott, M. Blazo and E. Stuchlík.

Like the diatoms, chrysophytes use silica, either as scales or as a protective wall for encysted resting stages. Their siliceous resting stages (stomatocysts) are common in lake sediment and the composition of the assemblages often reflects their surrounding physico-chemical environment (Smol 1995; Duff et al. 1997; Steward et al. 2000; Pla 2001; Zeeb and Smol 2001; Kamenik et al. 2001; Pla et al. 2003, 2005). As a result, chrysophyte cyst stages have been used in palaeolimnological studies to assist in environmental change reconstructions (Smol 1995; Zeeb and Smol 2001).

Using surface sediment samples from Lochnagar collected in the mid-1990s, 49 different morphotypes of stomatocyst were identified (S. Pla unpublished). This is a low number in comparison with other oligotrophic lakes. However, if sediment from the past 200 years is considered (see below) over 160 morphotypes can be found indicating that the diversity of these algae has declined in recent years. Dominant cysts in the surface sediments were unornamented (morphotypes S046 and S001; see Pla 2001 for classification codes), small cysts (diameter < 0.5 µm, S170, S005, S089a, S004 and S050) were also common and were similar to those found in oligotrophic lakes elsewhere (Duff et al. 1995; 1997; Pla 2001; Wilkinson et al. 2001; Pla et al. 2003; Pla and Anderson 2005).

The history of the aquatic vegetation

The microfossils preserved in sediments accumulated in the deep water areas of Lochnagar provide an excellent record of past species abundances and changes that can be used to infer aspects of the loch's environmental history. The diatom microfossil record for Lochnagar has already demonstrated how the species composition of sedimentary assemblages has changed to species more tolerant of increasing water acidity (Jones et al. 1993; Chapter 14: Monteith et al. this volume). The diatom record of acidification of Lochnagar is composed overwhelmingly of benthic species and is discussed elsewhere (Chapter 14: Monteith et al. this volume). The other group of siliceous algae common in loch sediments are the chrysophytes. These algae also provide excellent evidence about environmental changes in the loch as does the pollen record.

Chrysophytes

A biostratigraphic diagram constructed from a sediment core (NAG19) collected from the loch in 1997 (Figure 7) shows the down-core percentage frequency changes of particular morphotypes in the chrysophycean stomatocyst assemblages. The core is dated by its atmospheric contamination record (Chapter 17: Rose and Yang this volume) and the stratigraphy shows quite significant changes in stomatocyst composition during the past several hundred years. Most importantly the frequencies of several cysts increased in the late-19th century. The changes are quite sharp compared with those in the diatom record at this time (Chapter 14: Monteith et al. this volume). Also shown in Figure 7, is the first principal component (axis 1 scores) of variability derived from Principal Components Analysis (PCA) (see Pla 1999). This component is a numerical measure that encapsulates biostratigraphic changes into a single score. It indicates that the stomatocyst record is responding to significant environmental changes during the 19th century that parallel a decline in species diversity (Figure 7). Low first principal component scores correspond with increased acidity (Pla et al. 2003) and declining acidity of the loch during the recent past is also shown by benthic diatom record (Jones et al. 1993). However, in the last two decades of the 20th century, the first principal component scores (Figure 7) indicate a small recovery in assemblage composition as several cyst morphotypes common in early 19th century sediment show small abundance increases towards the core top.

There are probably several environmental processes driving the frequency changes in stomatocyst abundances in this Lochnagar core. Recent acidification is clearly a significant factor (Chapter 14: Monteith et al. this volume) but timing of the trend changes in stomatocysts is quite different from that displayed by the diatoms (cf. Jones et al. 1993). It is believed that diatoms (predominantly benthic in habit) and chrysophytes (planktonic in habit) are affected by different factors or have different responses to similar factors. Nitrate concentration in the water seems to be closely linked with the chrysophyte stomatocyst assemblages in several lake data sets (in the Pyrenees, Pla et al. 2003; in Greenland, Pla and Anderson 2005). In the dataset for Scotland (S. Pla, Loughborough University unpublished) both nitrate concentration and pH are correlated with CCA axis 1 suggesting that nitrate is derived from atmospheric deposition. Also, it has been suggested recently (Pla and Catalan 2005; Kamenik and Smith 2005) that stomatocysts indicate physical changes in lake water such as ice-cover duration and stratification patterns (e.g., thermocline development). These physical characteristics could account for the high seasonality shown by chrysophytes (Siver 1995). In Lake Redon, a high altitude (2240 m.a.s.l.) site in the Pyrenees (Pla and Catalan 2005), the second principal component of variability (CCA axis 2) of the stomatocyst record was correlated with summer temperature fluctuations ($r = 0.66$; average temperature of June, July and August, see Agusti-Panareda and Thompson 2000). In Lochnagar, a short duration thermocline can also develop during the summer (Chapter 5: Thompson et al. this volume) and it is possible that the stomatocyst record in the loch could also indicate trends in water column stability.

Pollen

The sediments of Lochnagar also contain pollen and other microfossil remains of aquatic plants (Patrick et al. 1989) but these show relatively little change during the recent period (Figure 8). The pollen biostratigraphy of a short core (NAG3) shows that frequencies of *Isoetes lacustris* have remained fairly constant in recent centuries but the abundances of *Sphagnum* spores have increased since the late-19th century. This increase could reflect recent catchment peat erosion but the lack of *Isoetes* change tends to suggest that aquatic sphagna have increased as a result of recent acidification. There are also single levels in core NAG3 that record occurrences of *Sparganium* and *Nymphaea*. The significance of these latter two records is unclear; they could indicate the temporary presences of these species in the loch or 'contamination' of the sediment record by extraneous pollen from elsewhere. The latter is possible for *Nymphaea* but *Sparganium* is known also to occur in older sediments from the loch. The vegetational history for Lochnagar over most of the Holocene period was examined in a longer core dated by ^{14}C AMS (Dalton et al. 2001; 2005). This record shows rather more floristic change with *Potamogeton*, *Sparganium* and *Equisetum* being occasionally present

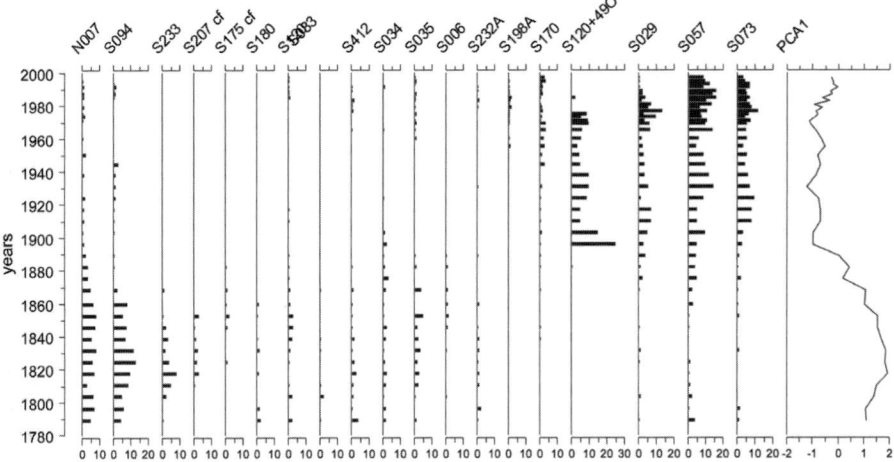

Figure 7. Summary biostratigraphic diagram showing percentage frequency distributions of chrysophycean stomatocysts in a short core (NAG19) collected from Lochnagar in 1997. Dates for sediment depths are calculated from the spheroidal carbonaceous particle record (Rose and Appleby 2005) and those prior to 1850 AD are by extrapolation. PCA axis 1 scores for each assemblage are shown on the right of the figure (see text).

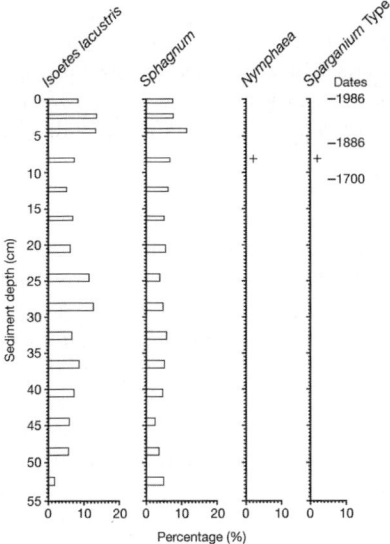

Figure 8. Aquatic vegetation history of Lochnagar. Sediment core profiles selected from the pollen percentage frequency diagram for a short core (NAG3)(Patrick et al. 1989). Note that dates are calculated from ^{210}Pb measurements and those prior to 1850 AD are by extrapolation.

during the past c. 8500 years (Figure 9). The core shows vegetation changes according to various time scales and the main features include a decline in *Isoetes lacustris* after about 7000 years BP. *Sphagnum* spores were always at low frequency. Remains of the aquatic green algae *Pediastrum* and *Botryococcus* were also present throughout most of the sequence. The former, however, disappeared from the record about 500 years ago but *Botryococcus* has persisted in low abundance since about 8000 years ago when it was relatively abundant. The early phase of high abundances of both these algae indicate rather higher nutrient and pH status than in the loch today.

Discussion

Scotland offers a wide range of freshwater loch-types and Lochnagar occupies an extreme situation where harsh conditions combined with high altitude and a granitic catchment play a major role in influencing the abundances and occurrences of aquatic plants. These conditions conspire to produce loch water with very low nutrients and mineral content that supports rather specialised but low species diversity aquatic plant communities.

Plant distributions

Despite the general floristic impoverishment of vascular aquatic plants in Lochnagar with only two species being currently present, the micro-algae are relatively diverse with over one hundred species of diatoms and chrysophytes. This relatively high species diversity for the micro-algae may be related to the ability of these organisms to exploit an array of niches in a nutrient-poor environment where competition by other major plant groups is low. As with the terrestrial flora of Lochnagar (Chapter 7: Birks this volume), some of the diatom species can be considered as characteristic of the alpine zone (e.g., *Psammothidium scoticum* and several other small *Psammothidium* species, see Flower and Jones 1989) and they are rarely found outside the Cairngorms region in the UK. Although Lochnagar offers an extreme and fairly unusual habitat example in Scotland, alpine lakes at considerably higher altitudes are common in mountainous regions elsewhere in Europe. However, many other species recorded at Lochnagar are not restricted to Cairngorm lochs and are known from other lakes with similar typology. For example, several lakes in the granitic Gredos Mountains of Central Spain possess several *Frustulia* and *Aulacoseira* taxa that occur in Lochnagar. The same is true for softwater lakes above 2000 m altitude in the Alps (e.g., Psenner and Schmidt 1992) and even in eastern Siberia (Flower et al. 1997a). Lake Redon in the Spanish Pyrenees is one alpine site that has been studied extensively and its diatom flora is similar to that of Lochnagar in that the dominant species are also *Aulacoseira* spp. (*A. lirata, A. alpigena*). However, because of geological differences and less organic acidity, the pH of Redon is rather higher than Lochnagar, at around pH 6.4 and other species, including the chrysophytes, reflect this lack of acidity (Pla 1999). Clearly many of the acid cold water benthic diatoms living in Lochnagar enjoy at least a pan-Eurasian distribution and some acidophilous taxa even extend to the southern hemisphere (Flower 2005).

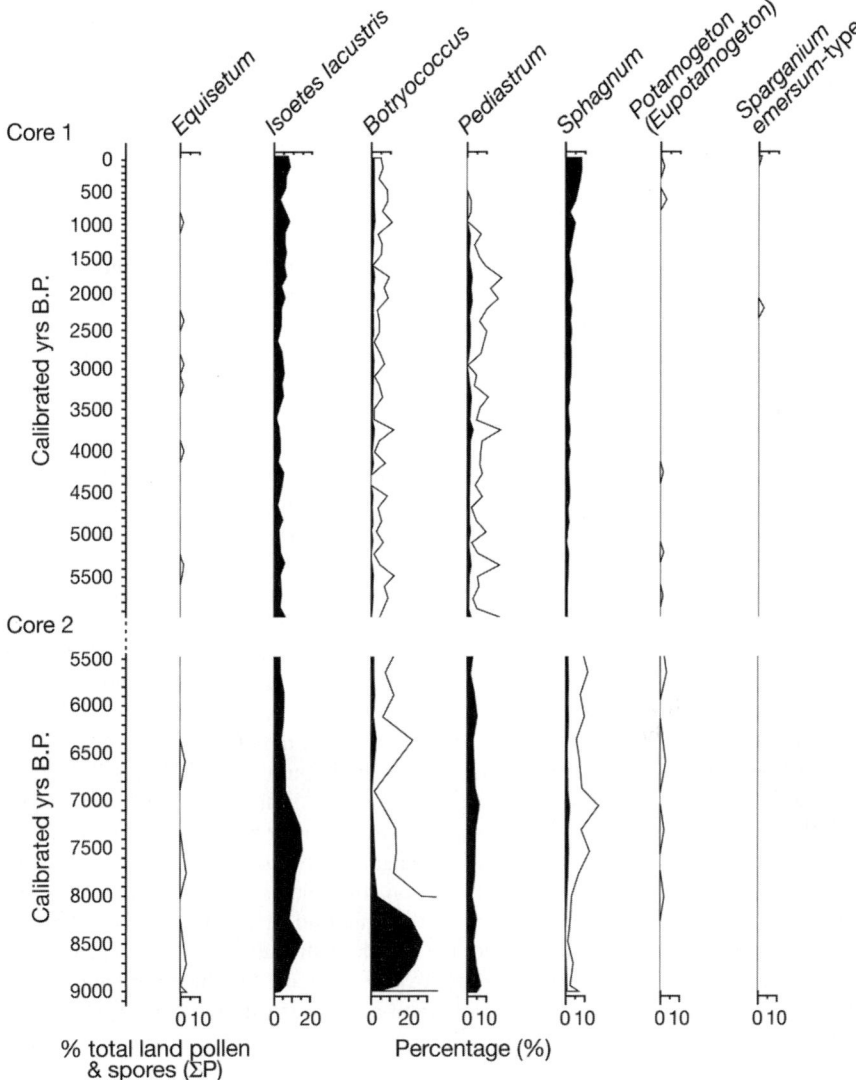

Figure 9. Aquatic vegetation history of Lochnagar. Sediment core profiles are selected from the plant microfossil percentage frequency diagrams produced from cores NAG27 and 30 (after Dalton et al. 2001). The profiles are constructed from two overlapping sediment cores (cores 1 and 2). Dates are calibrated radiocarbon years BP.

Presence of the two vascular aquatic plant species, *Isoetes lacustris* and *Juncus bulbosus* var. *fluitans*, in Lochnagar is unexceptional. In contrast to some of the diatom algae, neither of these species can be considered as indicative of alpine conditions *per se*; they occur in many acid lochs and lakes at a variety of altitudes in the UK (Monteith and Evans 2000; Palmer et al. 1992). Consequently, the alpine nature of Lochnagar is perhaps best exemplified by the absence of some typical oligotrophic aquatic macrophytes such as *Lobelia dortmanna* and *Littorella uniflora* (cf. Figure 1). These plants occur frequently in lochs elsewhere and are probably excluded from Lochnagar by physiological limitations. A general discussion of species richness in upland lakes exceeds the scope of this Chapter but observations at Lochnagar and other sites do question a simple altitudinal explanation of aquatic plant diversity (cf. Jones et al. 2003). *L. uniflora* and several other submerged aquatic plants grow in Sandy Loch at slightly higher altitude than Lochnagar implying that altitude-temperature effects on plant diversity can be conditioned locally by water quality and site aspects. Emergent aquatic plants are absent at both sites and Light (1974) attributed this *inter alia* to ice scour in Cairngorm lochs. The evaluation of environmental controls on diversity needs consideration of a wide geographical range of sites that display long environmental gradients. For example, in a different climatic zone, alkaline Lake Titicaca at 3800 m altitude in South America is generally ice free and possesses extensive marginal reed swamps.

Altitude and aspect undoubtedly influence aquatic plant diversity in Lochnagar but a range of other factors are implicated (cf. Jones et al. 2003) and these include water quality, light and local temperature effects. Temperature as well as light can influence plant reproduction mechanisms and/or seed germination and, although an ice-scour explanation may hold for the absent of emergent and very shallow rooted plants in Cairngorm lochs (Light 1974), the lack of aquatic plants below c. 3 m water depth in Lochnagar requires another explanation. This maximum depth limitation seems unrelated to light penetration *per se* (water transparency in the loch usually exceeds 9 m). Furthermore, in lakes of similar transparency but at lower altitudes, *Isoetes lacustris* can grow at twice the depth observed in Lochnagar (Don Monteith University College London unpublished). Temperature effects on the plants reproductive cycle are suspected as the most likely cause of growth limitation at depth. If this is correct, predicted climate change could extend the depth distribution of these plants and possibly others in the loch, especially if ice cover is reduced. On the other hand, global warming will threaten the persistence of arcto-alpine terrestrial plants generally in Scotland (Birks 1997).

Environmental quality

Despite being affected by acid deposition and exposed to a harsh climate, the current water acidity of Lochnagar is usually above pH 5 but below 6 (depending on year and season, Monteith and Evans 2000). This is not excessively low compared with some lochs elsewhere in Scotland (Palmer et al. 1994) but low pH does physiologically preclude many aquatic macrophytes (Spence 1964). Changes in CO_2 concentration in both water and sediment are thought to be important (Farmer 1990) and alkalinity in the pH 5-6 range offers little buffering capacity. The recent acidification of the loch by atmospheric pollution began during the latter part of the 19th century (Chapter 14:

Monteith et al. this volume) and has promoted floristic changes especially of benthic diatoms and planktonic chrysophytes. Consequently, some species have effectively disappeared from the aquatic flora (Jones et al. 1993). During UKAWMN monitoring of Lochnagar, significant further floristic changes have been detected in living communities of both the aquatic macrophytes and epilithic diatoms. The monitoring period partly coincides with the trend of diminishing atmospheric acid pollution which for the UK occurred since about 1980 (Mason 1990). The overall acidity of Lochnagar has, however, shown little sign of recovery during the 1990s (Chapter 14: Monteith et al. this volume). Several monitored upland lochs elsewhere in the UK do show evidence for recovery of aquatic plant communities (Monteith and Evans 2000). The subdued response of Lochnagar could indicate that other contaminants linked with atmospheric acid pollution such as nitrogen compounds are interfering with ecological recovery (Chapter 14: Monteith et al. this volume). Furthermore, climate change and reduced acid atmospheric pollution are linked with increasing concentrations of dissolved organic compounds in upland waters (Jones et al. 2003; Evans et al. 2005). Dissolved organic carbon in Lochnagar has almost doubled in concentration since 1988 (Evans et al. 2005) and is undoubtedly influencing the ecology and distribution of submerged aquatic vegetation. However, the continuing atmospheric contamination of the loch by heavy metals and persistent organic pollutants (Rose et al. 2001) cannot, as yet, be directly linked with any changes in plant distributions.

The overall implications are, that even at this remote alpine site, vegetation changes are occurring that are the result of a complex of environmental change drivers. These drivers include pollutants as well as the subtle effects brought about by climate change. The precise environmental change processes that are influencing plant growth in Lochnagar are currently unclear but the recent increases in *Juncus bulbosus* var. *fluitans* could be related to increased dissolved nitrogen compounds irrespective of any recovery in water pH. Until more is known about driver interactions within the catchment and within the water body of the loch, it is not possible to make confident predictions about future biological changes. It is likely that diminishing atmospheric contamination combined with a milder climate will promote loch water alkalinity (cf. Schindler et al. 1996) and encourage plant growth. However, because of internal recycling mechanisms, biological lag times and further changes in water transparency and quality, biological responses will be slow, difficult to predict and not necessarily 'desirable'.

Assessing vegetation change

The only sure way to track the future development of aquatic vegetation within Lochnagar is to continue and perhaps intensify the monitoring work currently undertaken as part of the UKAWMN. Without monitoring, the botanical changes occurring in the loch would remain largely undetected. Because aquatic plant sampling has been restricted mainly to annual assessment, there is very little information on seasonality or vegetational changes on shorter time scales. Such changes might be linked with plant reproductive strategies and ecological interactions under the influence of both intrinsic and extrinsic environmental change drivers. Furthermore, physiological responses of aquatic plants in Lochnagar to these drivers remain largely unresearched. The increasing depth range exhibited by *Isoetes lacustris* since 1988 is one example where physiology is undoubtedly linked with extrinsic environmental factors but mechanisms involving temperature are currently only speculative hypotheses. Monitoring is not only concerned with detecting the pace and nature of environmental change (as is well born out by the UKAWMN results for Lochnagar), it is also about identifying the processes of biological change and setting hypotheses about the driving causes.

At the more basic level perhaps, compiling species biodiversity inventories together with an adequate system for recording community composition and locational information provides the foundation for setting and evaluating species baselines. Rather than focussing on species composition of aquatic communities (cf. Rodwell 1995), investigations at Lochnagar demonstrate the importance of monitoring the locations and extent of individual species. This is epitomised by the remarkable expansion of *Juncus bulbosus* var. *fluitans* since 1988. More precise measurements on the pace and extent of species abundance changes are needed and these could be made using a differential Global Positioning System (dGPS) to monitor the cover of individual aquatic macrophyte species annually. Defining the extent of plant community compositional changes in time and space can contribute to regional biodiversity action plans, help select areas of biological significance and evidence environmental change processes in relation to pollution. These criteria are relevant to the European Union's Water Framework Directive (WFD, see Foster et al. 2001; Chapter 13: Rose et al. this volume) where ecological criteria are of particular importance for improved management of freshwaters generally.

Partly stemming from implications of the WFD, more attention is now being given to the restoration of damaged aquatic habitats in Scotland (e.g., Doughty et al. 2002). However, alpine lochs such as Lochnagar offer a special case where local management activities are largely irrelevant and restoring reference conditions precisely is probably now impossible because of climate change and continuing air pollution have re-set base-line reference conditions (cf. Battarbee et al. 2005b). Mitigating recent acidification effects in Lochnagar for example, by liming, is similarly not feasible logistically. Some forms of acidic atmospheric pollution are however diminishing (as a result of emission changes) and detecting signs of biological recovery is one aim of the UKAWMN programme (Monteith and Evans 2000). The value of such surveillance monitoring can be enhanced by use of sediment records. The presences of microfossils in pre-acidification sediments can help identify modern potential restoration targets (Flower et al. 1997b). Using both diatom and cladoceran remains, Simpson et al. (2005) compared species assemblages in pre-impacted (pre ~ 1800 AD) Lochnagar sediment (in core NAG6) with modern assemblages of these organisms in surface (contemporary) sediments in lakes elsewhere in the UK. This 'time for space' analogue matching technique indicated that two sites, Loch Toll an Lochan (57° 47.83' N; 5° 14.62'W) in northwest Scotland and Llyn Clyd in Wales, are closely matched with the pre-impacted Lochnagar assemblages. The analysis also implicated the modern aquatic macrophytes currently growing in Llyn Clyd (including *Nitella flexilis*, *Myriophyllum alterniflorum*) and in Loch Toll an Lochan (including *Sparganium angustifolia*, *Littorella uniflora*) as possible reference species for Lochnagar. These species are identified by virtue of their association with the precisely matched diatom and cladoceran assemblages in the surface sediments of these two lakes, indicated schematically in Figure 10.

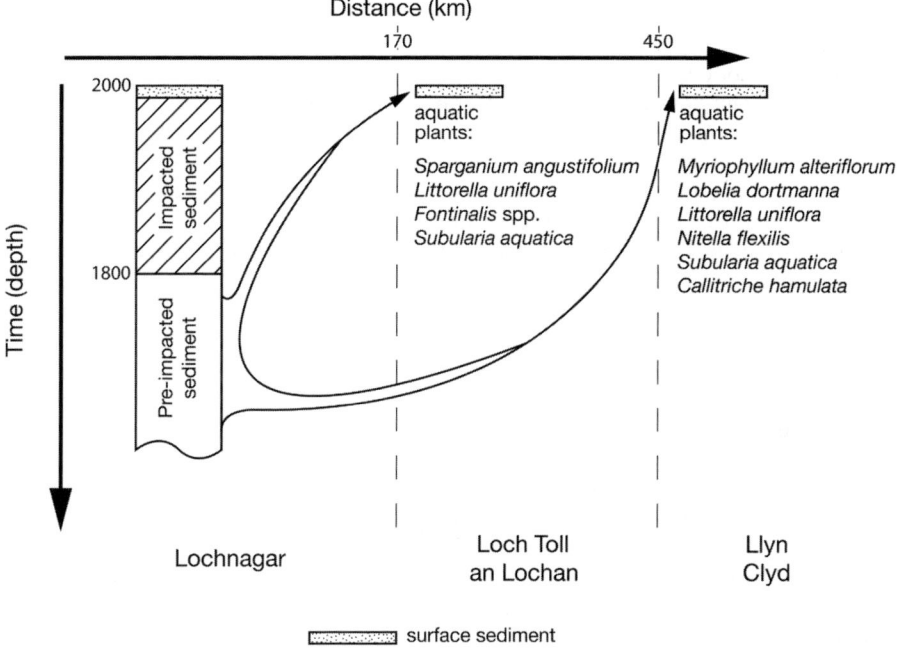

Figure 10. A schematic representation of analogue matching lake sediments. Fossil assemblages of aquatic organisms (diatoms and Cladocera preserve well in lake sediments) were used to compare pre-impacted Lochnagar sediment (pre- c. 1800 AD) with those modern assemblages found in surface sediments from lakes elsewhere. Numerical analysis (Simpson et al. 2005) showed that surface sediments from two sites, Loch Toll an Lochan (northwest Scotland) and Llyn Clyd (Wales), were most closely matched with pre-impacted assemblages in Lochnagar; the analysis also implicates several aquatic macrophytes as reference species for Lochnagar (see text). This Figure is adapted from Flower et al. (1997a).

Use of modern analogue matching for implicating the presence of aquatic plant species in pre-impacted Lochnagar can be further evaluated from sediment vegetation records. For Lochnagar we are fortunate in having a detailed Holocene vegetation history (Dalton et al. 2001; 2005) and by back-checking with the pollen record (cf. Figure 9) both *Sparganium* and *Littorella* are indicated to be formerly present in the loch. This tends to confirm the value of analogue matching but there is no evidence that several aquatic plant taxa (*Nitella* and *Myriophyllum*) implicated by the Llyn Clyd match were ever present in Lochnagar. Loch Toll an Lochan therefore seems to be the closest biologically matched reference site for Lochnagar although extending the surface sediment dataset with more potential modern analogue sites in the Highlands could change this designation. Furthermore, the analogue matching technique may not be appropriate for species with geographically restricted distributions. However, more specific comment concerns *Potamogeton* taxa. *Potamogeton* has not been recorded from Loch Toll an Lochan but the former presence of this taxon in Lochnagar is indicated by the sedimentary pollen record (cf. Figure 9). This is strong evidence for this plant formerly growing in Lochnagar but questions arise about the significance of the *Potamogeton* record. Does the presence of this pollen represent established communities or brief periods of opportunistic growth or transport of pollen from elsewhere? Unfortunately, individual *Potomogeton* species cannot be ascertained from the pollen record but the current distribution of *Potamogeton* taxa (such as *P. polygonifolius*) in upland UK sites indicates that Lochnagar is near their environmental limit (Preston and Croft 1997). Nevertheless, the pollen record is consistent with some *Potamogeton* plants persisting in or near Lochnagar during the past.

There are difficulties in establishing the precise nature of past plant communities in Lochnagar but it is clear from the fore-going that these communities have changed on a range of time-scales. The most recent and rapid botanical changes have occurred within the past two centuries mainly as a result of atmospheric pollution while earlier changes are more strongly linked with climate and ecosystem ontogeny. On a centennial time-scale, catchment-driven changes have been predominant (Dalton et al. 2005) and peat bog records for the Highland region indicate that bog surface wetness has varied markedly during the past 4500 years in sympathy with climate (Langdon and Barber 2005). Whether future atmospheric pollution and climate change interactions will favour the return of species formerly present in Lochnagar is unclear but is best assessed through long term biological monitoring. If any ecological restoration by direct intervention is ever contemplated for Lochnagar then reference sites identified by modern analogue matching techniques could conceivably serve as potential sources for re-introductions of aquatic plants and other aquatic organisms that do not leave fossil records. However, more detailed examination of both micro- and macrofossil remains in shallow and deep water sediment cores from Lochnagar could also yield more evidence about the aquatic plants and animals that formerly existed in the loch.

Summary

- Lochnagar possesses considerable biological merit as an alpine aquatic ecosystem.

- The loch possesses only two aquatic vascular plant species yet microphyte diversity is relatively high with over one hundred species of diatoms and chrysophytes being recorded.

- The low diversity of aquatic macrophytes probably reflects both water quality and more directly altitude related factors such as temperature and ice-scour effects.

- Environmental conditions at Lochnagar are undoubtedly too harsh for many species of aquatic macrophyte yet, since annual monitoring began in 1988, the cover distribution of one species (*J. bulbosus* var. *fluitans*) has expanded considerably.

- On a longer time-scale, microfossil records in sediment cores from the loch show that significant floristic changes have occurred as a result of climate change and more recently of atmospheric pollution.

- Nevertheless, Lochnagar remains a prime example of an alpine loch in the Scottish Highlands; it possesses immense value as a national amenity and is a national heritage ecosystem relevant to landscape conservation policy.

- As a result of the research invested in Lochnagar, particularly during the past twenty years, a strong foundation has been laid for understanding the relationships between its aquatic plant communities and environmental change processes. Continued surveillance of Lochnagar and similar sites are required to help inform national legislation initiatives and management policies that are needed to conserve the Highland environment.

Acknowledgements

We wish to thank all our colleagues in the ECRC for the various ways in which they have contributed to planning, facilitating and organising the various lines of research conducted at Lochnagar in the past two decades, especially Viv Jones, Nigel Cameron, Neil Rose, Simon Patrick, John Birks and Rick Battarbee. We are especially grateful to Evzen Stuchlik and his colleagues for the raw data on phytoplankton abundances in Lochnagar. The manuscript benefited from comments by Catherine Duigan and Nigel Willby. Jo and Molly Porter have provided invaluable field assistance with over 250 field visits to Lochnagar during the past decade. Funding for this work is from several sources including Ensis Ltd. (UCL) and DEFRA (Department for Environment, Food and Rural Affairs, UK). Aquatic plant monitoring at Lochnagar is carried out by the ECRC (UCL) and is funded through DEFRA. The Countryside Commission for Wales (CCW) and Scottish Natural Heritage (SNH) kindly arranged permission to access the JNCC Standing Waters Database for aquatic plants. Elanor McBay and Miles Irving kindly helped with producing the diagrams.

References

Agusti-Panareda A. and Thompson R. 2000. Retrodiction of air temperature at eleven remote alpine and arctic lakes in Europe from 1781 to 1997. J. Paleolim. 28: 23-36.

Arts G.H.P. 2002. Deterioration of Atlantic soft water macrophyte communities by acidification, eutrophication and alkalinisation. Aq. Bot. 73: 373–393.

Battarbee R.W., Jones V.J., Flower R.J., Cameron N.G., Bennion H., Carvalho L. and Juggins S. 2001. Diatoms. In: Smol J.P., Birks H.J.B. and Last W.M. (eds.) Tracking environmental change using lake sediments. Volume 3: Terrestrial, algal and siliceous indicators. Kluwer Academic Publishers, Dordrecht. pp 155–202.

Battarbee R.W., Curtis C.J. and Binney H.A. (eds.) 2005a. The Future of Britain's Upland Waters. The Proceedings of a Meeting held on 21st April 2004 at the Environmental Change Research Centre, UCL. Ensis Publishing. London. 53 pp.

Battarbee R.W., Anderson N.J., Jeppesen E. and Leavitt P.A. 2005b. Combining palaeolimnological and limnological approaches in assessing lake ecosystem response to nutrient reduction. Freshwat. Biol. 50: 1772–1780.

Bennion H., Simpson G., Battarbee R.W., Cameron N.G., Curtis C.J., Flower R.J., Hughes M., Jones V.J., Kernan M., Monteith D. T., Patrick S.T., Rose N.L., Sayer C.D. and Yang H. 2002. Environmental change in Scottish fresh waters. In: Usher M.B., Mackay E.C. and Curran J.C. (eds.) The state of Scotland's environment and natural heritage, The Stationery Office, Edinburgh. pp. 145–152

Birks H.J.B. 1973. Past and present vegetation of Skye. A palaeoecological study. Cambridge University Press, Cambridge. 415 pp.

Birks H.J.B. 1997. Scottish Biodiversity in historical context. In: Flemming L.V., Newton A.C., Vickery J.A. and Usher M.B. (eds.) Biodiversity in Scotland: Status, trends and initiatives. The Stationary Office, Edinburgh. 21–35 pp.

Birks H.J.B. This volume. Chapter 7. Flora and vegetation of Lochnagar – past, present and future. In: Rose N.L. (ed.) 2007. Lochnagar: The natural history of a mountain lake. Springer. Dordrecht.

Boston H.L. and Adams M.S. 1987. Productivity, growth and photosynthesis of two small 'isoetid' plants, Littorella uniflora and Isoetes macrospora. J. Ecol. 75: 333–350.

Brettum P. and Halvorsen G. 2004. The phytoplankton of Lake Atnsjoen, Norway – a long term investigation. Hydrobiol. 521: 141–147.

Brook A.J. 1964. The phytoplankton of the Scottish freshwater lochs. In: Burnett J.H. (ed.). The Vegetation of Scotland. Oliver and Boyd Ltd. London. pp. 290–300

Burnett J.H. 1964. The Vegetation of Scotland. Oliver and Boyd, Edinburgh. 613 pp.

Cox E.J. 1984. Observations on some benthic diatoms from North German lakes: The effect of substratum and light regime. Verh. Internat. Verein. Limnol. 22: 924–928.

Dalton C., Battarbee R.W., Birks H.J.B., Brooks S.J., Cameron N.G., Evershed R.P., Peglar S.M., Scott J.A. and Thompson R. 2001. Holocene lake sediment core sequences from Lochnagar, Cairngorm Mts., Scotland UK. Final Report for CHILL-10000. Environmental Change Research Report No. 77. University College London. London. 94 pp.

Dalton C., Birks H.J.B., Brooks S.J., Cameron N.G., Evershed R.P., Peglar S.M., Scott J.A. and Thompson R. 2005. A multi-proxy study of lake-development in response to catchment changes during the Holocene at Lochnagar, north-east Scotland. Palaeogeog. Palaeoclim. Palaeoecol. 221: 175–201.

Darling Fraser F. and Boyd Morton J. 1964. The Highlands and the Islands. Collins, London. 336 pp.

Doughty R.C., Boon P.J. and Maitland P. 2002. The State of Scotland's Freshwaters. In: Usher M.B., Mackay E.C. and Curran J.C. (eds.) The State of Scotland's Environment and Natural Heritage, The Stationery Office, Edinburgh. pp. 117–144.

Duff K.E., Zeeb B. and Smol J.P. 1995. Atlas of Chrysophycean Cysts. Kluwer Academic Publishers, Dordrecht. 189 pp.

Duff K.E., Zeeb B.A. and Smol J.P. 1997. Chrysophyte cyst biogeographical and ecological distributions: a synthesis. J. Biogeog. 24:791–812.

Duigan C. 2005. Why do we care about upland waters? In: Battarbee R.W., Curtis C.J. and Binney H.A. (eds.) The Future of Britian's Upland Waters. Proceedings of a meeting held 21st April 2004 at the Environmental Research Centre, UCL. Ensis Publishing London. pp 8–11.

Eloranta P. 1995. Biogeography of chrysophytes in Finish lakes. In: Sandgren C.D., Smol J.P. and Kristiansen J. (eds.) Chrysophyte algae. Ecology, phylogeny and development. Cambridge University Press, Cambridge. pp. 214–231

Engstrom D.R., Fritz S.C., Almendinger J. E. and Juggins S. 2000. Chemical and biological trends during lake evolution in recently deglaciated terrain. Nature 408: 161–166

Evans C.D., Monteith D.T., and Cooper D.M. 2005. Long-term increases in surface water dissolved organic carbon: observations, possible causes and environmental impacts. Environmental Pollution 137: 55–71.

Farmer A.M. 1990. The effects of acidification on aquatic macrophytes-a review. Environ. Pollut. 65: 219–240

Felip M., Bartumes F., Halac S. and Catalan J. 1999. Microbial plankton assemblages, composition and biomass, during two ice-free periods in a deep high mountain lake (Estany Redo, Pyrenees) J. Limnol. 58: 193–202.

Flemming L.V., Newton A.C., Vickery J.A. and Usher M. B. 1997. Biodiversity in Scotland: Status, Trends and Initiatives. The Stationary Office, Edinburgh. 309 pp.

Flower R.J. 2005. The taxonomy and ecological study of diatoms from freshwater habitats in the Falkland Islands, South Atlantic. Diatom Res. 20: 23–96

Flower R.J. and Jones, V.J. 1989. Taxonomic descriptions and occurrences of new *Achanthes* taxa in acid lakes in the UK. Diatom Res. 4: 227–239.

Flower R.J., Politov S.V., Rippey B., Rose N.L., Appleby P.G. and Stevenson A. C. 1997a. Sediment records of the extent and impact of atmospheric contamination from a remote Siberian highland lake. Holocene 7: 161–173.

Flower R.J., Juggins S. and Battarbee R.W. 1997b. Matching diatom assemblages in lake sediment cores and modern surface sediment samples: the implications for lake conservation and restoration with special reference to acidified systems. Hydrobiol. 344: 27–40.

Foster D., Wood A. and Griffiths M. 2001.The EC Water Framework Directive and its implications for the Environment Agency. Freshwat. Forum 16: 4–28.

Fott J., Blazo M., Stuchlík E., and Strunecky O. 1999. Phytoplankton in three Tatra Mountain lakes of different acidification status. J. Limnol. 58:107–116.

Gibson C.E., Anderson N.J. and Howarth E.Y. 2003. *Aulacoseira subarctica*: taxonomy, physiology, ecology and palaeoecology. Europ. J. Phycol. 38: 83–101.

Hughes M. This volume. Chapter 2. Physical characteristics of Lochnagar. In: Rose N.L. (ed.) 2007. Lochnagar: The natural history of a mountain lake. Springer. Dordrecht.

Hutchinson G.E. 1975. A Treatise on Limnology. Volume III. Limnological Botany. Wiley, New York. 660 pp.

Jenkins A., Reynard N., Hutchins M., Bonjean M. and Lees M. This volume. Chapter 9. Hydrology and hydrochemistry of Lochnagar. In: Rose N.L. (ed.) 2007. Lochnagar: The natural history of a mountain lake. Springer. Dordrecht.

Jones I.J., Wei Li and Maberly S.C. 2003. Area, altitude and aquatic plant diversity. Ecography 26: 411–420.

Jones V.J., Flower R.J., Appleby P.G., Natkanski J., Richardson N., Rippey B., Stevenson A.C. and Battarbee R.W. 1993. Evidence for the acidification and atmospheric contamination of lochs in the Cairngorm and Lochnagar areas of Scotland. J. Ecol. 81: 2–34.

Kamenik C. and Smith R. 2005. Chrysophyte resting stages: a tool for reconstructing winter/spring climate from Alpine lake sediments. Boreas 34: 477–489.

Langdon P.G. and Barber K.E. 2005. The climate of Scotland over the last 5000 years inferred from multiproxy peatland records: Inter-site correlations and regional variability. J. Quat. Sci. 20: 549–566.

Light J.J. 1975. Clear lakes and aquatic bryophytes in the mountains of Scotland. J. Ecol. 63: 937–943.

Mackereth F.J.H. 1957. Chemical analysis in ecology illustrated from Lake District tarns and lakes. I. Chemical analysis. Proc. Linn. Soc. Lond. 167: 159–164.

Mason B.J. 1990. The Surface Water Acidification Programme. Cambridge University Press, Cambridge. 562 pp.

Monteith D.T. and Evans C.D. (eds) 2000. UK Acid Waters Monitoring Network: 10 Year Report. Ensis Publishing, London 361 pp.

Monteith D.T., Evans C.D. and Dalton C. This volume. Chapter 14. Acidification of Lochnagar and prospects for recovery. In: Rose N.L. (ed.) 2007. Lochnagar: The natural history of a mountain lake. Springer. Dordrecht.

Murray J. and Pullar L. 1910. Bathymetrical survey of the Scottish fresh-water lochs. The Challenger Office, Edinburgh. 1420 pp.

Palmer M.A. 1992. A Botanical Classification of Standing Waters in Great Britain. Second edition. Joint Nature Conservation Committee, Peterborough. 44 pp.

Palmer M.A., Holmes N.T.H. and Bell S.L. 1994. Macrophytes. In: Maitland P.S., Boon P.J. and McLusky D.S. (eds.) The freshwaters of Scotland: A national resource of international significance. John Wiley and Sons Ltd. Chichester. 145–169 pp.

Patrick S.T., Flower R.J., Appleby P.G., Oldfield F., Rippey B., Stevenson A.C., Darley J. and Battarbee R.W. 1989. Palaeoecological evaluation of the recent acidification of Lochnagar,

Scotland. Palaeoecology Research Unit, University College London, Research Paper no. 34. 57 pp.

Pearsall W.H. 1921. The development of vegetation in English lakes considered in relation to the general evolution of glacial lakes and rock basins. Proc. Roy. Soc. Lond. B 92: 259–284.

Pla S. 1999. Chrysophycean cysts from the Pyrenees and their applicability as palaeoenvironmental indicators. Ph. D. thesis, University of Barcelona, Barcelona, Spain. 277 pp.

Pla S. 2001. Chrysophycean cysts from the Pyrenees. J. Cramer, Berlin-Stuttgart. 179 pp.

Pla S. and Anderson, N.J. 2005. Environmental factors correlated with chrysophyte cyst assemblages in low arctic lakes of south-west Greenland. J. Phycol. 41: 957–974

Pla S. and Catalan J. 2005. Chrysophyte cysts from lake sediments reveal the submillennial winter/spring climate variability in the northwestern Mediterranean region throughout the Holocene. Climate Dynamics 24: 263–278.

Pla S., Camarero L. and Catalan J. 2003. Chrysophyte cyst relationships to water chemistry in Pyrenean lakes (NE Spain) and their potential for environmental reconstruction. J. Paleolim. 30: 21–34.

Preston C.D. and Croft J.M. 1997. Aquatic Plants in Britain and Ireland. Harley Books Colchester. 365 pp.

Psenner R. and Schmidt R. 1992. Climate-driven pH control of remote alpine lakes and effects of acid deposition. Nature 356: 781–783.

Pugnetti A. and Bettineti R. 1999. Biomass and species structure of the phytoplankton of an high mountain lake (Lake Paione Superiore, Central Alps, Italy) J. Limnol. 58: 127–130.

Raven J. and Walters M. 1954. Mountain Flowers. New Naturalist Library, Collins, London. 240 pp.

Roelofs J.G.M., Brandrud T.E. and Smolders A.J.P. 1994. Massive expansion of Juncus bulbosus L. after liming of acidified SW Norwegian lakes. Aq. Bot. 48: 187–202.

Rodwell J.S. (ed.) 1995. British plant communities. Volume 4: Aquatic communities, swamps and tall-herb fens. Cambridge University Press, Cambridge. 283 pp.

Rörslett B. 1985. Death of submerged macrophytes – actual field observations and some implications. Aq. Bot. 22: 7–19.

Rörslett B. and Brettum P. 1989.The genus Isoëtes in Scandinavia: An ecological review and perspectives. Aq. Bot. 35: 223–261.

Rose N.L. This volume. Chapter 8. The sediments of Lochnagar: Distribution, accumulation and composition. In: Rose N.L. (ed.) 2007. Lochnagar: The natural history of a mountain lake. Springer. Dordrecht.

Rose N.L. and Appleby P.G. 2005. Regional applications of lake sediment dating by spheroidal carbonaceous particle analysis I: United Kingdom. J. Paleolim. 34: 349–361.

Rose N.L. and Yang H. This volume. Chapter 17. Temporal and spatial patterns of spheroidal carbonaceous particles (SCPs) in sediments, soils and deposition at Lochnagar. In: Rose N.L. (ed.) 2007. Lochnagar: The natural history of a mountain lake. Springer. Dordrecht.

Rose N.L., Backus S., Karlsson H. and Muir D.C.G. 2001. An historical record of toxaphene and its congeners in a remote lake in Western Europe. Environ. Sci. Technol. 35: 1312–1319.

Rose N.L., Metcalfe S.M., Benedictow A.C., Todd M., Nicholson J. This volume. Chapter 13. National, international and global sources of contamination at Lochnagar. In: Rose N.L. (ed.) 2007. Lochnagar: The natural history of a mountain lake. Springer. Dordrecht.

Rott E. 1988. Some aspects of the seasonal distribution of flagellates in mountain lakes. Hydrobiol. 161: 159–170

Round F.E. 1981. The Ecology of Algae. Cambridge University Press. 653 pp.

Schindler D.W. Bayley S.E., Parker B.R., Beaty K.G., Cruickshank D.R., Fee E.J., Schindler E.U. and Stainton M.P. 1996. The effects of climate warming on the properties of boreal lakes and streams at the Experimental Lakes Area, northwestern Ontario. Limnol. Oceanogr. 41: 1004–1017.

Schuurkes J.A.A.R., Kok C.J. and Den Hartog C. 1986. Ammonium and nitrate uptake by aquatic plants from poorly buffered and acidified waters. Aq. Bot. 24: 131–146.

Shilland E.M., Monteith D.T., Bonjean M., and Beaumont W.R.C. 2005. The United Kingdom Acid Waters Monitoring Network Data Report for 2004–2005 (year 17). Report to the Department for Environment, Food and Rural Affairs (Contract EPG 1/3/160). ENSIS Ltd., London. 188 pp.

Simpson G.L., Shilland E.M., Winterbottom J.M. and Keay J. 2005. Defining reference conditions for acidified waters using a modern analogue approach. Environ. Pollut. 137: 119–133.

Siver P.A. 1995. The distribution of chrysophytes along environmental gradients: Their use as biological indicators. In: Sandgren C.D., Smol J.P. and Kristiansen J. (eds.). Chrysophyte Algae. Ecology, Phylogeny and Development. Cambridge University Press, Cambridge. 232–268 pp.

Smol J. P. 1995. Application of chrysophytes to problems in paleoecology. In: Sandgren C., Smol J.P. and Kristiansen J. (eds.). Chrysophyte algae: Ecology, phylogeny and development. Cambridge University Press, Cambridge. pp. 303 29.

Spence D.H.N. 1964. The macrophyte vegetation of freshwater lochs, swamps and associated fens. In: Burnett J.H. (ed.) The vegetation of Scotland. Oliver and Boyd Ltd. London. 306–381 pp.

Steward K., Gregory-Eaves I., Zeeb B.A. and Smol J.P. 2000. Covariation among Alaskan chrysophyte stomatocyst assemblages and environmental gradients: a comparison with diatoms. Nord. J. Bot. 2: 357–68.

Svedäng M.U. 1990. The growth dynamics of *Juncus bulbosus* L.: A strategy to avoid competition? Aq. Bot. 37: 123–138.

Thompson R., Kettle H., Monteith D.T. and Rose N.L. This volume. Chapter 5. Lochnagar water-temperatures, climate and weather. In: Rose N.L. (ed.) 2007. Lochnagar: The natural history of a mountain lake. Springer. Dordrecht.

Wathne B.M., Patrick S.T., Monteith D.T. and Barth H. 1995. Acidification of Mountain Lakes: Palaeolimnology and Ecology – AL:PE. AL:PE 1 Report. Ecosystems Research Report Series, Number 9. European Commission, Luxembourg. 296 pp.

West W. and West G.S. 1909. The British freshwater phytoplankton with special reference to desmid plankton, and the British distribution of desmids. Proc. Roy. Soc. B81: 165–206.

Wetzel R.G. 1975. Limnology. W. B. Saunders Company, Philadelphia. 767 pp.

Wilkinson A.N., Zeeb A.B. and Smol J.P. 2001. Atlas of Chrysophycean Cysts. Volume II. Kluwer Academic Publishers, Dordrecht. 169 pp.

Zeeb B.A. and Smol J.P. 2001. Chrysophytes scales and cysts. In: Smol J.P., Birks H.J.B. and Last W.M. (eds.) Terrestrial, algal and siliceous indicators. Kluwer Academic Publishers, Dordrecht. 203–244 pp.

11. PATTERN AND PROCESS IN THE LOCHNAGAR FOOD WEB

GUY WOODWARD (g.woodward@qmul.ac.uk)
and KATRIN LAYER
School of Biological and Chemical Sciences
Queen Mary University of London
Mile End Road
London E1 4NS
United Kingdom

Key words: community closure, detritivory, food web, herbivory, invertebrate, predation, recovery, trophic interactions

Introduction

Palaeolimnological records show that Lochnagar started to acidify in the mid-19th century, as the Industrial Revolution progressed and intensified (Rose et al. 2004). Biomonitoring of the extant community has focussed on detecting responses to more recent abatement strategies, which have reduced acidifying emissions markedly since the 1970s (Monteith and Evans 2000; Monteith et al. 2005). To date, however, there is little evidence of chemical recovery: pH has remained relatively unchanged, possibly due to declines in sulphate being offset by increased nitrate deposition. Similarly, full basin fluxes of key trace metals have not declined and, in some instances, have even increased (Rose et al. 2004). This is likely to reflect continuing inputs of previously deposited metals from the catchment. Climatic change has been suggested as an explanation for these continuing inputs, via enhanced soil erosion and leaching from the catchment. Climate forecasts for the 21st century suggest accelerated temperature increases combined with a rise in winter precipitation (Chapter 18: Kettle and Thompson this volume). These predicted changes are likely to exacerbate the input of metals from the catchment, whereas higher temperatures might act antagonistically, serving to promote recovery from acidification (Rose et al. 2004). The effects on the biota could be profound, as not only are many metals toxic, but also rising temperatures will elevate basal metabolic rates, and this could have far-reaching effects across multiple scales of biological organisation, from the individual to the entire ecosystem (Woodward et al. 2005a).

Given the current absence of clear evidence for chemical recovery, any biological recovery seems unlikely, at least in the short-term, although in the longer term, perhaps

N.L. Rose (ed.), Lochnagar: The Natural History of a Mountain Lake
Developments in Paleoenvironmental Research, 231–252.
© *2007 Springer.*

over tens of decades, we should expect to see shifts in the community in response to an eventual reversal of acidification. Any eventual recovery, however, is unlikely to be an instantaneous response to pH; if new colonists are to persist, they will need to establish themselves within the food web and overcome any capacity to resist invasion that the system may possess. Thus, during the early stages of biological recovery, we should expect to see rare but increasingly frequent occurrences of acid-sensitive species, particularly during the summer months when pH is relatively high and aerial insects (which dominate the animal community) are dispersing overland. Consequently, the proximity and size of neighbouring source pools of such species is a potential rate-limiting step and, because dispersal abilities differ among species, so will their colonisation of the loch. In addition to these extrinsic factors, intrinsic characteristics of the loch's resident community might also play an important role in determining the rate of recovery. For instance, it has been suggested that the lack of biological recovery in some acidified freshwaters that have exhibited chemical recovery could be due to some kind of ecological inertia, associated with the internal dynamics of the system itself (Ledger and Hildrew 2005). If this is true, then we might observe a hysteresis in the trajectory of recovery, such that recovery does not simply equate with a reversal of the changes that occurred during acidification.

Knowledge of the structure of the extant community is a prerequisite for identifying the onset of signs of biological recovery. With this aim in mind, we sought to characterise the main features of the Lochnagar food web, with a particular emphasis on the invertebrates that occupy the intermediate trophic levels (other groups are dealt with in more detail elsewhere in this volume). As a first step, we aimed to place Lochnagar within the context of other systems from across a wide pH gradient, to identify commonalities and exceptions to the general patterns. We then focussed more specifically on the major properties of the Lochnagar food web, in terms of web size (i.e., species richness), trophic architecture (e.g., numbers of species and individuals at different trophic levels), spatial compartmentalisation (e.g., pelagic and benthic 'subwebs'), and the likely trophic basis of production (e.g., autochthonous *versus* allochthonous inputs). The data presented here are preliminary, but a necessary first step to developing a more integrated approach to understanding the Lochnagar food web and how it is likely to respond to environmental change. In summary, the principal questions addressed in this Chapter are: how is the extant community structured, how might it respond to anticipated increases in pH, and over what time scales might we expect to see evidence of biological recovery?

Methods

The role of pH in structuring the community was investigated by comparing the sample scores from the first-axis of an ordination of the benthic invertebrate community with the loch's acid-neutralising capacity over the entire time series, from 1988 onwards (after Monteith et al. 2005). In addition, the frequency of occurrence of benthic macroinvertebrate indicator species was assessed by plotting their independently-derived pH optima, calculated from weighted-averaging regressions for individual taxa in an extensive study carried out in Finland (from Hämäläinen and Huttunen 1996), against their percent frequency of occurrence in the Lochnagar food web over time (after Woodward et al. 2002).

Data were collated from a range of sources to construct a summary food web. Zooplankton and phytoplankton data were drawn from the EMERGE dataset (data supplied courtesy of Martin Kernan, ECRC), and fish, macroinvertebrate and diatom data were taken from the more extensive UKAWMN dataset (see Monteith and Evans 2000 for details of sampling methods). As the pelagic and benthic samples were collected using different methods, the data were used mainly to explore broad patterns in community structure and web architecture (e.g., construction of a summary web from pooled data cf. Woodward et al. 2005b), rather than more fine-grained analyses (e.g., consumer-resource dynamics of individual feeding links).

Recent research has highlighted the importance of body-size as a determinand of food web structure and dynamics (Cohen et al. 2003; Brose et al. 2005; Woodward et al. 2005a; b; c). To explore these relationships in Lochnagar, body masses of the different members of the food web were measured directly, as in the case of the data for trout, zooplankton and phytoplankton, or taken from literature reviews of potential mass for macroinvertebrates (i.e., final instar mass) and diatoms. Regression equations were used to predict body mass from maximal body dimensions, where raw mass data were not available (e.g., Brose et al. 2005; Edington and Hildrew 1995; Elliot 1996; Elliott et al. 1988; Friday 1988; Hynes 1993; Wallace et al. 2003; Woodward et al. 2005a). Functional feeding groups (FFGs), *sensu* Cummins and Klug (1979), were also assigned using data extracted from the published literature. Similarly, data from the literature were used to infer the presence of individual feeding links between pairs of species, in order to construct a preliminary food web for the loch (Warren 1989; Woodward and Hildrew 2001; Schmid-Araya et al. 2002a, b; Brose et al. 2005; Edwards 2005). As consumers in acid waters tend to be extremely generalist and feed on virtually any prey that is smaller than themselves (e.g., Warren 1989; Woodward and Hildrew 2001), it is reasonable to assume that if consumer-resource species pairs that have been described from other systems also coexist in Lochnagar, then the feeding link between them will also be present. Although no direct observations were made of individual feeding links, this is common practice in many food web studies (Hall and Raffaelli 1993). Archived invertebrate samples are available for dissection, should the postulated structure of the food web need to be validated more rigorously at a later date.

Results

The macroinvertebrate community of the loch contained 35 taxa over the entire time series (1988-2004; Table 1). This seemingly low biodiversity was typical of a more general relationship between species richness and pH seen in other standing waters in the UK (Figure 1). There was no obvious relationship between species richness and abundance in Lochnagar (Figure 2), although several taxa were not resolved to species and this undoubtedly resulted in underestimates of biodiversity to some extent (cf. Figures 2a and b). Chironomids in particular were very abundant, but were not resolved any further than this single grouping. The benthic invertebrate assemblage was dominated numerically by either chironomid midge larvae or stoneflies, depending on the year of sampling, with predatory beetles and caddisflies occupying the intermediate trophic levels, and brown trout at the top of the web. Among the macroinvertebrates there were orders of magnitude differences in the abundance of the different functional

Table 1. The macroinvertebrate fauna of Lochnagar, including functional feeding group (FFG) classifications (secondary feeding groups are shown in parentheses for taxa that belong to more than one group). Regression equations used to predict dry weights of taxa within Lochnagar are shown, together with units used for linear body dimensions (HW = Head-capsule width; BL = Total body length). Morphologically similar taxa or higher taxonomic levels, shown in square parentheses, were used where equations were unavailable for Lochnagar taxa. The source of each equation is denoted by a letter at the end of each row (see key below). r^2 values are given where available.

FFG	Taxon	y	x	Regression equation	r^2	
	Trichoptera (Caddisflies)					
G(C)	*Anabolia nervosa* [cased caddis]	ln(mg)	lnHW	1.30 + 3.62(x)	0.82	a
P	*Plectrocnemia conspersa*	$\log_{10}(\mu g)$	$\log_{10}$HW	2.58 + 2.8(x)	-	g
P	*Polycentropus* sp.	ln(mg)	lnHW	1.30 + 3.62(x)	0.87	a
S	*Chaetoperyx villosa* [case-bearing caddis]	ln(mg)	lnHW	1.30 + 3.62(x)	0.82	a
G	*Drusus annulatus* [case-bearing caddis]	ln(mg)	lnHW	1.30 + 3.62(x)	0.82	a
S(P)	*Halesus* sp. [case-bearing caddis]	ln(mg)	lnHW	1.30 + 3.62(x)	0.82	a
S(P)	Limnephilidae [case-bearing caddis]	ln(mg)	lnHW	1.30 + 3.62(x)	0.82	a
S(P)	*Limnephilus* sp. [case-bearing caddis]	ln(mg)	lnHW	1.30 + 3.62(x)	0.82	a
	Plecoptera (Stoneflies)					
S	*Brachyptera risi* [*Leuctra* sp.]	ln(mg)	lnHW	-5.11 + 2.46(x)	0.88	a
S	*Capnia* sp. [*Leuctra* sp.]	ln(mg)	lnBL	-5.11 + 2.46(x)	0.88	a
P	*Chloroperla torrentium* [*Siphonoperla* sp.]	ln(mg)	lnHW	-0.27 + 2.7(x)	0.8	b
P	*Chloroperla tripunctata*	ln(mg)	lnHW	-0.27 + 2.7(x)	0.8	b
P	*Diura bicaudata*	$\log_{10}(\mu g)$	$\log_{10}$HW	0.16 + 3.66(x)	0.89	c
P	*Isoperla grammatica*	$\log_{10}(\mu g)$	$\log_{10}$HW	0.28 + 2.66(x)	0.68	c
S	*Leuctra fusca* [*Leuctra* sp.]	ln(mg)	lnHW	-5.11 + 2.46(x)	0.88	a
S	*Leuctra hippopus* [*Leuctra* sp.]	ln(mg)	lnHW	-5.11 + 2.46(x)	0.88	a
S	*Leuctra inermis* [*Leuctra* sp.]	ln(mg)	lnHW	-5.11 + 2.46(x)	0.88	a
S	*Leuctra nigra* [*Leuctra* sp.]	ln(mg)	lnHW	-5.11 + 2.46(x)	0.88	a
S	*Nemoura* sp. [*Nemurella pictetii*]	$\log_{10}(\mu g)$	$\log_{10}$HW	2.71 + 3.13(x)	0.8	e
S(G)	*Nemurella pictetii*	$\log_{10}(\mu g)$	$\log_{10}$HW	2.71 + 3.13(x)	0.8	e
S(G)	*Protonemura* sp. [*Nemurella pictetii*]	$\log_{10}(\mu g)$	$\log_{10}$HW	2.71 + 3.13(x)	0.8	e
	Ephemeroptera (Mayflies)					
G	Leptophlebiidae	ln(mg)	lnHW	-0.83 + 4.25(x)	0.86	b
G	*Paraleptophlebia* sp. [Leptophlebiidae]	ln(mg)	lnHW	-0.83 + 4.25(x)	0.86	b
	Coleoptera (Beetles)					
S	*Oulimnius tuberculatus* [adult Col.]	ln(mg)	lnBL	-2.01 + 3.23(x)	0.97	d
P	*Agabus guttatus* [adult Col.]	ln(mg)	lnBL	-2.01 + 3.23(x)	0.86	b
P	*Helophorus* sp. [adult Col.]	ln(mg)	lnBL	-2.01 + 3.23(x)	0.86	b
P	*Hydroporus palustris* [adult Col.]	ln(mg)	lnBL	-2.01 + 3.23(x)	0.86	b
P	*Oreodytes davisii* [adult Col.]	ln(mg)	lnBL	-2.01 + 3.23(x)	0.97	d
P	*Oreodytes sanmarkii* [adult Col.]	ln(mg)	lnBL	-2.01 + 3.23(x)	0.97	d
P	*Potamonectes depressus* [adult Col.]	ln(mg)	lnBL	-2.01 + 3.23(x)	0.97	d
P	*Potamonectes griseostriatus* [adult Col.]	ln(mg)	lnBL	-2.01 + 3.23(x)	0.97	d
	Megaloptera (Alderflies)					
P	*Sialis lutaria*	$\log_{10}(\mu g)$	$\log_{10}$HW	2.68 + 2.90(x)	-	h

Table 1 (cont.)

FFG	Taxon	y	x	Regression equation	r^2	
	Tipulidae (Craneflies)					
S(P)	Tipulidae	ln(mg)	lnBL	-5.3 + 2.36(x)	-	h
	Chironomidae (Midges)					
C(G,P)	Chironomidae	$\log_{10}(\mu g)$	$\log_{10}HW$	3.17 + 2.3(x)	-	h
	Oligochaetae (Segmented worms)					
C	Oligochaetae	g		$\pi r^2 l\,(1.05))/4$	-	f

a: Baumgärtner and Rothhaupt (2003); b: Burgherr and Meyer (1997); c: Edwards (2005); d: Towers et al. (1994); e: Hildrew and Townsend (1982); f: S. Rundle (unpublished data); g: Hildrew and Townsend (1976); h: Woodward et al. (2005b)

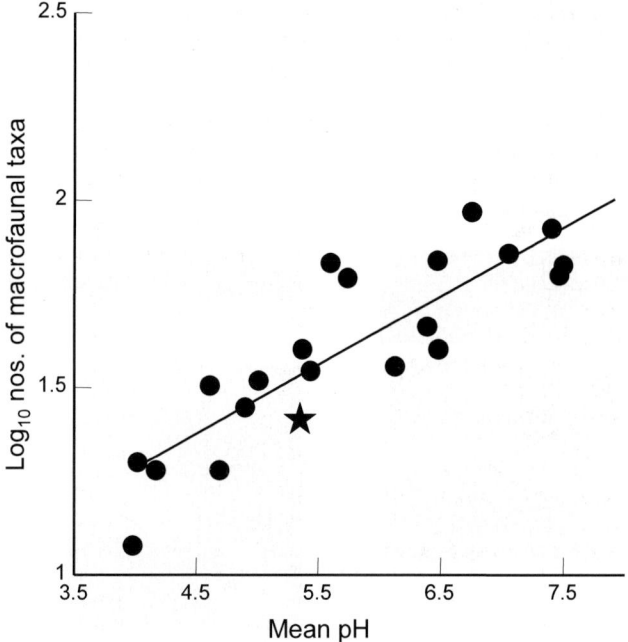

Figure 1. Macrofaunal species richness plotted as a function of pH in 21 freshwaters. The Lochnagar summary food web (denoted by the star symbol) contains 27 macrofaunal species, within the confidence limits predicted by the general relationship for the remaining 20 sites (Regression equation: $y=0.185x + 0.54$; $r^2=0.76$; $n=20$).

feeding groups (Figure 3). The macroinvertebrate benthic assemblage was characterised almost exclusively by detritivores (shredders and collectors; Figure 3). The shredders were dominated numerically by the stonefly, *Capnia sp.*, and the collectors were characterised by the deposit-feeding oligochaetes. Chironomids were excluded from Figure 3, because they were poorly resolved taxonomically, but it is likely that most of these were also collectors (described later). There were very few grazers and grazer- collectors: a single species of mayfly, *Paraleptophlebia* sp.,

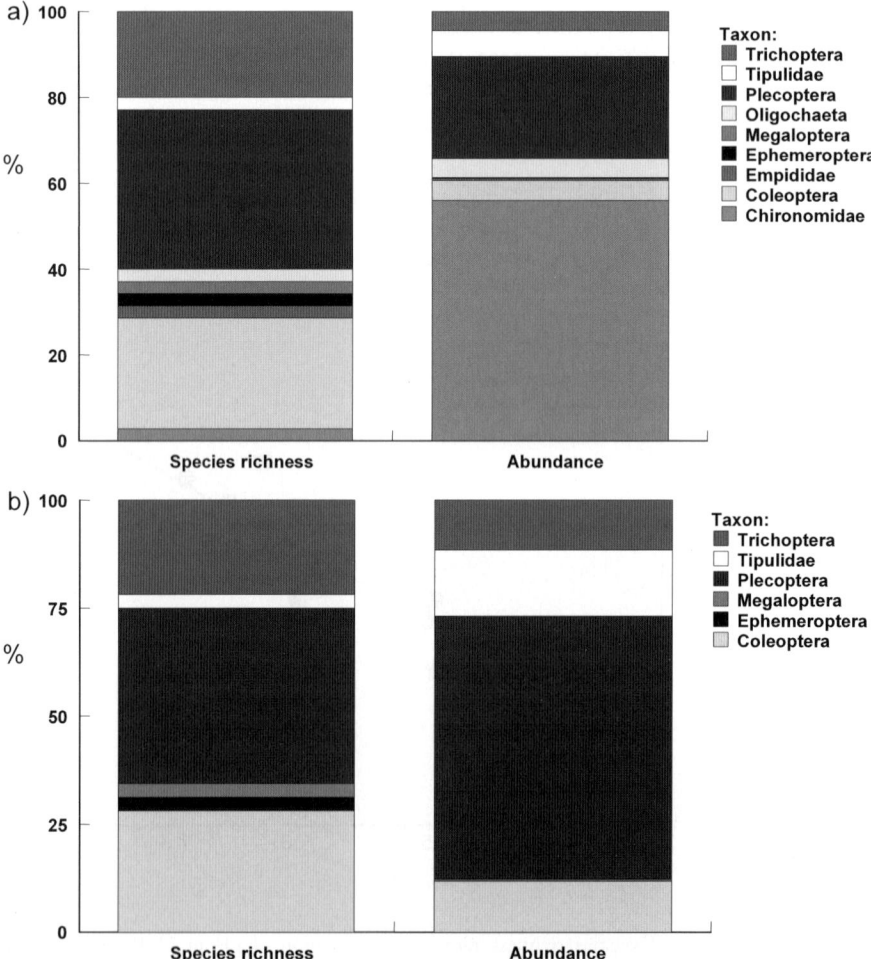

Figure 2. Relative taxon richness and numerical abundance for the Lochnagar benthic macroinvertebrate assemblage. In (a) all taxa are included, even though several groups are poorly resolved taxonomically (chironomids and oligochaetes, for instance, are each grouped into single elements). In (b) only groups that have been resolved to species are included.

was found occasionally, but it was vanishingly rare, presumably reflecting the loch's poor quality algal food resources, which appear insufficient to sustain a significant assemblage of herbivores. There was no clear progressive change in pH over time, and the regular occurrence of Nemouridae stoneflies and other acid tolerant taxa (e.g., *Plectrocnemia conspersa*) suggested that acidity had not ameliorated significantly since the onset of sampling in 1988. The profound effects of acidity on the composition of the macroinvertebrate community were evident, as revealed by the very tight correlation between sample scores from the first axis of a PCA ordination and the acid-neutralising capacity (ANC) of the loch, which fluctuated in synchrony over time (Figure 4). These temporal changes in ANC resulted in shifts in the relative abundance of certain invertebrate groups over time: for example, in 1989-1991 the chironomids, which accounted for 56% of the numerical abundance of benthic invertebrates when averaged over the entire time series, were virtually absent. Conversely, in 1993 and 1994, the stonefly, *Capnia sp.*, which was the numerically dominant species on many other occasions, was effectively absent. Some of the rarer species were also absent on certain sampling occasions: this could be an artefact

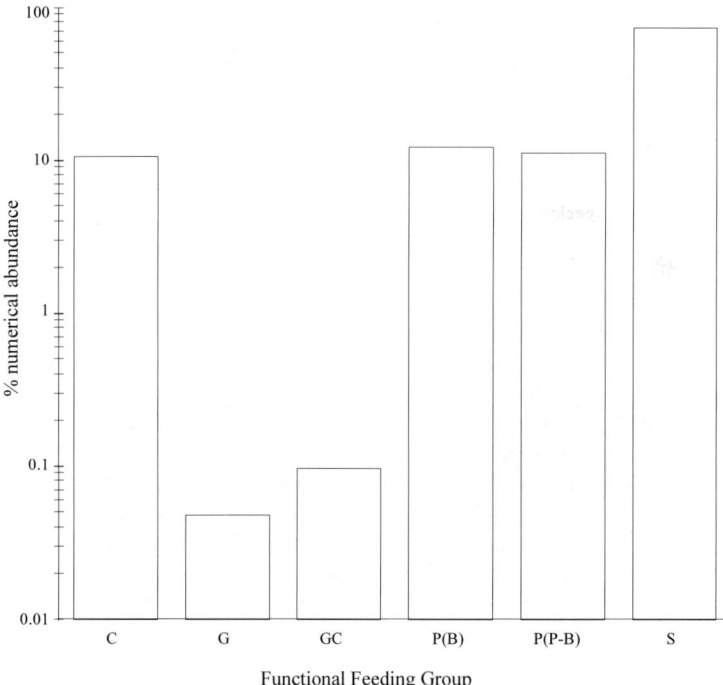

Figure 3. Distribution of relative numerical abundance among the different functional feeding groups of macroinvertebrates, averaged over the entire time series (P(B) = benthic predators; P(P-B) = pelagic-benthic predators, G = grazer, C = collector, GC = grazer-collector, S = shredder). Collectors and shredders are detritivorous, whereas grazers are predominantly herbivorous algal grazers.

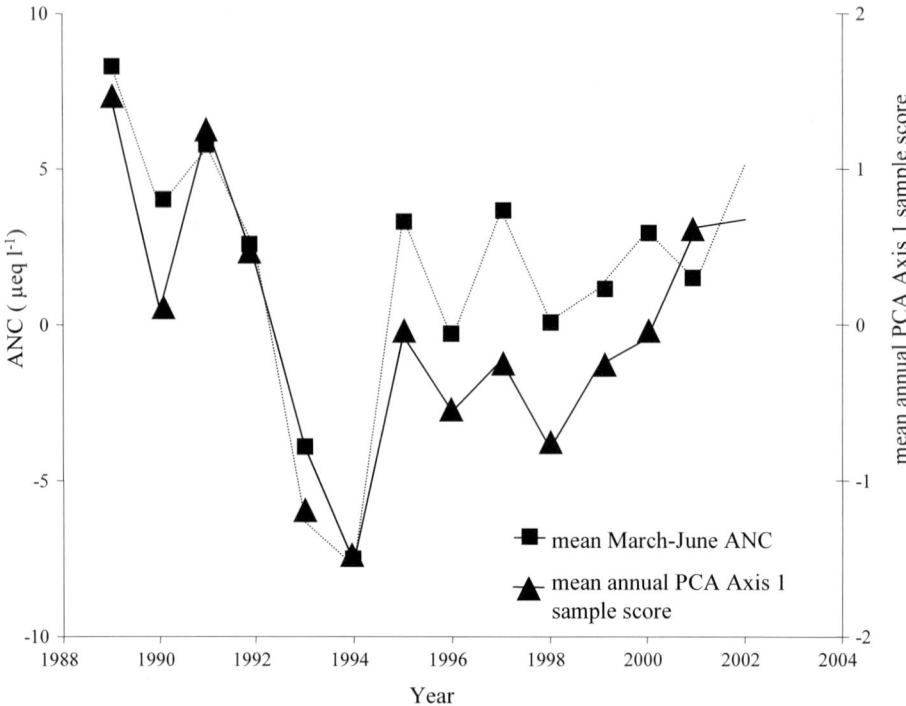

Figure 4. Acid-neutralising capacity (ANC) and PCA Axis 1 scores for the benthic macroinvertebrate community between 1989-2002 (data courtesy of Don Monteith, ECRC).

of rarefaction, as these taxa were generally not missing from the record for more than two consecutive years. However, the strong negative correlation between percentage occurrence and independently-derived pH optima, suggested that at least some of the rarer, more acid-sensitive species, such as *Paraleptophlebia* sp, were repeatedly invading the system but were unable to persist (Figure 5).

The summary community food web contained 212 species in total, which were subdivided spatially into 40 pelagic species, 162 benthic species, and 10 pelagic-benthic species (Figure 6). This latter group, which linked the pelagic and benthic subwebs, included trout, at the top of the web, and the loch's beetle assemblage of eight predators and one primary consumer, *Oulimnius tuberculatus*. Within the pelagic subweb there were four common species of zooplankton, one of which was a predatory copepod, and 36 species of phytoplankton. There was clear size-structuring within the food web, with declines in both diversity and abundance as body-mass (and trophic status) increased (Figures 7 and 8). The principal direction of energy flux within the food web was from smaller, more abundant resource species to larger, rarer consumers, such that food resources were concentrated into progressively fewer individuals and species (Figure 7).

This pattern was also evident within the pelagic subweb, with a negative power-law relationship between body-mass and abundance among the zooplankton and phytoplankton (Figure 8). Maximum food chains lengths were short within the pelagic subweb (four links). Chains were longer in the benthic subweb (up to nine links), although most feeding paths included only 3-4 species (Figures 6 and 7).

Discussion

At macroecological scales species richness is often constrained by external environmental drivers, as revealed by the power law relationship between hydrogen ion concentrations and species richness (Figure 1; cf. Petchey et al. 2004). Lochnagar fits this general relationship, with its low biodiversity being typical of an upland oligotrophic, acidified lake. It also has relatively high numbers of large invertebrate predators and acid-tolerant indicator species, and algal resources are likely to be of limited availability and of poor food quality. The food web is highly interconnected, food chains are short, and trout, at the top of the food web, are scarce, slow growing and in poor body condition, which actually deteriorates with age (Chapter 12: Rosseland et al. this volume), reflecting patterns that are typical of other acid systems (Warren 1989; Woodward and Hildrew 2001; Cohen et al. 2003).

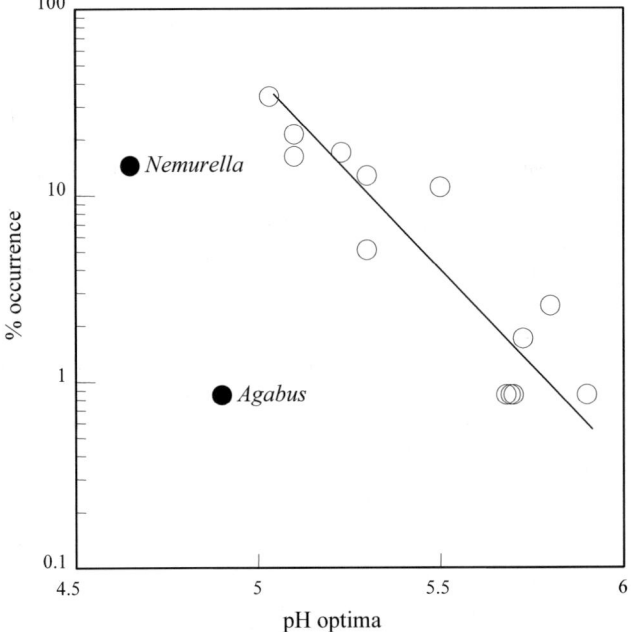

Figure 5. Percentage occurrence per sampling occasion for benthic macroinvertebrates, plotted as a function of independently-derived pH optima (see Methods). The two taxa shown as solid circles were identified as outliers and excluded from the analysis: *Agabus* is a large predator and therefore likely to be rare due to its trophic position and *Nemurella* appears to have started to invade only recently as it has replaced the more acid sensitive *Nemoura* species to become the dominant Nemouridae stonefly. (Regression equation: $y=-2.20x + 12.6$; $r^2=0.80$)

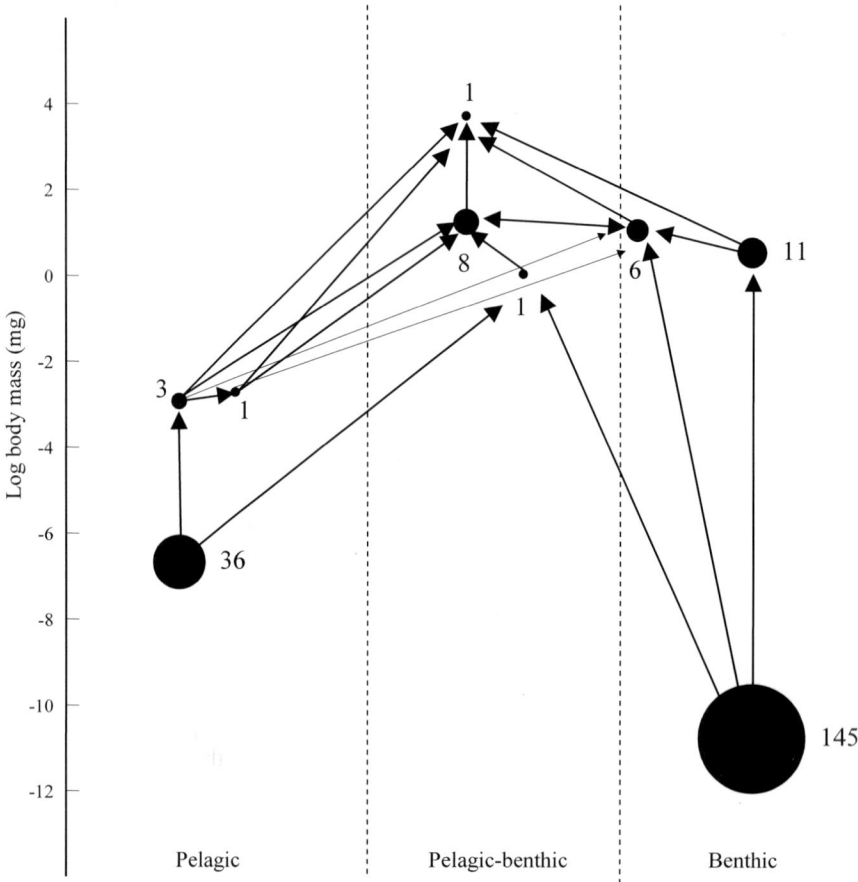

Figure 6. Schematic food web for Lochnagar, with species grouped into broad trophic elements, as defined by body mass, diet and microhabitat use. The *y*-axis is the log_{10} average body mass (mg dry mass) per taxon. The area of the circles represents the taxon richness for each trophic element. The arrows show the direction of energy flux, from resources to consumers. In total the food web contains 212 species, which are subdivided spatially into 40 solely pelagic species, 162 solely benthic species, and 10 pelagic-benthic species which span the two 'subwebs'.

Most of the loch's species are benthic, reflecting the greater availability of both food and refugia within this microhabitat, relative to the pelagic zone, which contains less than one quarter of the total species complement. The lower trophic levels in the benthos are the most diverse component of the food web, with the 145 diatom species alone accounting for more than all the other species summed together. Benthic filter-feeders are absent, presumably due to the highly oligotrophic nature of the loch and the lack of significant inputs of running water. Grazers were also virtually absent, with the exception of small numbers of larvae of the mayfly *Paraleptophlebia* sp. Although their taxonomy is poorly resolved in the current dataset, the chironomids in Lochnagar are likely to be mostly collectors, with a few omnivorous predatory chironomids (Tanypodinae), as in other acidified food webs (e.g., Woodward et al. 2005b). Grazing chironomids should be rare, given the virtual absence of grazers among the other benthic invertebrate groups and the likely poor quality of the algal resource. The larger macrophyte plants within the loch are unlikely to be consumed by aquatic invertebrate 'herbivores', because the macrophyte-piercer functional feeding group is absent and living tissues of these primary producers are well-defended against many other consumers. The macrophytes are therefore likely to enter the aquatic food webs only via detrital pathways, following winter die-back, as is typical of many freshwaters (Webster and Benfield 1986; but see Elger and Willby 2003). The benthic shredder community of stonefly and caddisfly larvae are therefore likely to be the major processors of plant litter.

Within the Lochnagar food web, energy flows predominantly from small, abundant prey to larger, rarer consumers, and species richness also declines as body size (and trophic status) increases (cf. Cohen et al. 2003; Woodward et al. 2005a). Essentially, this means that energy becomes concentrated within progressively fewer individuals and species as it moves through the food web, conforming to Elton's early ideas relating to the "pyramid of numbers" (Elton 1927). Despite the early recognition of such patterns, detailed empirical evidence has only emerged relatively recently and formal mathematical exploration of these patterns is still in its infancy (Warren 2005). It appears, however, that such trends might be common, especially in freshwaters, where most of the data have been collected to date. It would seem likely, then, that Lochnagar could provide a potentially important insight into more general ecological phenomena, and would benefit from further research along these lines.

At the top of the Lochnagar food web, brown trout are potential predators of all of the macroinvertebrate groups at lower trophic levels, in addition to the pelagic zooplankton and terrestrial insects taken from the water surface. Trout are highly opportunistic predators and, depending on food availability, will switch readily between benthic, pelagic, and surface feeding (Chapter 12: Rosseland et al. this volume). Feeding links from trout to 21 of the 40 animal species within the loch were found in the metadata compiled by Brose et al. (2005) and Edwards (2005), although in reality it is likely that the diet of Lochnagar trout is even broader, and all of the animal species in the web are likely to be potential prey. It is difficult to make generalisations about trout feeding in Lochnagar, as only a single sample of 18 fish collected in July 2001 has been used to examine diet directly in the loch, and of these three had empty guts and only six were above 30cm length (Chapter 12: Rosseland et al. this volume). In other oligotrophic lakes, however, brown trout are known to favour the littoral zone rather than the open water (Cavalli et al. 1998), so it is possible that they have stronger interactions with the benthic subweb than with pelagic prey species: indeed of the 15 individuals with identifiable prey in their guts, more than 50% of their gut contents, by volume, were derived from the benthos (Chapter 12: Rosseland et al. this volume). In addition, ontogenetic shifts are likely, with smaller trout potentially being primarily benthic predators, and larger individuals feeding on terrestrial prey and smaller conspecifics (Cavalli et al. 1998). Although no direct evidence of cannibalism was found during the July 2001 sampling, the number of individuals examined was very small and about one third of those caught were in a suitable size range to be cannibalistic (Chapter 12: Rosseland et al. this volume).

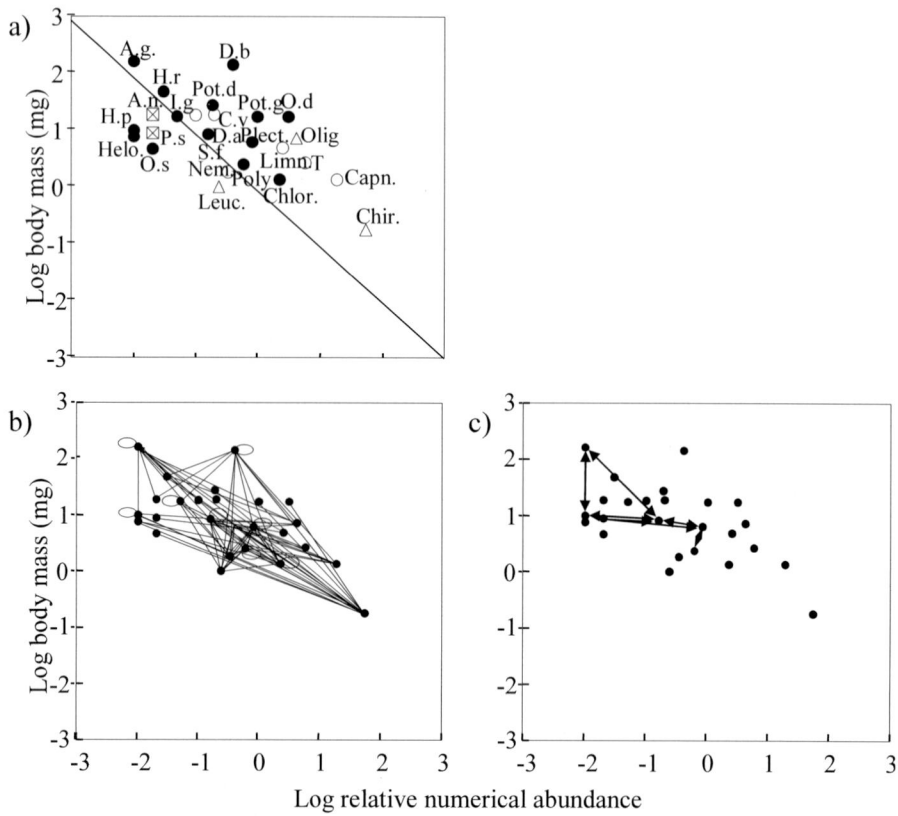

Figure 7. Size structuring within the benthic macroinvertebrate portion of the food web, illustrating the log-log decline in abundance as body-mass increases. In the upper panel (a), the symbols represent functional feeding groups of the different taxa: solid circles denote predators, open circles denote shredders, triangles denote collector-gatherers and squares denote grazers. Taxa are: *Agabus guttatus* (A.g.); *Anabolia nervosa* (A.n); *Capnia* sp. (Capn.); *Chaetopteryx villosa* (C.v); Chironomidae (Chir); *Chloroperla* sp. (Chlor); *Diura bicaudata* (D.b); *Drusus annulatus* (D.a); *Halesus radiatus* (H.r); *Helophorus* sp. (Helo.); *Hydroporus palustris* (H.p); *Isoperla grammatica* (I.g); *Leuctra* sp. (Leuc); *Limnephilus* sp. (Limn); *Nemoura* sp. (Nem); Oligochaetae (Olig); *Oreodytes davisii* (O.d); *Oreodytes sanmarkii* (O.s); *Paraleptophlebia* sp. (P.s); *Polycentropus* sp. (Poly); *Potamonectes depressus* (Pot.d); *Potamonectes griseostriatus* (Pot.g); *Sialis lutaria* (S.l); and Tipulidae (T). Putative feeding links between taxa were derived from literature reviews (see Methods) and are shown in the lower panels (b and c). Panel b) shows the principal direction of energy flux within the food web from smaller, more abundant taxa to larger, rarer taxa, with an increasing concentration of resources into progressively fewer species. Ellipses denote cannibalism. In panel c), unidirectional arrows indicate links where the usual direction of energy flux is reversed (i.e., from larger prey to smaller predator), and double-headed arrows denote feeding loops (*a* eats *b* eats *a*).

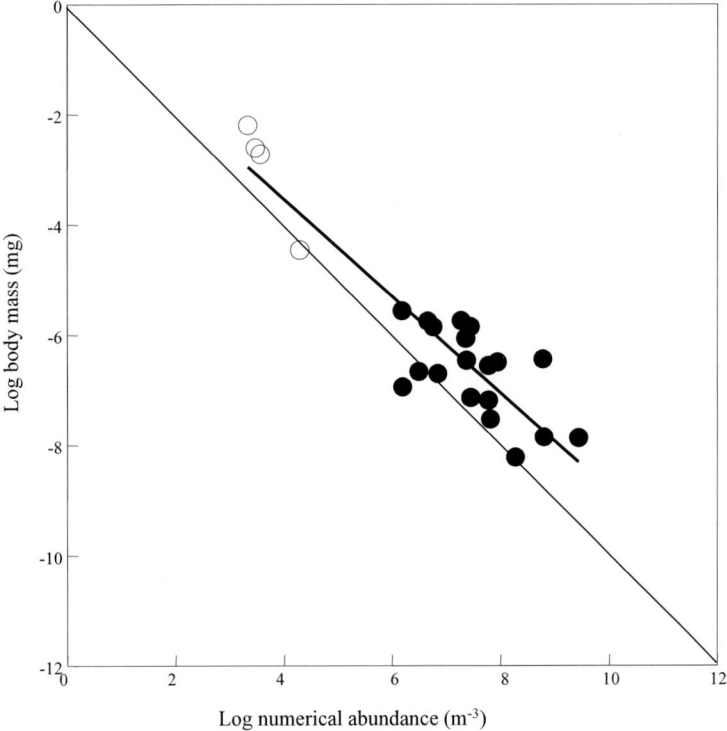

Figure 8. Allometric power law scaling in the size-structure of the planktonic subweb. The mean body mass of each zooplankton (open circles) and phytoplankton (closed circles) species is plotted against numerical abundance on log-log axes. (y = -0.88x - 0.03; r^2=0.83)

At the intermediate trophic levels, the predatory beetles in the loch will feed on both the pelagic zooplankton and the benthic macroinvertebrate assemblage, and will be preyed upon themselves by trout. Among the benthic invertebrate predators, few are likely to be able to capture the adult beetles, with the possible exceptions of the two species of polycentropodid net-spinning caddis. The larval forms of the beetles would, however, be likely prey for some of the larger benthic predators, so complex feeding loops are likely (i.e., larvae of species *a* are eaten by species *b*, which is itself eaten by adults of species *a*). It is also likely, however, that at certain times of the year, particularly when generations overlap, large individuals of 'small' predators will prey upon 'large' prey species in their early instars, and this will further increase the number of pairwise links between species within the web. This life-history omnivory is common in invertebrate-dominated food webs, and is likely to give rise to seasonal reversals in the usual rank order of trophic status when generations overlap (Holt and Polis 1997). Similarly, cannibalism can be common when there is a large size disparity between cohorts or instars within a population: this is true for *Plectrocnemia* and *Sialis* species, which are both present in Lochnagar. These reversals or loops in energy flux are likely to be most prevalent within the benthic subweb, because of the high degree of overlap between the body-sizes of consumers and resources.

In addition to animal-only omnivory, 'true' omnivory, whereby predators consume non-animal prey, is common among some of the loch's predatory stoneflies, such as *Chloroperla* and *Isoperla* species, which can act as facultative grazers (Edwards 2005). However, the scarcity of benthic algae in Lochnagar and the rarity of specialist grazers suggest that this is likely to be a trivial energy source for potential omnivores. Some of the loch's predators are also facultative detritivores, and this type of omnivory is typically exhibited as ontogenetic shifts (Woodward and Hildrew 2002b). For instance, the small *Chloroperla* and *Isoperla* stoneflies are predominantly detritivorous in their early instars, becoming progressively more predatory as they grow (Edwards 2005; Woodward unpublished data). Similarly, many supposedly 'detritivorous' Limnephilidae caddis, such as *Halesus* sp., which are found in the loch, can become voracious predators in the final instar (Warren 1989; Malmqvist 1994; Brose et al. 2005).

In summary then, the consumers in Lochnagar are generalists at all trophic levels, and many of the predators are omnivorous to some degree: this opportunistic feeding is perhaps not surprising, given the scarcity and poor quality of food resources within the loch. As in other acid systems, it seems likely that the consumers will eat most potential food items that they can ingest, and will not display the degree of feeding specialisation seen in other, more productive systems. Consequently, the functional feeding group classifications that are commonly applied to freshwater invertebrates, *sensu* Cummins and Klug (1979), are not as rigid as they may seem. It should be borne in mind, however, that these categories were originally created to describe feeding modes, rather than diet *per se* (as is commonly, but erroneously assumed) (Ledger and Hildrew 2002a; b). Nonetheless, despite the blurring of the boundaries between categories, in terms of diet, differences in abundance among the groups spanned orders of magnitude within Lochnagar, and the food web was dominated by detritivores (either shredders of leaf-litter, or the smaller collectors of fine particulates) and predators, with very few true grazers, which tend to specialise on feeding on algal biofilms.

So what is the main determinant of diet, given the apparent prevalence of generalism within the food web? There is increasing evidence that unlike many of their terrestrial counterparts, few freshwater predators are true specialists and most are opportunistic, with their diet being determined largely by the relative body-size of their prey (e.g., Warren 1989; 1996; Cohen et al. 2003; Woodward et al. 2005a; b; Woodward and Hildrew 2002a; b). The Lochnagar predators fit this generalisation, as revealed by the literature survey of their feeding relationships, and they are capable of preying upon virtually every other animal species smaller than themselves within the food web. Recent research has shown that, on average, consumers tend to be between 1-4 orders of magnitude larger than their prey in terms of body mass, although the range can be considerably greater (Woodward et al. 2005a; Brose et al. 2005): this ratio would effectively place every prey species in Lochnagar within the range of every larger macroinvertebrate predator in the food web.

The trophic generalism exhibited by the predators is applicable to many of the invertebrate primary consumers, which also have catholic diets. This trophic level within the Lochnagar web is characterised by a large contingent of generalist detritivores and herbivore-detritivores, although body-size constraints on feeding might be less obvious here. The stoneflies in particular operate as facultative grazers in acidified systems in the absence of specialist grazers, such as mayflies and snails (e.g., Ledger and Hildrew 2002a; b; 2005). The dominance of generalists at all trophic levels will inevitably result in a highly interconnected, reticulate food web, as is evident in Figure 7. Similar patterns have been reported from many seemingly disparate systems, including soil food webs, estuaries, and other acidified lakes and streams (Williams and Martinez 2000; Woodward et al. 2005c). Such complex networks are likely to be dynamically unstable if the interactions between individual species are strong (i.e., a substantial *per capita* effect of a predator species on the growth rate of a prey population; May 1973). However, if interactions are weak, as seems to be the case in most instances, this complexity might actually serve to stabilise the system (McCann et al. 1998). So, if a food web is highly interconnected, and most of those links are weak, it should be dynamically stable: i.e., resilient and/or resistant to perturbations.

Can we make any inferences about the strength of interactions within Lochnagar, and the system's potential stability? There are several lines of evidence that suggest that most interactions within the food web might indeed be weak. For instance, donor-controlled food webs, based on detrital inputs, appear to be relatively stable (DeAngelis 1975). This is because the consumers do not influence the rate of resource supply, so interaction strengths at the base of the food web are effectively zero (McCann et al. 1998). The structure of the benthic community in Lochnagar suggests that most of the energy inputs to the food web are derived from detritus, rather than primary production, because specialist grazers are vanishingly rare and most of the primary consumers are either shredders or collectors of detritus. The possibility that Lochnagar's food web is based largely on external detrital subsidies is perhaps not as surprising as it might at first seem, as even the much larger and deeper Loch Ness derives much of its energy from external subsidies of detritus, in the form of peat particles (Jones et al. 2001). It would be highly informative to examine the isotopic fractionation of $\delta^{13}C$ of members of the food web, to establish whether this is also the case in Lochnagar. We might expect that once pH starts to rise during the recovery process, the food web will become increasingly autochthonous, and this should be reflected by shifts in the carbon isotope ratios of consumers.

Despite the interactions between primary consumers and basal resources being donor-controlled, predator-prey interactions higher in the food web do not necessarily follow this pattern. However, among the invertebrate predators within Lochnagar, any one species might be too rare to have a significant impact on individual prey species on its own. This is also likely to be the case for the vertebrate predators, as the low densities of trout are probably insufficient to induce the strong cascading effects of fish predation that have been reported in more productive systems (e.g., Power 1990; Scheffer 1998; Jones and Sayer 2003). Consequently, rather than the tight pairwise interactions normally associated with traditional Lotka-Volterra predator-prey dynamics, the entire suite of generalist predators is likely to interact with the prey assemblage in a far more diffuse way in Lochnagar.

A prerequisite for dynamical stability is high persistence of community structure and, despite the interannual fluctuations in community composition in response to ANC, Lochnagar's community is relatively persistent, at least when compared with many other systems, such as productive lowland lakes (e.g., Scheffer 1998), with no clear progressive shift in community composition since the onset of sampling in 1988. Indeed, the fact that the community returns to an almost identical state in years when ANC is similar might suggest some form of density-dependent regulation of species populations, such that variations in composition are restricted to occur within a narrow range. In common with other acidified freshwaters, much of the temporal variation occurs at relatively short time scales, rather than at intergenerational scales, with the percentage variance in community composition attributable to between year effects being comparable to that due to seasonal variation (Monteith and Evans 2000; Woodward et al. 2002).

The prospects for biological recovery

Species loss in response to stressors is typically non-random and often follows a recognisable and predictable trajectory (Raffaelli 2004), and this sequence forms the basis of many bioassessment metrics, such as AWIC (Acid Waters Indicator Community; Davy-Bowker et al. 2003). Large, rare species, and particularly those at the higher trophic levels, are often the first to be lost when a system is stressed, and this alters the size-spectrum and the distribution of interaction strengths within the food web (Woodward et al. 2005a). These changes have potentially profound implications for ecosystem functioning at all scales of biological organisation, from the individual to the entire food web, particularly because these larger, more vulnerable, species tend to exert disproportionately strong influences on a given ecosystem (Woodward et al. 2005a).

Perhaps unsurprisingly, since it is likely to lag behind the still incomplete chemical recovery of Lochnagar, there is little evidence of biological recovery to date, although there are between-year fluctuations in portions of the invertebrate assemblage that mirror shifts in ANC. These changes appear to be driven largely by climatic factors, and represent shifts in the relative abundances of resident species, rather than the invasion of new colonists. However, in terms of assessing the prospects for its potential recovery in the future, we need to consider how aspects of the extant community might influence its future responses to environmental change. There are several reasons as to why we might expect a time-lag or hysteresis in the loch's recovery, and these different hypotheses are not necessarily mutually exclusive. First, chemical recovery is likely to be incomplete for a significant period because stored pollutants can still enter the system (e.g., Rose et al. 2004), and extremely acid events are often stronger determinants of community responses than average conditions (e.g., Hämäläinen and Huttunen 1996). Second, given the relative inaccessibility of the site, the recolonisation process will vary markedly among taxa, reflecting their different dispersal capabilities and constraints (Bunn and Hughes 1997; Wilcock et al. 2001). Third, there might be some form of biological inertia inherent within the ecosystem itself, such that the configuration and strength of interactions among species confers resistance to potential invaders.

These three rate-limiting steps for recovery can, to some extent, be viewed as a temporal sequence, in that chemical amelioration must progress to a certain degree before potential colonists are able to survive, and that in order for them to persist and flourish they must overcome any inertia to invasion that might exist within the food web. The evidence for chemical inertia has already been discussed in detail in this volume and elsewhere (Evans and Monteith 2001; Rose et al. 2004), but the potential constraints on biological recovery are less well-known, particularly this concept of inertia, which could make trajectories of response to rising pH difficult to identify. A recent paper by Ledger and Hildrew (2005) raised the possibility of resistance to recovery in acidified systems.

Although the theoretical reasoning underpinning this suggestion focussed on the idea of community closure, from a competition-niche theory perspective, the structure and dynamics of the food web are also likely to play an important role. Community closure, which is largely couched in the language of competitive interactions, suggests that saturation is approached at higher pH as species become packed evermore tightly in niche space, with a relaxation of constraints on niche width in depauperate (i.e., acidified) systems permitting a degree of competitive release. One potential example is the loss of specialist grazers (e.g., snails and mayflies) from acidified systems. Because an algal resource is still retained in acid conditions (despite often declining markedly in food quality and abundance), the loss of these consumers can be partially compensated for by an increase in facultative algal grazing among the remaining guild of generalist 'herbivore-detritivores' (Woodward and Hildrew 2002a; Ledger and Hildrew 2005). The presence of these generalist consumers might then make it difficult for specialists to reinvade, following an amelioration of acidity, particularly as they are subsidised by other food sources (e.g., detritus) and thus at higher densities than if they were feeding on algae alone.

Although there is some evidence in support of the idea of community closure, it must be borne in mind that aquatic systems are not directly comparable with many of the terrestrial systems that have driven the development of competition theory, and other types of interaction might also be important. For instance, freshwater food webs have an important vertical dimension, such that they are often characterised by predator-prey interactions, rather than purely horizontal competitive interactions, and this has important consequences for food web dynamics and ecosystem functioning (Petchey et al. 2004). One example of this is found in the 'mesopredator release' (*sensu* Courchamp et al. 1999) of large invertebrate predators frequently observed in acidified freshwaters (Woodward and Hildrew 2002a). Essentially, as fish populations are suppressed or extirpated at low pH, a guild of large-bodied invertebrate predators supplants them at the top of the food web. These mesopredators, which are themselves suppressed at high pH by fish predation, include *Sialis* and *Plectrocnemia* species, both of which are found in Lochnagar. Lentic mesopredators also include many pelagic-benthic species, usually represented by Hemiptera and Coleoptera, and the latter in particular are common components of Lochnagar's predator assemblage. In systems dominated by predator-prey interactions, competition for enemy-free space can produce effects that are indistinguishable from traditional resource competition: indeed, this phenomenon is also known as "apparent competition" (*sensu* Holt 1977). Experimental manipulations, mathematical modelling and longer-term monitoring of changes in the composition of the Lochnagar food web, in the face of the eventual anticipated rises in pH, will provide insight into the process of recovery and whether acidified webs might be inherently resistant to change.

Future directions

Clearly, if we are to improve insight into the potential for biological recovery within the loch, the sophistication of the biomonitoring techniques will need to be enhanced. Given that this is the first attempt to construct a model of the Lochnagar food web based on data that were collected by different researchers with different goals in mind, we are cautious about making too many inferences at this stage, as more detailed dietary analysis (e.g., Chapter 12: Rosseland et al. this volume) and experimental manipulations are required. As a first step, we recommend closer integration of the different taxonomic-based studies (e.g., dissection and gut content analysis of invertebrates to characterise diets via direct observation), and the standardisation of methods so that, for instance, densities of all organisms can be expressed as absolute abundance, or preferably biomass, per unit area or volume. Also, the meiofauna have not been addressed in detail to date, due to a lack of data, and there will undoubtedly be numerous representatives of this very speciose group within both the pelagic and benthic portions of the web. Some taxa are permanent meiofauna, such as cladocerans and rotifers, which spend their entire lifecycle within this size-class, whereas others are only temporary members, passing into macrofaunal size-classes as they grow. The former group are represented by the microcrustacea within the pelagic web, and the latter group includes the chironomids, which are the most abundant 'macrofaunal' group within the loch in most years. Improving the taxonomic resolution of these smaller organisms is likely to provide important new insights into the biology of Lochnagar, and the chironomids are an obvious place to start.

We also advocate the quantification of ecosystem processes (e.g., rates of primary production and decomposition), to provide a more holistic view of the system's ecology: to date, the ecological work has focussed solely on the structural attributes of the community. Further, in order to estimate the system's capacity for biological recovery (and to ascertain the likely trajectory) we will need to assess the ability of acid-sensitive species to arrive, establish populations, and to persist. As a first step, we can assess the loch's degree of isolation from potential colonists by using microsatellite markers for a range of different taxa already present in Lochnagar and its environs (Wilcock et al. 2001). To complement these genetic data, ecological surveys could be carried out for focal taxa (e.g., light trapping for caddis; sticky traps for mayflies) and the two datasets could then be combined and analysed using GIS techniques. To test the inertia hypothesis, invasion experiments could be carried out using field-based mesocosms containing potential colonists and representatives of the ambient benthos. Stable isotope analysis of the main species within the food web offers a potentially powerful means of detecting biological recovery, as we should predict that as pH increases there should be an increased reliance on autochthonous production and a lengthening of food chains, which can be detected by changes in the ratios of carbon and nitrogen isotopes, respectively. These empirical and experimental approaches could then be integrated with new mathematical models that are able to take the system's own internal dynamics into account, particularly with respect to body size (e.g., Emmerson and Raffaelli 2004; Emmerson et al. 2005; Montoya et al. 2005). In summary, the food web of Lochnagar offers an ideal system with which to monitor the process of recovery, and with which to develop and test models of community change in response to stressors: the challenge is now to implement the necessary research required to attain this goal.

Summary

- The Lochnagar food web is typical of an acidified, oligotrophic upland lake. There are few species and numerical densities are low at all trophic levels, reflecting the system's low productivity.

- Although shifts in the system's acid neutralising capacity dominate the between-year temporal shifts in the community, there is little evidence of any sustained and progressive biological recovery from acidification.
- The food web can be subdivided into two spatial compartments (the pelagic and benthic 'subwebs') that are linked to one another by a subset of species, including the top predator, brown trout.
- Within the food web, energy flows from smaller, more abundant species to larger, rarer species, and it is divided among progressively fewer species.
- The consumer assemblage is characterised by trophic generalists, and the structure of the food web suggests that detritus, perhaps in the form of peat particles, is the primary energy source, rather than autochthonous production. The algal assemblage might form an occasional supplementary food source for the facultative grazers among the primary consumers, but it appears to be insufficient to support significant populations of specialist grazers.
- The generalist nature of the consumers and the potential for omnivory are likely to create a highly reticulate, interconnected food web, with a large number of donor-controlled feeding links, which could combine to make the extant community dynamically stable and resistant to invasion, and this might slow the pace of future biological recovery.

Acknowledgements

We would like to thank Martin Kernan and Don Monteith for supplying us with additional data. We would also like to thank Neil Rose, Mark Ledger and an anonymous referee for their helpful comments, which greatly improved the manuscript. We would also like to thank the University of London Central Research Fund for providing financial assistance to GW.

References

Baumgärtner D. and Rothhaupt K.-O. 2003. Predictive length-dry mass regressions for freshwater invertebrates in a pre-Alpine lake littoral. International Review of Hydrobiology 88: 453–463.

Brose U., Berlow E.L., Jonsson T., Banasek-Richter C., Bersier L.-F., Blanchard J.L., Brey T., Carpenter S.R., Cattin Blandenier M.-F., Cohen J.E., Cushing L., Ali Dawah H., Dell T., Edwards F., Harper-Smith S., Jacob U., Knapp R.A., Ledger M.E., Martinez N.D., Memmott J., Mintenbeck K., Pinnegar J.K., Rall B.R., Rayner T., Ruess L., Ulrich W., Warren P., Williams R., Woodward G. and Yodzis P. 2005. Body sizes of consumers and their resources. Ecology 86: 2545–2546.

Bunn S.E. and Hughes J.M. 1997. Dispersal and recruitment in streams: evidence from genetic studies. Journal of the North American Benthological Society 16: 338–346.

Burgherr P. and Meyer E.I. 1997. Regression analysis of linear body mass dimensions vs. dry mass in stream macroinvertebrates. Archiv für Hydrobiologie 139: 101–112.

Cavalli L., Chappaz R., and Gilles A. 1998. Diet of arctic charr (*Salvelinus alpinus* (L.)) and brown trout (*Salmo trutta* L.) in sympatry in two high altitude alpine lakes. Hydrobiologia 386: 9–17.

Cohen J.L., Jonsson T. and Carpenter S.R. 2003. Ecological community description using the food web, species abundance, and body size. Proc. Nat. Acad. Sci. 100: 1781–1786.

Courchamp F., Langlais M. and Sugihara G. 1999. Cats protecting birds: modelling the mesopredator release effect. J. Animal Ecol. 68: 292–293.

Cummins K.W. and Klug M.J. 1979. Feeding ecology of stream invertebrates. Ann. Rev. Ecol. Syst. 10: 147–172.

Davy-Bowker J., Furse M.T., Murphy J.F., Clarke R.T., Wiggers R and Vincent H.M. 2003. Development of the AWIC (Acid Water Indicator Community) macroinvertebrate family and species level scoring systems, Monitoring Acid Waters Phase 1 Report R&D Technical Report P2/090/4.

DeAngelis D.L. 1975. Stability and connectance in food web models. Ecology 56: 238–243.

Edington J.M. and Hildrew A.G. 1995. Caseless caddis larvae. Freshwater Biological Association Scientific Publication No. 53, Titus Wilson & Son, Kendal, 134 pp.

Edwards F.K. 2005. Fish presence and the ecology of stream invertebrate predators. PhD thesis. The University of Edinburgh, UK. 243 pp.

Elger A. and Willby N.J. 2003. Leaf dry matter content as an integrative expression of plant palatability: the case of freshwater macrophytes. Functional Ecol. 17: 58–65.

Elliott J.M. 1996. British freshwater Megaloptera and Neuroptera: a key with ecological notes. Freshwater Biological Association Scientific Publication No. 54, Titus Wilson & Son, Kendal, 52 pp.

Elliott J.M., Humpesch U.H. and Macan T.T. 1988. Larvae of the British Ephemeroptera. Freshwater Biological Association Scientific Publication No. 49, Titus Wilson and Son, Kendal, 145 pp.

Elton C.S. 1927. Animal Ecology. Sidgwick and Jackson, London. 209 pp.

Emmerson M.C. and Raffaelli D.G. 2004. Body size, patterns of interaction strength and the stability of a real food web. J. Animal Ecol. 73: 399–409.

Emmerson M.C., Montoya J.M. and Woodward G. 2005. Body size, interaction strength and food web dynamics. In: de Ruiter P.C., Wolters V. and Moore J.C. (eds.) Dynamic Food Webs: Multispecies assemblages, ecosystem development, and environmental change. Academic Press, San Diego, pp. 167–178.

Evans C.D. and Monteith D.T. 2001. Chemical trends at lakes and streams in the UK Acid Waters Monitoring Network, 1988–2000: Evidence for recent recovery at a national scale. Hydrol. Earth Sys. Sci. 5: 351–366.

Friday L.E. 1988. A key to the adults of British water beetles. Field Studies, Vol. 7 No. 1, Field Studies Council Publication 189. 151 pp.

Hall S.J. and Raffaelli D. 1993. Food webs: theory and reality. Adv. Ecol. Res. 24: 187–239.

Hämäläinen H. and Huttunen P. 1996. Inferring the minimum pH of streams from macroinvertebrates using weighted averaging regression and calibration. Freshwat. Biol. 36: 697–709.

Hildrew A.G. and Townsend C.R. 1976. The distribution of two predators and their prey in an iron-rich stream. J. Animal Ecol. 45: 41–57.

Hildrew A.G. and Townsend C.R. 1982. Predators and prey in a patchy environment: a freshwater study. J. Animal Ecol. 51: 797–815.

Holt R.D. 1977. Predation, apparent competition, and the structure of prey communities. Theoret. Pop. Biol. 12: 197–229.

Holt R.D. and Polis G.A. 1997. A theoretical framework for intraguild predation. Am. Naturalist 147: 396–423.

Hynes H.B.N. 1993. A key to the adults and nymphs of the British stoneflies. Freshwater Biological Association Scientific Publication No. 17, Titus Wilson & Son, Kendal. 90 pp.

Jones J.I. and Sayer C.D. 2003. Does the fish-invertebrate-periphyton cascade precipitate plant loss in shallow lakes? Ecol. 84: 2155–2167.

Jones R.I., Grey J., Quarmby C. and Sleep D. 2001. Sources and fluxes of inorganic carbon in a deep, oligotrophic lake (Loch Ness, Scotland). Global Biogeochem. 15: 863–870.

Kettle H. and Thompson R. This volume. Chapter 18. Future climate predictions for Lochnagar. In: Rose, N.L. (ed.) 2007. Lochnagar: The natural history of a mountain lake. Springer. Dordrecht.

Ledger M.E. and Hildrew A.G. 2000a. Herbivory in an acid stream. Freshwat. Biol. 43: 545–556.

Ledger M.E. and Hildrew A.G. 2000b. Resource depression by a trophic generalist in an acid stream. Oikos 90: 271–278.

Ledger M.E. and Hildrew A.G. 2005. The ecology of acidification and recovery: Changes in herbivore-algal food web linkages across a stream pH gradient. Environ. Pollut. 137: 103–118.

McCann K., Hastings A. and Huxel G.R. 1998. Weak trophic interactions and the balance of nature. Nature 395: 794–798.

Malmqvist B. 1994. Preimaginal blackflies (Diptera: Simuliidae) and their predators in a central Scandinavian lake outlet stream. Ann. Zool. Fennici. 31: 245–255.

May R.M. 1973. Stability and Complexity in Model Ecosystems. Princeton University Press, Princeton. 236 pp.

Monteith D.T. and Evans C.D. (eds.) 2000. UK Acid Waters Monitoring Network: 10 Year Report. Analysis and Interpretation of Results, April 1988-March 1998. Ensis Ltd, London. 364 pp.

Monteith D.T., Hildrew A.G., Flower R.J., Raven P.J., Beaumont W.R.B., Collen P., Kreiser A.M., Shilland E.M. and Winterbottom J.H. 2005. Biological responses to the chemical recovery of acidified fresh waters in the UK. Environ. Pollut. 137: 83–101.

Montoya J.M., Emmerson M.C., Solé R.V. and Woodward G. 2005. Perturbations and Indirect Effects in Complex Food Webs. In: de Ruiter P.C., Wolters V. and Moore J.C. (eds.), Dynamic Food Webs: Multispecies assemblages, ecosystem development, and environmental change. Academic Press, San Diego, pp. 369–380.

Petchey O.L., Downing A.L., Mittelbach G.G., Persson L., Steiner C.F., Warren P.H. and Woodward G. 2004. Species loss and the structure and functioning of multitrophic aquatic systems. Oikos 104: 467–478.

Power M.E. 1990. Effects of fish in river food webs. Science 250: 811–814.

Raffaelli D.G. 2004. How extinction patterns affect ecosystems. Science 306: 1141–1142.

Rose N.L., Monteith D.T., Kettle H., Thompson R., Yang H. and Muir, D. 2004. A consideration of potential confounding factors limiting chemical and biological recovery at Lochnagar, a remote mountain loch in Scotland. J. Limnol. 63: 63–76.

Rosseland B.O., Rognerud S., Collen P., Grimalt J.O., Vives I., Massabuau J-C., Lackner R., Hofer R., Raddum G.G., Fjellheim A., Harriman R. and Piña B. This volume. Chapter 12. Brown trout in Lochnagar: Population and contamination by metals and organic micropollutants. In: Rose N.L. (ed.) 2007. Lochnagar: The natural history of a mountain lake. Springer. Dordrecht.

Scheffer M. 1998. Ecology of Shallow Lakes. Chapman & Hall, London. 357 pp.

Schmid-Araya J.M., Hildrew A.G., Robertson A., Schmid P.E and Winterbottom J. 2002a. The importance of meiofauna in food webs: Evidence from an acid stream. Ecology 83, 1271–1285.

Schmid-Araya J.M., Schmid P.E., Robertson A., Winterbottom J.H., Gjerløv C. and Hildrew A.G. 2002b. Connectance in stream food webs. J. Animal Ecol. 71: 1056–1062.

Towers D.J., Henderson I.M. and Veltman C.J. 1994. Predicting dry weight of New Zealand aquatic macroinvertebrates from linear dimensions. New Zealand J. Mar. Freshwat. Res. 28: 159–166.

Wallace I.D., Wallace B. and Philipson G.N. 2003. Keys to the case-bearing caddis larvae of Britain and Ireland. Freshwater Biological Association Scientific Publication No. 61, Titus Wilson and Son, Kendal. 259 pp.

Warren P.H. 1989. Spatial and temporal variation in the structure of a freshwater food web. Oikos 55: 299–311.

Warren P.H. 1996. Structural constraints on food web assembly. In: Hochberg M.E., Clobert J. and Barbault R. (eds.) Aspects of the genesis and maintenance of biological diversity,. Oxford University Press, Oxford, UK. pp. 142–161.

Warren P.H. 2005. Wearing Elton's Wellingtons: Why body size still matters in food webs. In: de Ruiter P.C., Wolters V. and Moore J.C. (eds.) Dynamic food webs: Multispecies assemblages, ecosystem development, and environmental change. Academic Press, San Diego, pp. 128–136.

Webster J.R. and Benfield E.F. 1986. Vascular plant breakdown in freshwater ecosystems. Annual Rev. Ecol. Systematics 17: 567–594.

Wilcock H.R., Hildrew A.G. and Nichols R.A. 2001. Genetic differentiation of a European caddisfly: Past and present gene flow among fragmented larval habitats. Molec. Ecol. 10: 1821–1834.

Williams R.J. and Martinez N.D. 2000. Simple rules yield complex food webs. Nature 404: 180–183.

Woodward G. and Hildrew A.G. 2001. Invasion of a stream food web by a new top predator. J. Animal Ecol. 70: 273–288.

Woodward G. and Hildrew A.G. 2002a. Food web structure in riverine landscapes. Freshwat. Biol. 47: 777–798.

Woodward G. and Hildrew A.G. 2002b. Body-size determinants of niche overlap and intraguild predation within a complex food web. J. Animal Ecol. 71: 1063–1074.

Woodward G., Jones J.I. and Hildrew A.G. 2002. Community persistence in Broadstone Stream (U.K.) over three decades. Freshwat. Biol. 47: 1419–1435.

Woodward G., Thompson R., Townsend C.R. and Hildrew A.G. 2005a. Pattern and process in food webs: Evidence from running waters. In: Belgrano A., Scharler U., Dunne J. and Ulanowicz B. (eds.) Aquatic food webs: An ecosystem approach. Oxford University Press. pp. 51–66.

Woodward G., Speirs D.C. and Hildrew A.G. 2005b. Quantification and resolution of a complex, size-structured food web. Adv. Ecol. Res. 36: 85–135.

Woodward G., Ebenman B., Emmerson M., Montoya J.M., Olesen J.M., Valido A. and Warren P.H. 2005c. Body size in ecological networks. Trends Ecol. Evol. 20: 402–409.

12. BROWN TROUT IN LOCHNAGAR: POPULATION AND CONTAMINATION BY METALS AND ORGANIC MICROPOLLUTANTS

BJØRN OLAV ROSSELAND (bjorn.rosseland@umb.no)
Department of Ecology and Natural Resource Management,
Norwegian University of Life Sciences
P.O. Box 5003, N-1432 Aas
Norway

SIGURD ROGNERUD
Norwegian Institute of Water Research, NIVA Branch Office East
Sandvikaveien 4,
N-2312 Ottestad
Norway

PETER COLLEN
FRS Freshwater Laboratory
Faskally, Pitlochry,
Perthshire, PH16 5LB
United Kingdom

JOAN O. GRIMALT
Department of Environmental Chemistry,
Institute of Chemical and Environmental Research (CSIC)
Jordi Girona, 18
08034-Barcelona
Spain

INGRID VIVES
Department of Environmental Chemistry,
Institute of Chemical and Environmental Research (CSIC)
Jordi Girona, 18
08034-Barcelona
Spain

JEAN-CHARLES MASSABUAU
Laboratoire d'Ecophysiologie et Ecotoxicologie des Systèmes Aquatiques
UMR 5805, CNRS & Univ. Bordeaux 1
Place du Dr Peyneau, 33 120 Arcachon
France

253

N.L. Rose (ed.), Lochnagar: The Natural History of a Mountain Lake
Developments in Paleoenvironmental Research, 253–285.
© 2007 *Springer.*

REINHARD LACKNER
Institute of Zoology, University of Innsbruck
Technikerstr. 25
A-6020 Innsbruck
Austria

RUDOLF HOFER
Institute of Zoology, University of Innsbruck
Technikerstr. 25
A-6020 Innsbruck
Austria

GUNNAR G. RADDUM
Institute of Biology, University of Bergen
Thormøhlensgt. 49
N-5006 Bergen
Norway

ARNE FJELLHEIM
Stavanger Museum
Musegata 16
N-4010 Stavanger
Norway

RON HARRIMAN
FRS Freshwater Laboratory
Faskally, Pitlochry,
Perthshire, PH16 5LB
United Kingdom

BENJAMIN PIÑA
Department of Molecular and Cellular Biology,
Institute of Chemical and Environmental Research (CSIC)
Jordi Girona, 18
08034-Barcelona
Spain

Key words: brominated flame retardants, brown trout, ecotoxicology, endocrine disruption, fish physiology, fish population, heavy metals, mercury, mountain lake, organic pollutants, POPs, Salmo trutta, stable isotopes

Introduction

To our knowledge, only two fish species have been found in Lochnagar; brown trout (*Salmo trutta*) and the European eel (*Anguilla anguilla*). The presence of eels in Lochnagar was first mentioned by Nethersole-Thomson and Watson (1974) and this observation, as well as the capture of an eel at the site in July 2001 by Fisheries Research Services (FRS) (Peter Collen, FRS pers. commun.), is described by Gardiner and Mackay (2002). The earliest description we have found of the brown trout population in Lochnagar dates back to 1891, when McConnochie (1891) writes in his book on Lochnagar; "The loch swarms with trout of a fair size, but they are generally stiffish to take, unless a slight breeze agitates the water. Permission to fish is of course necessary. The trout are so numerous that at times the surface of the loch has the appearance of "boiling" with them. They were first placed there in 1851". We have found no later record of stocking at the site and the Balmoral Estates Office is unaware of any such activity (Peter Ord, Balmoral Estates Office pers. commun.).

The brown trout population in Lochnagar has been indirectly monitored since 1988 by the FRS, as a part of the UK Acid Waters Monitoring Network (UK AWMN). Indirectly in this context means that few if any complete standard test fishing operations have taken place in the loch itself, but the trout population inhabiting the outlet stream from Lochnagar has been electrofished annually at three permanent stretches. The complete dataset from two of these stretches is presented here. As part of the EU funded AL: PE 2 project, a small sample of trout (n = 10) were analysed for heavy metals and persistent organic pollutants (POPs) (Rosseland et al. 1997) while in 2000 as part of the subsequent EMERGE project, a further 18 fish were taken for tissue and blood samples as part of an extensive and detailed analytical programme. This work followed strict protocols (Rosseland et al. 2001) to allow direct comparisons with data from fish in other European mountain lakes and to ensure quality control was maintained. These data form the basis for the majority of this Chapter.

The fish population in Lochnagar

Historical data from electrofishing the lake outlet stream

The trout population in the outlet stream from Lochnagar has been monitored annually by electrofishing three 50m sections. Each section is electrofished three times, in a sequential removal method, to provide a reliable estimate of the number of individuals per m^2. The results from two of the three sections (Middle and Lower) are shown in Figure 1. The Figure shows that there have been changes in density both between sections and between years over the period of the study. The year classes represented in these two sections have also varied considerably between years. Closest to the Lochnagar outflow stream (Middle section), four year classes (0+ to 3+) have been found. The oldest year class (3+), however, has been lacking in the years 1992, 1995-98, 2001-2002 and 2004. In 2004, the oldest year class found on this stretch was 1+ (Figure 2). At the Lower section, up to five year classes have been found, with a 4 year old trout found in 2001 (Figure 2). However, the 3+ year class was missing in the years 1996, 2000-2001 and 2004. These data suggest that the water quality and/or the physical conditions (drought or high temperatures etc.) may have been variable over the study period. Despite such variations, there is no sign of recruitment failure in the outlet stream from Lochnagar.

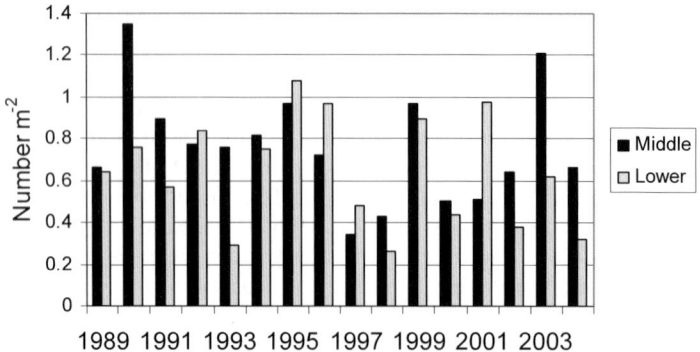

Figure 1. Brown trout densities (all age classes) in two 50m sections of the outlet stream from Lochnagar. (Data from: FRS Freshwater Laboratory.)

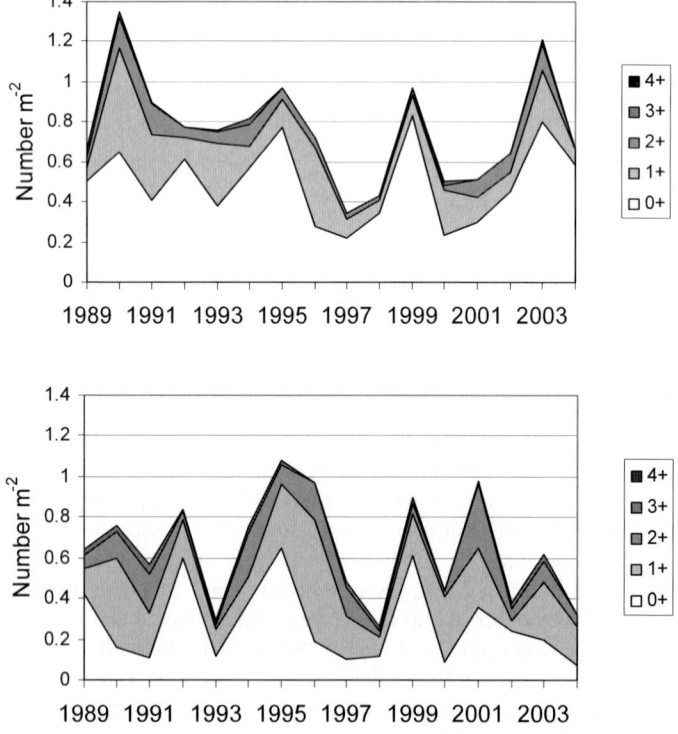

Figure 2. Densities of the different year classes from 1989 to 2004 in the outlet stream from Lochnagar, at the Middle (upper) and Lower (lower) section. Data from: FRS Freshwater Laboratory.

Population data from Lochnagar

Because no standard test fishing with gillnet series has been performed at Lochnagar, the results from the fly rod catches in 2001 are the only data to assess population characteristics. As fly fishing will never reflect a representative age structure, and definitively not the youngest age classes, we have to use the age and size at catch data to reconstruct growth patterns in the population (Figure 3). The age of the 18 brown trout, of which 72% were males, was determined by the use of otoliths. As shown in Figure 3, the specimens represented age classes from 3 to 11 years of age. Based on these data, the population seems to have a linear growth up to the age of 5 years, and thereafter growth stagnates.

The Condition Factor (CF = (length in cm)^3/(weight in g) x100) (where CF = 1 is a normal "well shaped" trout) for the specimens are shown in Figure 4. Lochnagar trout generally had a low CF, and a tendency for a reduced CF with age (not statistically significant). This reduction both in growth and CF might either be due to

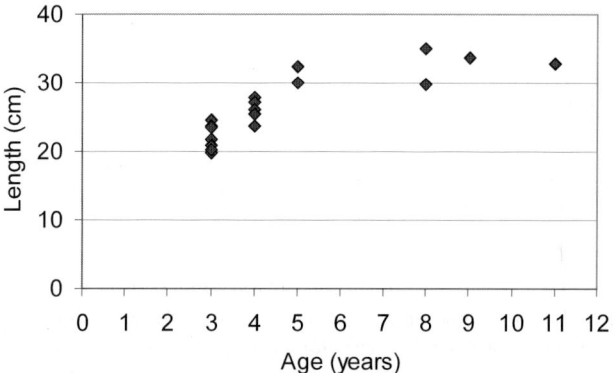

Figure 3. The age and length of the 18 brown trout caught by fly rod in Lochnagar in July 2001.

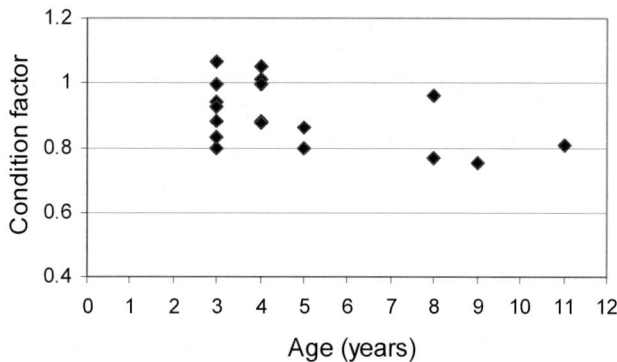

Figure 4. The Condition Factor (CF) of brown trout as a function of age in Lochnagar in July 2001.

(a) a shortage of food relative to the existing fish biomass in the loch (stunted population), or (b) that the physiological cost of living in the loch due to a marginal acidic water quality results in a higher energy cost for survival and thus a reduced allocation of surplus energy used for growth (Rosseland et al.1980).

Age at spawning

Based on the 18 brown trout, of which five were females and 13 males, first age of spawning appears to be 3 years for both females and males as no post-spawners younger than 4 years were found (Table 1).

Table 1. The maturity stage of brown trout in relation to age in Lochnagar in July 2001. The number of trout in each category is given.

	Age	1	2	3	4	5	6	7	8	9	10	11
Immature	I-II			3	2							
Recruit spawners	III-IV			4	2							
Post spawners resting	VII/II				1	1						
Post spawners spawning	VII/III					1			2	1		1

Food

Stomach content
The degrees of stomach fullness were estimated and ranged from 0 (empty) to 5 (full). Three brown trout had no food in the stomach at the time of catch (Figure 5). All stomach samples were preserved in ethanol and later sorted and identified using a binocular microscope. The composition of food items in the stomach was estimated according to the 'point's method' of Hynes (1950). Of the 15 analysed stomach contents, Cladocera (small crustaceans with 99% as *Eurycercus lamellatus*), Tipulidae (crane flies) and other Diptera (true flies) represented 30%, 28% and 20% by volume, respectively (Figure 6). Chironomids (non-biting midges), which constitute an important food source in many high mountain lakes (Schnell and Barlaup 1992), represented only 11% by volume in Lochnagar at the time of sampling. Chironomids are generally more important to brown trout as food in alpine lakes during winter and spring (Pechlaner and Zaderer 1985) and this decreases during the summer (Lien 1978). This may explain the relatively low abundance in the fish stomachs from Lochnagar at the time of sampling. Plecoptera (stoneflies), Tricoptera (caddisflies), other aquatic invertebrates and terrestrial insects, together represented 11% by volume. No fish remains were found in the stomach samples.

Flesh colour
Diet will also influence a trout's flesh colour. If the diet contains small crustaceans with carotenoids in their exoskeleton, the flesh colour will turn from white to pink and red. The percentage of fish with flesh colours white, pink and red, were 39, 50 and 11%, respectively. Only 3 and 4 year old fish had red flesh (Figure 7).

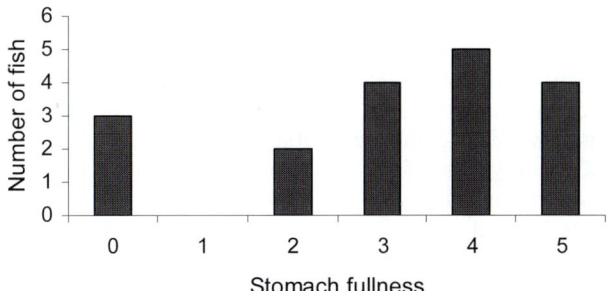

Figure 5. Stomach fullness of the brown trout caught by fly rod in Lochnagar in July 2001.

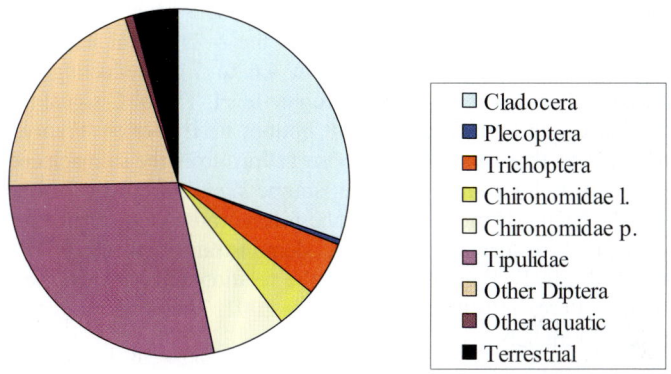

Figure 6. Analyses of the stomach content (volume %) of 15 brown trout caught by fly rod in Lochnagar in July 2001.

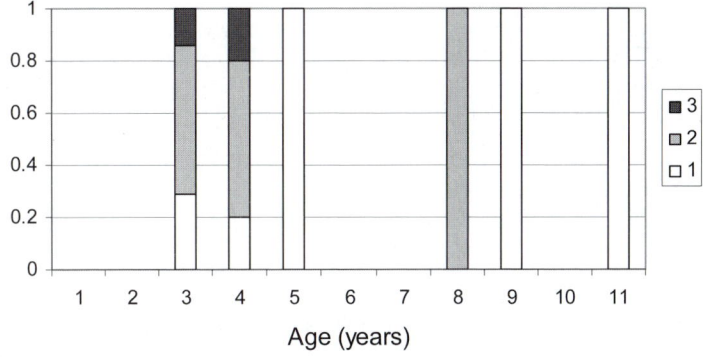

Figure 7. Flesh colour of brown trout (white 1, pink 2 and red 3) caught by fly-rod in Lochnagar in July 2001, expressed as proportion within each year class.

Measuring the position in the food chain by $\delta^{15}N$ analysis and the carbon source by $\delta^{13}C$ analysis.

Diet can vary significantly both temporally and among individuals of the same species (Kidd et al. 1995). A more integrated way of studying food web structure in relation to the age of the fish involves measurements of nitrogen and carbon isotopes to trace food-web linkages. Generally, there is a significant enrichment of the heavy isotope of nitrogen (^{15}N) relative to light isotope (^{14}N) through the food web (Peterson and Fry 1987). Hence, the ratio of $^{15}N:^{14}N$ ($\delta^{15}N$) provides a time-integrated and continuous measure of the relative trophic position of consumers (Kiriluk et al. 1995). The fractionation of ^{13}C in the food chain is much less, and the ratio between $^{13}C:^{12}C$ ($\delta^{13}C$) is more indicative of the carbon sources of the assimilated diet than trophic position (DeNiro and Epstein 1978; Fry and Sherr 1984; France 1995). Thus, a dual isotope approach can indicate structures of the food-web essential for explaining concentrations of contaminants in fish.

The method of isotopic measurements for muscle of brown trout is described in Rognerud et al. (2002). Figure 8 shows that the $\delta^{15}N$ signal has a mean $\pm$ standard deviation (SD) value of 5.96 $\pm$ 0.58 (from 4.6 to 7.0), reflecting a non-piscivorous population with all fish within the same trophic level. The $\delta^{13}C$ signal (-25.4 to -22.2) is typical for fish eating herbivorous or soft bottom invertebrates (France 1997; Karlsson and Byström 2005), thus supporting the observations from the stomach content analysis.

Figure 9 shows, for comparison, the isotope signals from Lochnagar together with lakes in the EU MOLAR project, where the piscivorous Arctic charr (*Salvelinus alpinus* L.) from Lake Arresjøen on Svalbard, demonstrate a mean $\delta^{15}N$ signal of 14.1 (Rognerud et al. 2002). Figure 10 shows the relationship between the individual $\delta^{15}N$ signal and age of brown trout in Lochnagar in 2001. There are no trends towards a shift in trophic level with age.

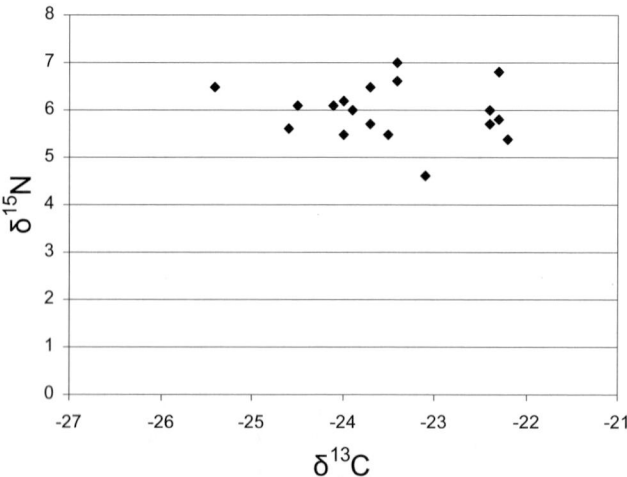

Figure 8. Relationship between stable-carbon and -nitrogen isotope ratios of individual fish from Lochnagar in 2001.

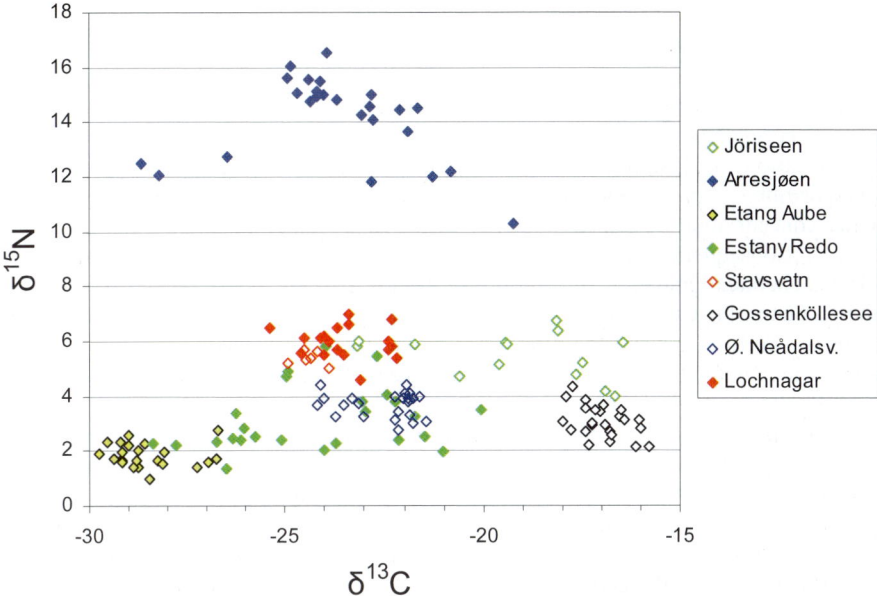

Figure 9. Relationship between stable-carbon and -nitrogen isotope ratios of individual fish from lakes studied in the MOLAR project (Rognerud et al. 2002) with Lochnagar data added.

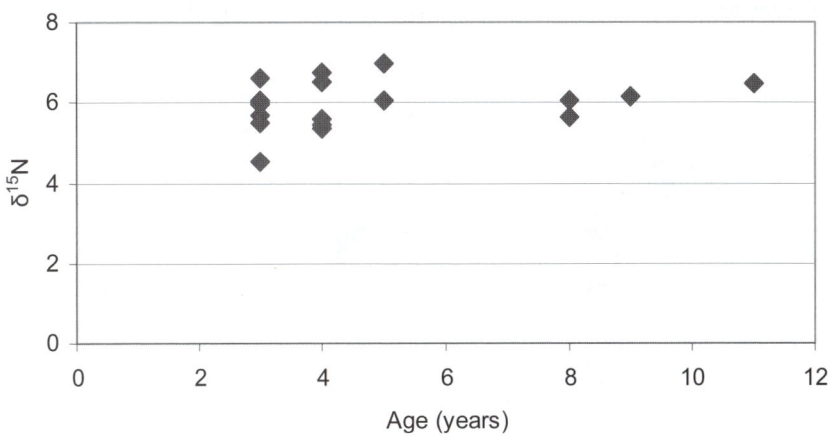

Figure 10. Relationship between stable nitrogen isotope ratios of individual fish against age from Lochnagar in 2001.

Physiological status and osmoregulation in relation to water chemistry

It is well known that in remote mountain lakes, the ionic concentration of the water can be so low that these sites are among the more extreme biotopes that fish can colonise. Indeed, when only minute amounts of water salts are present (e.g., conductivity < 10 μS cm^{-1}), ion regulation is an outstanding challenge. It was therefore important to study if the situation in Lochnagar should be considered as extreme in terms of ionic composition.   For any aquatic animal living in freshwater the basic problem is that its inner concentration of solutes is higher than that of the external environment (Figure 11).   Consequently, water tends to enter and solutes tend to leak following their electro-chemical gradients. In addition solutes are also lost by excreted urine. To solve the problem a first strategy is to reduce membrane permeability, but gills always remain a major site of water and ion movement due to the need to maintain a thin and large surface area with efficient gas diffusion properties. A second strategy is then to counterbalance the ion leak by an equivalent flow in the opposite direction but the problem becomes severe when the water salt concentration reaches extremely low values. This is one of the limits of life in freshwater, and the situation is even more critical in the presence of contaminants that can alter gill physiology by increasing water-blood barrier distances and tissue permeability at gill level. Therefore, fish living under low ionic strength conditions are sensitive organisms for biomonitoring of any stressors affecting osmoregulation and / or respiration. As an example, lakes at high altitude and with low ionic contents were the first to lose fish populations in Norway during the early period of acidification (Muniz and Leivestad 1980).

In Table 2, we compare the water ionic compositions with the ability of trout caught by fly-rod, and thus struggling for life, to maintain their blood concentration of sodium, [Na]$_{blood}$, and chloride, [Cl]$_{blood}$ thus avoiding natremia and chloremia. These ions are the main osmotic components of the extracellular compartment in all aquatic animals. Importantly, these values are quite comparable to those of most teleost fish, either living in freshwater or the sea (Kirschner 1991). Thus, it must be concluded that in

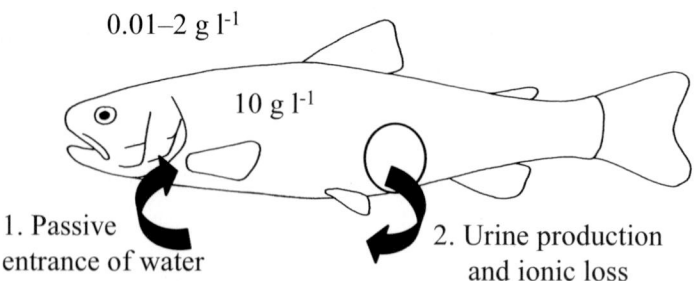

0.01–2 g l^{-1}

10 g l^{-1}

1. Passive entrance of water

2. Urine production and ionic loss

3. Re-uptake of the ionic loss by active pumping at gill level

Figure 11. A basic problem for a fish living in freshwater is ion regulation. The animal must maintain a proper concentration of ions in their body fluid when the concentration difference between water and blood tends to dissipate the concentration difference. Any significant deviation is incompatible with life. This is worst in low mineralised water.

Lochnagar, fish are perfectly able to maintain a normal and standard ion regulation even if, at a first glance, the concentrations of solutes in the water appear rather low. In fact, a comparison with 23 other mountain lakes sampled around Europe shows that Lochnagar does not have excessively low ionic strength (Figure 12) and therefore this is not a fundamental problem for the fish population present in Lochnagar.

Table 2. Comparison of concentrations of sodium and chloride in the water of Lochnagar and the blood of fish (n = 16). The blood concentrations are normal by comparison to fish living in plain water with higher water ionic concentrations.

	Sodium concentration	Chloride concentration
Water (mmol·L^{-1})	0.093	0.080
Blood (mmol·L^{-1})	132.6 ± 7.9	119.1 ± 2.6

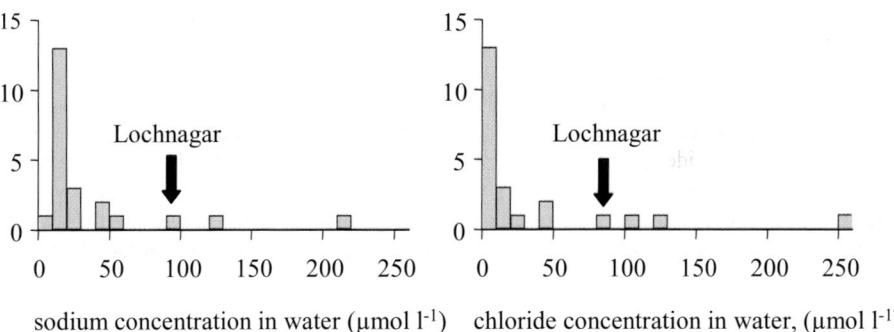

sodium concentration in water (µmol l^{-1})   chloride concentration in water, (µmol l^{-1})

Figure 12. Distribution of sodium (left) and chloride (right) concentrations in waters of 23 remote mountain lakes sampled in Europe (Wathne and Rosseland 2000). The Na and Cl concentrations observed in Lochnagar are in the upper part of the distribution showing that, by comparison to other mountain lakes, Lochnagar cannot be considered an extremely low mineralised water body.

Metal concentration in fish organs

Bioavailable metals taken up by the fish gill

The gill is considered to be an important site for direct toxic effects to metals in high concentrations, for sub-lethal effects at lower metal concentrations, and, along with uptake from food, an important point of entry into the organism for both essential trace elements (Cu, Zn, Se, Mn, Fe) and non-essential elements (Al, As, Cd, Cr, Ni, Pb). Although essential in trace amounts, excess water concentration of positively charged free ions and low molecular weight metal species exert toxic effects. Positively charged metal species form bonds with negatively charged phospholipids on the gill epithelium and glycoproteins in the mucus layer (Exley et al. 1996), and bioconcentrate relative to the water. Some metals will bind to the gill only in relation to the lake water concentration, whereas others will bioaccumulate with exposure time (age).

The binding of excess metals causes inflammation, swelling, and sometimes irradiation of the secondary lamella, initiating a series of resisting mechanisms. The acute toxicity of aluminium is well described, and serves as a model of acute metal toxicity (Rosseland and Staurnes 1994). Two distinct toxic mechanisms of Al-gill interactions are described, and their relative importance varies with the physico-chemical properties of the environment. Ion regulatory disturbances are the main cause of death and physiological disturbance at pH lower than 5, while respiratory problems dominate at higher pH. Loss of plasma ions in freshwater is caused by increased permeability of the gill with a resulting increase in ion efflux, and disruption of ion uptake mechanisms (reduced NaK-ATPase activity) (Staurnes et al. 1984; Rosseland et al. 1992). Metals can bind to different sites at the gill membrane, and thereby compete with important ions necessary for homeostasis. Several models exist for describing the magnitude and effect of metal binding to gills (reviewed by Niyogi and Wood 2004): the 'biotic ligand model (BLM)', the 'gill surface interaction model (GSIM)' and the 'free ion activity model (FIAM)'. The first of these models has evolved from the other two. The common concept of these models is that free metal ions compete with naturally occurring cations for binding sites at the gill (biotic ligands). Complexation by abiotic ligands (e.g., dissolved organic matter, carbonates, sulphides) strongly affects this competition. The sites of toxic action at the gill surface are the specific binding sites, presumed to be active ion transport pathways. These models are based on an equilibrium geochemical modelling framework. The binding affinity (log K) and binding capacity (Bmax) describes the toxicity of each metal, with toxicity increasing with log K (Figure 13).

In acid freshwaters, it is well established that positively charged aluminium species (inorganic Al, labile Al) cause chronically detrimental effects in fish. Aluminium exerts its toxic properties by being accumulated on or into fish gill tissues (Muniz and Leivestad 1980; Rosseland and Staurnes 1994; Rosseland et al. 1999; Gensemer and Playle 1999; Kroglund et al. 2001; Røyset et al. 2005). In our studies, the second gill arch on the right side of the brown trout was used as a biomarker for gill reactive metals. Gills were analysed after a method described in Røyset et al. (2005), and expressed as μg metal g^{-1} gill wet weight (ww). For comparison, many data found in the literature express metal concentrations based on gill dry weight (dw), which is found by multiplying wet weight by five (a standard factor used by e.g., Norwegian Institute for Water Research). Mean gill concentrations of aluminium, arsenic (As), cadmium (Cd), chromium (Cr), copper (Cu), iron (Fe), manganese (Mn), nickel (Ni), lead (Pb), selenium (Se) and zinc (Zn) for ten Lochnagar trout are given in Table 3.

Gill aluminium
In neutral or non-acidified waters, gill Al concentration in Atlantic salmon is always < 1-2 μg g^{-1} gill (ww) or < 5-10 μg g^{-1} gill (dw) (Rosseland et al. 2000, Kroglund and Rosseland 2004). In brown trout, a gill Al concentration of 4.9 ± 2.7 μg g^{-1} gill (dw) was found on reference fish from neutral waters, whereas exposed to acidified waters for 1-2 weeks, the trout could still osmoregulate normally but had an increased gill Al

level of 50 µg g^{-1} gill (dw) (Røyset et al. 2005). The concentration of gill Al in Lochnagar was 21.6 ± 4.8 µg g^{-1} gill (ww) (Table 3) and shows no bioaccumulation with age (Figure 14). However, compared to other mountain lakes in the EU EMERGE project, Lochnagar trout had the highest gill Al concentration (Rosseland et al. 2003)(Figure 15).

Table 3. Gill metal concentration in µg g^{-1} gill (ww) of 10 brown trout from Lochnagar in July 2001. The analyses were performed by NIVA.

	Al	As	Cd	Cr	Cu	Fe	Mn	Ni	Pb	Se	Zn
Mean	21.6	0.2	2.1	0.6	0.4	55.7	4.2	0.3	3.9	2.4	78.9
SD	4.8	0.3	1.0	0.1	0.1	13.6	0.8	0.1	0.7	0.6	13.6

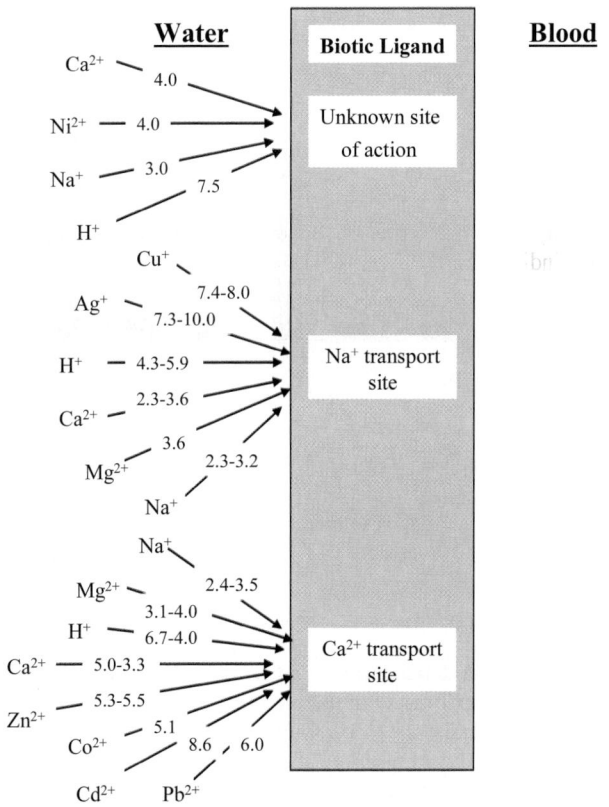

Figure 13. Binding affinities (log K) for free ions of different metals and environmental cations interacting at different sites of toxic action on the biotic ligand (i.e., gill surface). Adapted from: Niyogi and Wood (2004).

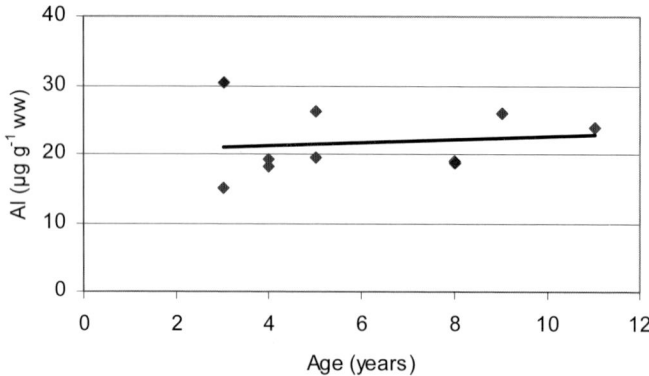

Figure 14. Gill Al concentration in brown trout in Lochnagar in July 2001, in relation to age $(y = 0.23x + 20.25, R^2 = 0.081)$.

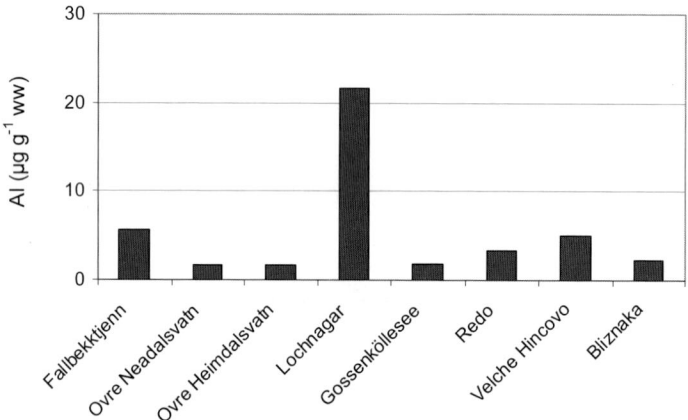

Figure 15. Mean gill Al concentration in salmonids from EMERGE lakes (After: Rosseland et al. 2003).

Other gill metals

Figures 16 – 18 show the concentration of other gill metals in brown trout from Lochnagar. As with Al, no bioaccumulation with age seems to occur. Few, if any, reference data on these metals are to be found in the literature, except for iron which has a comparable reference concentration of 10 - 30 $\mu gFe\ g^{-1}$ gill (ww) with the highest concentrations in iron contaminated waters of 100 $\mu gFe\ g^{-1}$ gill (ww) (Rosseland et al. 2000; 2004). As the water Al and Fe concentrations are positively correlated, a high gill Al and gill Fe was to be expected in Lochnagar. Figure 19 shows that the mean concentration of gill Pb and Cd were highest in brown trout from Lochnagar, compared to the other fish populations in mountain lakes studied in the EMERGE project.

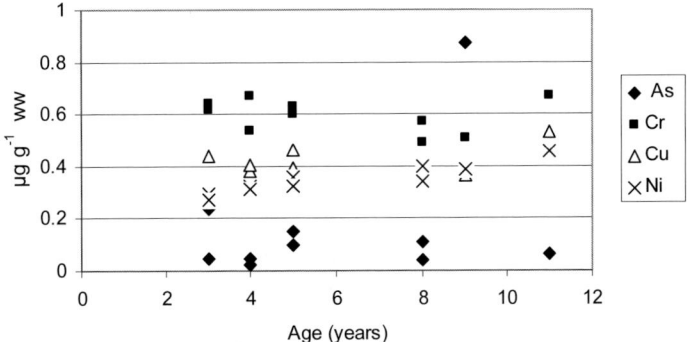

Figure 16. Gill As, Cr, Cu and Ni concentration in brown trout in Lochnagar in July 2001, in relation to age.

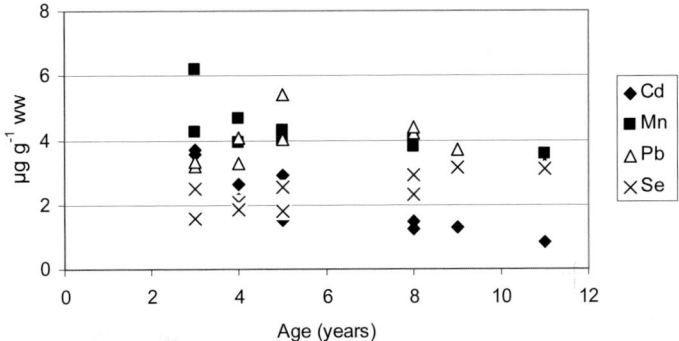

Figure 17. Gill Cd, Mn, Pb and Se concentration in brown trout in Lochnagar in July 2001, in relation to age.

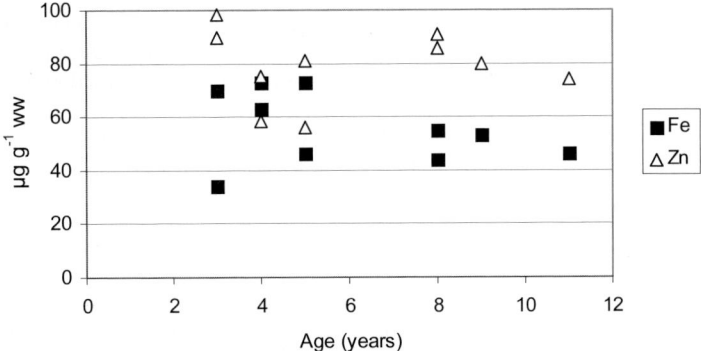

Figure 18. Gill Fe and Zn concentration in brown trout in Lochnagar in July 2001 in relation to age.

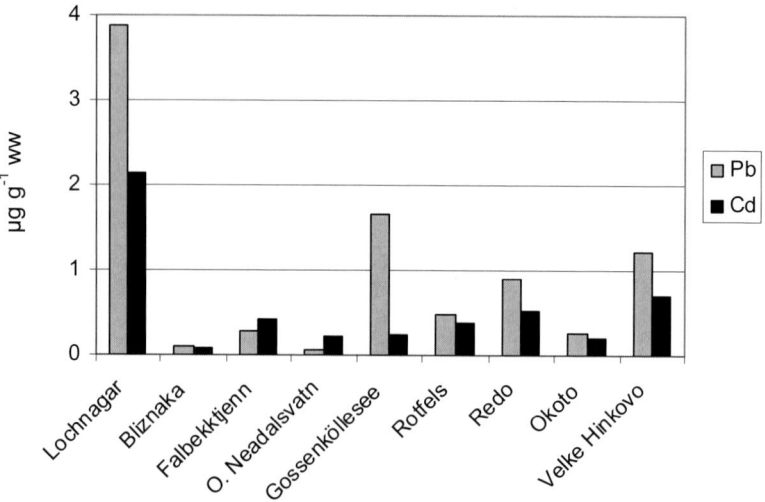

Figure 19. Mean concentration of Pb and Cd on gills of salmonids in lakes from the EU EMERGE project.

Bioavailable metals in liver and kidney

During the European mountain lake project AL:PE 2, ten fish from each of 15 lakes including Lochnagar were sampled to determine whether micro-pollutants could be found in fish from these remote sites. The test fishing in Lochnagar took place in July 1993, and five fish were analysed for heavy metals and five fish for organic pollutants. In the 2-4 year old fish, weighing between 41 and 296 g, the Cd and Pb in the liver ranged from 1.33 – 2.58 and 0.52 – 0.77 µg g^{-1} (ww), respectively. Lochnagar was found to be the fourth most polluted lake for both metals, as illustrated for Cd in Figure 20 (Rosseland et al. 1997; 1999).

In the subsequent EMERGE project, the priority was to measure organic pollutants in fish liver and hence the kidney was analysed for metals. Table 4 shows the mean and standard deviation of selected elements for the 18 brown trout sampled by fly fishing in

Table 4. Total concentrations (mean and SD) of kidney As, Cd, Pb and Se in µg g^{-1} (ww) from 18 brown trout sampled in July 2001, and the linear regression to length (L) and age (A).

	As	Cd	Pb	Se
Mean	0.15	19.56	12.82	7.82
SD	0.13	6.00	5.65	4.80
Lin. Reg. L	$y = 0.019x - 0.36$	$y = 0.9143x - 4.71$	$y = 0.8568x - 9.93$	$y = 0.7318x - 11.61$
	$R^2 = 0.466$	$R^2 = 0.541$	$R^2 = 0.535$	$R^2 = 0.541$
Lin. Reg. A	$y = 0.0501x - 0.097$	$y = 1.3715x + 12.93$	$y = 2.0574x + 2.87$	$y = 1.7332x - 0.56$
	$R^2 = 0.838$	$R^2 = 0.315$	$R^2 = 0.8$	$R^2 = 0.786$

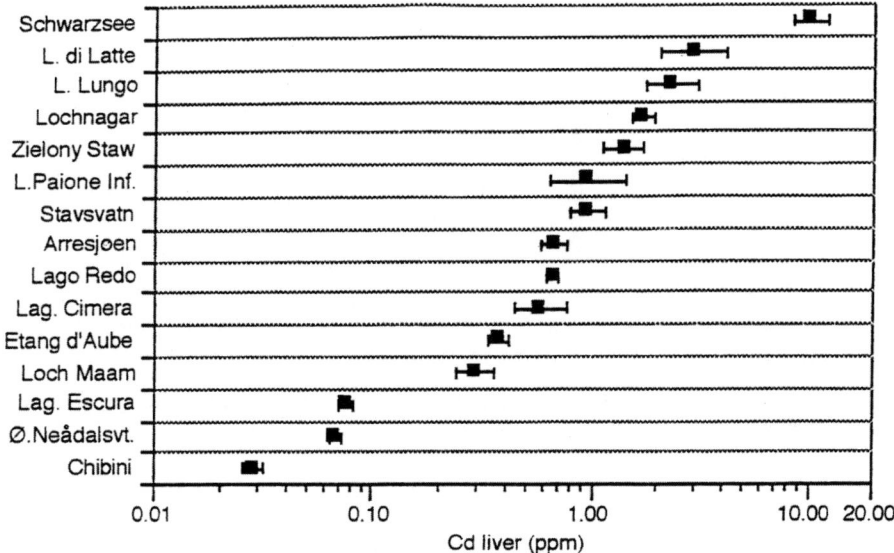

Figure 20. Concentration (with standard errors) of cadmium in fish (Salmonids) liver from the various AL:PE 2 sites, with no adjustments for differences in fish parameters (age, size) or fish species (Rosseland et al. 1997, 1999).

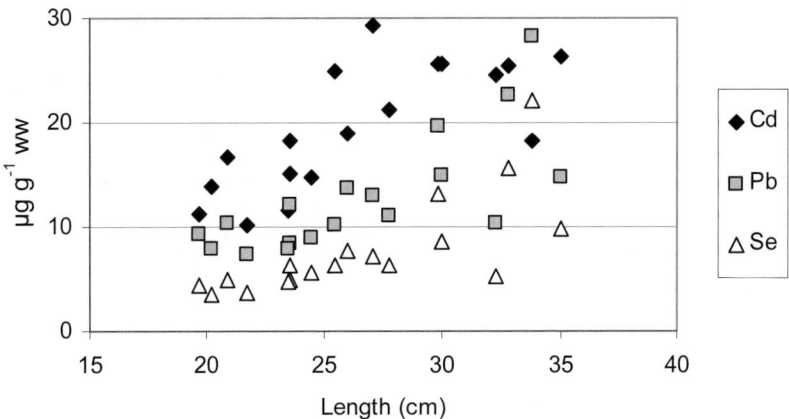

Figure 21. Concentration of cadmium, lead and selenium in kidney in relation to length of brown trout sampled in Lochnagar in July 2001.

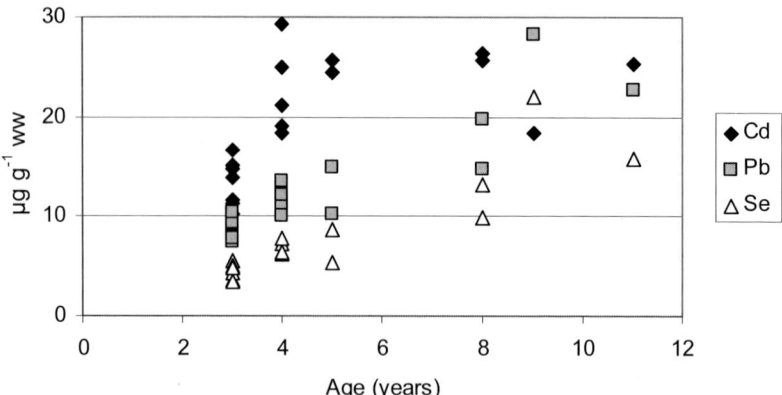

Figure 22. Concentration of cadmium, lead and selenium in kidney in relation to age of brown trout sampled in Lochnagar in July 2001.

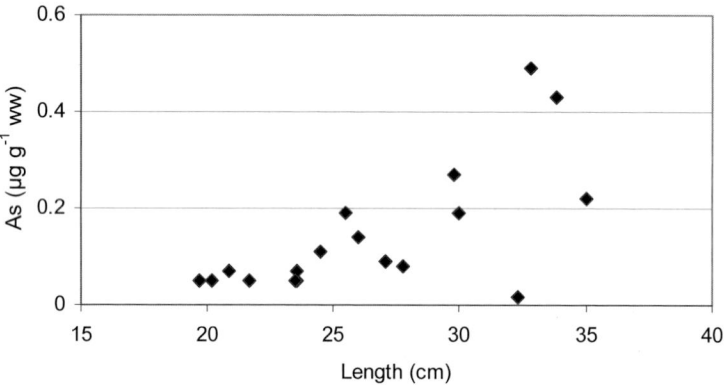

Figure 23. Concentration of arsenic in kidney in relation to size (length) of brown trout sampled in Lochnagar in July 2001.

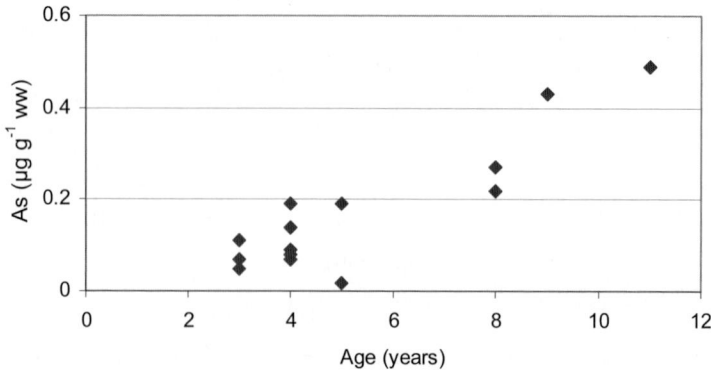

Figure 24. Concentration of arsenic in kidney in relation to age of brown trout sampled in Lochnagar in July 2001.

July 2001. Figures 21 to 24 show the concentrations (total) as a function of length and age, respectively. All metals showed a bioaccumulation both with increased length and age of the fish. Compared to other European alpine lakes, the levels of Cd and Pb in Lochnagar, was the highest of all brown trout populations sampled. Figure 25 and 26 shows a comparison of Cd and Pb concentrations in a 'standard fish' of 250g. Only the Slovakian lake Velche Hincovo in the Tatra Mountains had a kidney Cd concentration comparable with that of Lochnagar, but the Lochnagar trout had the highest concentration of kidney Pb (Figure 26). Figure 27 shows the mean kidney concentration of Se in the 12 EMERGE lakes (all Salmonidae species), where only Velche Hincovo had higher values than Lochnagar.

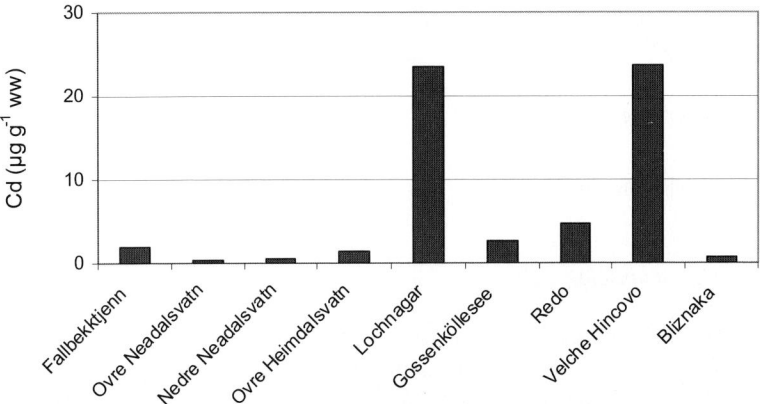

Figure 25. Concentration of kidney cadmium in standardised 250g brown trout from nine EMERGE lakes (Rosseland et al. 2003).

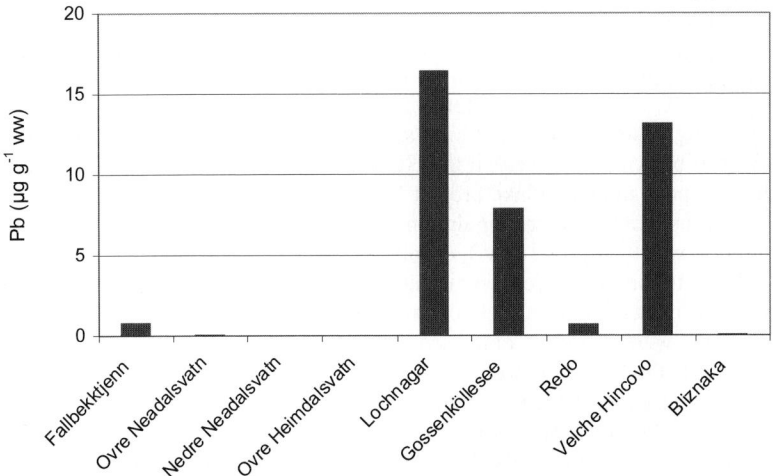

Figure 26. Concentration of lead in kidney in a standardised 250g brown trout in nine EMERGE lakes (Rosseland et al. 2003).

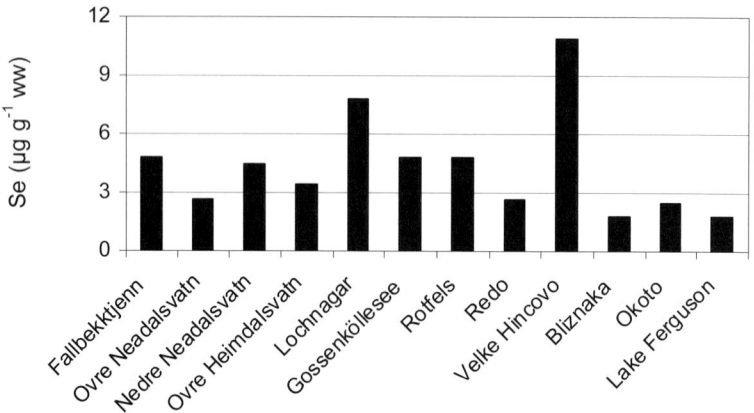

Figure 27. Mean kidney Se in salmonids from 12 EMERGE lakes.

Mercury (Hg) in muscle

The 1998 UNECE protocol targets three particularly harmful metals: cadmium, lead and mercury. Of these, mercury (as methyl Hg) is the only heavy metal that can both bioaccumulate and biomagnify in a food chain, such that if there are fish-eating specimens (piscivorous or cannibal) in a fish population, the Hg concentration might be increased by a factor of 2-3 for a fish of the same age. Mercury is an increasing problem worldwide, as many lakes have been found to have levels of Hg in fish exceeding the dietary recommendations of the World Health Organisation (WHO) of 0.3 µg g^{-1} (ppm). The United States Environmental Protection Agency (US EPA) is even more restrictive, and now recommends 0.18 ppm Hg as a Federal Advisory limit for consumption of mercury-contaminated fish (US EPA 2003). The recommendations from FAO/WHO (2003) on Provisional Tolerable Weekly Intake (PTWI) of 1.6 µgHg kg^{-1} body weight (bw), means that a person of 70 kg should not consume more than a 500 g fillet of brown trout a week if the Hg concentration in that fish is 0.2 µg g^{-1} (ww).

In the European mountain lake project MOLAR, the highest Hg levels were found in the piscivorous Arctic charr population in Arresjøen, Svalbard, where the highest concentration was 0.44 µg g^{-1} (ww), despite Hg concentrations in the sediment and lake water exhibiting the lowest concentrations of all analysed lakes (Rognerud et al. 2002). In 1993, five brown trout from Lochnagar were analysed for Hg in muscle and the concentrations were found to range from 0.04 to 0.08 µg g^{-1} (ww) (Rosseland et al. 1997). In 2001, the Hg levels were higher still, ranging from 0.035 to 0.23 µg g^{-1} (ww) (Figure 28). This upper value thus exceeds the WHO dietary recommendation.

As shown by the nitrogen isotope ratio (δ^{15}N) (Figures 8 – 10), the brown trout in Lochnagar are a non-piscivorous population. Therefore, the Hg data in Figure 29 show that the Hg accumulation in Lochnagar is the highest of all non-piscivorous populations studied. This means that the bioaccumulation of mercury in Lochnagar must reflect a

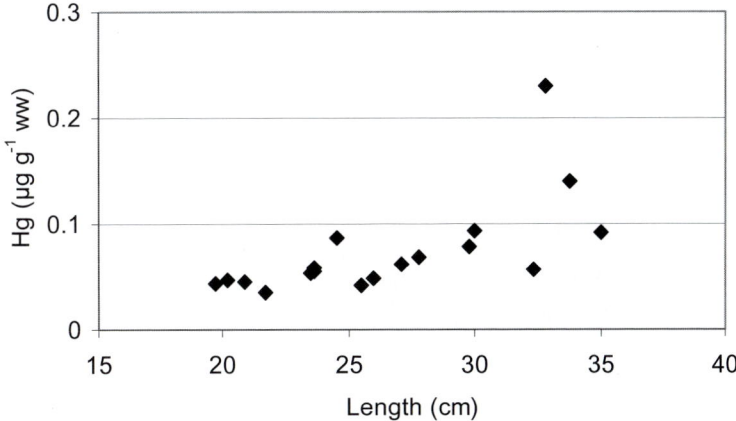

Figure 28. Concentration of Hg in brown trout from Lochnagar in July 2001, in relation to length.

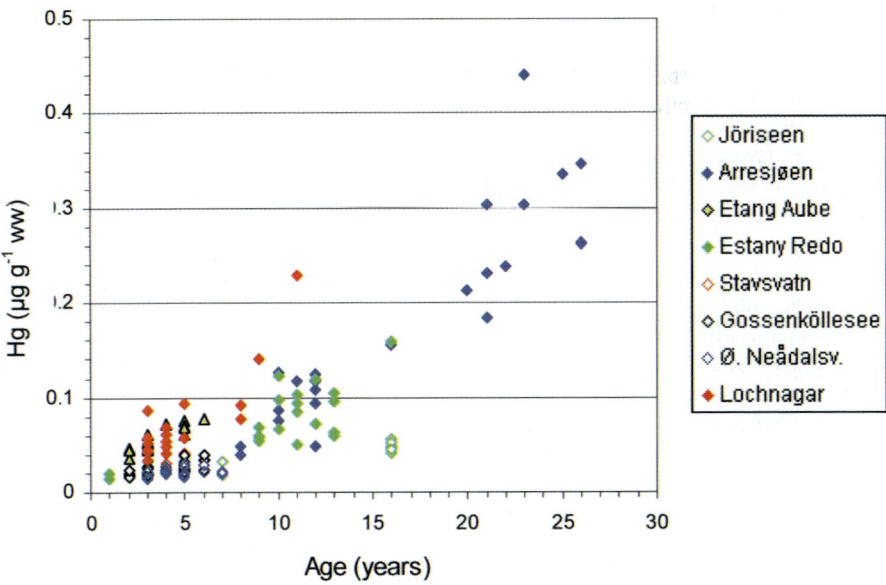

Figure 29. Concentration of Hg as a function of age in brown trout from Lochnagar in July 2001, compared to the concentration in lakes analysed in the MOLAR project (Rognerud et al. 2002). Note the high values for the piscivorous Arresjøen population.

high mercury transportation to the loch as methyl mercury concentration in the water is known to be low (Neil Rose, University College London unpublished data).

Persistent organic pollutants (POPs) in liver and muscle and polycyclic aromatic hydrocarbons (PAH) in liver

Organochlorine compounds (OC)

Organochlorine compounds (OC), such as polychlorinated biphenyls (PCBs), hexachlorocyclohexanes (HCHs), hexachlorobenzene (HCB) and DDTs, are ubiquitous contaminants on our planet. Once released into the environment they may be transported over long distances (Wania and Mackay 1996) and be incorporated into many biogeochemical cycles without undergoing degradation (Chapter 16: Muir and Rose this volume). Despite the discontinued use of many of them, their presence in the ecosystems of cold and remote sites, such as in sediments and fish in high mountain lakes, has been documented (Rosseland et al. 1999; Grimalt et al. 2001; Vilanova et al. 2001a;b) where they are enriched and 'trapped' by condensation due to low temperatures (Grimalt et al. 2001). Moreover, organochlorine compounds tend to accumulate in organic tissues due to their lipophilicity and resistance to degradation (de Voogt and Brinkman 1989). The combination of these two effects may eventually result in concentration levels that are toxic for the organisms living in these sites. The accumulation of OC in fish tissues may result from direct water intake (bioconcentration) and/or from prey ingestion (biomagnification).

From the sampling in Lochnagar in July 1993, five fish, weighing between 108 and 211g, were analysed for PCBs, HCB, HCH, and DDT. The fish had concentrations of total PCBs between 2.01 to 5.77 ng g^{-1} (ww), a mean ± standard deviation of 3.48 ± 1.84 ng g^{-1} (ww), and between 134.8 and 289 ng g^{-1} lipid weight (lw) (mean ± standard deviation of 187 ± 82 ng g^{-1} lipid weight (lw)) (Rosseland et al. 1997). These levels of major chlorinated compounds in Lochnagar were considered to be low in comparison to other high mountain lakes. The observed concentrations were as expected when considering the relatively low altitude and relatively high annual average temperature of Lochnagar within the dataset. PCBs were the compounds in highest concentrations, while the concentrations of total DDTs ranged among the lowest in the mountain lake series (Rosseland et al. 1997). The average concentrations of OCs in the five brown trout sampled in Lochnagar in 2001 are shown in Table 5.

In the EMERGE project, analyses of organic micropollutants in liver and PAH in bile confirmed earlier findings that these contaminants are to be found in all arctic and alpine lakes (Rosseland et al. 1999; Rognerud et al. 2002; Grimalt et al. 2001; Vives et al. 2004a; b; 2005). The higher PCB congeners (> PCB 153) dominated in EMERGE lakes, and the highest concentrations were found in the Tatras, Pyrenees and Rila Mountains (Bulgaria), whereas the concentrations in Lochnagar were lower. The same situation was observed for DDTs, with pp'-DDE having the highest concentration in fish muscle (Table 5). Both PCBs and DDTs have well known biomagnification properties (Rognerud et al. 2002). These general results from Lochnagar and the EMERGE study are comparable to the concentrations found for the same species in other European high mountain lakes and in fish from low altitude freshwater systems

Table 5. Mean and standard deviation (SD) of different organic compounds in muscle of five brown trout from Lochnagar in July 2001. Concentrations in ng g^{-1} (ww).

	α-HCH	HCB	γ-HCH	PCB28	PCB52	PCB101
Mean	0.01	0.08	0.17	0.08	0.08	0.22
SD	0.01	0.01	0.07	0.11	0.07	0.08

	PCB118	PCB138	PCB153	PCB180	DDE	DDT
Mean	0.24	0.42	0.5	0.38	0.96	0.15
SD	0.09	0.24	0.3	0.22	0.49	0.05

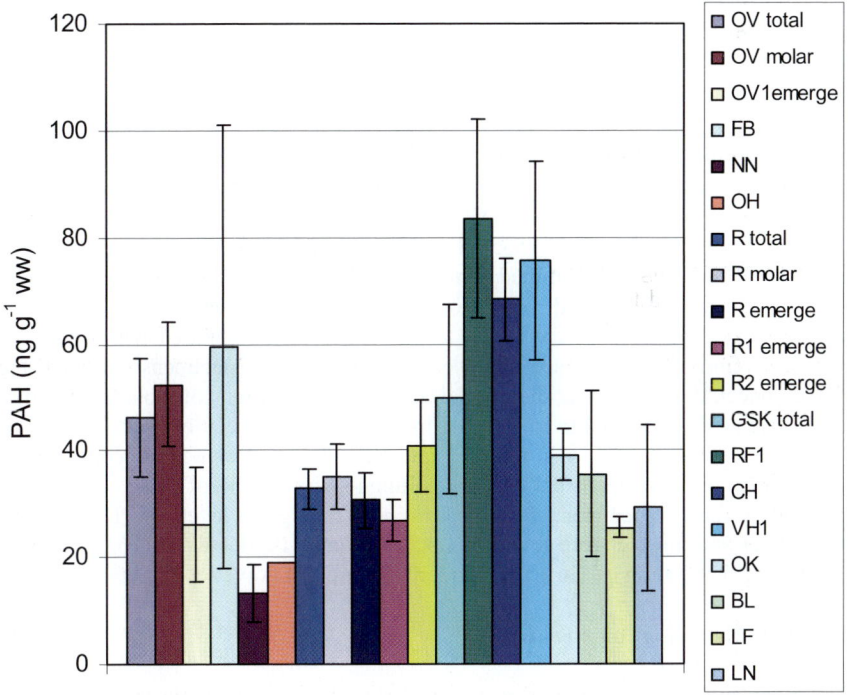

Figure 30. Sum PAH in liver of salmonids in EMERGE lakes. High concentrations in brown trout were found in lakes in the Tatra Mountains, the Tyrol and in Neadalen, Norway, especially Lake Fallbekktjenn (FB, left). Levels in Lochnagar (LN, far right) were close to the average of the EMERGE lakes. For abbreviations, see text. The numbers adjacent to some abbreviations refer to the sampling series for lakes sampled at different times.

(Swackhamer and Hites 1988; Andersson et al. 1988; Leiker et al. 1991; Grimalt et al. 2001).

Polycyclic Aromatic Hydrocarbon (PAH)

Polycyclic Aromatic Hydrocarbon (PAHs) were also analysed in the liver of trout from 61 fish distributed among eight lakes encompassing Greenland (Lake Ferguson LF), the Pyrenees (Redo R), the Alps (Gossenköllesee GSK; Rotfels RF), Scotland (Lochnagar LN), Norway (Øvre Neådalsvatn ON; Nedre Neådalsvatn NN; Ovre Heimdalsvatn OH; Fallbekktjenn FB), Rila (Okoto OK; Bliznaka, BL) and the Tatra Mountains (Velche Hinkovo VH; Cherny Lake, CH). The observed relative distribution of PAHs was quite uniform in all fish, being dominated by phenanthrene followed by fluorene, fluoranthene and pyrene. The mean concentrations of total PAH were distributed over a narrow interval encompassing between 8.8 ng g^{-1} ww in Fallbekktjenn, Norway and 44 ng g^{-1} ww in Rotfels in the Alps (Figure 30). Within these data, biological factors, for example, age and condition, cannot explain the differences in concentration between lakes. PAH are therefore highly metabolised upon incorporation into fish. This result is consistent with the high concentrations of hydroxy-PAH in fish bile.

Brominated flame retardants

Another group of organohalogen pollutants that was investigated in these high mountain lakes were the polybrominated diphenyl ethers (PBDE) that are widely used as flame retardants. The major BDE congeners in all cases were BDE-47 and -99, followed by BDE-100, BDE-153, BDE-154 and BDE-28. These compounds were found in all the samples examined. Their average concentrations 110 - 1300 and 69 - 730 pg g^{-1} (ww) in liver and muscle or 2400 - 40000 and 2900 - 41000 pg g^{-1} lipid weight (lw), respectively, were in the lower range when compared with those of fish from less remote locations (see Vives et al. 2004c). Male specimens exhibited higher PBDE concentrations in liver than females. The concentrations of most PBDEs in liver were correlated with fish age ($p < 0.01$). In contrast, no correlation was found between PBDE concentrations in fish muscle and age. This difference is consistent with observations of other persistent organic pollutants since some of these compounds are excreted by female specimens during spawning. The highest levels of PBDEs (ww and lw) in liver and muscle were found in Lochnagar at 11000 and 1200 pg g^{-1} (ww) and 366000 and 177000 pg g^{-1} (lw), respectively (Figure 31) (Vives et al. 2004c).

Endocrine disruption compounds

Studies of endocrine disruption compounds (i.e., those that mimic the female steroid hormone estradiol) in fish were performed by recombinant yeast assays (RYA) based on engineered yeast strains that harbour two foreign genetic elements: a human oestrogen receptor and a reporter gene whose expression is made dependent on the presence of oestrogens and whose final product concentration is easy to quantify. This is a simplified version of the mechanism by which natural oestrogens operate in vertebrates. Exposure of the Lochnagar fish muscle extracts containing the organochlorine compounds showed that the level of oestrogenicity in Lochnagar samples is above the average for European high mountain lakes (Figure 32).

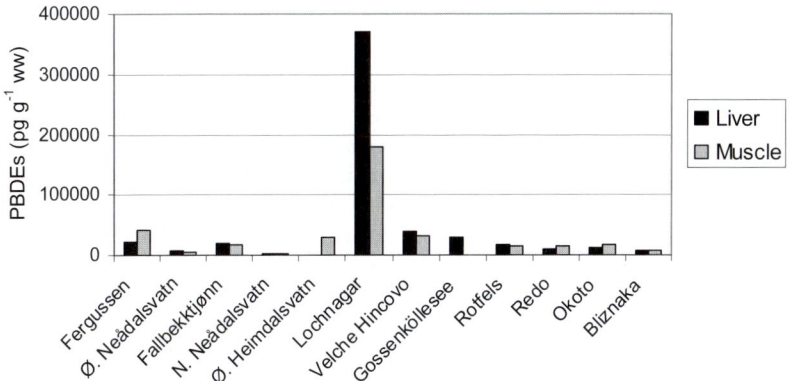

Figure 31. Sum PBDEs in liver and muscle of salmonids sampled in the EMERGE project. Data from Vives et al. 2004c.

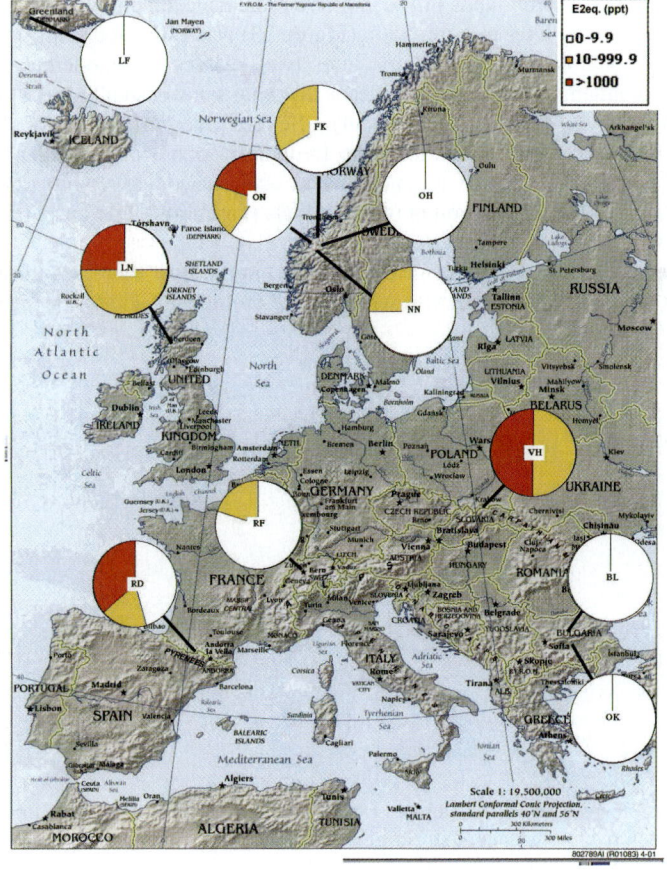

Figure 32. Summary of the estrogenic effects observed in fish as measured by RYA.

Liver histology as an indicator of oxidative stress

Both trace metals and organic pollutants can, through free radical formation, induce oxidative stress. This can cause damage to cell membranes and a series of other negative effects to single cells and organs including DNA breakage leading, for example, to cancer. A series of 'anti-oxidant' reactions are developed in most organisms, including fish. Several enzymatic and gene expression methods are available to measure the anti-oxidant activity. The main organ for detoxifying these compounds is the liver, and in severe cases of detoxification, damage to liver cells can be observed.

Histological slides were prepared from liver samples of all fish collected in the EMERGE project and evaluated under the microscope. Liver inflammation and accumulation of melanomacrophages were the most prominent histological alterations varying between fish populations of EMERGE lakes. Melanomacrophages are macrophages containing melanine granula incorporated from melanophores (pigment cells). As macrophages are cells feeding on foreign material and cell debris, increased numbers of macrophages indicate a faster turnover of tissue, which in turn may have been caused by toxic compounds producing free radicals leading to cell death.

The degree of focal liver inflammation (Figure 33) was estimated semi-quantitatively dividing this pathological alteration into two classes (1=moderate, 2=severe). 'Moderate liver inflammation' (Class 1) means that only some or small focal inflammations were found on liver sections (Figure 34). 'Severe liver inflammations' (Class 2) means that large focal inflammations were frequently found, in particular around bile ducts but also in the liver parenchyma. Focal leucocyte (white blood cells) accumulations were observed around blood vessels (vasculitis), bile ducts (choleangitis) and in the liver parenchyma (hepatitis).

The number of melanomacrophages (Figure 33) in the liver increases during disease, intoxication and starvation, but also with the age of the fish. Thus, the age corrected

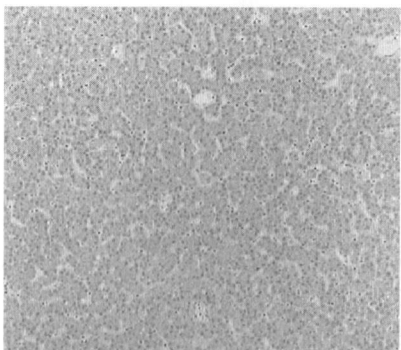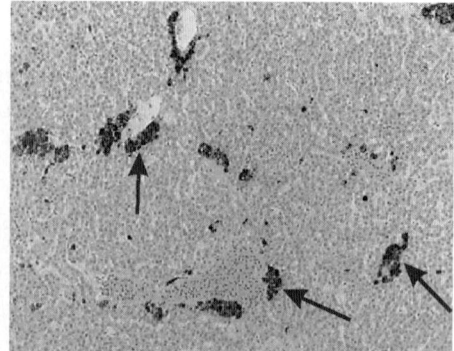

Figure 33. Histological section of a brown trout liver. Left panel: liver from a normal (healthy) fish. Right panel: liver with a high number of melanomacrophages accumulated around veins and bile ducts (arrows). (Magnification = 100x)

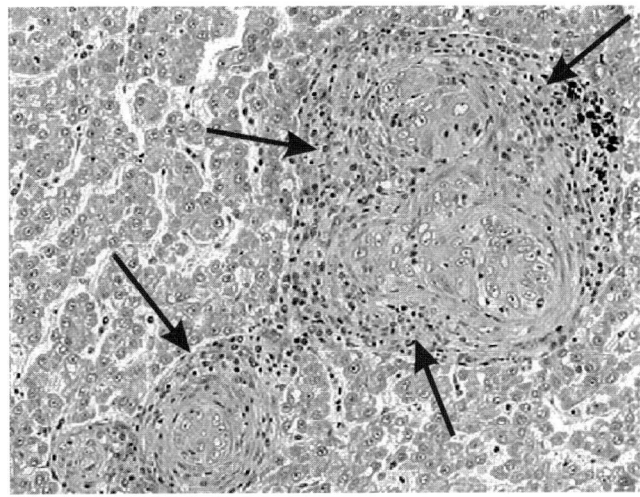

Figure 34. Brown trout liver with focal inflammation (arrows) around bile ducts. (Magnification = 200x)

number of melanomacrophages per mm^2 of liver slides has been estimated for a standard six year old fish, calculated from the age / melanomacrophage relationship of each population.

Melanomacrophages (MM)

A typical increase in the number of hepatic melanomacrophages (MM) with the age of brown trout were found only in a few populations (Velke Hincovo, Lochnagar, Fallbekktjern) while in the majority of populations the range of age classes was too small or the individual variation too high to establish a positive correlation between age and MM numbers. Brown trout populations of Lochnagar, Velke Hincovo, Ovre Heimdalsvatn, and Nedre Neådalsvatn had the highest MM density, the lowest was found in Gossenköllesee and probably also in Bliznaka in which only three year old fish were caught (no age correction possible) (Figure 35). No correlation was found between MM density in the liver and the condition factor of fish.

Focal liver inflammation

More than 70% of brown trout from Ovre Heimdalsvatn and Velke Hincovo displayed focal leucocyte accumulations around veins, bile ducts, and in the liver parenchyma (Figure 36). A relatively high percentage of trout with focal liver inflammation were also found in both lakes of Neådalen, whereas the percentage found in Lochnagar was relatively low. Liver inflammation did not correlate with the age of the fish.

In eight out of nine brown trout populations the abundance of MM correlated positively with the percentage of fish with focal liver inflammation. Only brown trout livers from Lochnagar displayed a high abundance of MM but a low rate of inflammation (Figure 37). The reason for this is unknown.

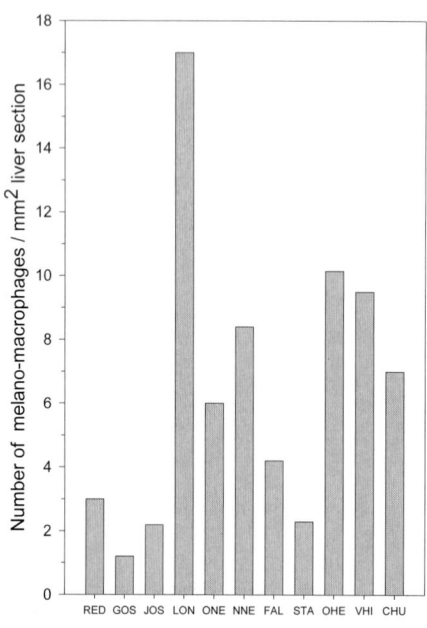

RED Redo
GOS Gossenköllesee
JOS* Jörisee
ONE Ovre Neadalsvatn
NNE Nedre Neadalsvatn
OHE Ovre Heimdalsvatn
FAL Fallbekkfjern
STA* Stavsvatn
VHI Velke Hincovo
LON Lochnagar
CHU* Chuna

Figure 35. Number of melanomacrophages per mm^2 of liver section in brown trout corrected to a six year old standard fish (see text). Lakes that were part of MOLAR project are indicated with *.

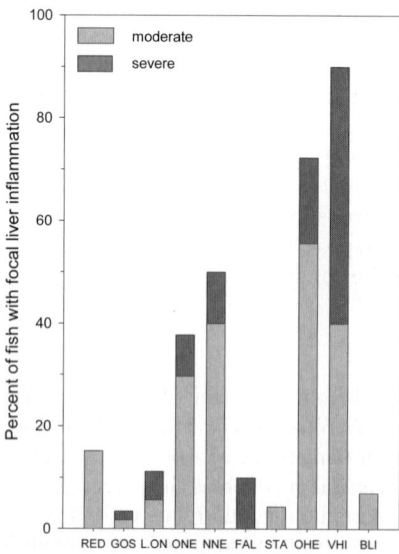

Figure 36. Percentage of brown trout with moderate and severe focal inflammation of the liver. For site abbreviations, see Figure 35.

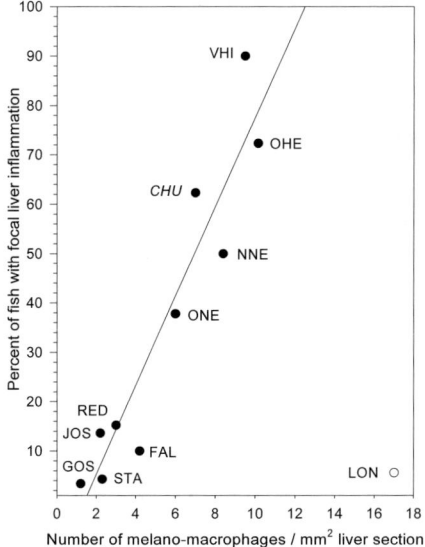

Figure 37. Correlation between the relative abundance of melanomacrophages (MM) in the liver and the percentage of brown trout with focal liver inflammation. The population of Lochnagar (LON; bottom right) did not fit this pattern.

Summary

- Electrofishing data show that the population of brown trout in the Lochnagar outflow stream has gone through years of marginal water quality and/or physically challenging conditions. Several years with trout year classes absent normally is an indicator of stress.
- Data from 18 brown trout caught by fly fishing in July 2001 form the main basis for characterising the fish population in Lochnagar. From otolith age interpretation, it seems that stagnation in growth took place after an age of five years.
- The condition factor (CF) also indicates that the surplus energy for growth was lacking in older fish, as CF fell at the same age as growth ceased. This indicates either a high density (stunted) population, or a population stressed by marginal water quality.
- An increased metabolism, due to compensatory mechanisms to maintain a normal ion regulation or enforced detoxification activities, can reduce the surplus energy normally used for storage of energy (high CF) and growth. Post-spawners are more sensitive to acid waters than younger stages. As the brown trout in Lochnagar start to spawn at three years of age, this might add to the cost of survival for the older fish in the population. Typically, some fish were not spawning every year.
- The brown trout showed a normal range of stomach filling at the time of catch (from empty to full). The food reflected the isotope analyses ($\delta^{15}N$ and $\delta^{13}C$),

namely that none had sign of fish debris in the stomach, and that the invertebrates mostly represented herbivorous or soft bottom groups which can also inhabit the deeper part of the lake.

- Based on blood samples, the plasma concentrations of the major osmotic ions Na^+ and Cl^- were within the lower normal range for brown trout, indicating a successful osmoregulation at the time of sampling. However, the ionic composition of the Lochnagar water is not as extreme as other high mountain lakes and this has helped the brown trout in Lochnagar to maintain their osmotic homeostasis.

- The metal accumulation on the gills, however, indicates that especially Al, Pb and Cd are high in Lochnagar, compared to other lakes studied. None of these metals, including Cr, Cu, Ni, Mn, Zn, Fe, As and Se, bioaccumulated on the gills as a function of age. The specific level of Al on the gills was $20 - 30$ times higher than background level for non-acidified lakes, but the amount was still within a level where brown trout could regulate their ion balance. For comparison, Atlantic salmon would have suffered ion-regulatory problems with the same amount of Al on their gills.

- Analyses of the kidney showed clear bioaccumulation with size (length) and age of all metals analysed: Pb, Cd, As and Se. Compared to the levels in other EMERGE lakes, Lochnagar had the highest level of Pb and Cd, and the second highest of selenium. Selenium, however, acts as a protective metal by binding and detoxifying Hg and other toxic metals in fish. The high Se concentration in kidney from brown trout in Lochnagar could be caused by this.

- Mercury is the only heavy metal that has both bioaccumulation and biomagnification properties. The brown trout population in Lochnagar is non-piscivorous but had the highest bioaccumulation rate of all non-piscivorous fish in the studied mountain lakes. Although only one of the oldest brown trout exceeded 0.2 μgHg g^{-1} muscle (ww), old fish in Lochnagar will have a contamination level of Hg approaching the dietary advice level given by WHO and US EPA. The high Hg accumulation probably reflects a high Hg load to the loch.

- Persistent organic pollutants (POPs) were in low levels when compared to other alpine lakes in agreement with the low altitude and relatively high annual average temperature. However, the concentration of polybrominated diphenyl ethers (PBDE) in Lochnagar were the highest of all EMERGE lakes. The distinct high concentrations of these compounds in fish from Lochnagar reflect inputs from unknown pollution sources.

- Histological examination of liver tissue of the brown trout in Lochnagar revealed clear signs of toxic effects. Lochnagar fish livers had the highest frequency of melanomacrophages mm^{-2} when compared to brown trout populations in other mountain lakes.

- Since no standard test fishing with a series of gillnets in autumn has ever been performed in Lochnagar, we are unable to characterise the population structure and year-class strength. The data presented here on the level of pollution load and subsequent effects on the fish, however, are reliable and robust. These findings confirm that the pollution load of heavy metals and POPs to the brown trout in Lochnagar exceed critical levels for effects i.e., in liver tissue.

- To be able to study whether the pollution loads have affected the population structure and critical parameters for sustainable production, a standard test fishing with a gillnet series following international protocols is highly recommended for Lochnagar.

Acknowledgements

We wish to thank Janey Keay, Alistair McCartney, Saul Mackay, Dave Stewart and Andy Walker for their valuable assistance in the field work at Lochnagar. The study was a part of the EU Project EMERGE, contract EVK1-CT-1999-00032.

References

Andersson O., Linder C.E., Olsson M., Reutergardh L., Uvemo U.B. and Wideqvist U. 1988. Spatial differences and temporal trends of organochlorine compounds in biota from the Northwestern Hemisphere. Arch. Environ. Contam. Toxicol. 17: 755–765.

DeNiro M. J. and Epstein S. 1978. Influence of diet on the distribution of carbon isotopes in animals. Geochim. Cosmochim. Acta 42: 495–506.

de Voogt P. and Brinkman U. 1989. Production, properties and usage of polychlorinated biphenyls. In: Kimbrough R.D. and Jensen A.A.. (eds.) Halogenated biphenyls, naphthalenes, dibenzodioxins and related products. Elsevier Science Publishers B.V. (Biomedical Division), Amsterdam, The Netherlands. pp. 3–45.

Exley C., Wicks A.J., Hubert R.B. and Birchall J.D. 1996. Kinetic constraints in acute aluminium toxicity in the rainbow trout (*Oncorhynchus mykiss*). J. Theoret. Biol. 179: 25–31.

FAO/WHO 2003. Summary and conclusions of the sixty-first meeting of the Joint FAO/WHO Expert Committee on Food Additives (JECFA) Rome. 10-19 June 2003. (www.fao.org/ es/esn/jecfa/index_en.stm, www.who.int/pcs/jecfa/jecfa.htm)

France R. L. 1995. Differentiation between littoral and pelagic food webs in lakes using stable carbon isotopes. Limnol. Oceanogr. 40: 1310–1313.

France R.L. 1997. Stable carbon and nitrogen isotopic evidence for ecotonal coupling between boreal forests and fishes. Ecol. Freshwat. Fish. 6: 78-87.

Fry B. and Sherr E.B. 1984. $\delta^{13}C$ measurements as indicators of carbon flow in marine and freshwater ecosystems. Contrib. Mar. Sci. 27: 13–27.

Gardiner R. and Mackay D.W. 2002. Fish Populations. In: Gimingham C. (ed.) Ecology, land use and conservation of the Cairngorms. Packard Publishing Ltd. Chichester. pp. 148-159.

Gensemer R.W. and Playle R.C. 1999. The bioavailability and toxicity of aluminium in aquatic environments. In: Logan T.J. (ed.) Critical reviews in environmental science and technology. 29: 315 –450.

Grimalt J.O., Fernandez P., Berdie L., Vilanova R., Catalan J., Psenner R., Hofer R., Appleby P.G., Rosseland B.O., Lien L., Massabuau J.C. and Battarbee R.W. 2001. Selective trapping of organochlorine compounds in mountain lakes of temperate areas. Environ. Sci. Technol. 35: 2690 - 2697.

Hynes H.B.N. 1950. The food of freshwater sticklebacks (*Gasterosteus aculeatus* and *Pygosteus pungitius*), with a review of methods used in the studies of the food of fishes. J. Animal Ecol. 19: 36 - 58.

Karlsson J., and Byström P. 2005. Littoral energy mobilization dominates energy supply for top consumers in subarctic lakes. Limnol. Oceanogr. 50: 538-543.

Kidd K.A., Schindler D.W., Muir D.C.G., Lockhart W.L. and Hesslein R.H. 1995. High concentrations of toxaphene in fish from a subarctic lake. Science 269: 240–242.

Kiriluk R. M., Servos M. R., Whittle D. M., Cabana G. and Rasmussen J.R. 1995. Using ratios of stable nitrogen and carbon isotopes to characterize the biomagnification of DDE, mirex, and PCB in a lake Ontario pelagic food web. 1995. Can. J. Fish. Aquat. Sci. 52: 2660–2674.

Kirschner L.B. 1991. Water and ions. In: Ladd Prosser C. (ed.), Environmental and Metabolic Animal Physiology. Wiley-Liss Inc., New-York. pp. 13–107.

Kroglund F. and Rosseland B.O. 2004. Effects of acidification on Atlantic salmon parr and smolt quality. Norwegian Institute for Water Research Report 4797-2004, ISBN 82-577-4475-1. 45 pp. (In Norwegian with English summary)

Kroglund F., Teien H-C., Rosseland B.O. and Salbu B. 2001. Time and pH-dependent detoxification of aluminium in mixing zones between acid and non-acid rivers. Wat. Air Soil Pollut. 130: 905–910.

Leiker T.J., Rostad C.E., Barnes C.R. and Pereira W.E. 1991. A reconnaissance study of halogenated organic compounds in catfish from the lower Mississippi River and its major tributaries. Chemosphere 23: 817–829.

Lien L. 1978. The energy budget of the brown trout population of Øvre Heimdalsvatn. In: Vik R. (ed.) The Lake Øvre Heimdalsvatn, a subalpine freshwater ecosystem - Holarctic Ecol. 1: 279–300.

McConnochie A.I. 1891. Lochnagar. D.Wyllie and Son. Aberdeen. 188 pp.

Muir D.C.G. and Rose N.L. This volume. Chapter 16. Persistent organic pollutants in the sediments of Lochnagar. In: Rose N.L. (ed.) 2007. Lochnagar: The natural history of a mountain lake. Springer. Dordrecht.

Muniz I.P. and Leivestad H. 1980. Acidification-effects on freshwater fish. In: Drablos D. and Tollan A, (eds.) SNSF Project, Proc. Int. Conf. Ecological Impact of Acid Precipitation, 11–14 March 1980. Norway: Sandefjord; 1980. pp. 84–92.

Nethersole-Thompson D. and Watson A. 1974. The Cairngorms: Their natural history and scenery. Collins, London. pp. 77–80.

Niyogi S. and Wood C.M. 2004. Biotic ligand model, a flexible tool for developing site-specific water quality guidelines for metals. Environ. Sci. Technol. 38: 6177–6192.

Pechlaner R. and Zaderer P. 1985. Interrelations between brown trout and chironomids in the alpine lake Gossenköllesee. Verh. Internat. Verein. Limnol. 22: 2620–2627.

Peterson B. J. and Fry B. 1987. Stable isotopes in ecosystem studies. Ann. Rev. Ecol. Syst. 18: 293–320.

Rognerud S., Grimalt J.O., Rosseland B.O., Fernández P., Hofer R., Lackner R., Lauritzen B., Lien L., Massabuau J.C. and Vilanova R. 2002. Mercury, and organochlorine contamination in brown trout (Salmo trutta) and Arctic charr (Salvelinus alpinus) from high mountain lakes in Europe and the Svalbard archipelago. Wat. Air Soil Pollut. Focus 2: 209–232.

Rosseland B.O. and Staurnes M. 1994. Physiological mechanisms for toxic effects and resistance to acidic water: An ecophysiological and ecotoxicological approach. In: Steinberg C.E.W. and Wright R.W. (eds.) Acidification of freshwater ecosystems: Implications for the future. John Wiley and Sons, Ltd. pp 227–246,

Rosseland B.O., Sevaldrud I., Svalastog D. and Muniz I.P. 1980. Studies on freshwater fish populations - effects of acidification on reproduction, population structure, growth and food selection. In: Tollan A. and Drabløs D. (eds.) Ecological impact of acid precipitation. SNSF Project, Proc. Int. Conf. Ecological Impact of Acid Precipitation, 11–14 March 1980. Sandefjord, Norway. pp 336–337

Rosseland B.O., Blakar I., Bulger A., Kroglund F., Kvellestad A., Lydersen E., Oughton D.H., Salbu B., Staurnes M. and Vogt R. 1992. The mixing zone between limed and acidic river waters: Complex aluminium chemistry and extreme toxicity for Salmonids. Environ. Pollut. 78: 3–8.

Rosseland B.O., Grimalt J., Lien L., Hofer R., Massabuau J-C., Morisson B., Rodriguez A., Moiseenko T., Galas J. and Birks J.B. 1997. Chapter 4 Fish. Population structure and concentrations of heavy metals and organic micropollutants. In: Wathne B., Patrick S. and Cameron N. (eds.) AL:PE – Acidification of mountain lakes: Palaeolimnology and ecology.

Part 2 – Remote mountain lakes as indicators of air pollution and climate change. NIVA Report 3638–97 pp. 4–1 to 4–73.

Rosseland B.O., Massabuau J-C., Grimalt J., Hofer R., Lackner R., Rognerud S. and Lien L. 1999. The ecophysiology and ecotoxicology of fishes as a tool for monitoring and management strategy of high mountain lakes and lakes and rivers in acidified areas. Zoology 190: 90–100.

Rosseland B.O., Maroni K., Salbu B. and Rosten T. 2000. Water Quality Project '99. Results from a survey of 53 salmon smolt farms in spring 1999 on raw water, production water, tank water and aluminium and iron in gill tissue of Atlantic salmon smolt. Compendium KPMG, NIVA, LAK/IBK-NLH (Restricted) (In Norwegian).

Rosseland B.O., Massabuau J-C., Grimalt J., Hofer R., Lackner R., Raddum G., Rognerud S., Vives I. 2001. Fish ecotoxicology, The EMERGE fish sampling manual for live fish. The EMERGE Project (European Mountain lake Ecosystems: Regionalisation, diaGnostic and socio-economic valuation). (http://www.mountain-lakes.org/emerge/methods/29.pdf)

Rosseland B.O., Massabuau J.C., Hofer R., Grimalt J. Rognerud S., Lackner R., Vives I., Ventura M., Stuchlik E., Harriman R., Collen P., Raddum G.G., Fjellheim A. and Trichkova T. 2003. EMERGE WP5. Ecotoxicology. In: Patrick S. (ed.) EMERGE final report. pp 42–51.

Rosseland B.O., Rosten T., Salbu B., Kristensen T., Atland A., Heier L.S., Teien H-C., Maroni K. and Bæverfjord G. 2004 Water quality and its effects on aquaculture operations. Profet workshop, Dublin, Ireland. (http://www.feap.info/news/RTD/profet/irl/rosseland_en.asp)

Røyset O., Rosseland B.O., Kristensen T., Kroglund F., Garmo Ø.A. and Steines E. 2005. Diffusive gradient in thin films sampler predict stress in brown trout (*Salmo trutta* L.) exposed to aluminium in acid fresh waters. Environ. Sci. Technol. 39: 1167–1174.

Schnell Ø. A and Barlaup B. T. 1992. Non-biting midges - An important food resource for brown trout and Arctic charr in reservoirs. - In: The Fish symposium 1992. Norwegian Electricity Association, Oslo, pp. 347 - 364 (in Norwegian).

Staurnes M., Blix P. and Reite O.B. 1993. Effects of acid water and aluminum on parr–smolt transformation and seawater tolerance in Atlantic salmon, *Salmo salar*. Can. J. Fish. Aquat. Sci. 50: 1816–1827.

Swackhamer D.L. and Hites R.A. 1988. Occurrence and bioaccumulation of organochlorine compounds in fishes from Siskiwit lake, Isle Royale, Lake Superior. Environ. Sci. Technol. 22: 543–548.

US-EPA 2003. Draft Report on the Environment 2003.Technical Document.Chapter 2.5 Consumption of Fish and Shellfish.

Vilanova R, Fernández P, Martınez C and Grimalt J. 2001a. Polycyclic aromatic hydrocarbons in remote mountain lake waters. Water Res. 35: 3916 –3926.

Vilanova R.M., Fernández P., Grimalt J.O. 2001b. Polychlorinated biphenyl partitioning in the waters of a remote mountain lake. Sci. Total Environ. 279: 51–62.

Vives I, Grimalt J.O, Catalan J, Rosseland B.O. and Battarbee R.W. 2004a. Influence of altitude and age in the accumulation of organochlorine compounds in fish from high mountain lakes. Environ. Sci. Technol. 38: 690–698

Vives I., Grimalt J.O., Fernández P. and Rosseland B.O. 2004b. Polycyclic aromatic hydrocarbons in fish from remote and high mountain lakes in Europe and Greenland. Sci. Tot. Environ. 324: 67–77.

Vives I., Grimalt J.O., Lacorte S., Guillamón M, Barceó D and Rosseland B.O. 2004c. Polybromodiphenyl Ether Flame Retardants in fish from lakes in European High Mountains and Greenland. Environ. Sci. Technol. 38: 2338 –2344.

Vives I., Grimalt J.O., Ventura M., Catalan J. and Rosseland B.O. 2005. Age dependence of the accumulation of organochlorine pollutants in brown trout (*Salmo trutta*) from a remote high mountain lake (Redó, Pyrenees). Environ. Pollut. 133: 343–350.

Wania F. and Mackay D. 1996. Tracking the distribution of persistent organic pollutants. Environ. Sci. Technol. 30: 390A–396A.

Wathne B.M. and Rosseland B.O. (eds.) 2000. MOLAR Final Report. Measuring and modelling the dynamic response of remote mountain lake systems to environmental change: A programme of Mountain Lake Research - MOLAR. NIVA Report 96061–1.

PART III: ANTHROPOGENIC IMPACTS FROM
ATMOSPHERIC POLLUTANT DEPOSITION

13. NATIONAL, INTERNATIONAL AND GLOBAL SOURCES OF CONTAMINATION AT LOCHNAGAR

NEIL L. ROSE (nrose@geog.ucl.ac.uk)
Environmental Change Research Centre
University College London
Pearson Building, Gower Street,
London WC1E 6BT
United Kingdom

SARAH E. METCALFE
School of Geography,
The University of Nottingham,
University Park,
Nottingham NG7 2RD
United Kingdom

ANNA C. BENEDICTOW
Air Pollution Section
Norwegian Meteorological Institute
P.O.Box 43 Blindern
0313 Oslo
Norway

MARTIN TODD
Department of Geography
University College London
Pearson Building, Gower Street,
London WC1E 6BT
United Kingdom

and

JIM NICHOLSON

Key words: back-trajectories, EMEP, HARM, source apportionment, sulphur deposition

N.L. Rose (ed.), Lochnagar: The Natural History of a Mountain Lake
Developments in Paleoenvironmental Research, 289–315.
© *2007 Springer.*

Introduction

Atmospherically transported pollutants have reached all parts of the globe and nowhere can now be classed as truly pristine. Both the Arctic and Antarctic are known to have been impacted by trace metals (e.g., Boutron 1982; Hermanson 1993; Wolff and Suttie 1994; Wolff et al. 1999), persistent organic pollutants (POPs) (e.g., Macdonald et al. 2000; Borghini et al. 2005) and fossil-fuel derived particulates (e.g., Murphey and Hogan 1992; Rose et al. 2004a) and the lake sediment record from these areas has demonstrated that fluxes have increased over recent decades. Semi-volatile POPs transported in the atmosphere are preferentially deposited in colder regions and this effects polar regions by movement latitudinally (Wania and Mackay 1996) as well areas that are colder by virtue of their altitude (Grimalt et al. 2001). As a consequence, these contaminants have been shown to have bioaccumulated to significant levels in the sediments (e.g., Fernández et al. 1999; 2000) and ecosystems (e.g., Vives et al. 2004a; b) of mountain lakes. Given these recent increases in input, and the potential for climate change to exacerbate the release of previously deposited contaminants from catchment soils to freshwaters (Yang et al. 2002a; Rose et al. 2004b) and enhance their atmospheric distribution, the identification of the sources of pollutants impacting remote areas is both critical and urgent.

Source apportionment is an essential part of environmental protection as it enables emissions reductions to be targeted more effectively and it has, therefore, become fundamental to national and international policy. To this end, source identification was a driving force behind the establishment of the UNECE Convention on Long Range Transboundary Air Pollution and is one of the main objectives of EMEP (Co-operative Programme for Monitoring and Evaluation of the Long Range Transmission of Air Pollutants in Europe) (e.g., EMEP 1996; Ilyin et al. 2001). More recently, the intercontinental transport of pollution has become of increasing concern (Berntsen et al. 1999; Jaffe et al. 1999; 2003a; b; Bailey et al. 2000; Wilkening et al. 2000) and here again source apportionment has a critical part to play in identifying the relative roles that both national and imported atmospheric pollutants may take in damage to environmental, and also human, health. Furthermore, the occurrence of meteorological conditions resulting in the transport of pollutants over more than 1000 km with little dilution (Scorer 1992) implies that even in remote areas where wet deposition is considerably more important than dry deposition, episodic inputs of pollutants at high concentrations can occur (e.g., Davies et al. 1984) rather than the more diffuse, lower concentration inputs usually associated with these areas (e.g., Rose et al. 2004a). Therefore, the distance of remote areas from pollution sources is now no longer seen as sufficient for their protection.

In these respects, Lochnagar is no different from any other remote site where contamination sources are solely atmospheric in origin. Contemporary and historical sources are both widespread and various and there is evidence for both episodic and long-term deposition from national, continental and global sources. Here, we use a combination of empirical and modelling approaches with diverse data to attempt to identify the sources of contamination affecting Lochnagar on a range of spatial scales from Scottish to global.

Modelling source attribution

Modelling annual average depositions

Direct measurements can only provide limited data with respect to the origin of pollutants, so atmospheric transport models play a key role. Here, we use output from two different dispersion models to identify the sources of pollution reaching Lochnagar on annual timescales. Ideally, the same model would be used for all pollutants and all scales, but this was not possible.

(i) The Hull Acid Rain Model (HARM) is a receptor-orientated statistical Lagrangian model describing the coupled behaviour of SO_2, NO_X, HCl and NH_3 in air parcels which arrive at locations within the United Kingdom (Metcalfe et al. 1995) having travelled across the EMEP grid. It incorporates orographic enhancement to improve the representation of wet deposition and is thus ideal for use at remote upland sites such as Lochnagar. This model has been extensively validated against data from the UK's monitoring networks and the best available estimates of deposition, and has been shown to be able to reproduce the overall amount and spatial distribution of deposition of potentially acidifying species of sulphur and nitrogen. HARM 11.5 is used to identify the contribution of Scottish and other UK sources to the deposition of sulphur and nitrogen at the site.

(ii) To model the contributions from terrestrial sources within European countries and shipping within European waters, deposition values are extracted from Source-Receptor (SR) calculations undertaken with the Unified EMEP model rev.1.6.12 presented in EMEP Status Report 1/03 Part III (Tarrasón et al. 2003) and emissions documented in EMEP/MSC-W Note 1/2002 (Vestreng and Klein 2002). The EMEP model is driven by a full meteorological model, while HARM employs a highly simplified climatology and straight line trajectories. In the EMEP Unified models, calculations of SR relationships are carried out by perturbing emissions from individual countries, or from groups of sources, in separate model runs. SR matrices are generated for each country by reducing emissions of one or more precursors by a given percentage. SO_X emissions are the main contributor to oxidised sulphur deposition, but NO_Y and NH_3 emissions also affect its deposition. Further information about the described SR calculations can be found in Tarrasón et al. (2003). It should be noted that although total deposition from the EMEP model for some pollutants is close to estimates based on measurements, it significantly underestimates wet deposition of sulphur, oxidised and reduced nitrogen to upland Britain (NEGTAP 2001; Fowler et al. 2004). The absence of orographic enhancement from the EMEP model, included in UK measurement-based maps and in HARM, is key in this respect. The best estimates of deposition at Lochnagar (provided by CEH Edinburgh), indicate that wet deposition of sulphur, oxidised and reduced nitrogen greatly exceeds dry deposition but the EMEP model is unable to reproduce this. As a result, great care needs to taken in interpreting and especially combining output from different models, particularly given the significant differences in treatment between HARM and the EMEP model.

Back trajectory analysis of episodic deposition

Back trajectory analysis was used to quantify transport pathways and therefore likely pollution sources for periods of peak spheroidal carbonaceous particle (SCP) deposition at Lochnagar (Chapter 17: Rose and Yang this volume) for the years 2000 – 2003 inclusive. Periods of analysis for each calendar year were selected by determining the 14 day period with maximum observed SCP deposition. Trajectories were calculated using three dimensional wind fields from the ECMWF operational analyses which have a 1.25 degree resolution and temporal frequency of six hours. The parcel positions were updated every hour. Back trajectories were released from Lochnagar at five vertical levels (950, 850, 700, 600 and 500 hPa), since the height of pollution transport is unknown, every 12 hours over the 14 day period covered by each SCP sample. The back trajectories were calculated for seven days following release. The data were processed by calculating the four dimensional (latitude, longitude, height, time) densities of trajectories over each 14-day period. To gain the total trajectory densities the four dimensional vector for each of the trajectories released from Lochnagar over the period were summed.

Scottish sources

The evidence for impacts from Scottish sources at Lochnagar is mainly derived from modelling studies, as the distinction between Scottish and non-Scottish UK sources is difficult to determine empirically. Modelling predicts Scotland to be a net exporter of oxidised nitrogen (i.e., more is emitted from Scottish sources than is deposited across the country), a net importer of most trace metals and reduced nitrogen, whilst imports and exports of SO_X are approximately in balance (Fowler et al. 2002; McDonald et al. 2002). Here, HARM is used as an approach to provide data for sources of deposition derived from Scottish power stations and refineries, based on 1999 emissions, the last year for which these data were available to us. The locations of major point sources used in the modelling exercise, relative to Lochnagar, are illustrated in Figure 1.

Table 1 shows the HARM modelled contributions of the four major industrial fossil-fuel combustion sources in Scotland (see Figure 1) to deposition at Lochnagar in 1999. For both wet and dry deposition of sulphur and NO_Y-N, Longannet, the 2300 MW coal-fired power station on the Firth of Forth is by far the major contributor, representing 48 – 60 % of the total Scottish contribution to inputs of these pollutants. Cockenzie, a 1100 MW coal-fired power station contributes 17 – 19 % of the sulphur and 21 – 24% of the NO_Y-N, whilst the oil refinery at Grangemouth represents 19 – 20% and c. 11% respectively. Cockenzie and Grangemouth are also located on the Firth of Forth c. 110 km from Lochnagar. Peterhead, a 1550 MW dual-fired (oil / natural gas; but formerly just oil) power station located on the east coast of Scotland 45 km north of Aberdeen (c. 100 km from Lochnagar) contributes a much lower fraction of sulphur (3 %). This is probably due to a combination of two factors. First, the dual-firing of Peterhead by oil

Table 1. Total sulphur and oxidised nitrogen deposition at Lochnagar from Scottish sources for 1999 as modelled by HARM. All values in kg ha^{-1} yr^{-1} 'Scotland %' represents the percentage of total deposition derived from Scottish sources. Numbers in parentheses represent percentage derived from the individual source.

	Peterhead	Longannet	Grangemouth	Cockenzie	Total	Scotland %
Wet SO$_X$	0.007 (3)	0.165 (60)	0.056 (20)	0.046 (17)	0.274	7
Dry SO$_X$	0.009 (3)	0.193 (60)	0.060 (19)	0.060 (19)	0.322	17
Wet NO$_Y$	0.018 (17)	0.050 (48)	0.011 (11)	0.025 (24.0)	0.104	3
Dry NO$_Y$	0.005 (13)	0.021 (55)	0.004 (11)	0.008 (21.0)	0.038	11

and natural gas results in reduced sulphur emissions and second, the sources on the Firth of Forth are 'upwind' of Lochnagar as compared to Peterhead which is northeast and therefore downwind. The Firth of Forth thus represents the major Scottish source area for these acidic pollutants to Lochnagar, and Longannet, Cockenzie and Grangemouth contribute an estimated 62.9 of the 64.7 kg of sulphur deposited onto the loch and its catchment area from Scottish sources in 1999 (Pie A in Figure 2). However, put into context, Scottish sources contribute only a small fraction of the total deposition (all sources) at Lochnagar. Table 1 shows that the modelled total of these four major Scottish sources only contribute 7% and 17% of wet and dry sulphur respectively. Similarly, total Scottish inputs of wet and dry oxidised nitrogen are 3% and 11% respectively.

Empirical evidence for the impact of pollutants from Scottish sources at Lochnagar is restricted to studies on fly-ash particles. Fly-ash is the particulate material emitted along with flue-gases resulting from fossil-fuel combustion and as such is indicative of contamination derived from this source-type (Rose 2001). It comprises two particle types, spheroidal carbonaceous particles (SCPs) (Chapter 17: Rose and Yang this volume) and inorganic ash spheres (IASs). Research on the spatial distribution of SCPs (Rose and Juggins 1994) and IASs (Rose 1996) across Scotland, including Lochnagar, have been undertaken using lake surface sediments. Concentrations of both SCPs and IASs at Lochnagar were typical of this area of Scotland, showing no undue local enhancement. These data show Lochnagar to be contaminated from sources burning fossil-fuels at industrial temperatures, but do not provide any indication as to where these sources are located. Further information can be obtained, however, by using a ratio of the two particle-types and from the characterisation of SCPs.

SCP characterisation was developed in the 1990s to allocate SCPs to their source fuel-type and was based on the semi-quantitative chemistry of individual particles, measured by energy dispersive spectroscopy linked to scanning electron microscopy (SEM-EDS) (Rose et al. 1996). This was applied to SCPs extracted from surface sediments across the UK including Lochnagar. Scottish fossil-fuel combustion is dominated by coal and therefore, as expected, the SCPs from the surface sediments at Lochnagar were found to be dominated by those from the coal fuel-type. However, in common with other sites from the east of Scotland there was a higher than average allocation of oil-derived particles. Given the distribution of industrial oil combustion sources in Scotland, and that the 'high' oil-SCP region was centred on the east coast north of Aberdeen, it seems reasonable to assume that the main source of the

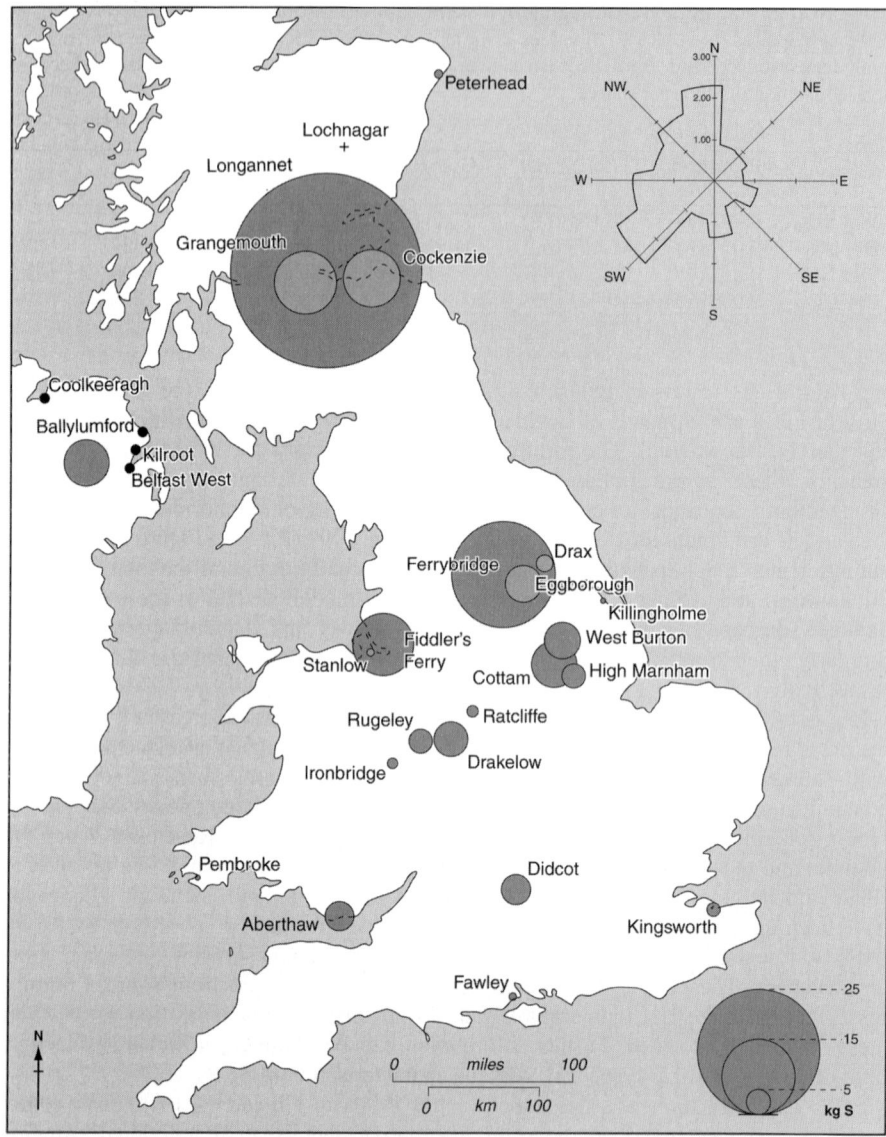

Figure 1. Major point sources in the UK affecting Lochnagar. Proportional circles represent the deposition of sulphur in kg onto Lochnagar and its catchment area in 2000 from that source. Circle in Northern Ireland represents the combined deposition from the four sources shown. Wind rose used in HARM also shown. Wind rose data are percent frequency by 5 degree sector aggregated to 30 degrees.

oil-derived SCPs found at Lochnagar was the then oil-fired Peterhead power station (Rose and Harlock 1998). This is supported by IAS: SCP ratio data. In the UK, IASs are almost solely derived from coal combustion, and so a simple ratio of the two particle types can provide an estimate of the relative inputs from coal and oil. The IAS:SCP ratio for Lochnagar showed a level of oil-derived contamination intermediate between that of sites located close to oil sources and a value typical of a site dominated by coal (high IAS:SCP ratio; Rose 1996). This confirms a measurable input from industrial oil combustion despite dominance from coal sources.

UK sources

Historically, the UK has been a major net exporter of pollutants (NEGTAP 2001). This is not only due to its long industrial history, but also because of its location to the west, and hence 'upwind', of mainland Europe. Given the predominant wind directions from the west and southwest (see wind-rose on Figure 1), it might be expected that transport of pollutants, from Europe to the UK, would be low and this is discussed in the section below. If this is the case, and Lochnagar also only receives low levels of contamination from Scottish sources, then the main source of contamination at the site must be non-Scottish UK sources.

 Indirect evidence for this is available from the sediment record. The historical trends of trace metals (Yang et al. 2002a; b; Chapter 15: Tipping et al. this volume), SCPs (Rose and Monteith 2005; Chapter 17: Rose and Yang this volume) and acidification (Chapter 14: Monteith et al. this volume) show good agreement with national trends which, considering the low inputs from Scottish sources, suggests a considerable impact from beyond the immediate region. Similarly, the detection of a millennial-scale (possibly up to 6000 years) increase in Hg concentrations in the sediment record at Lochnagar (Yang and Rose 2003) suggests contamination from beyond Scotland, although contributions from early domestic peat combustion cannot be ruled out (cf. Meharg and Killham 2003). Furthermore, a significant ($p < 0.01$) reduction in the concentration of sulphate has been recorded in Lochnagar waters by the UK Acid Waters Monitoring Network over the last 15 years (Davies et al. 2005; Chapter 9: Jenkins et al. this volume). This reduction, in response to those of SO_X emissions and deposition, is due to a number of factors. These include the decline in heavy industry, the change in fuel-usage in power stations from coal and oil to natural gas and the introduction of flue-gas desulphurisation (FGD) at two of the UK's largest coal-fired power stations at Drax and Ratcliffe-on-Soar in 1995/6. While the decline in heavy industry has been severe, the switch to natural gas has been less significant in Scotland due to the role of hydro-electric power. However, both power stations retro-fitted with FGD are located at least 400 km away (Figure 1), so any lessening of deposition at Lochnagar as a result of this installation provides an indication of the importance of this industrial area as a source region for Lochnagar. This is discussed further below. Although such evidence may suggest UK sources, many of the same patterns are observed not only in the UK but across Europe. Therefore, whilst the scale of deposition and impact at Lochnagar implies sources beyond Scotland, it does not necessarily implicate non-Scottish UK sources alone. In order to obtain further evidence regarding these sources it is necessary to return to modelling.

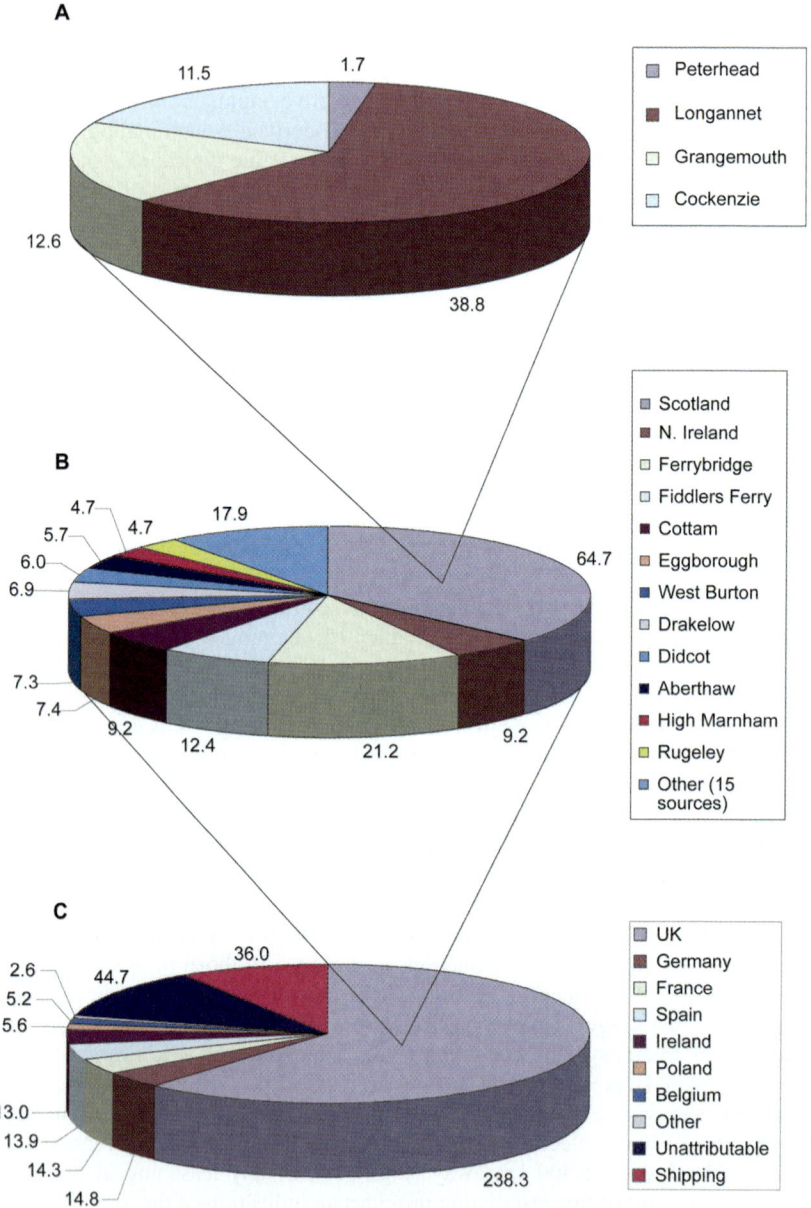

Figure 2. Contributions of sulphur to the total deposition over the combined loch and catchment area at Lochnagar from (a) Scottish sources for 1999 (HARM); (b) sources in UK for 2000 (England and Wales) and 1999 (Scotland and Northern Ireland) (HARM) and (c) European sources for 2000 (EMEP). Pies show proportion from each source to the total from that region. Figures on each pie are kg of sulphur deposited at Lochnagar from that source in that year.

Source attribution using long range transport models

The contributions of sources located in England and Wales to the wet and dry deposition of sulphur at Lochnagar as modelled by HARM for 2000 (not 1999 as for Scottish sources) are given in Table 2. The top six individual sources, representing over 60% of the England and Wales total, are all located in the north of England or the English Midlands (Figure 1) identifying this as a major source area for contaminants deposited at Lochnagar (Figure 1). Deposition from power stations in Northern Ireland has also been modelled, using 1999 emissions, and are also included in the Table. The combined inputs of sulphur from the Northern Ireland power stations are shown to be equivalent to those from Cottam power station (Table 2) and hence make a significant contribution to sulphur deposition at Lochnagar. The location of Northern Ireland upwind of the site makes this contribution unsurprising.

Table 2. Contribution of wet and dry sulphur to Lochnagar from power station and refinery sources in England and Wales for 2000 as modelled by HARM. Values in kg ha^{-1} yr^{-1}. Load represents kg sulphur deposited across the Lochnagar catchment and attributed to that source for 2000. The percentage contribution from each source to the total for England and Wales and the cumulative percentage are also shown. Combined results from Northern Ireland power stations (1999 emissions) are shown for comparison. [Ref. = refinery]

	Wet S	Dry S	Total SO$_X$	Load	Contribution %	Cumulative %
Ferrybridge	0.11	0.09	0.20	21.2	20.4	20.4
Fiddlers Ferry	0.07	0.04	0.11	12.4	11.9	32.4
Cottam	0.05	0.04	0.09	9.2	8.9	41.3
Eggborough	0.04	0.03	0.07	7.4	7.1	48.4
West Burton	0.04	0.03	0.07	7.3	7.0	55.5
Drakelow	0.04	0.02	0.06	6.9	6.7	62.2
Didcot	0.04	0.02	0.06	6.0	5.8	67.9
Aberthaw	0.04	0.02	0.05	5.8	5.6	73.5
High Marnham	0.03	0.02	0.04	4.8	4.6	78.1
Rugeley	0.03	0.02	0.04	4.8	4.6	82.7
Drax	0.02	0.01	0.03	3.4	3.2	86.0
Kingsnorth	0.02	0.01	0.03	2.7	2.6	88.6
Ratcliffe	0.01	0.01	0.02	2.3	2.2	90.8
Ironbridge	0.01	0.01	0.02	2.1	2.0	92.8
Fawley Ref.	0.01	< 0.01	0.01	1.5	1.5	94.2
Stanlow Ref.	0.01	< 0.01	0.01	1.5	1.5	95.7
Killingholme Ref.	< 0.01	< 0.01	001	1.0	0.9	96.6
Pembroke Ref.	< 0.01	< 0.01	0.01	1.0	0.9	97.6
Tilbury	< 0.01	< 0.01	0.01	0.7	0.6	98.2
S. Killingholme Ref.	< 0.01	< 0.01	0.01	0.5	0.5	98.7
Milford Haven Ref.	< 0.01	< 0.01	< 0.01	0.4	0.4	99.2
North Tees Ref.	< 0.01	< 0.01	< 0.01	0.4	0.4	99.6
Littlebrook	< 0.1	0.001	0.002	0.2	0.2	99.8
Fawley	< 0.1	0	0.001	0.1	0.1	99.9
Coryton	< 0.1	0	0.001	0.1	0.1	100.0
N. Ireland power stations	0.05	0.04	0.09	9.2		

Other sites contributing to the total are even more remote. Didcot coal-fired power station in Oxfordshire (c. 610 km), Fawley oil refinery and power station on the south coast of England (c. 700 km) and Pembroke and Milford Haven oil refineries in South Wales (c. 600 km) all contribute to the deposition at Lochnagar albeit in decreasing quantities and possibly with large uncertainties. Although the results from English and Welsh sources (2000) are not directly comparable with those from Scotland and Northern Ireland (1999), it can be seen that their combined contribution to sulphur deposition at Lochnagar is considerably higher than that for Scottish sources and they also provide a larger impact than the closer, but smaller Northern Ireland sources. HARM predicts that sources in England and Wales contributed 103.5 kg sulphur to the combined loch and catchment area at Lochnagar in 2000, whilst the 1999 total for Scotland alone was 64.7 kg. Assuming these totals did not drastically change over one year, then sources in England and Wales contributed almost 60% of the total sulphur derived from the UK, and sources in Northern Ireland a further 5% (Pie B in Figure 2). This highlights the importance of the north of England and English Midlands region as a pollutant source for the site.

McDonald et al. (2002), using the FRAME-HM model, also indicate that non-Scottish UK sources are a major source of deposition of heavy metals. The impact from Scottish sources was focussed on a small region centred on the Firth of Forth, while non-Scottish UK sources were found to be important only in southern Scotland. The model indicates that the largest input component for the region around Lochnagar was found to be from beyond the UK (see below).

The importance of the Midlands and north of England as a source for Lochnagar is further emphasised by considering the impact of FGD on the sources of SO_X. Table 3 shows the HARM modelled contributions for Lochnagar from England and Wales sources for 1994, prior to FGD installation. Comparison of Table 3 with Table 2 illustrates the effect of FGD retrofitting on the importance of Drax and Ratcliffe-on-Soar for sulphur inputs to Lochnagar. Prior to FGD, Drax was the single most important UK source to Lochnagar, contributing over 9% of the total UK load and over 12% of the England and Wales total. Ratcliffe-on-Soar was the sixth most important England and Wales source contributing 6.5%. Post-FGD, Drax drops to 11th and Ratcliffe to 13th and their contributions to the England and Wales load at Lochnagar for 2000 are 2 to 3% respectively. This represents a reduction at Lochnagar of almost 60 kg yr[-1] from these two sources alone and serves to highlight the significance of FGD in the UK's sulphur reduction programme. However, while there is significant evidence for northern England being a major source for Lochnagar, there is also considerable variability in pollution sources and this is demonstrated by using back trajectory or (parcel trajectory) analysis.

Source attribution using back trajectory analysis

Atmospheric parcel trajectories have been widely used as a basic method of describing the pathways followed by pollutants (Merrill and Kim 2004) and as a tool for tracing particulates back to specific sources (e.g., for identifying likely sources of Saharan dust deposited in Spain; Rodriguez et al. 2001). Here, back trajectories released from the Lochnagar region during periods of peak SCP deposition are used to quantify transport pathways and therefore likely pollution sources. SCP fluxes have been measured in

Table 3. Contribution of total sulphur deposition to Lochnagar from England and Wales point sources for 1994 as modelled by HARM. Only sources contributing >1% are shown. Values in kg ha^{-1}. Load represents kg sulphur deposited across the Lochnagar catchment and attributed to that source for 1994. The percentage contribution from each source to the total is also shown.

	kg ha^{-1}	Load	Contribution
Drax	0.39	42.4	9.1
Ferrybridge C	0.30	32.1	6.9
West Burton	0.27	29.4	6.3
Eggborough	0.24	25.8	5.5
Cottam	0.22	23.9	5.1
Ratcliffe-on-Soar	0.21	23.0	4.9
Blyth B	0.20	21.6	4.6
Rugeley B	0.19	20.1	4.3
Thorpe Marsh	0.15	15.7	3.4
Didcot	0.14	15.6	3.3
Fiddlers Ferry	0.12	12.8	2.7
Ironbridge	0.08	8.9	1.9
Willington B	0.07	7.9	1.7
High Marnham	0.07	7.5	1.6
Skelton Grange	0.06	6.5	1.4
Kingsnorth	0.06	6.2	1.3
Drakelow	0.05	5.5	1.2
Tilbury	0.05	5.2	1.1
Total	2.8	309.9	
Total all sources	4.3	467.1	

deposition samples from Lochnagar on a fortnightly basis since 1996 and show considerable temporal variability in inputs (Chapter 17: Rose and Yang this volume). The 14-day sample period containing the maximum SCP deposition for each calendar year from 2000 – 2003 inclusive was identified for inclusion in this analysis. Figure 3a shows the trajectory density, integrated over all vertical levels, 168 hours (seven days) after release for the 7th to 21st May 2003, the period of maximum SCP flux at Lochnagar in that year. As Figure 3a shows, the main transport pathway to Lochnagar is from the west, with air originating over the west Atlantic (centred on 52° N) and passing over Northern Ireland and southwest Scotland. The trajectory densities show remarkably little dispersion, indicating a consistent pathway during this period, quite insensitive to parcel release height. There are no major SCP sources in southwest Scotland and high SCP concentrations in lake sediments from this region are thought to be due to transport from sources at intermediate distances in many directions (Rose and Harlock 1998). Therefore it seems likely that the high flux of SCPs deposited during this period of May 2003 is from power station sources in Northern Ireland. While Northern Ireland has been cited as a source for contaminants deposited in Galloway, this is the first direct evidence for impact from these sources in northeast Scotland.

By contrast, the three other periods of peak SCP input at Lochnagar, from 2000 to 2002 (Figure 3b, c and d), are each derived from different source areas. For the SCP peak deposited between 10th and 24th April 2002 (Figure 3b) the major source area is a wide region of the north Atlantic with subsequent transport over the north of England. Hence the SCPs deposited at this time are probably derived from the northern England source area previously identified as important for Lochnagar. For those SCPs deposited in the 2001 peak between 2nd and 16th May (Figure 3c), there is a wide range of pathways although transport from the southeast (i.e., from across the North Sea and into Europe) is prominent. Most curiously, those deposited in the peak of 8th to 22nd March 2000 appear to pass over no land until they cross northern Scotland (Figure 3d). The source of the SCPs in this latter 'episode' therefore remains unclear although there could be some influence from northwestern Spain (e.g., As Pontes; one of the largest sulphur dioxide emitters in the 'former' EU 15). In summary, back trajectory analysis suggests that SCPs deposited at Lochnagar are derived from a range of sources across the UK and also possibly from continental Europe. It is also clear that models which employ straight-line trajectories will be unable to reproduce some of these patterns of behaviour.

Evidence from the literature provides other examples where back trajectory analysis has been used to trace the origin of depositional episodes in this region of Scotland to sources in the UK and beyond. Davies et al. (1984) reported a 5cm black snowfall in a c. 200 km^2 area around the Cairngorm Mountains on 20th February 1984. The snow in this episode was acidic (pH 3.0) with a large particulate component consisting of c. 29% carbon and particles identifiable as coal fly-ash. The back-trajectory showed a track over north and central England suggesting that this was the major source of the particulates scavenged by the snow. However, contributions from further back along the track, possibly as far as Eastern Europe, were not excluded. Such large and acidic episodes are known to have considerable impacts on streamwater chemistry (e.g., Tranter et al. 1988) and are not new phenomena. The Rev. James Rust (1864) reported a series of three "black rain showers" in 1862 and 1863 affecting areas of the Scottish east coast just north of Aberdeen and a further episode in Edinburgh. Brimblecombe et al. (1986) used back trajectories to determine the possible sources of Rust's episodes. Although aware of the "shortcomings of the meteorological data available for the mid-19th century", they showed that the trajectories of the former episodes took them south and west across central England, while the latter again tracked through central England but then moved east to Europe within two days. These studies provide further evidence for the Midlands and north of England as a major source of pollutants to Scotland over a 150 year period, but contributions from Europe also appear to be quite likely.

European sources

Source attribution for deposition at Lochnagar in 2000 using the EMEP model has been carried out in two modes. For sulphur deposition, data are catchment specific while for nitrogen, values represent deposition for the standard 50 x 50 km grid square within which Lochnagar lies. The main contributors to sulphur deposition are listed in Table 4. It should be noted that the EMEP modelled total sulphur deposition of 3.58 kg S ha yr^{-1} is significantly less than the latest CEH estimate (NEGTAP 2001) for the Lochnagar grid cell of 11.2 kg S ha yr^{-1} (for 1998 – 2000) and this discrepancy should be borne in mind when considering the EMEP output. It can be seen that, as mentioned in the previous section, the UK is by far the greatest contributor of SO$_X$ to Lochnagar with over 238 kg deposited to the loch and catchment area in 2000, representing 61% of

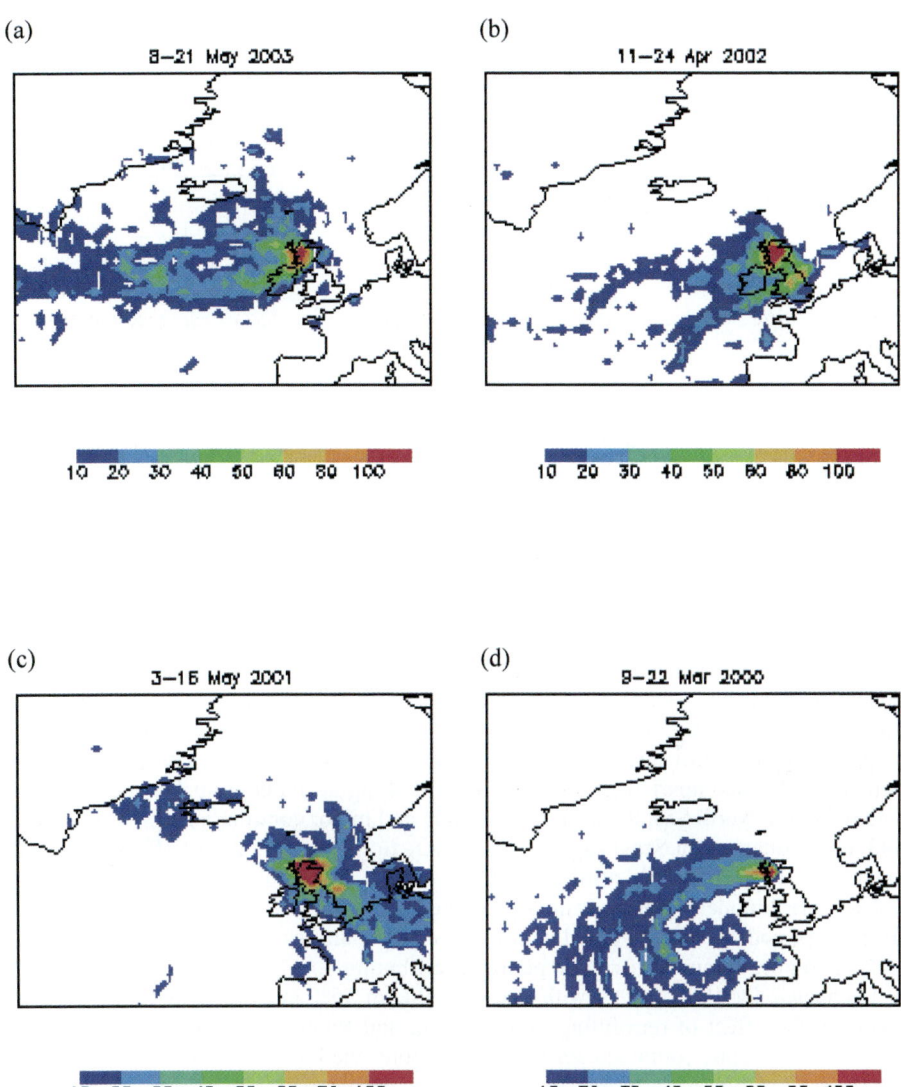

Figure 3. Back trajectory densities for parcels released from Lochnagar every 12 hours over the 14-day period represented by maximum annual spheroidal carbonaceous particle (SCP) deposition for (a) 7th-21st May 2003; (b) 10th – 24th April 2002; (c) 2nd – 16th May 2001; (d) 8th – 22nd March 2000.

Table 4. Contribution of sulphur to Lochnagar from European and shipping sources for 2000 as modelled by EMEP. Values in kg ha^{-1}. Load represents kg sulphur deposited across the Lochnagar catchment and attributed to that source for 2000. The percentage contribution from each source to the total is also shown.

	kg ha^{-1}	Load	Contribution
UK	2.20	238.3	61.3
Germany	0.14	14.8	3.8
France	0.13	14.3	3.7
Spain	0.13	13.9	3.6
Ireland	0.12	13.0	3.4
Poland	0.05	5.6	1.5
Belgium	0.05	5.2	1.3
Other	0.02	2.6	0.7
Unattributable	0.41	44.7	11.5
Shipping	0.33	36.0	9.3
Total	3.58	388.41	

the total load (or 69% of that attributable to sources). All other individual countries contribute less than 4% of the total and only Germany, France, Spain and Ireland provide more than 2%. These international contributions are shown in Figure 2 (Pie C). Some perspective is gained as to the scale of these inputs when it is considered that the UK coal-fired power station at Didcot, 600 km away from Lochnagar (Figure 1), would appear to contribute about as much sulphur deposition to the site (Table 2) as the emissions from either Poland or Belgium. Interestingly, EMEP output indicates that shipping in the North Sea in 2000 contributed more sulphur to Lochnagar than any country in Europe apart from the UK. A total shipping input from European waters (including the North Sea, the northeast Atlantic and minor contributions from the Baltic and the Mediterranean Seas) supplied 36 kg SO_X to Lochnagar, almost 10% of the total and almost 12% of the total attributable to sources.

The EMEP model is able to estimate European contributions to the load at Lochnagar over the period 1985 – 1996 and data for selected countries are given in Table 5. The UK, of course, dominates over the period, contributing 60 – 72% of the SO_X deposition at Lochnagar for 1985 – 1995 with a lower value of 55% in 1996. This lower value is probably the effect of retrofitting FGD to Drax and Ratcliffe-on-Soar in 1995/6. It is notable that, despite continued emissions reductions, the UK fraction increases again by 2000, presumably as a result of emissions controls being implemented in other European countries. Indeed, the depositional trends for all European nations across the period seems to be one of general decline, although the magnitude of changes over the decade varies considerably and a few, such as Germany, Poland and the Czech and Slovak Republics show maximum contributions to Lochnagar in the final year of the time-series. By contrast, the contributions from shipping appear to change little across the period such that the percentage contribution from this source to Lochnagar increases from 4.5% in 1985 to 7.3% in 1996. If the data from 2000 are also considered, then the result is the trend shown in Figure 4 and it would appear that since the mid-1990s the relative contribution of sulphur from shipping sources to Lochnagar has more than doubled. Modelling by the EU's 'Clean Air for Europe' (CAFE)

Table 5. Annual sulphur deposition (kg) across the combined Lochnagar catchment and loch area from selected European countries and marine sources 1985 – 1996 as modelled by EMEP. [Cz and Slovak = Czechoslovakia and later Czech Republic and Slovakia; Germany = combined East and west Germany and later unified Germany; Nat. Ocean = natural marine sources].

	1985	1987	1989	1991	1993	1995	1996
Belgium	4.74	3.12	6.46	4.59	3.37	2.03	3.22
Cz and Slovak	11.10	5.26	1.11	5.75	13.01	7.63	17.17
France	8.74	12.98	9.42	15.36	11.98	6.90	7.47
Germany	60.51	44.90	14.71	21.51	33.45	11.87	24.36
Ireland	16.36	16.10	26.24	32.04	19.92	22.16	18.49
Poland	7.91	6.70	0.87	3.34	11.61	8.35	20.91
Spain	7.49	1.39	3.95	5.62	6.79	3.67	3.51
UK	505.04	519.00	490.54	584.27	438.27	339.58	285.87
Baltic Sea	1.59	0.71	0.67	1.01	1.07	0.98	1.52
North Sea	12.33	10.31	10.68	13.07	13.65	10.03	12.50
Atlantic.	17.75	13.55	21.26	22.54	16.58	19.70	17.47
Med. Sea	0.00	0.00	0.00	0.00	0.00	0.00	1.61
Nat. Ocean	18.75	19.34	23.00	24.04	16.24	19.23	16.15
Attributed Total	698.03	669.45	620.46	744.08	595.05	463.55	449.23
Unattributed	90.02	81.46	116.62	103.86	78.86	80.15	70.89
TOTAL	788.04	750.91	737.08	847.94	673.91	543.70	520.12

(http://europa.eu.int/comm/environment/air/cafe/index.htm) programme predicts that, even after accounting for MARPOL limits on the sulphur content of marine fuels, emissions of SO_2 from international shipping in European waters are expected to increase by 45% between 2000 and 2020. Further, with the continued decline of SO_2 emissions from terrestrial sources, emissions from international shipping in European waters is expected to surpass the total terrestrial emissions from the 25 EU member states by 2020. It therefore seems likely that this source of sulphur will become increasingly important for deposition at Lochnagar over the next two decades.

Data for European contributions of other pollutants are less site-specific and EMEP have supplied data for oxidised and reduced nitrogen for the 50 x 50 km grid square within which Lochnagar lies. These are shown in Tables 6 and 7 respectively. As with sulphur, EMEP modelled depositions are considerably lower than estimated by CEH (NEGTAP 2001), which are 9.2 kg N ha yr^{-1} for oxidised nitrogen and 7.7 kg N ha yr^{-1} for reduced N. For both forms of nitrogen, the EMEP model indicates that the UK is by far the largest source with 47% and 71% of the total NO_Y-N and NH-N respectively. The other main contributing countries are much the same, although Germany, the Netherlands and Norway are more important for NO_Y inputs and Ireland for NH-N. Removing the Scottish contribution from the rest of the UK, shows that Scotland would be the sixth highest contributing nation to this 50 x 50 km square for NO_Y and fourth for NH-N. Marine sources are again important for oxidised N. The North Sea and northeast Atlantic contribute more than another country apart from the UK and France and the combined shipping contribution is almost 15%. As might be expected, the contribution of shipping to reduced nitrogen deposition is very small.

Whilst the work of Davies et al. (1984) and Brimblecombe et al. (1986) hint at a possible European contribution to pollution episodes in Scotland (see above), other studies show more direct evidence. Both Barnes and Eggleton (1977) and Buchanan et al. (2002) showed significant transport from mainland Europe, with the latter stating that during 1995-1996, PM_{10} (particulate matter with an aerodynamic diameter less than 10 μm) concentrations "increased by 11 μg m^{-3} (compared with annual mean) when air mass back-trajectories were from Eastern Europe" whilst Beverland et al. (2000) came to similar conclusions for Leuchars, near Dundee, on the east coast of Scotland.

One of the best known cases of pollution transport from Eastern Europe to the UK resulted from the accident at the Chernobyl nuclear power plant, 125 km north of Kiev in Ukraine, on 26th April 1986. The release of radioactive materials lasted nearly ten days and these were transported over a wide area of Europe. The lack of rainfall in the Ukraine and surrounding parts of the (former) Soviet Union resulted in a significant fraction of volatile fission products, such as ^{131}I, ^{132}Te and ^{137}Cs, remaining in the atmosphere until interception by rainfall. The plume passed over the UK on 2nd – 4th May and rainfall patterns over that period resulted in the main deposition occurring in west Scotland, northwest England, north Wales and Northern Ireland (Clark and Smith 1988). By comparison, the northeast of Scotland was affected only slightly, with an estimated total ^{137}Cs deposition across the area of less than 5 kBq m^{-2} (Clark and Smith 1988). However, this was sufficient to be detectable in sediment cores taken from Lochnagar later in 1986 (NAG3) and 1991 (NAG6), but not in a core taken in 1996 (NAG8). In both NAG3 and NAG6, a peak in the concentration of ^{137}Cs in the surface sediments was ascribed to Chernobyl. This was distinguishable from other sources of ^{137}Cs by the presence of a second peak of this radionuclide at a lower sediment depth, contemporaneous with a peak in ^{241}Am (Figure 5) and hence indicative of global weapons testing (see below). ^{134}Cs was also released from Chernobyl but whilst this has been recorded in some lake sediment cores it has not been detected in any from Lochnagar, presumably as deposition was only limited.

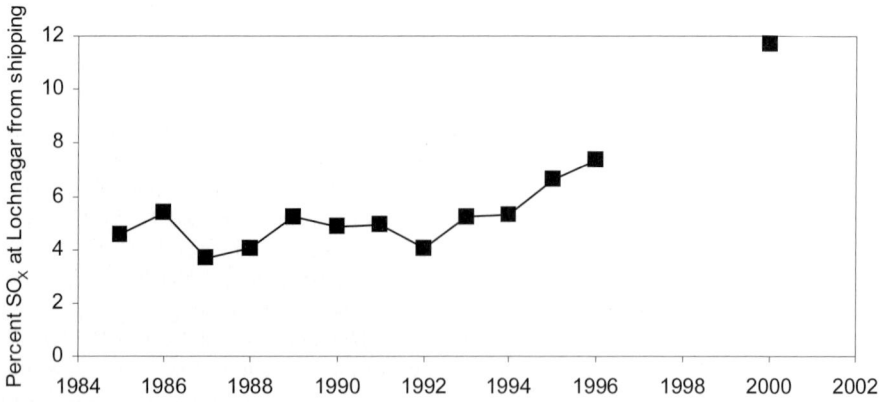

Figure 4. Percent of total sulphur deposited at Lochnagar derived from shipping in European waters 1985 – 2000. Data from EMEP modelling.

Table 6. Contribution of NO_Y-N to Lochnagar from European countries and shipping sources for 2000 as modelled by EMEP. Values in kg ha^{-1}. Load represents kg nitrogen deposited across the Lochnagar catchment and attributed to that source for 2000. The percentage contribution from each source to the total is also shown.

	kg ha^{-1}	Load	% Contribution
UK	1.60	174.03	47.6
France	0.28	29.95	8.2
Germany	0.19	20.40	5.6
Netherlands	0.14	14.76	4.0
Ireland	0.12	12.59	3.4
Spain	0.11	12.15	3.3
Belgium	0.08	8.68	2.4
Norway	0.04	3.91	1.1
Poland	0.02	2.60	0.7
Denmark	0.02	2.60	0.7
Portugal	0.02	2.60	0.7
Czech Rep	0.02	2.17	0.6
Sweden	0.01	0.87	0.2
Italy	0.01	0.87	0.2
Total shipping	0.50	54.25	14.8
Unattributable	0.22	23.44	6.4
TOTAL	3.37	365.86	

Table 7. Contribution of NH-N to Lochnagar from European and marine sources for 2000 as modelled by EMEP. Values in kg ha^{-1}. Load represents kg nitrogen deposited across the Lochnagar catchment and attributed to that source for 2000. The percentage contribution from each source to the total is also shown.

	kg ha^{-1}	Load	% Contribution
UK	2.740	297.29	71.2
France	0.364	39.49	9.5
Ireland	0.220	23.87	5.7
Germany	0.136	14.76	3.5
Spain	0.080	8.68	2.1
Netherlands	0.072	7.81	1.9
Belgium	0.056	6.08	1.5
Poland	0.020	2.17	0.5
Portugal	0.020	2.17	0.5
Denmark	0.016	1.74	0.4
Czech Rep	0.012	1.30	0.3
Italy	0.012	1.30	0.3
Other	0.024	2.58	0.6
Total shipping	0.060	6.51	1.6
Unattributable	0.016	1.74	0.4
TOTAL	3.848	417.51	

The Lochnagar sediment record provides further evidence for inputs from Eastern Europe using the temporal record of toxaphene (Rose et al. 2001; Chapter 16: Muir and Rose this volume). Toxaphene is a pesticide that was widely used following the ban on DDT in the 1970s, particularly as an insecticide in the cotton-growing industry and as a piscicide to remove undesirable fish. However, it has neither been produced nor used in the UK and has since been placed on the United Nations Economic Commission for Europe (UNECE) list of banned persistent organic compounds. The historical record of toxaphene in the sediments of Lochnagar therefore solely reflects long-range transport to the site. The details of the toxaphene record are discussed further in Chapter 16 (Muir and Rose this volume), but the double peak in the profile is thought to represent inputs from both eastern Europe (notably the former East Germany) and an intercontinental or global pattern as a result of production in the United States (Rose et al. 2001).

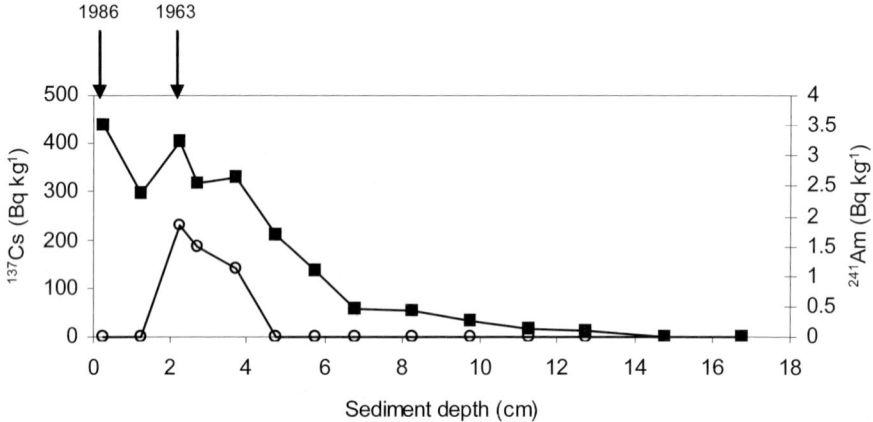

Figure 5. Sediment record of radionuclide fallout in core NAG3. [137]Cs shown by solid symbols (■) and [241]Am by open symbols (○). Coincident peaks in ^{137}Cs and [241]Am indicate the peak in global weapons testing in 1963; a later peak in ^{137}Cs alone indicates deposition from the Chernobyl accident in 1986.

Intercontinental and global sources

In recent years the intercontinental movement of pollutants has become of increasing concern and while much of the literature relates to trans-Pacific movement from Asia's expanding industries to the west coast of the USA (Berntsen et al. 1999; Jaffe et al. 1999; 2003a;b; Bailey et al. 2000; Wilkening et al. 2000) there is also evidence for movement from Europe to Asia (Wild et al. 2004; Pochanart et al. 2003), from North America to Europe (Holloway et al. 2003; Stohl et al. 2003; Huntrieser et al. 2005) and the movement of Saharan dusts and Chinese loess from Africa and Asia to Europe (Grousset et al. 2003).

At Lochnagar there are three main strands of evidence pertaining to intercontinental or global contamination. The first of these is discussed in more detail in Chapter 16 (Muir and Rose this volume) and concerns the peak in sediment toxaphene concentrations in the mid-1970s thought to relate to US production and usage. The second is from the sediment record of ^{137}Cs and ^{241}Am. These artificial radionuclides were first introduced into the environment on a global scale in 1954 by the atmospheric testing of thermonuclear weapons. Fallout of these radionuclides from this source reached a maximum value in 1963 and then declined sharply following the treaty in that year banning further atmospheric tests. The sediment records of these isotopes are used widely, along with the naturally occurring radionuclide ^{210}Pb, in constructing chronologies for sediment cores. As such, the record of both ^{137}Cs and ^{241}Am has been identified in a number of sediment cores from Lochnagar thereby identifying contamination from this source at the site (Figure 5). However, with a half-life of 30.1 years the sediment record of ^{137}Cs from atmospheric weapons testing should become undetectable at Lochnagar by the end of the 22nd century and the record from Chernobyl a few decades later. The half-life of ^{241}Am, on the other hand, is 432 years and so the impact from nuclear weapons testing should remain detectable in the sediment record of Lochnagar for at least another 3000 years.

The third line of evidence for global impacts at Lochnagar is more indirect and concerns the effects of global climate change. Whilst most research is, quite rightly, focussed on the future impacts of continuing climatic changes there are indications that impacts are already being recorded. Current changes and future impacts are discussed in more detail elsewhere in this book and summarised in the final chapter (Chapter 19: Rose and Battarbee this volume). However, in brief, direct evidence includes increasing air and water temperatures (Chapter 5: Thompson et al; Chapter 18: Kettle and Thompson) and the loss of continuous winter snow and ice cover (Chapter 5: Thompson et al. this volume). Indirect evidence for observed changes which may implicate climate change include elevated soil erosion (Chapter 6: Helliwell et al. this volume), increasing water concentrations of dissolved organic carbon (DOC) in the loch water (Freeman et al. 2001; Chapter 14: Monteith et al. this volume), the possible confounding of recovery from acidification (Rose et al. 2004b; Chapter 14: Monteith et al. this chapter) and the elevated inputs of previously deposited trace metals from catchment soils to the loch and sediment record (Yang et al. 2002a; Rose et al. 2004b; Chapter 15: Tipping et al. this volume).

Discussion

For historical, political and geographical reasons, mountainous areas often form national boundaries, and mountain lakes are therefore frequently subject to atmospherically transported pollutants from multi-national sources. Depending on source proximity and prevailing meteorological conditions, sources from outside the 'host' nation can even provide greater inputs than the country in which the lake is geographically located. Mountain lakes in the southern French Alps, close to the Italian border, are good examples of this as they receive considerable inputs from the industrial regions of northeast Italy (e.g., Rose et al. 1999).

For mountain areas in the UK, the situation is simpler. Due to its position to the west of mainland Europe the UK has been, and continues to be, the source of much of the pollutants deposited within its own borders. The evidence for pollutant sources specifically impacting Lochnagar support this and show that, while there are a large number of sources on a broad range of geographical scales contributing to the pollutant load, the main source area would appear to be the north, and Midlands, of England. There is also evidence, however, for considerable impact from very long range. The concentrations and accumulation rates of toxaphene in the lake sediments of Lochnagar reached levels equivalent to those of the Great Lakes of North America, where toxaphene has both been produced and widely used as a pesticide (Rose et al. 2001) and where significant riverine inputs might be expected to have elevated levels over and above that from atmospheric deposition alone. The reasons for these high levels in Lochnagar remain unclear, principally due to the lack of comparable data from elsewhere in the UK and Europe, and it could be that accumulation rates and concentrations of the compound are significantly elevated above those in Lochnagar in regions of southern and Eastern Europe closer to sources. If this is the case there are considerable implications for human and environmental health in these regions.

Given the diversity of sources, the protection of fragile mountain lake ecosystems, such as that at Lochnagar, requires local and national legislation as well as international co-operation on emissions policies. Hence, while the threats come from a range of spatial scales, from regional to global, so must the protection. As Doughty et al. (2002) state "The main driving forces for the future are political decisions concerning the environment taken at global (e.g., climate change), European (e.g., EU Directives), United Kingdom (e.g. Wildlife and Countryside Acts,), Scotland (e.g. Green Paper on freshwater fish) and local council (e.g., Local Biodiversity Action Plans (LBAPs)) levels. European legislation in particular is now a strong steering force for the way in which Scotland's fresh waters are conserved and managed".

For Lochnagar, at the local level, there are no major sources of air pollution and no specific control plans. However, at the Scottish scale, emissions control legislation such as the Pollution Prevention and Control (Scotland) Regulations (SSI 2000/323), the Large Combustion Plants (Scotland) Regulations (SSI/2002/493) and the Solvent Emissions (Scotland) Regulations (SSI 2004/26) are in place and, at the UK level, there is a long history of legislation including the Control of Pollution Act (1974), the Environment Protection Act 1995, the Pollution Prevention and Control Acts of 1974 (water), 1981 (air) and 1999, as well as the Clean Air Acts of 1956, 1968 and 1993. Further, restructuring of the energy supply industry for Northern Ireland, moving away from its historical reliance on coal-fired local generation, is likely to reduce the impact of this source on deposition at Lochnagar.

At the European scale, emissions have been restricted through various EU Directives (most recently the National Emissions Ceilings Directive). The Convention on the Conservation of European Wildlife and Natural Habitats (the 'Bern Convention'; 82/72/EEC), the European Councils Directive on Integrated Pollution Prevention and Control (the IPPC Directive; 96/61/EC) and more recently the European Union's Water Framework Directive (WFD) (2000/60/EC) also set the legislative framework for environmental protection. The WFD came into force in 2000 and became law in Scotland at the end of 2003 through the Water Environment and Water Services (Scotland) Act. It establishes an integrated approach to the protection, improvement and sustainable use of Europe's rivers, lakes, estuaries, coastal waters and groundwater and aims to achieve good ecological quality in all relevant waters within 15 years. It requires that biological, hydromorphological and chemical elements of water quality should be based on the degree to which present day conditions deviate from those expected in the absence of significant anthropogenic influence, termed reference conditions. The WFD specifically requires the determination of reference conditions for different waterbody types in order to identify sites of 'High' status, i.e., those where the various elements correspond to undisturbed conditions. The four categories of 'Good', 'Moderate', 'Poor' and 'Bad' status are defined according to the degree of deviation from the reference state. In the absence of long-term data, the WFD states that reference conditions based on modelling may be derived using hindcasting methods, and the study of lake sediment records is one such approach (Pollard and Huxham 1998; European Union 2000).

Beyond Europe, the UNECE Convention on Long-Range Transboundary Air Pollution (LRTAP) was established in 1979 and via a series of Protocols has been instrumental in the negotiated reduction of a range of air pollutants emitted to the atmosphere. Initially, these focussed mainly on acidifying pollutants, but in 2003 the Aarhus Protocols on heavy metals and persistent organic pollutants (POPs) also came into force while the Protocol setting 2010 emission ceilings for sulphur, NO_Y, VOCs and ammonia came into force in 2005. Although the hemispherical treaty controlling the intercontinental transport of pollutants envisaged by Holloway et al. (2003) has yet to become a reality, the UNECE LRTAP has also recently set up a Task Force on Hemispherical Transport of Air Pollution (http://www.htap.org/index.htm) (TF HTAP) to develop a fuller understanding of the problem, focussing on fine particles, acidifying deposition, mercury, POPs and the precursors of ozone. Furthermore, models dealing with the hemispherical transport of toxic substances using the MSCE-Hem-HM and MSCE-Hem-POP models for trace metals (including Hg) and POPs respectively are now becoming more developed (Travnikov 2001; Travnikov and Ryaboshapko 2002; Gusev et al. 2005) while on a global scale, some treaties are already in place such as the Kyoto Protocol to the United Nations Framework Convention on Climate Change (UNFCCC) which entered into force in 2005 following the ratification by sufficient countries to account for more than 55% of the total carbon dioxide emissions for 1990.

There is, therefore, already a wealth of legislation in place that should be protecting fragile ecosystems such as that at Lochnagar. Yet, as can be seen from Chapters 14 to 17, despite major reductions in emissions over the last few decades there is a continuing impact from atmospherically deposited contaminants. This ongoing threat is due to the potentially massive store of pollutants in the catchment soils, deposited from the atmosphere as a result of 150 – 200 years of industrial emissions and whose release may be accelerated as a result of a changing climate. While legislation is in place to attempt to reduce, or prevent, further emissions of acidifying gases, trace metals and POPs at source, thereby lowering the inputs from the atmosphere, so pollutants released from catchments may become more significant. Already, it has been calculated that transport from the catchment soils at Lochnagar is responsible for 78% and 91% of the Hg and Pb inputs to the loch, respectively (Yang et al. 2002a) and that hundreds of years worth of deposition, at 1998 levels, remain in the soils. Exacerbation of this process will therefore maintain, and possibly increase, pollutant inputs to the loch regardless of emissions controls policies implemented in European, UK or Scottish Parliaments. Further protection from increased catchment inputs at Lochnagar and other sites like it will therefore depend on a global commitment to tackle climate change.

Finally, there is a need to know that the implementation of pollution controls, at whatever geographical scale, is effective at protecting vulnerable areas. There is, therefore, a need for ongoing monitoring of a range of parameters at a series of lake types. The work described in this book represents a significant investment of resources and the Lochnagar dataset is probably unique because of it. However, as the impacts of climate change are increasingly identified through a range of ecological compartments, these data represent a valuable baseline against which to observe and measure future impacts, but this can only be done by a continued commitment to science at the site and other lakes and lochs like it. Furthermore, while palaeolimnological data have provided the long-term context for observations of anthropogenically-driven environmental change at Lochnagar and the UK AWMN through 18 years of high quality biological and chemical data is now providing evidence for some recovery from acidification and 'new' impacts from climatic changes, there is a further need for accurate, validated models to predict future changes. Only through the use of a combination of contemporary monitoring, palaeo-techniques and modelling can full human-environment interactions be identified (Dearing et al. 2006a; b) and appropriate precautions be taken against future impacts.

Summary

- Modelling and empirical data provide evidence for a range of pollutant sources impacting Lochnagar across broad geographical scales.
- Within Scotland, the main source area of contamination at Lochnagar is the Firth of Forth and, in particular, the coal-fired power station at Longannet.
- The HARM model shows sources in England and Wales contributed >60% of the total sulphur derived from within Great Britain. The north of England and the Midlands appear to be the major source areas. This is supported by back-trajectory analysis.
- EMEP modelling indicates that the UK is the major source of acidifying pollutants to Lochnagar, although the limitations of the EMEP model in reproducing site-specific deposition must be considered.
- Shipping within European waters makes a significant contribution to sulphur deposition at Lochnagar and this is likely to increase as marine emissions are forecast to exceed European terrestrial emissions by 2020. Shipping is also a significant source of oxidised nitrogen at Lochnagar.
- Evidence for contamination of Lochnagar from European sources includes back-trajectory analysis of black snow events and the sediment record of both toxaphene from Eastern Europe and the Chernobyl accident in Ukraine in 1986.
- Evidence for intercontinental and global sources impacting Lochnagar include, the sediment record of toxaphene of North American origin, the record of ^{137}Cs and ^{241}Am from atmospheric testing of thermo-nuclear weapons and direct and indirect evidence for global climate change.
- As sources of contamination at Lochnagar are from a broad geographical range, so protection must address this range (local, national and international).

- The availability of appropriate models can make a vital contribution to source attribution and the assessment of emissions reduction policies.
- Despite current legislation, a major source of present and future contamination comes from previously deposited pollutants, stored in catchment soils and available to be released as a result of climate change.
- Current datasets provide an excellent baseline against which to measure future changes in contamination status at Lochnagar. There is, therefore, a need for a commitment to future monitoring to determine both how well implemented policy protects remote fragile ecosystems and as an early warning for future changes at less sensitive sites as a result of climate change impacts.

Acknowledgements

We wish to thank Peter Appleby for the use of his radionuclide data and Duncan Whyatt and Dick Derwent for their helpful comments on an earlier version of the manuscript.

References

Bailey R., Barrie L.A., Halsall C.J., Fellin P. and Muir D.C.G. 2000. Atmospheric organochlorine pesticides in the Western Canadian Arctic: Evidence of trans-Pacific transport. J. Geophys. Res. 105: 11805–11811.

Barnes R.A. and Eggleton A.E.J. 1977. The transport of atmospheric pollutants across the North Sea and English Channel. Atmos. Environ. 11: 879–892.

Berntsen T.K., Karlsdottir S. and Jaffe D.A. 1999. Influence of Asian emissions on the composition of air reaching the Northwestern United States. Geophys. Res. Lett. 26: 2171–2174.

Beverland I.J., Tunes T., Sozanska M., Elton R.A., Agius R.M. and Heal M.R. 2000. Effect of long-range transport on local PM_{10} concentrations in the UK. Internat. J. Environ. Health Res. 10: 229–238

Borghini F., Grimalt J.O., Sanchez-Hernandez J.C. and Bargagli R. 2005. Organochlorine pollutants in soils and mosses from Victoria Land (Antarctica). Chemosphere. 58: 271–278.

Boutron C.F. 1982. Atmospheric trace metals in the snow layers deposited at the South Pole from 1928 to 1977. Atmos. Environ.16: 2451–2459.

Brimblecombe P., Davies T. and Tranter M. 1986. Nineteenth century black Scottish showers. Atmos. Environ. 1986; 20: 1053–1057.

Buchanan C.M., Beverland I.J. and Heal M.R. 2002. The influence of weather-type and long-range transport on airborne particle concentrations in Edinburgh, UK. Atmos. Environ. 36: 5343–5354.

Clark M.J. and Smith F.B. 1988. Wet and dry deposition of Chernobyl releases. Nature. 332: 245–249.

Davies J.J.L., Jenkins A., Monteith D.T., Evans C.D. and Cooper D.M. 2005. Trends in surface water chemistry of acidified UK Freshwaters, 1988–2002. Environ. Pollut. 137: 27–39.

Davies T.D., Abrahams P.W., Tranter M., Blackwood I., Brimblecombe P. and Vincent C.E. 1984. Black acidic snow in the remote Scottish highlands. Nature. 312: 58–61.

Dearing J.A., Battarbee R.W., Dikau R., Larocque I. and Oldfield F. 2006a. Human-environment interactions: learning from the past. Region. Environ. Change. 6: 1–16.

Dearing J.A., Battarbee R.W., Dikau R., Larocque I. and Oldfield F. 2006b. Human-environment interactions: towards synthesis and simulation. Region. Environ. Change. 6: 115–123.

Doughty C.R., Boon P.J and Maitland P.S. 2002. The state of Scotland's fresh waters. In: Usher M.B., Mackey E.C. and Curran J.C. (eds.) The State of Scotland's Environment and Natural Heritage. The Stationary Office, Edinburgh. pp.117–144.

EMEP. 1996. Transboundary air pollution in Europe. EMEP/MSC-W Report 1/96. Part One. Estimated dispersion of acidifying agents and of near surface ozone. 152 pp.

European Union. 2000. Establishing a framework for community action in the field of water policy. Directive of the European Parliament and of the Council 2000/60/EC. PE-CONS 3639/1/00 REV 1, Luxembourg.

Fernández P., Vilanova R.M. and Grimalt J.O. 1999. Sediment fluxes of polycyclic aromatic hydrocarbons in European high altitude mountain lakes. Environ. Sci. Technol. 33: 3716–3722.

Fernández P., Vilanova R.M., Martinez C., Appleby P. and Grimalt J.O. 2000. The historical record of atmospheric pyrolytic pollution over Europe registered in the sedimentary PAH from remote mountain lakes. Environ. Sci. Technol. 34: 1906–1913.

Fowler D., McFadyen G.G., Ellis N.E. and MacGregor W.G. 2002. Air pollution and atmospheric deposition in Scotland: Environmental and natural heritage trends. In: Usher M.B., Mackey E.C. and Curran J.C. (eds.) The State of Scotland's Environment and Natural Heritage. The Stationary Office, Edinburgh. pp. 89–110.

Fowler D., Smith R.I. and Muller J. 2004. United Kingdom contribution to the EMEP assessment. In: Bartnicki J. and Lövblad G. (eds.) EMEP Assessment Part II. National contributions. Norwegian Meteorological Institute, Oslo. 250 pp.

Freeman C., Evans C.D., Monteith D.T., Reynolds B. and Fenner N. 2001. Export of organic carbon from peat soils. Nature. 412: 785.

Grimalt J.O., Fernández P., Berdie L., Vilanova R.M., Catalan J., Psenner R., Hofer R., Appleby P.G., Rosseland B.O., Lien L., Massabuau J.C. and Battarbee R.W. 2001. Selective trapping of organochlorine compounds in mountain lakes of temperate areas. Environ. Sci. Technol. 35: 2690–2697

Grousset F.E., Ginoux P., Bory A. and Biscaye P.E. 2003. Case study of a Chinese dust plume reaching the French Alps. Geophys. Res. Lett. 30: 1273–1277.

Gusev A., Mantseva E., Shatalov V. and Travnikov O. 2005. Hemispheric model (MSCE-Hem) of persistent toxic substances dispersion in the environment. MSC-E Technical note 11/2005. Co-operative Programme for Monitoring and Evaluation of the Long-Range Transmission of Air Pollutants in Europe (EMEP). Meteorological Synthesizing Centre – East.

Helliwell R.C., Lilly A. and Bell J.S. This volume. Chapter 6. The development, distribution and properties of soils in the Lochnagar catchment and their influence on soil water chemistry. In: Rose N.L. (ed.) 2007. Lochnagar: The natural history of a mountain lake. Springer. Dordrecht.

Hermanson M.H. 1993. Historical accumulation of atmospherically derived pollutant trace metals in the Arctic as measured in dated sediment cores. Wat. Sci. Technol. 28: 33–41.

Holloway T., Fiore A. and Hastings M.G. 2003. Intercontinental transport of air pollution: Will emerging science lead to a new hemispheric treaty. Environ. Sci. Technol. 37: 4535–4542.

Huntrieser H., Schlager H., Forster C., Stohl A., Aufmhoff H., Arnold F., Scheel H.E., Campana M., Gilge S., Eixmann R. and Cooper O. 2005. Intercontinental pollution transport from North America to Europe: Experimental evidence from airborne measurements and surface observations. J. Geophys. Res. – Atmospheres. 110: D1. 1283–1305

Ilyin I., Ryaboshapko A., Afinogenova O., Berg T. and Hjellbrekke A-G. 2001. Evaluation of transboundary transport of heavy metals in 1999. Trend analysis. EMEP / MSC-E Report 3/2001. 129 pp.

Jaffe D.A., Anderson T., Covert D., Kotchenruther R., Trost B., Danielson J., Simpson W., Berntsen T., Karlsdottir S. Blake D., Harris J., Carmichael G. and Uno I. 1999. Transport of Asian Air Pollution to North America. Geophys. Res. Lett. 26: 711–714.

Jaffe D., McKendry I., Anderson T. and Price H. 2003a. Six 'new' episodes of trans-Pacific transport of air pollutants. Atmos. Environ. 37: 391–404.

Jaffe D., Snow J. and Cooper O. 2003b. The April 2001 Asian dust events: Transport and substantial impact on surface particulate matter concentrations across the United States. EOS, Nov 18th.

Jenkins A., Reynard N., Hutchins M., Bonjean M. and Lees M. This volume. Chapter 9. Hydrology and hydrochemistry of Lochnagar. In: Rose N.L. (ed.) 2007. Lochnagar: The natural history of a mountain lake. Springer. Dordrecht.

Kettle H. and Thompson R. This volume. Chapter 18. Future climate predictions for Lochnagar. In: Rose, N.L. (ed.) 2007. Lochnagar: The natural history of a mountain lake. Springer. Dordrecht.

MacDonald R.W., Barrie L.A., Bidleman T.F., Diamond M.L., Gregor D.J., Semkin R.G., Strachan W.M.J., Li Y.F., Wania F., Alaee M., Alexeeva L.B., Backus S.M., Bailey R., Bewers J.M., Gobeil C., Halsall C.J., Harner T., Hoff J.T., Jantunen L.M.M., Lockhart W.L., Mackay D., Muir D.C.G., Pudykiewicz J., Reimer K.J., Smith J.N., Stern G.A., Schroeder W.H., Wagemann R. and Yunker M.B. 2000. Contaminants in the Canadian Arctic: 5 years of progress in understanding sources, occurrence and pathways. Sci. Tot. Environ. 254: 93–234.

McDonald A.G., Nemitz E., Dragosits U., Sutton M.A. and Fowler D. 2002. Modelling heavy metal deposition across Scotland. In: Usher, M.B., Mackey, E.C. and Curran, J.C. (eds.) The State of Scotland's Environment and Natural Heritage. The Stationary Office, Edinburgh. pp. 111–116.

Meharg A.A. and Killham K. 2003. A pre-industrial source of dioxins and furans. Nature. 421: 909–910.

Merrill J.T. and Kim J. 2004. Meteorological events and transport patterns in ACE-Asia. J. Geophys. Res. 109 (D19): art. no. D19S18.

Metcalfe S.E., Whyatt J.D. and Derwent R.G. 1995. A comparison of model and observed network estimates of sulphur deposition across Great Britain for 1990 and its likely source attribution. Quart. J. Roy. Meteorol. Soc. 121: 1387–1411.

Monteith D.T., Evans C.D. and Dalton C. This volume. Chapter 14. Acidification of Lochnagar and prospects for recovery. In: Rose N.L. (ed.) 2007. Lochnagar: The natural history of a mountain lake. Springer. Dordrecht.

Muir D.C.G. and Rose N.L. This volume. Chapter 16. Persistent organic pollutants in the sediments of Lochnagar. In: Rose N.L. (ed.) 2007. Lochnagar: The natural history of a mountain lake. Springer. Dordrecht.

Murphey B.B. and Hogan A.W. 1992. Meteorological transport of continental soot to Antarctica? Geophys. Res. Lett. 19: 33–36.

NEGTAP (National Expert Group on Transboundary Air Pollution) 2001. Transboundary air pollution: Acidification, eutrophication and ground-level ozone in the UK. Prepared on behalf of the UK Department for Environment, Food and Rural Affairs (DEFRA) and the devolved administrations. 314 pp.

Pochanart P. Akimoto H., Kajii Y., Potemkin V.M. and Khodzher T.V.2003. Regional background ozone and carbon monoxide variations in remote Siberia / East Asia. J. Geophys. Res. – Atmospheres. 108: D1. 4010–4028

Pollard P. and Huxham M. 1998. The European Water Framework Directive: A new era in the management of aquatic ecosystem health? Aquat. Conservat.: Marine and Freshwater Ecosystems. 8: 773 –792.

Rodriguez S., Querol X., Alastuey A., Kallos G. and Kakaliagou O. 2001. Saharan dust contributions to PM10 and TSP levels in Southern and Eastern Spain. Atmos. Environ. 35: 2433–2447.

Rose N.L. 1996. Inorganic ash spheres as pollution tracers. Environ. Pollut. 91: 245–252.

Rose N.L. 2001. Fly-ash particles. In: Last W.M. and Smol J.P. (eds.). Tracking environmental change using lake sediments. Volume 2: Physical and Geochemical methods. Kluwer Academic Publishers, Dordrecht. pp. 319–349.

Rose N.L. and Battarbee R.W. This volume. Chapter 19. Past and future environmental change at Lochnagar and the impacts of a changing climate. In: Rose N.L. (ed.) 2007. Lochnagar: The natural history of a mountain lake. Springer. Dordrecht.

Rose N.L. and Harlock S. 1998. The spatial distribution of characterised fly-ash particles and trace metals in lake sediments and catchment mosses in the United Kingdom. Wat. Air Soil Pollut.106: 287–308.

Rose N.L. and Juggins S. 1994. A spatial relationship between carbonaceous particles in lake sediments and sulphur deposition. Atmos. Environ. 28: 177–183.

Rose N.L. and Monteith D.T. 2005. Temporal trends in spheroidal carbonaceous particle deposition derived from annual sediment traps and lake sediment cores and their relationship with non-marine sulphate. Environ. Pollut. 137: 151–163.

Rose N.L. and Yang H. This volume. Chapter 17. Temporal and spatial patterns of spheroidal carbonaceous particles (SCPs) in sediments, soils and deposition at Lochnagar. In: Rose N.L. (ed.) 2007. Lochnagar: The natural history of a mountain lake. Springer. Dordrecht.

Rose N.L., Juggins S. and Watt J. 1996. Fuel-type characterisation of carbonaceous fly-ash particles using EDS-derived surface chemistries and its application to particles extracted from lake sediments. Proc. Roy. Soc. Lond. Series A. 452: 881–907.

Rose N.L., Harlock S. and Appleby P.G. 1999. The spatial and temporal distributions of spheroidal carbonaceous fly-ash particles (SCP) in the sediment records of European mountain lakes. Wat. Air Soil Pollut.113: 1–32.

Rose N.L., Backus S., Karlsson H. and Muir D.C.G. 2001. An historical record of toxaphene and its congeners in a remote lake in Western Europe. Environ. Sci. Technol. 35: 1312–1319.

Rose N.L., Rose C.L., Boyle J.F and Appleby P.G. 2004a. Lake sediment evidence for local and remote sources of atmospherically deposited pollutants on Svalbard. J. Paleolim. 31: 499–513

Rose N.L., Monteith D.T., Kettle H., Thompson R., Yang H. and Muir D.C.G. 2004b. A consideration of potential confounding factors limiting chemical and biological recovery at Lochnagar, a remote mountain loch in Scotland. J. Limnol. 63: 63–76.

Rust R.J. 1864. The Scottish black rain showers and pumicestone shoals of the years 1862 and 1863. William Blackwood and Sons, Edinburgh. 64 pp.

Scorer R.S. 1992. Deposition of concentrated pollution at large distance. Atmos. Environ. 26A: 793–805.

Stohl A., Huntrieser H., Richter A., Beirle S., Cooper O.R., Eckhardt S., Forster C., James P., Spichtinger N., Wenig M., Wagner T., Burrows J.P. and Platt U. 2003. Rapid intercontinental air pollution transport associated with a meteorological bomb. Atmos. Chem. Phys. 3: 969–985.

Tarrasón L., Jonson J.E., Fagerli H., Benedictow A., Wind P., Simpson D. and Klein H. 2003. Transboundary Acidification, Eutrophication and Ground Level Ozone in Europe. EMEP Status Report 1/2003, Part III: Source-Receptor Relationships. The Norwegian Meteorological Institute, Oslo, Norway. 102 pp.

Thompson R., Kettle H., Monteith D.T. and Rose N.L. This volume. Chapter 5. Lochnagar water-temperatures, climate and weather. In: Rose N.L. (ed.) 2007. Lochnagar: The natural history of a mountain lake. Springer. Dordrecht.

Tipping E., Yang H., Lawlor A.J., Rose N.L. and Shotbolt L. This volume. Chapter 15. Trace metals in the catchment, lake and sediments of Lochnagar: measurements and modelling. In: Rose N.L. (ed.) Lochnagar: The natural history of a mountain lake. Springer.

Tranter M., Abrahams P., Blackwood I., Brimblecombe P. and Davies T.D. 1988. The impact of a single black snowfall on streamwater chemistry in the Scottish Highlands. Nature 332: 826–829.

Travnikov O. 2001. Hemispheric model of airborne pollutant transport. MSC-E Technical note 8/2001. Co-operative Programme for Monitoring and Evaluation of the Long-Range Transmission of Air Pollutants in Europe (EMEP). Meteorological Synthesizing Centre – East. 37 pp.

Travnikov O. and Ryaboshapko A. 2002. Modelling of mercury hemispheric transport and depositions. MSC-E Technical report 6/2002. Co-operative Programme for Monitoring and Evaluation of the Long-Range Transmission of Air Pollutants in Europe (EMEP). Meteorological Synthesizing Centre – East. 69 pp.

Vestreng V. and Klein H. 2002. Emission data reported to UNECE/EMEP: Quality assurance and trend analysis and presentation of WebDab. Technical Report Note 1/2002, Meteorological Synthesizing Centre – West, The Norwegian Meteorological Institute, Oslo, Norway. 102 pp.

Vives I., Grimalt J.O., Catalan J., Rosseland B.O. and Battarbee R.W. 2004a. Influence of altitude and age in the accumulation of organochlorine compounds in fish from high mountain lakes. Environ. Sci. Technol. 38: 690–698.

Vives I., Grimalt J.O., Lacorte S., Guillamón M., Barceló D, and Rosseland B.O. 2004b. Polybromodiphenyl ether flame retardants in fish from lakes in European high mountains and Greenland. Environ. Sci. Technol. 38: 2338–2344.

Wania F. and Mackay D. 1996. Tracking the distribution of persistent organic pollutants. Environ. Sci. Technol. 30: 390A–396A

Wild O., Pochanart P. and Akimoto H. 2004. Trans-Eurasian transport of ozone and its precursors. J. Geophys. Res. – Atmos. 109: D11. 11286–11302

Wilkening K.E., Barrie L.A. and Engle M. 2000. Trans-Pacific Air Pollution. Science 290: 65–67.

Wolff E.W. and Suttie E.D. 1994. Antarctic snow record of southern hemisphere lead pollution. Geophys. Res. Lett. 21: 781–784.

Wolff E.W., Suttie E.D. and Peel DA. 1999. Antarctic snow record of cadmium, copper and zinc content during the twentieth century. Atmos. Environ.33: 1535–1541.

Yang H and Rose, N.L. 2003. Distribution of mercury in six lake sediment cores across the UK. Sci.Tot. Environ. 304: 391–404.

Yang H., Rose N.L., Battarbee R.W. and Boyle J.F. 2002a. Mercury and lead budgets for Lochnagar, a Scottish mountain lake and its catchment. Environ. Sci. Technol. 36: 1383–1388.

Yang H., Rose N.L., Battarbee R.W. and Monteith D.T. 2002b. Trace metal distribution in the sediments of the whole lake basin for Lochnagar, Scotland: A palaeolimnological assessment. Hydrobiol. 479: 51–61.

14. ACIDIFICATION OF LOCHNAGAR AND PROSPECTS FOR RECOVERY

DONALD T. MONTEITH (dmonteit@geog.ucl.ac.uk)
Environmental Change Research Centre
University College London
Pearson Building, Gower Street
London WC1E 6BT
United Kingdom

CHRIS D. EVANS
Centre for Ecology and Hydrology
Orton Building
Deiniol Road
Bangor LL57 2UP
United Kingdom

and

CATHERINE DALTON
Department of Geography
Mary Immaculate College
University of Limerick
South Circular Road
Limerick
Ireland

Key words: acidification, aquatic macrophytes, brown trout, diatoms, macroinvertebrates, nitrate, palaeoecology, recovery, sulphate

Introduction

Mountain lakes such as Lochnagar are commonly perceived to be some of our most undisturbed, or pristine, ecosystems. Despite their geographical isolation, however, geological and climatological characteristics often render them particularly sensitive to harm from atmospheric pollutants. Perhaps the most ecologically damaging effect of atmospheric contamination, acidification via 'acid rain', has received substantial scientific and political attention. Indeed, concern over acidification prompted much of

317

N.L. Rose (ed.), Lochnagar: The Natural History of a Mountain Lake
Developments in Paleoenvironmental Research, 317–344.
© 2007 *Springer*.

the early scientific research and subsequent monitoring of Lochnagar which underpin other Chapters within this volume. In this Chapter we provide a brief review of the current understanding of acidification processes and effects, summarise the research which has revealed the extent of the problem at Lochnagar, and consider to what extent recent and predicted improvements in air quality may stimulate recovery of the water chemistry of the loch and its plant and animal communities.

The chemical composition of lake water and its acidification by acid pollutants

Viewed remotely, mountain lakes may convey the appearance of simple precipitation collectors. However, as Jenkins et al. (Chapter 9: this volume) and Helliwell et al. (Chapter 6: this volume) demonstrate, the chemical quality of mountain lake water, even for relatively simple hydrological systems such as Lochnagar, is intrinsically linked to hydrochemical processes occurring within and beneath the sparse catchment soils.

Precipitation is slightly acidic naturally, as it contains carbonic acid which exists in the equilibrium between atmospheric carbon dioxide (CO_2) and water. As precipitation percolates through soil it becomes further enriched in CO_2 and organic compounds formed by the partial decomposition of plant and animal remains. This increases the level of carbonic and organic acids (represented analytically by the measurement of dissolved organic carbon, or DOC) which react with rock minerals in a process known as weathering to produce carbonates and bicarbonates in addition to increasing the levels of ionised bases (or base cations) including calcium.

In calcareous regions weathering results in a significant generation of bicarbonate and carbonate ions which raise soilwater pH. In contrast, where geology is relatively resistant to the weathering process (e.g., the biotite granites of Lochnagar - Chapter 3: Goodman this volume), rates are relatively slow and soilwater is acidic. Even in these relatively sensitive systems, however, weathering can generate sufficient base cations to buffer soils and groundwater, and hence lake water, against the natural acidity of precipitation. In these circumstances some interchange of positively charged acid hydrogen ions and base cations occurs at soil exchange sites (negatively charged regions of clay and organic particles), but overall the 'base-cation saturation' of the soil remains relatively stable. As water leaves the soil, a reduction in partial pressure results in the degassing of CO_2 and, consequently, to a solution enriched with bicarbonate. Providing there are no substantial changes in the acidity of precipitation, land use or major shifts in climate, the chemistry of lake water in such catchments might be expected to remain in a relatively stable base cation-bicarbonate equilibrium over time scales of decades to hundreds of years at least.

Over the past two centuries, however, the acidity of precipitation has increased substantially as a result of the emission of sulphur and nitrogen compounds into the atmosphere by industry (particularly coal and oil burning power stations), vehicular exhausts and agriculture. Following oxidation in the atmosphere, sulphurous and nitrous oxides react with water droplets to form sulphuric and nitric acids which reach the land surface as 'acid rain' or occult deposition (i.e., water in fog, dew, frost or rime), or are deposited in gaseous or particulate form to the land surface, i.e., as 'dry deposition'. Ammonia, which is derived mainly from agricultural sources, may also act as an acidifying agent, i.e., a supplier of hydrogen ions, if taken up directly by catchment vegetation or through conversion to nitrate by soil bacteria (Reuss and Johnson 1986).

The resulting elevated hydrogen ion concentration in percolating water accelerates the displacement of base cations from exchange sites. If this process exceeds the weathering rate the 'critical load' for the catchment is said to be exceeded and the soil begins to acidify (Chapter 6: Helliwell et al. this volume). Meanwhile, at least initially, the base cation concentration of the water entering lakes and streams increases. As the soil acidifies aluminium and iron oxyhydroxides become more abundant. It has been argued that this can result in the increasing adsorption of phosphate, resulting in a reduction in the export of phosphorus, which often limits primary production in upland waters (Jansson et al. 1986). However, there remains some debate as to the importance of acidification as an agent of oligotrophication (Olsson and Petterson 1993).

With time, the depletion of the base cation store renders the soil less able to buffer further inputs of acid. As water leaves the soil, the hydrogen ion concentration may be sufficiently high that the remaining inorganic carbon stays in the form of CO_2, and all bicarbonate alkalinity is lost. Thus the water received by streams and lakes becomes progressively enriched in hydrogen and aluminium ions, and impoverished with regard to bicarbonate and, possibly, phosphorus.

The importance of acidity for lake ecosystems

Spatially and temporally the pH, alkalinity, base cation, aluminium and nutrient concentration of lake water are all closely interlinked for the reasons given above. Geographically these factors represent a major environmental gradient, ranging from acidic, nutrient poor, low ionic strength lakes, such as Lochnagar, to alkaline nutrient rich systems, and this governs a variety of ecological characteristics. For example Palmer et al. (1992), in a UK-wide lake survey, found that aquatic plant community types were dependent on alkalinity, pH and conductivity, while Allott and Monteith (1999) demonstrated that the species composition of an integrated biological dataset, consisting of diatoms, aquatic macrophytes, littoral macroinvertebrates, and open water zooplankton from 31 Welsh lakes varied along a gradient represented by acidity, conductivity, and phosphorus. Petchy et al. (2004) demonstrated that lake biodiversity varies along a pH gradient, and showed clear links between species richness for a range of biological groups and acidity (and see Chapter 11: Woodward and Layer this volume).

Acidity influences aquatic biota in various ways, even within the relatively narrow range exhibited by acid sensitive lakes in the UK today (i.e., approximately pH 5.0 – 6.5). For example, aquatic plant productivity, particularly in more remote upland environments, is often limited by the availability of phosphorus. Certain algae which thrive in acidic conditions (i.e., some diatom species) have been shown to possess acid phosphatase enzymes which enable them to utilise organic phosphorus when the soluble reactive form is deficient (Smith 1990).

Dissolved inorganic carbon (DIC), required for aquatic photosynthesis, is often at limiting concentrations in lakes. Many aquatic plants are adapted to utilising specific forms, i.e., carbonate, bicarbonate or CO_2, which exist in pH dependent equilibrium.

The preferred form in circumneutral lakes (i.e., with a pH of around 7.0) is bicarbonate which is taken up directly by the submerged leaves of many elodeid species (plants with elongate, branching growth forms). The elodeid growth form represents an adaptation to DIC limitation, by minimising the aqueous boundary layer which forms a resistance to DIC movement from the open water into plant cells. Below a pH of around 5.5 bicarbonate is almost completely replaced by CO_2, but while water column concentrations of CO_2 tend to be strongly limiting, higher concentrations are available within organic sediments and in the water column outside periods of maximum productivity, i.e., during winter and early spring.

Vestergaard and Sand-Jensen (2000) observed that while alkaline lakes in Denmark tend to be dominated by elodeids, low alkalinity lakes are dominated by isoetids (prostrate, slow growing, rosette forming species) and bryophytes (mosses and liverworts). For the same dataset Maberly and Madsen (2002) found that the ability of species to use bicarbonate within each of five plant-defined lake groups was strongly related to the average alkalinity of the lakes they represented. While Maberly and Madsen (2002) were able to demonstrate that certain species, such as the water milfoil (*Myriophyllum alterniflorum*) were flexible in terms of their ability to utilise bicarbonate or CO_2 from the water column, it is clear that this and most other elodeid species are generally absent from severely acidified lakes.

A major exception to the general observations on elodeid distribution is the rush *Juncus bulbosus* var. *fluitans*, which is adapted to more acidic conditions, but thrives particularly in waters where dissolved CO_2 is more abundant. Svedäng (1992) found that this plant was able to grow at relatively low temperatures in the early spring and was thus able to sequester DIC at a time when competitive demand was relatively low. Isoetid species are adapted to deriving CO_2 from the lake sediment, and some use Crassulacean Acid Metabolism (CAM) pathways to fix carbon during the night when there is less biological demand. The vascular plant assemblage of Lochnagar today is limited to one isoetid species, *Isoetes lacustris*, and *J. bulbosus* var. *fluitans*, and therefore reflects the chronic acidity of the system, although climatic factors may also prevent the establishment of other species (see Chapter 10: Flower et al. this volume).

Many aquatic animals are acutely sensitive to elevated levels of hydrogen and inorganic aluminium ions in water. Fish and aquatic invertebrates need to continually replace vital ions (i.e., chloride, and the base cations sodium, potassium and calcium) (see Chapter 12: Rosseland et al. this volume) which are lost via diffusion across permeable membranes, and particularly gills where blood plasma and the external water are separated by a very thin respiratory epithelium. Active ion transport is achieved using ion pump mechanisms and the potential replenishment rate is dependent on external concentrations. Where water is rich in hydrogen and aluminium ions, relative to base cations, the former may be taken up preferentially resulting in potentially lethal internal ion inbalances. This process was once considered the prime cause of fish death in acid waters (Potts and McWilliams 1991). High internal hydrogen and aluminium concentrations may also render the gill even more permeable to sodium, perhaps as a result of the loss of calcium from the gill surface (McWilliams 1983).

Twitchen (1987) found that sodium uptake in the stonefly species *Leuctra moselyi*, which is characteristic of circumneutral waters, was substantially reduced in comparison with the acid tolerant species *Amphinemura sulcicollis* when both were exposed to acid water. He attributed this to differences in the relative affinities of the two species for sodium and hydrogen ions. Benthic macrocrustaceans (e.g., *Gammarus* sp.), zooplanktonic microcrustaceans (e.g., *Daphnia* sp.), and molluscs are particularly susceptible to low ionic concentrations, perhaps due to their relatively permeable membranes. Even air breathing macroinvertebrates are potentially vulnerable to acid toxicity, as these too depend on specialised tissues (in this case 'chloride cells') to regulate ion transport. Vangenechten et al. (1991) observed that corixid (water boatmen) species which are found in acid waters showed an adaptation to sodium ion regulation which allows relatively high uptake even when hydrogen ion levels are high. As a general rule the effects of high hydrogen and aluminium concentrations are reduced by elevated concentrations of calcium and/or DOC, but both of the latter are low in Lochnagar. The macroinvertebrate fauna of Lochnagar today is thus restricted mostly to a small number of hardy, acid tolerant stonefly and chironomid species (Chapter 11: Woodward and Layer this volume).

In addition to its effects on osmoregulatory disruption, the precipitation of aluminium on fish gill membranes also causes mucous clogging, intracellular aluminium accumulation, and high levels of haemocrit due to a reduction in blood plasma volume (Rosseland and Staurnes 1994; Chapter 12: Rosseland et al. this volume). Respiratory problems may lead to hypoxia, resulting in a decrease in blood pH and bicarbonate and an increase in lactate, normally with lethal consequences. The salmonid species, Atlantic salmon (*Salmo salar*) and brown trout (*Salmo trutta*), often the top predators in naturally acid waters in the UK and Scandinavia, are sensitive to these effects. Importantly, however, inorganic aluminium concentrations are normally only found at benign concentrations when pH is depressed by organic (i.e., naturally derived) acids only.

Through various direct physiological impacts on species, acidity will also influence various ecosystem functions, such as primary productivity, detrital decomposition by bacteria and fungi (which make organic matter more available to detritivorous invertebrates), predation rates, etc. (Chapter 11: Woodward and Layer this volume). Acidification therefore results in physiological stress, species loss, and wider changes to ecosystem-scale processes. Furthermore, there is increasing evidence to suggest that acidification is accompanied by a reduction in the concentration of DOC which provides the brown staining of lake waters within peaty catchments (Evans et al. 2005). Acidification might therefore be expected to increase photic depth, i.e., the maximum depth within a lake at which photosynthesis can occur, while, perhaps, reducing the protection to near surface dwelling plants and animals against the potentially harmful effects of UV-B radiation.

Palaeolimnological reconstruction of lake acidity

Natural long-term acidification

According to classical limnological theory lake environments mature through time, through a process known as ontogeny, from infertile to increasingly productive stages, and finally through to extinction by the process of infilling (Pearsall 1920). However, with the development of palaeolimnological science during the 20th century it has become clear that lake evolution can take various trajectories depending on location, geology, climate and land use, among other factors (e.g., Engstrom et al. 2000). Hard water lakes formed by glacial action, such as Lochnagar, may actually often undergo a process of chronic oligotrophication (i.e., a progressive loss of productivity and increase in acidity). Freshly exposed glacial till offers a readily soluble source of minerals, including base cations and phosphorus, a lack of which often limits primary production in these environments today. As a result this lake-type may experience an early 'eutrophic' alkaline stage, in which planktonic species thrive and productivity is relatively high, followed by a gradual decline in aquatic fertility as acid soils develop, mineral availability declines and the acidity of run-off increases.

Evidence for this process comes primarily from diatom-based studies of sediment cores. Diatoms, a group of siliceous algae, are ubiquitous primary producers in lakes and flowing waters. Since the time of Hustedt (1930s) individual species have been known to show strong preferences for waters of a particular, and narrow, pH range. Diatom cell walls are siliceous and preserve well in lake sediments. Therefore, the acidity of a lake can be reconstructed though time from changes in the species composition from deep (i.e., old) through to surficial (most recent) sediments (Battarbee 1984; Battarbee and Charles 1987).

Reconstructions of lake water pH are based on the relative abundance of individual species across a pH gradient in a 'training set' of surface sediment (i.e., modern) diatom samples and water chemistry. Early methods (e.g., Charles 1985; Birks 1987) were based on multiple regression models and the grouping of species into pH categories (e.g., Flower 1986). These were superceded by maximum likelihood (ML) and weighted averaging (WA) regression and calibration (ter Braak and van Dam 1989), which were shown to provide higher levels of prediction. WA is based on the weighted average of pH optima of all species occurring in a sediment sample, with the pH optima of species in turn determined by the average pH of the lakes in which they occur weighted by the taxon's relative abundance.

Renberg (1990), in a palaeolimnological study of the lake Lilla Öresjön in southwest Sweden, demonstrated that it had experienced an almost continuous decline in pH, from over pH 7.2 (12600 before present (BP)) to around pH 5.2 by 2300 BP. Similar pH trajectories have been reconstructed for Devoke Water in the English Lake District and Loch Sionascaig in northwest Scotland (Atkinson and Haworth 1990), and Kråkenes Lake in western Norway (Bradshaw et al. 2000). The Round Loch of Glenhead in southwest Scotland also shows gradual long-term acidification although from a more acidic initial phase (Jones et al. 1989; Birks et al. 1990) reflecting the relatively poor weathering of its granitic catchment.

Recently, Dalton et al. (2005) investigated the Holocene history of Lochnagar in a detailed multi-proxy palaeolimnological study. Sediment cores (with overlapping depths) were carbon-dated by accelerated mass spectrometry (AMS) and the age of the most recent sediments was determined by linear extrapolation from 1295 years BP, at a sediment depth of 30cm, to the modern surface. Diatom analysis revealed that in the early Holocene (c. 9000 – 8000 BP ± 80) the loch was dominated by *Aulacoseira perglabra* and *A. distans* var. *nivalis*. These are benthic-tychoplanktonic species, i.e., those normally attached to submerged surfaces but occasionally carried into open water. According to a diatom-pH transfer function developed for European mountain lakes (Cameron et al. 1999) this assemblage indicated a pH of around 6.5 (the root mean standard error of prediction for this model was 0.33 pH units).

Diatom assemblages then indicate that there was a gradual long-term decline in pH throughout much of the Holocene (see Figure 1 for the post 6000 BP pH reconstruction), although this was not monotonic. The early dominant species were gradually replaced by the more benthic *A. lirata* over the period 8000-6000 BP, reflecting a decline in alkalinity, as pH declined to circa pH 6.0. This was followed by oscillations in pH at millennial and shorter frequencies. *Fragilaria* taxa increased between 5500-4000 BP, indicating a return to more alkaline conditions. This period was also considered to be warmer, on the basis of chironomid remains, and subject to more human activity, as indicated by an increase in grass and heathland pollen at the expense of woodland types. After 4000 BP *Fragilaria vaucheriaea* was replaced gradually by the more acid tolerant *F. virescens* var. *exigua,* and this remained common until around the time represented by the most recent AMS date (i.e., approximately 1300 BP). This suggested a decline and subsequent stabilisation of pH at around pH 6.2, although inferred pH for individual sediment samples ranged from pH 5.8 to 6.6.

In the sediment samples immediately above the most recently AMS-dated sample diatom inferred pH declined sharply from around pH 6.2 to 5.5. On the basis of the available dating this was interpreted by Dalton et al. (2005) as the beginning of a natural acidification phase which continued for most of the last millennium almost to the present day. However, we have re-assessed the chronology for the upper samples using ^{210}Pb dating (a more sensitive dating technique but restricted to the last 150 years)

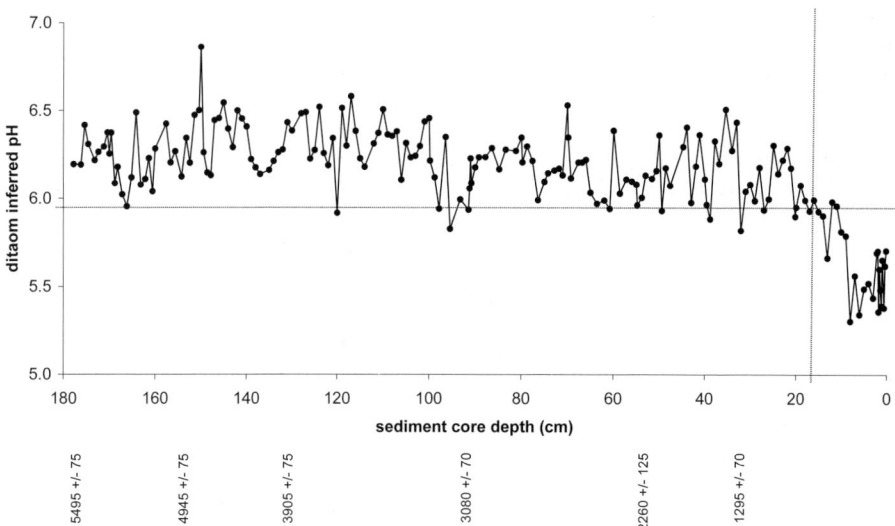

Figure 1. Diatom inferred pH reconstruction of the AMS ^{14}C dated sediment core NAG27 from Lochnagar (from Dalton et al. 2005). Cross-wires identify the possible onset of the anthropogenic acidification stage, and the concomitant inferred pH at this point, based on comparison with ^{210}Pb-based studies of other cores.

and with reference to spheroidal carbonaceous particle (SCP) profiles, which provide an independent indicator of industrial contamination (Chapter 17: Rose and Yang this volume). This comparison suggests that the loch actually underwent two distinct phases of pH decline over the period represented by the top 30 cm of the core. The first phase, in which change was relatively gentle, was marked by an increase in n-alkanes (lipid biomarkers) and organic matter content of the sediments. This is consistent with a period of increased delivery of eroded peat to the loch, a feature common to many upland lakes around the UK, although normally thought to occur more recently, i.e., 200-300 years ago (Stevenson et al. 1990). In the second phase, diatom inferred pH declined more sharply, as acidophilic species increased rapidly and the acid sensitive species *F. vaucheriaea* was lost. The onset of this latter stage (at around 18 cm depth) coincided with the earliest evidence for SCPs in sediments, while species changes are comparable to those described in other palaeoecological studies of Lochnagar and attributed to the period of anthropogenic acidification.

The recent acidification of Lochnagar by anthropogenic pollutants

Diatom species changes in the upper 18cm of the NAG27 core (Dalton et al. 2005) are similar to those described for a ^{210}Pb-dated sediment core taken from Lochnagar by Jones et al. (1993). In the latter study *Achnanthes marginulata* began to expand shortly beneath the lowest dateable horizon (dated to 1877 AD). The increase in *A. marginulata* coincided with the first detection of SCPs and a significant increase in the sediment concentration of zinc and lead (also an indication of atmospheric contamination). From this point in the core to the surface the decline in diatom inferred pH mirrored the continued accumulation of these industrially derived pollutants, providing strong evidence of an anthropogenic acidification process.

There are slight differences in inferred pH values between the two studies which reflects differences in statistical methods. Jones et al. (1993), used two approaches: (i) WA, based on the Surface Waters Acidification Project (SWAP) training set (Birks et al. 1990) and (ii) a multiple regression technique based on coefficients determined by Flower (1986) because of concerns that the flora represented in the sediment cores of some sites might not be sufficiently represented in the SWAP training set. Dalton et al. (2005) on the other hand, used a WA model based on a training set compiled specifically for application to European mountain lakes by Cameron et al. (1999), who also provided a pH reconstruction of a core for Lochnagar. This training set overcomes the problem of 'sufficient analogues', the model has a relatively low root mean squared error of prediction, and the reconstructed pH for the sediment surface agrees closely with the measured pH of the loch at around the time the core was taken. Overall, therefore we believe that the reconstruction used by Dalton et al. (2005) provides the best available estimate of the historical pH trajectory of Lochnagar, although it is important to recognise that all such models carry a degree of uncertainty (see for example Battarbee et al. 2005).

Figure 2 provides a diatom stratigraphy for the ^{210}Pb-dated sediment core NAG6, taken from Lochnagar in 1990 and described by Patrick et al. (1995). The diatom inferred pH reconstruction, using the approach of Dalton et al. (2005) and the training set of Cameron et al. (1999), is illustrated in Figure 3. This indicates a pH of between 5.8 - 5.9, prior to the onset of

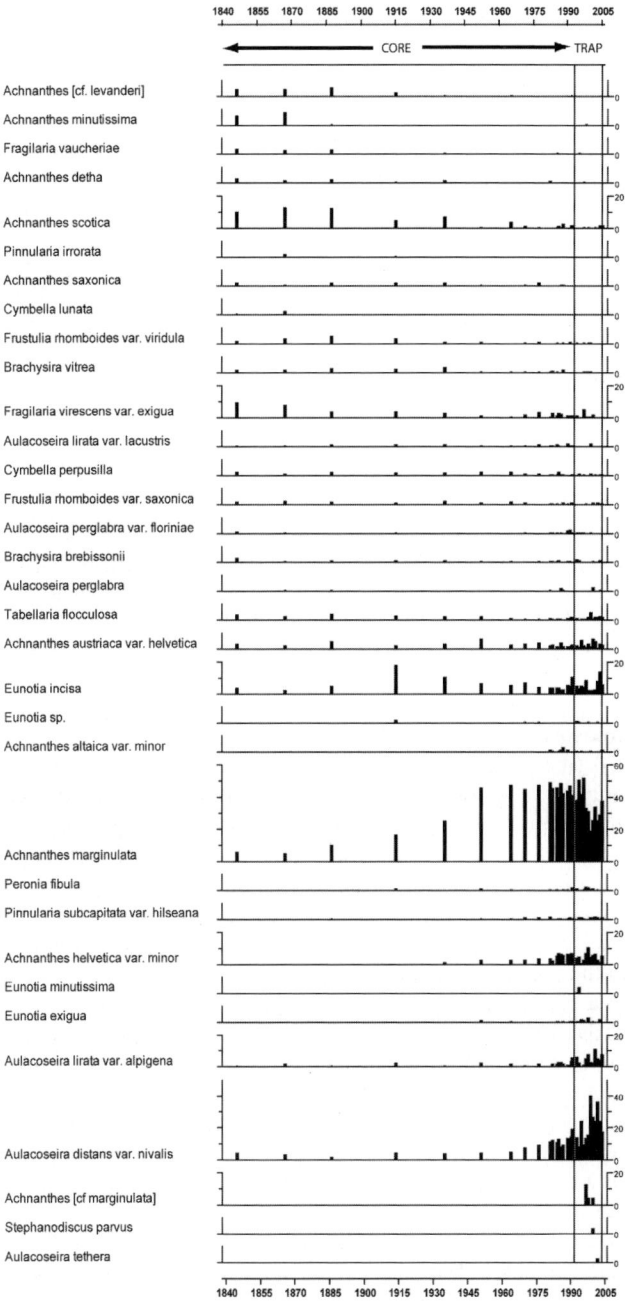

Figure 2. Diatom stratigraphy for the ^{210}Pb dated sediment core NAG6. Sediment trap samples collected since 1991 are represented to the right of the vertical line.

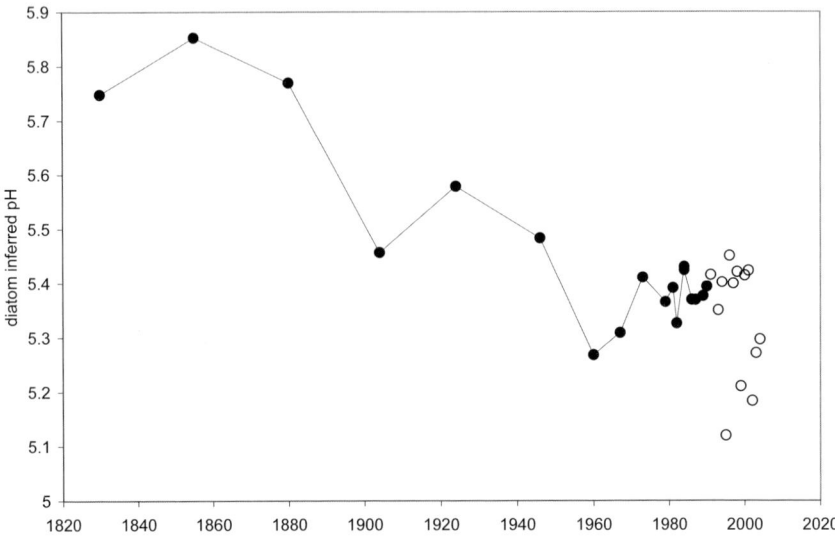

Figure 3. Diatom inferred pH reconstruction of ^{210}Pb-dated sediment core NAG27 from Lochnagar (black circles), based on a two component ALPE-training set WAPLS model (Cameron et al. 1999). Open circles represent inferred pH of diatom samples from annually collected sediment traps over the period 1991-2004.

significant atmospheric contamination, followed by rapid acidification down to pH 5.3 by 1960 -1970 (the period of greatest sulphur emissions and deposition in the UK). There then follows a slight improvement (of ~0.1 pH unit) until the most recent sediment sample, representing the time the core was taken i.e., 1990.

A pH range of 5.8 – 5.9 would therefore appear to be the current palaeoecological best estimate of the pre-industrial 'reference condition' for the loch. This implies that prior to the impact of acid deposition the loch would have been characterised by a significant level of bicarbonate alkalinity and a negligible concentration of inorganic aluminium, both crucial factors biologically.

If we adopt a revised chronology for core NAG27, on the basis of ^{210}Pb dating and SCP analysis, the multi-proxy data provide further insights into ecological responses to anthropogenic acidification at Lochnagar.   Pollen and spore analyses offer little evidence of change in aquatic macrophytes. The abundance of *Isoetes lacustris*, one of two vascular plants in the loch today, does not appear to have been strongly affected. There is evidence for an increase in the moss *Sphagnum* sp., but while *Sphagnum auriculatum* is a common aquatic species in the loch and is associated with acid lakes, it is not possible to distinguish its spores from those of other terrestrial species in the genus. The clearest evidence for a biological response (other than the diatoms) comes from chironomid remains. While the total number of individuals does not appear to have changed significantly, *Micropsectra insignilobis* and *Tanytarsus lugens* types have become significantly rarer while the proportion of *Heterotrissocladius marcidus*-type has increased substantially. However, there is little indication that any previously abundant chironomid species has become extinct in the loch over the last two centuries.

Evidence for recovery from acidification from environmental monitoring

Acid deposition

Situated toward the eastern side of Scotland, Lochnagar has historically received a substantially higher pollutant load than lochs at similar latitudes to the west. This is evident from data collated by the UK Acid Deposition Network (ADN, which was established in 1986 by the UK Department of the Environment and is now run by NETCEN on behalf of the Department for Environment, Food and Rural Affairs). Initially, Glen Dye was the closest wet deposition site to Lochnagar. Between 1987-1988 the precipitation weighted annual mean concentrations of non-marine sulphate (i.e., sulphate not attributable to seasalt) and nitrate at Glen Dye were 48-49 μeq l⁻¹ and 31-32 μeq l⁻¹ respectively (NETCEN 2004). By contrast, concentrations at the Allt a'Mharcaidh site to the west of the Cairngorms were 20-24 μeq l⁻¹ and 10-12 μeq l⁻¹ respectively.

Over the period 1986-2001, Fowler et al (2005) estimate that the non-marine sulphate concentration in wet deposition at Glen Dye declined by 1.46 μeq l⁻¹ yr⁻¹, while the nitrate concentration remained essentially unchanged (Figure 4). The scale of reduction in non-marine sulphate in precipitation is consistent with national-scale trends and reflects efforts by UK and foreign governments to reduce acid emissions and thereby abide by internationally agreed air quality protocols (NEGTAP 2001). In the UK this has been achieved predominantly through a move towards using more natural gas and the fitting of new flue gas desulphurisation (FGD) technology on existing coal burning power stations by the energy industry (Chapter 13: Rose et al. this volume).

The Glen Dye monitoring site is not ideally situated to represent conditions at Lochnagar, being 40 km away and 600 m lower in altitude. In 1999, however, the ADN incorporated a new wet deposition station within the Lochnagar catchment. For 2002 the precipitation weighted annual concentrations of non-marine sulphate and nitrate in wet deposition for the new site were 27 μeq l⁻¹ and 29 μeq l⁻¹ respectively, only fractionally higher than those recorded for Glen Dye in the same year. This suggests that the longer term Glen Dye record may provide a reasonable surrogate for depositional changes at Lochnagar (see Figure 4). All indications are, therefore, that Lochnagar has experienced a substantial reduction in acid inputs over the last 15 or more years as a direct result of the reduction in sulphur emissions.

Water chemistry

In 1988, the Department of Environment established the UK Acid Waters Monitoring Network (AWMN), to investigate the influence of emissions abatement strategy on acid sensitive freshwater ecosystems (Monteith and Evans 2005). Largely on the basis of palaeoecological evidence that the loch had acidified, Lochnagar was one of 11 lakes included in the network, which also included 11 streams. Since then AWMN water samples have been taken on a seasonal frequency from the loch outflow. 'Unstable'

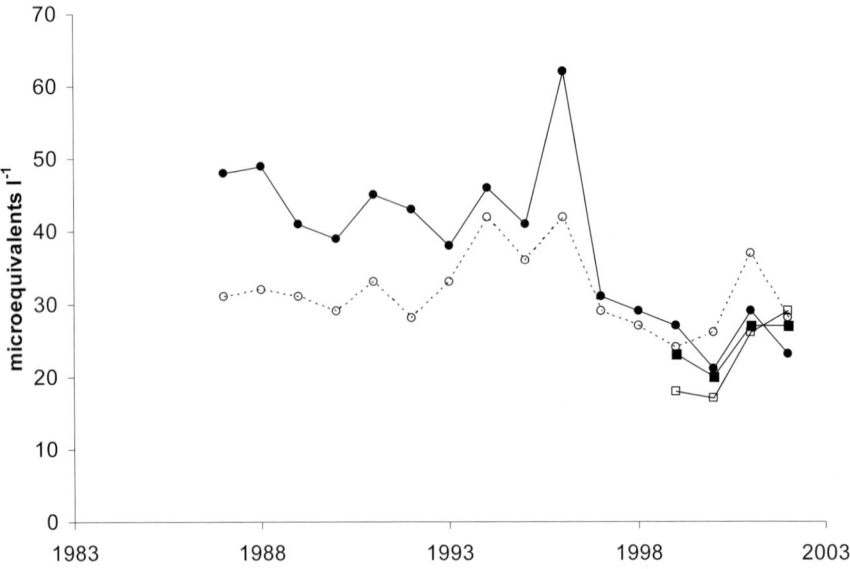

Figure 4. Time series of precipitation weighted annual mean concentrations of non-marine sulphate (filled symbols) and nitrate (open symbols), for the Glen Dye (circles) and Lochnagar (squares) wet deposition stations (data from NETCEN 2004).

determinands, which tend to change rapidly with time, such as pH and alkalinity, have been measured by the Freshwater Research Services (FRS) laboratory in Pitlochry, while to ensure national consistency, the more stable ions have been analysed by the AWMN central laboratory at CEH, Wallingford. For the purposes of this Chapter AWMN data are supplemented by data for three additional water samples analysed by FRS between 1980 – 1983, and by data for higher frequency (fortnightly) sampling conducted since 1996 on behalf of a series of EU funded projects (Chapter 1: Rose this volume), with all measurements again by FRS.

The resulting water chemistry time series for Lochnagar illustrate that the reduction in non-marine sulphate deposition has had a clear effect on surface water non-marine sulphate concentration (Figure 5), but that the impact on acidity is perhaps not as clear as might have been anticipated at the onset of monitoring.

Non-marine sulphate concentration in the outflow water fell markedly from over 100 µeq l^{-1} for the two 1980 samples to an average of c. 50 µeq l^{-1} for the first ten years of the AWMN (1988-1997). Concentrations declined further between 1998 and 2000 and since then have stabilised with values ranging from 36 – 45 µeq l^{-1}. Generally, concentrations in run-off are significantly higher than those measured in deposition but this is to be expected as a result of slight evaporative loss, and since dry deposition is not fully accounted for in bulk deposition samples. Cooper (2005), however, found slight discrepancies when attempting to model hypothetical run-off concentrations using the Glen Dye deposition record.

Despite the decline in non-marine sulphate concentration, the major driver of acidification, trends in the major toxic response variables, i.e., hydrogen and aluminium ions, and acid neutralising capacity (ANC: a parameter which summarises the balance between the sum of base cations and strong acid anions), are less obvious (Figure 5c). This is due primarily to an increase in nitrate (the second acidifying anion) which is of similar magnitude to the decline in non-marine sulphate (Figure 5a). Mean pH actually fell from around 5.4 to around 5.1 during the early 1990s (Figure 5b) when nitrate concentrations rose most sharply. Concomitantly, inorganic aluminium concentrations climbed dramatically; between 1993 and 1996 25 % of samples showed concentrations of over 80 μg l^{-1}, i.e., higher than the hypothetical threshold of 75 μg l^{-1} argued by Rosseland et al. (1990) to represent a level above which biological effects can become severe.

Since 1996, nitrate concentrations have stabilised while non-marine sulphate concentrations have continued to fall, and the net effect has been a slight overall reduction in the equivalent sum of these two strong acid anions. pH and inorganic aluminium levels have slowly returned to those recorded in the late-1980s and early-1990s. Only in 2005 did acidity levels fall beneath those of 1992, the least acidic year of monitoring up to this time. Between June 2005 and the time of the most recently available data (November 2005), pH remained above 5.5 and inorganic aluminium below 20 μg l^{-1} in each of the two weekly samples. The latter is lower than the concentration of 25 μg l^{-1} proposed by Rosseland et al. (1990) as a level beneath which effects on aquatic fauna would be negligible. However, definitive evidence of long-term improvement in water chemistry only becomes clear when the post-1988 monitoring record is considered in the context of the earliest available data. The minimum pH measurement of 4.9 was recorded in 1983; but unfortunately aluminium data are not available for this earlier period.

Despite the recent encouraging evidence for improvement in water quality, the apparent 'confounding' effect of nitrate on recovery gives cause for concern, particularly since there is no indication of an increase in nitrogen deposition over the period. Curtis et al. (2005) predicted that, as non-marine sulphate continues to decline nationally, nitrate would become the dominant agent of acidity for many AWMN sites by 2010 and at Lochnagar by 2021. However, they did not take into account the tendency for increasing trends in nitrate which would bring forward this date.

The cause of the recent increase in nitrate in Lochnagar is poorly understood. However, it is clear that currently more nitrogen is deposited on the catchment than is exported to the loch. In regions where pollutant nitrogen deposition is low, nitrogen is often limiting for terrestrial primary production and nitrate leaching tends to be negligible (Tamm 1992). However, Skeffington and Wilson (1988) suggested that elevated atmospheric inputs can result in rising levels in catchment soils to a point where they become saturated and significant leaching begins. Stoddard (1994) identified four stages of leaching ranging from year-round retention by soils to year-round leaching and proposed that this represented a temporal progression of sites impacted by nitrogen deposition. On this basis the nitrate time-series in Figure 5a might be considered to represent a progression from seasonal to year-round leaching. Comparison with other AWMN sites, the Round Loch of Glenhead and Loch Grannoch in Galloway, and Loch Chon in the Trossachs, revealed similar temporal patterns

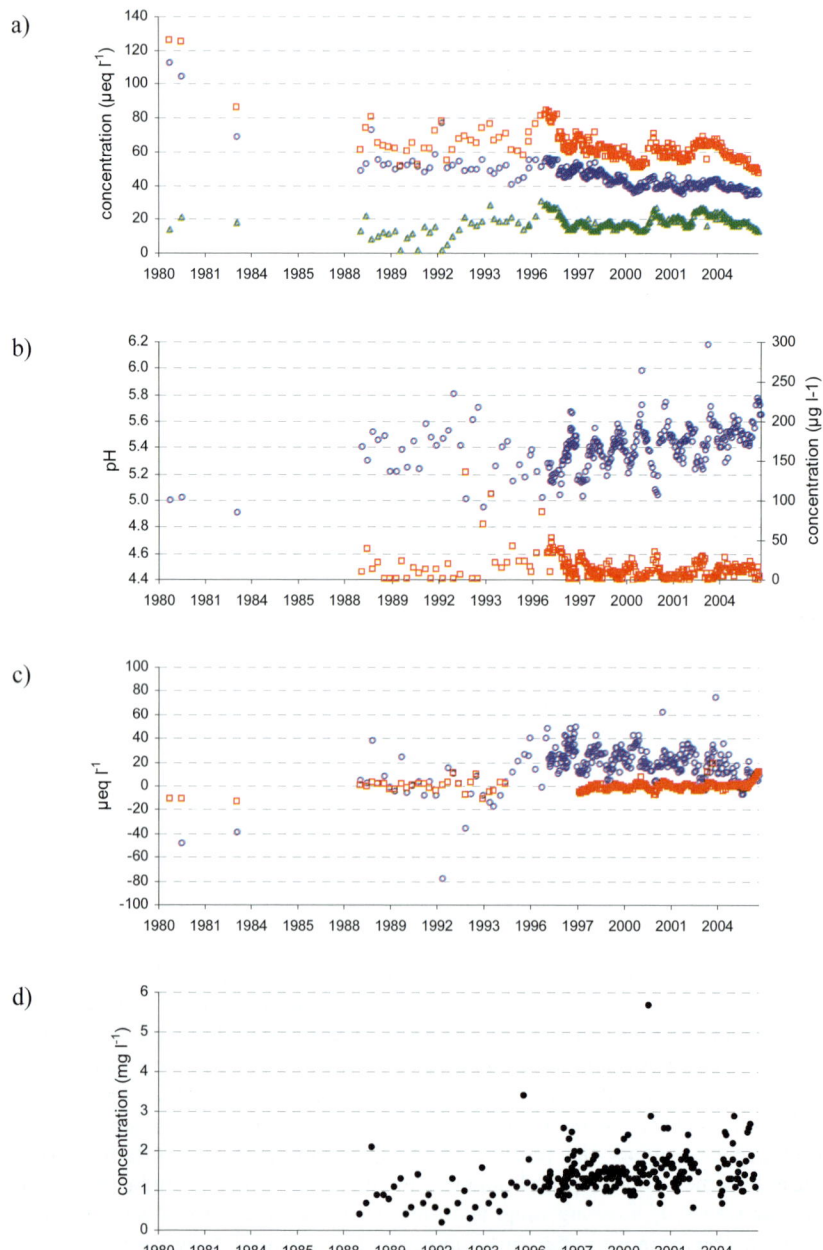

Figure 5. Water chemistry of samples taken from the outflow of Lochnagar since 1980. a) non-marine sulphate (blue); nitrate (green); sum of nitrate + non-marine sulphate (red)); b) pH (blue); labile monomeric aluminium (red); c) Acid Neutralising Capacity (blue); dual endpoint alkalinity (red); d) Dissolved Organic Carbon (no data prior to 1988)

(Davies et al. 2005). All sites were characterised by nitrate concentrations close to or below the limit of detection prior to 1993 after which all showed persistently elevated levels. However, this synchronous behaviour is at odds with the hypothetical process of 'nitrogen saturation' which should be site specific, and dependent on soil type, soil cover, nitrogen deposition and mineralisation rates. Furthermore, at least one of the three early samples (i.e., from December 1980) had a relatively elevated nitrate concentration, indicating that the post-1990 values might not be uniquely high over the longer term.

Nitrate time series from the AWMN and other monitoring records exhibit marked interannual variability which has been attributed to effects of summer drought (Reynolds et al. 1992), summer temperature (Mitchell et al. 1996) and cold winters. Monteith et al. (2000) demonstrated a particularly strong correlation between the state of the North Atlantic Oscillation, a major control on climate in this region (see Chapter 5: Thompson et al. this volume), and peak annual nitrate concentrations, which normally occur during spring, for AWMN sites. While the mechanisms behind the relationship are still unclear, Monteith et al. (2000) proposed that nitrate leaching might be promoted by biocidal effects of soil-freezing on plant roots thus releasing nitrogenous compounds for mineralisation, and a reduced uptake of nitrogen by soil bacteria. These hypotheses are supported by an apparent inverse relationship between nitrate concentration and a regional measure of mean December to March air temperature (Figure 6) over the monitoring period. While there is one obvious outlier in this relationship, representing the winter of 1991, the overall pattern suggests that the low nitrate concentrations in the first few years of monitoring may be linked to the anomalously warm winter conditions at the beginning of the record, when mean December to March Central England Temperatures for 1989 and 1990 were the warmest since 1834. The recent apparently year-round elevated concentrations in nitrate may therefore reflect the norm in recent decades, and the loch may have first experienced the characteristics of 'nitrogen saturation' before monitoring began.

Monitoring data demonstrate marked long-term variation in base-cation concentrations such as calcium, which largely follow variation in the sum of the concentration of acid anions. Also of note is a significant increase in the concentration of DOC (Figure 5d). This tendency has been observed at all sites on the AWMN (Freeman et al. 2001) and in many other regions of Europe and North America (Skjelkvåle et al. 2005). Various mechanisms have been proposed to explain this phenomenon, but most centre on climatic effects of rising air temperatures or increasing frequency of drought. Statistical analyses of the AWMN time series however, suggests that these trends may be linked to decreases in sulphur deposition (Stoddard et al. 2003; Evans et al. 2005), possibly as a result of its influence on soil acidity (see for example Clark et al. 2005) or ionic strength (Tipping and Hurley 1988). The long-term consequences of the current increase in DOC for the acidity of Lochnagar will only become clear once we have a better understanding of processes. If organic acidity is rising as a direct consequence of the declining 'mineral' acidity of deposition, hydrogen ion concentrations may not decline as much in response to declining inputs of strong acid anions as was once expected. Nevertheless, as DOC is represented mainly by weak acids, which only partly dissociate in solution, the net effect of this process should not hinder a rise in ANC or decline in inorganic aluminium and overall the change in chemistry should still be biologically beneficial.

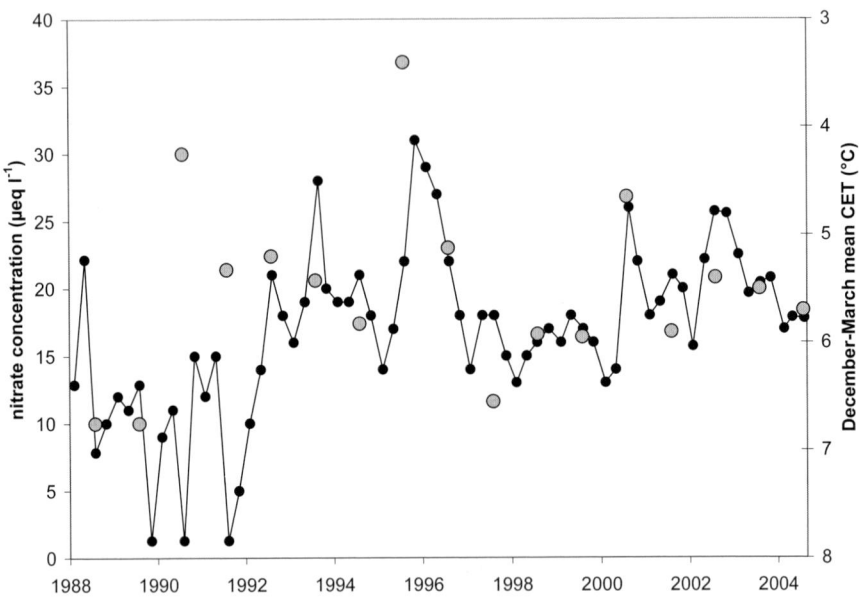

Figure 6. Time series of nitrate concentration (black circles) and mean December to March Central England Temperature (grey circles) for the period 1988-2005.

Biology

The AWMN biological monitoring of Lochnagar involves the annual assessment of epilithic diatom, sediment trap diatom, littoral macroinvertebrate, and salmonid populations, and the bi-annual assessment of aquatic macrophyte communities. Monteith et al. (2005) as part of a Network-wide assessment of biological trends found that, unlike many lakes where sulphate concentrations have declined, there was no evidence for statistically significant biological changes over time (1988-2002) in Lochnagar. Given the absence of any overall recovery in acidity over the same period this is not surprising. The monitoring data do, however, illustrate the current acid-biological status of the loch and the possible biological influences of interannual variability in acidity, and there is some evidence to suggest that over the most recent period (i.e., 2001 - 2004) certain species are beginning to benefit from recent chemical improvements.

Diatoms
The diatom flora of the loch is considered in greater detail by Flower et al. (Chapter 10: this volume). In most years epilithon samples have been dominated by the acidophilous *Achnanthes marginulata*, the species which showed a sharp increase in abundance from the mid-19th century as the loch acidified. Other important constituents of the epilithon

include *Tabellaria flocculosa, Eunotia incisa* and *Aulacoseira distans* var. *nivalis. E. incisa* shows strong interannual variability but no overall trend, but *A. distans* var. *nivalis* and *T. flocculosa* have clearly increased in abundance since around 1999. There is also tentative evidence for a recent fall in abundance of *A. marginulata*. Perhaps the most noteworthy observation is the recent detection of *Fragilaria vaucheriae* in one of the three diatom epilithon monitoring locations. This species was an important constituent of the sediment fossil flora of the loch over much of the Holocene but became undetectable in post-1900 sediments.

In addition to epilithon monitoring, diatoms have also been assessed in recently deposited sediment, caught by sediment traps (tubes suspended in deep water) which are retrieved annually each summer. While the epilithon record provides our best indication of temporal variability in the living flora of the loch, material from sediment traps is more directly comparable to the palaeoecological record because (a) it is also essentially a sedimentary fossil assemblage and (b) the location of the traps is close to the zone from which the cores are taken, and should better represent species growing in deeper water, including deep water epilithic, epipelic (mud dwelling) and tychoplanktonic taxa. The species composition of sediment trap samples is placed in the context of the NAG6 core assemblages in Figure 2. Figure 7 provides a temporal comparison of water pH and the abundance of key species in epilithic and trap samples, and thus an insight into possible lags between changing water chemistry, diatom production and diatom export to the sediment. The trough in pH between 1993-1995, which is most apparent for spring samples and is associated with the apparent step-change in nitrate, is concomitant with a peak in *A. marginulata* in the summer sampled epilithon in these years. The sediment trap samples (Figure 2) show a similar peak in this species between 1994-1996, i.e., displaced by one year. Furthermore, the sediment trap samples, unlike the epilithon, provide some indication of a recent downturn in the relative abundance of this indicator of acidification. Proportions of *A. marginulata* in 1999, 2000, 2002 and 2003 are all lower than the minimum abundance of 30% recorded in the previous seven annual samples.

In a longer-term context the NAG6 core shows *A. marginulata* at between 40-50% abundance in the upper layers. There is tentative evidence for a very slight reduction from peak abundances in samples dated at around 1980, and a gradual increase in diatom inferred pH is apparent slightly earlier, with the lowest values around 1960-1970 (Figure 3). Overall, therefore, a comparison of diatom assemblages in the NAG6 core and in sediment trap samples suggests that chemical and biological recovery may already have been underway before AWMN monitoring was initiated, and this is consistent with observations that sulphur deposition in the UK reached its peak around 1970 (NEGTAP 2001). There is, as yet, no evidence in the modern sediment of a significant presence of *F. vaucheriae,* which maintained abundances of c. 5% throughout most of the Holocene. While this species is locally abundant in the loch once more, a more substantial chemical improvement may be necessary before its fossils are detectable in sediment trap material.

To date there is still one obvious inconsistency in the argument that recent changes in the species composition of diatoms are indicative of biological recovery. Comparison of diatom trajectories in the NAG6 core and sediment trap samples show that while several of the more abundant species are now showing trends which are the reverse of those which occurred during acidification, the abundance of *A. distans* var. *nivalis*,

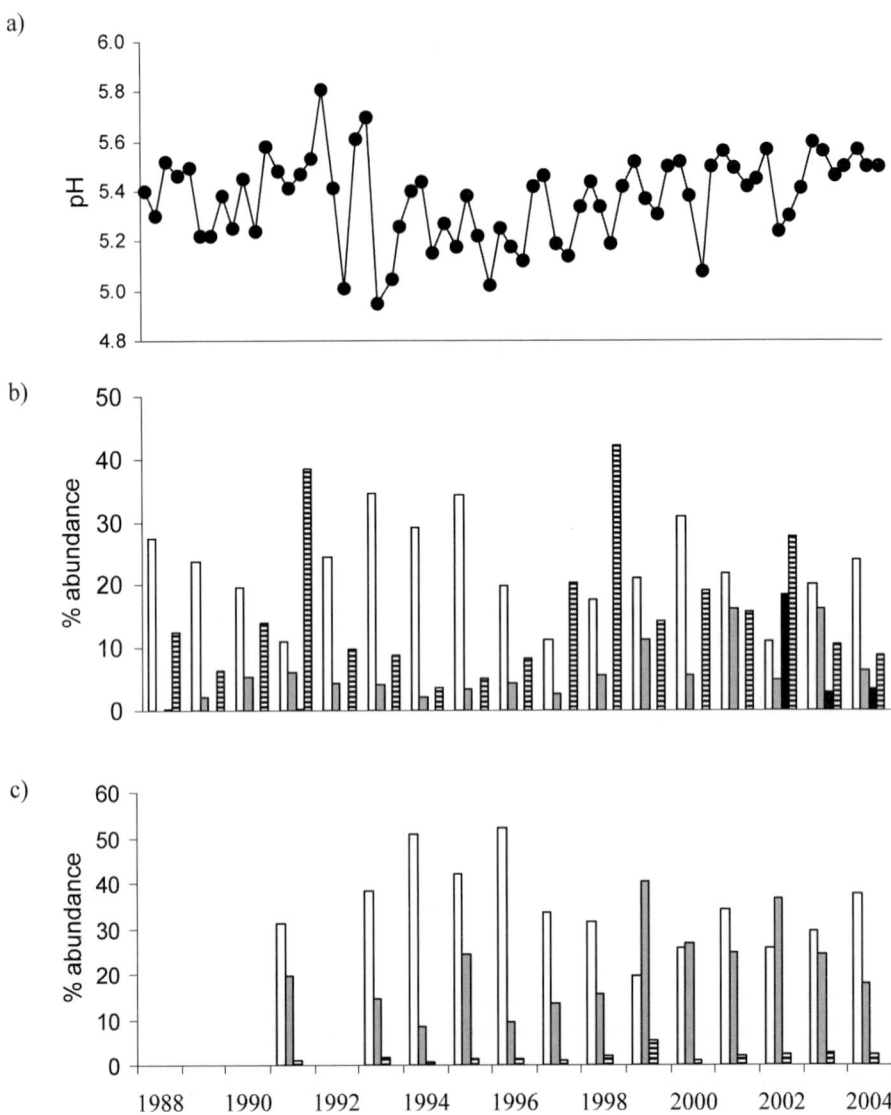

Figure 7. Time series for: (a) outflow pH, (b) % frequencies of four abundant diatom species in epilithon samples; *Achnanthes marginulata* (white bars); *Aulacoseira distans* var. *nivalis* (grey bars); *Tabellaria flocculosa* (black bars); *Fragilaria vaucheriae* (hatched bars); (c) Percentage frequencies of the species in (b) in the sediment trap samples.

which increased during acidification, has continued to increase in trap samples over the monitoring period (Figure 2). The recent trend in this species, leading to unprecedented proportions in the sediment, can be accounted for by a range of hypotheses which have different consequences for the long-term ecological recovery of Lochnagar. For example, its proportional increase might simply result from a declining flux of *A.marginulata*, and a lagged response of other acid sensitive taxa to improving chemical conditions. If this is the case, proportions should begin to fall again as these other taxa become re-established. Alternatively, the continued increase in *A. distans* var. *nivalis* might instead reflect a response to some other change in either chemistry, such as the increased availability of nitrate, or climate; for example, progressively earlier ice-off dates (see Chapter 5: Thompson et al. this volume).

Aquatic macrophytes
The aquatic macrophytes of Lochnagar are considered in detail by Flower et al. (Chapter 10: this volume). Essentially the vascular flora comprises two of the very few species which are adapted to survive in acidified UK lakes, the isoetid *Isoetes lacustris* and the rush *Juncus bulbosus* var. *fluitans*. Both species show some indication of changing abundance over the past 15 years but in neither case is it possible to link these directly to changing acidity. The maximum depth at which *I. lacustris* occurs has increased from around 2 m to 3 m since 1988, even though water colour has increased slightly as a result of the elevated dissolved organic carbon concentration. However, water transparency remains high; a Secchi disc suspended from a boat is often visible to a depth of 10 m or more. It is feasible that an increasing availability of nutrients (possibly recovery related) or an increase in early spring temperatures, as a result of earlier ice-off date, has facilitated greater leaf production rates, enabling plants to survive at greater depths.

   J. bulbosus var. *fluitans* was relatively scarce during the early years of monitoring but has become abundant, particularly along the east side of the loch, over the last 15 years (see Figure 3 in Chapter 10: Flower et al this volume). This may reflect a response to an enhanced availability of CO_2 due to an earlier ice-off date increasing the spring niche, or due to the partial photo-oxidation of DOC which has been increasing in concentration over the period of monitoring.

Littoral macroinvertebrates
The macroinvertebrate community of Lochnagar has been assessed by colleagues from Queen Mary University of London (QMUL), using a kick-sampling procedure on littoral habitats around the loch margin each May. The structure of this relatively species-impoverished community is described in detail by Woodward and Layer (Chapter 11: this volume). Chironomid species dominate the assemblage in most years, although in the less acidic years the acid tolerant stonefly, *Capnia* sp. is the most numerous. The balance between these two species drives the principal component of variability in the entire assemblage and can be related to the mean ANC of the March and June water samples for the same year (Chapter 11: Woodward and Layer this volume). *Capnia* sp. were rarest between 1993-1994, the period of the apparent step change in nitrate, lowest pH, highest inorganic aluminium levels, and dominance of *Achnanthes marginulata* in epilithic diatom samples. Over the same period there is no indication of any change in the overall abundance of individuals. To date it is not clear

whether this relationship results from direct toxic effects, varying quality or availability of food, changes in predation pressure, or co-variant biological and chemical responses to varying climate.

Salmonids
While we have not monitored the salmonid population of the loch *per se*, the outflow has been electrofished annually since 1989 by the FRS laboratory, Pitlochry. Characteristics of the brown trout population of the outflow and the loch are considered in more detail by Rosseland et al. (Chapter 12: this volume). AWMN time series demonstrate moderate and fluctuating numbers of fish in the loch outflow with evidence for recruitment every year since monitoring began. There is evidence for acid stress on the population from age class data, however, which demonstrates that few fish survive for more than three years. This is broadly consistent with Norwegian observations that, in the early stages of recovery the majority of fish die after the first year of spawning. Repeated spawning (i.e., spawning in successive years) is common in non-acidified systems but individuals are particularly sensitive to acidity immediately after spawning due to the particularly heavy demand on energy resources (Rosseland et al. 1980; Rosseland 1986). Rosseland et al. (Chapter 12: this volume) also show, from the outcome of a single recent rod-catch survey of the loch, that the population is stunted, i.e., after 5-6 years of age there tends to be no further length increment with time, while gill concentrations of aluminium are high compared to other populations studied from mountain lakes across Europe. To date we have no indication that the within-lake or outflow populations have benefited from the recent decline in acidity.

Future prediction

The dynamic biogeochemical model MAGIC (Cosby et al. 2001; Chapter 6: Helliwell et al. this volume) has been applied to Lochnagar on a number of occasions (e.g., Jenkins and Cullen 2001; Aherne et al. in press). Parameterisation of the model to two representative soil horizons within the catchment, and the simulated change in soil base status resulting from acidification, is described in the chapter by Helliwell et al. (Chapter 6: this volume). Here, the same source data have been used to calibrate the model to a single 'lumped' soil box, with the intention of simulating the average annual chemistry of loch water. The model was calibrated to the mean observed loch chemistry for the period 1989-1998, and to measured soil chemistry, by adjusting the weathering input of base cations and initial soil base status. Nitrogen 'saturation' of the soil was simulated in response to the gradual nitrogen enrichment of the soil (i.e., reduced soil C: N ratio) by nitrogen deposition, leading to reduced net immobilisation of incoming nitrogen over time. The model was run using historic reconstructions of pollutant sulphur and nitrogen deposition from 1850, and forecasts based on current emissions legislation (see Chapter 6: Helliwell et al. this volume).

  Results of the model simulation are shown in Figure 8. Sulphate, the dominant acidifying anion, is treated as conservative in the model, and the close fit between modelled and observed annual sulphate concentrations suggests that this assumption is valid. By contrast, the other major acidifying anion, nitrate, is highly variable between years. The model does predict a gradual increase in nitrate concentrations over time as

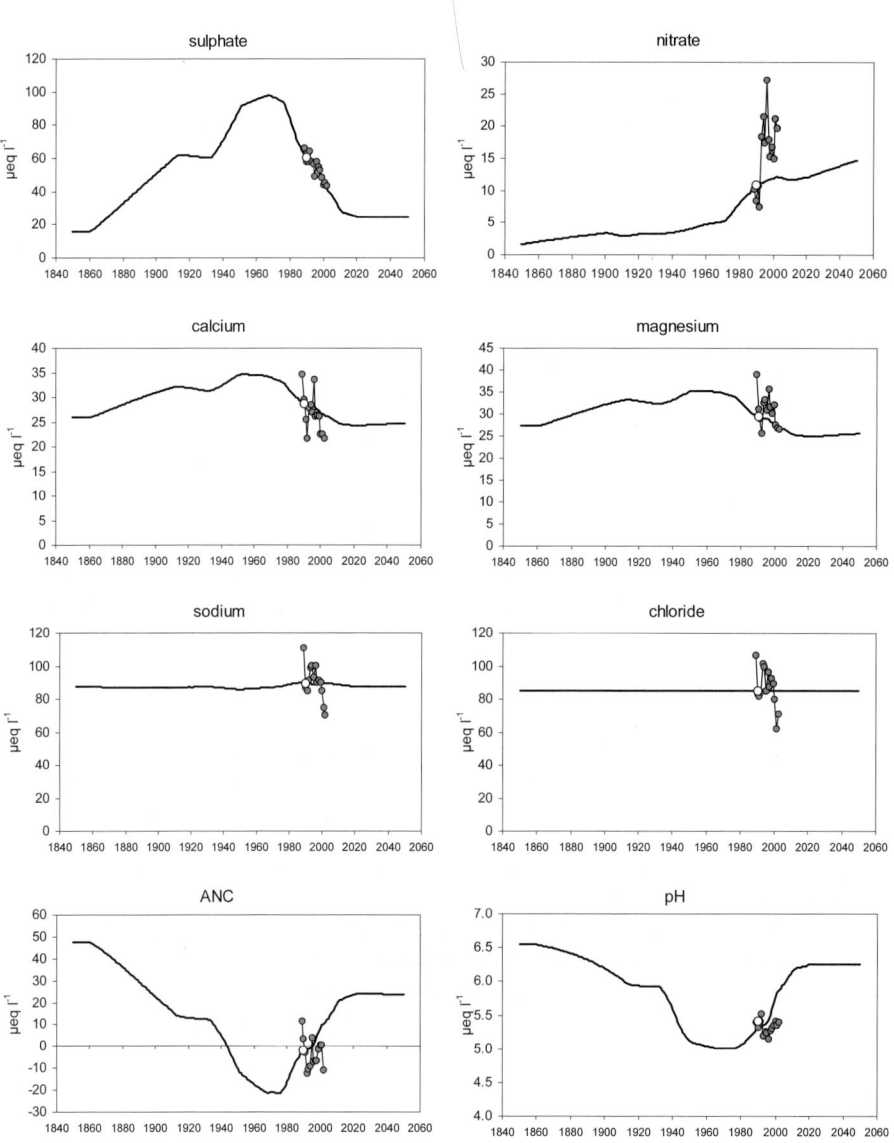

Figure 8. MAGIC parameters and model predictions. White dots represent MAGIC optimisation targets for 1990. Grey dots represent annual means of observed data.

soils become more nitrogen-enriched, and this is expected to continue in future despite some reduction in nitrogen deposition. However, the model does not reproduce the high level of variation observed. As discussed earlier, it is possible that the overall increase in nitrate at Lochnagar could indicate the relatively rapid onset of nitrogen saturation, but it is considered more likely that much of this short-term variability is related to climatic fluctuations, rather than a long-term increasing trend. The MAGIC simulation can therefore be considered an estimate of the underlying long-term increase in nitrate concentration at the site, on which (unmodelled) short-term climatic fluctuations are superimposed.

Short-term variations in nitrate will influence the concurrent behaviour of base cations such as calcium, and of the acid cations hydrogen and aluminium. Variations in the deposition of ions from marine sources have also been shown to cause year-to-year fluctuations in these ions in both UK monitoring data (e.g., Evans et al. 2001) and model simulations (e.g., Evans 2005). However, long-term variation in both base and acid cations should ultimately be driven by long-term change in sulphate and nitrate concentration. Climatic variations were not included in the current simulation so short-term variations, and most notably the apparent step-change acidification due to increased nitrate concentrations in 1993, are not captured by the model. However, the underlying decreasing trends in calcium and aluminium, and increases in ANC and pH, appear to be well simulated generally.

With regard to loch reference conditions, MAGIC retrodicts a pre-industrial pH of around 6.5, somewhat higher than the palaeolimnological estimate. MAGIC background pH estimates are sensitive to modelled concentrations of the weak acids (notably carbonic acid) that determine pH in the absence of strong mineral acids derived from acid deposition. Additionally, it has been suggested that DOC concentrations, and associated organic acidity, may be greater in the absence of strong mineral acids (Krug and Frink 1983; Evans et al. 2005); if organic acid concentrations were higher during reference conditions, then reference pH would be lower, closer to the palaeolimnological estimate. The same uncertainties apply to some extent to future predictions, as levels of mineral acid input decrease further. In the current simulation, loch pH is predicted to rise fairly rapidly before stabilising at around 6.2, although continued nitrogen saturation could cause a degree of re-acidification in the long term. ANC, which is less sensitive to assumptions about carbonic and organic acids, is predicted to have been around 45 μeq l^{-1} in pre-industrial times, falling to a severely acid -20 μeq l^{-1} in the 1960s and 1970s, and recovering to just above 20 μeq l^{-1} (the 'critical threshold' ANC used to define acceptable chemical status for fish in current critical loads estimates) by 2020.

Summary

- The water of Lochnagar has become increasingly acidic ever since the loch's formation at the end of the last glaciation, largely as a result of the development of its acidic soils and the export of base cations. However, while this natural process resulted in a pH of around 6.0, the loch water immediately prior to the onset of the industrial revolution would still have contained significant levels of bicarbonate and negligible levels of inorganic aluminium.

Chemical conditions at the time would therefore have been suitable for an acid sensitive montane aquatic flora and fauna, including a healthy fish population.

- Lochnagar is vulnerable to the effects of 'acid rain' due to its geological sensitivity, altitude and location on the relatively polluted eastern side of Scotland. Anthropogenic acidification, through the deposition of sulphur and nitrogen, resulted in a rapid decline in water quality. By the time sulphur deposition peaked, around 1970, the loch water would have become very acidic (average pH below 5.4), devoid of bicarbonate and contaminated with toxic concentrations of inorganic aluminium. Palaeoecological records demonstrate that this had a marked effect on the aquatic microflora and chironomid composition of the loch. The modern flora and fauna is characterised by relatively few taxa, all of which are adapted to highly acidic conditions, while the brown trout populations of the loch and the loch outflow are physiologically stressed and stunted.

- The effects of substantial reductions in non-marine sulphate deposition over the course of monitoring (1988-present) have been largely negated by an increase in nitrate over the same period. However the sum of the concentrations of these strong acid anions has fallen recently.

- Over much of 2005 pH remained above 5.5, alkalinity remained positive (i.e., some bicarbonate present) and inorganic aluminium levels remained below a theoretical biological damage threshold, for the first time since monitoring began.

- Whether the acidity of Lochnagar will recover further in response to anticipated reductions in sulphur deposition depends on the future tendency for nitrate release from catchment soils. If soils have reached a stable condition, and nitrogen deposition loads do not increase in the future, current nitrate concentrations may reflect the norm over the next few decades. If so, then the continued decline in sulphate concentration should be accompanied by a slow increase in pH, a restoration of some bicarbonate alkalinity and a reduction in inorganic aluminium to non-toxic concentrations. This should improve the health and diversity of organisms in the loch and its outflow significantly.

- While the future ecological health of the loch would seem to be dependent on nitrogen dynamics at the site, our ability to define realistic ecological reference targets for the 'natural' level of acidity of Lochnagar is currently hampered by an inadequate understanding of the mechanisms behind the recent, and substantial, increase in DOC concentration. Furthermore, our understanding of temporal and spatial dynamics, and physiological requirements of individual species and their trophic interactions needs to improve before we can predict the extent to which the ecosystem might return towards a pre-industrial condition and identify factors which might delay or prevent such recovery.

- Few mountain lakes worldwide have been monitored with the frequency, ecological scope or duration of the ongoing work at Lochnagar. Within the UK this work is undoubtedly exceptional, but it is reasonable to assume that Lochnagar is not unique, but rather is representative of other high altitude, acid sensitive lakes in montane regions of the UK, and beyond that have been similarly affected by acidification over recent decades. The study of Lochnagar therefore, provides an insight into the ecological functioning and environmental vulnerability of these important elements of our natural heritage. However, a continuation of monitoring and research is vital if we are to further develop approaches to their restoration and protection into the future.

Acknowledgements

The UK Acid Waters Monitoring Network (AWMN) and the Acid Deposition Network are funded by DEFRA. We particularly wish to thank: Jo and Mollie Porter, staff at FRS Pitlochry, and all others who have been involved in the sampling and analysis on behalf of the AWMN and associated projects over the last 18 years; Ewan Shilland for provision of data and Figures; and Jiří Kopáček, Chris Curtis, Roger Flower and Neil Rose for numerous helpful recommendations for improvements to the manuscript.

References

Aherne J., Helliwell R.C., Lilly A., Ferrier R.C. and Jenkins A. in press. Dynamic modelling of the acidification process: The two soil-layer approach. Wat. Air Soil Pollut.: Focus.

Allott T.E.H. and Monteith D.T. 1999. Classification of lakes in Wales for conservation using integrated biological data. 1999. A report to the Countryside Council for Wales by ENSIS Ltd. CCW Science Report No. 314. Environmental Change Research Centre, University College London, Research Report No. 53. 154 pp.

Atkinson K.M. and Haworth E.Y. 1990. Devoke Water and Loch Scionascaig: Recent environmental changes and the post-glacial overview. Phil. Trans. R. Soc. Lond. Series B-Biological Sciences. 327: 349–355.

Battarbee R.W.1984. Diatom analysis and the acidification of lakes. Phil. Trans. R. Soc. Lond. B 305, 451–477.

Battarbee R.W. and Charles D.F. 1987. The use of diatom assemblages in lakes sediments as a means of assessing the timing, trends and causes of lake acidification. Prog. Phys. Geog. 11: 552–580.

Battarbee R.W., Monteith D.T., Simpson G.L., Juggins S., Evans C.D. and Jenkins A. 2005. Reconstructing pre-acidification pH for an acidified Scottish loch: A comparison of palaeolimnological and modelling approaches. Environ. Pollut. 137: 135–150.

Birks H.J.B. 1987. Methods for pH-calibration and reconstruction from palaeolimnological data: Procedures, problems, potential techniques. Proceedings of the Surface Water Acidification Project (SWAP) mid-term review conference. Bergen 22-26 June 1987, pp. 370–380.

Birks H.J.B, Line J.M., Juggins S., Stevenson A.C. and ter Braak C.J.F. 1990. Diatoms and pH Reconstruction. Phil. Trans.Roy.Soc. Lond. Series B-Biological Sciences. 327: 263–278.

Bradshaw E.G., Jones V.J., Birks H.J. and Birks H.H. 2000. Diatom responses to late-glacial and early-Holocene environmental changes at Krakenes, western Norway. J. Paleolim. 23: 21–34.

Cameron N.G., Birks H.J.B., Birks H.H., Jones V.J., Berges F., Catalan J., Flower R.J., Garcia J., Kawecka B., Koinig K.A., Marchetto A., Sánchez-Castillo P., Schmidt R., Šiško M., Solovieva N., Štefková E. and Toro M. 1999. Surface-sediment and epilithic diatom pH calibration sets for remote European mountain lakes (AL:PE Project) and their comparison with the Surface Waters Acidification Programme (SWAP) calibration set. J. Paleolim. 22: 291–317.

Charles D.F. 1985. Relationships between surface sediment diatom assemblages and lake water characteristics in Adirondack lakes. Ecol. 66: 994–1011.

Clark J.M., Chapman P.J., Adamson J.K. and Lane S.N. 2005. Influence of drought induced acidification on the mobility of dissolved organic carbon in a peat soil. Global Change Biol. 11: 791–809.

Cooper D.M. 2005. Evidence of S and N deposition signals at the United Kingdom Acid Waters Monitoring Network sites. Environ. Pollut. 137: 41–54.

Cosby B.J., Ferrier R.C., Jenkins A. and Wright R.F. 2001. Modelling the effects of acid deposition: refinements, adjustments and inclusion of nitrogen dynamics in the MAGIC model. Hydrol. Earth Sys. Sci. 5: 499–517.

Curtis C.J., Evans C.D., Helliwell R. and Monteith D.T. 2005. Nitrate leaching as a confounding factor in chemical recovery from acidification in UK upland waters. Environ. Pollut 137: 73–82.

Dalton C., Birks H.J.B., Brooks S.J., Cameron N.G., Evershed R.P., Peglar S.M., Scott J.A. and Thompson R. 2005. A multi-proxy study of lake-development in response to catchment changes during the Holocene at Lochnagar, north-east Scotland. Palaeogeog. Palaeoclim. Palaeoecol. 221: 175–201.

Davies J.J.L., Jenkins A., Monteith D.T., Evans C.D. and Cooper D.M. 2005. Trends in surface water chemistry of acidified UK fresh waters, 1988–2002. Environ. Pollut. 137: 27–40.

Engstrom D.R., Fritz S.C., Almendinger J.E. and Juggins. S. 2000. Chemical and biological trends during lake evolution in recently deglaciated terrain. Nature 408: 161–166.

Evans C.D. 2005. Modelling the effects of climate change on an acidic upland stream. Biogeochem. 74: 21–46.

Evans C.D., Monteith D.T. and Harriman R. 2001. Long-term variability in the deposition of marine ions at west coast sites in the UK Acid Waters Monitoring Network: impacts on surface water chemistry and significance for trend determination. Sci. Tot. Environ. 265: 115–129.

Evans C.D., Monteith D.T. and Cooper D.M. 2005. Long-term increases in surface water dissolved organic carbon: Observations, possible causes and environmental impacts. Environ. Pollut. 137: 55–72.

Flower R.J. 1986. The relationship between surface sediment diatom assemblages and pH in 33 Galloway lakes: some regression models for reconstructing pH and their application to sediment cores. Hydrobiol. 143: 93–103.

Flower R.J., Monteith D.T., Tyler J. Shilland E. and Pla S. This volume. Chapter 10. The aquatic flora of Lochnagar. In: Rose N.L. (ed.) 2007. Lochnagar: The natural history of a mountain lake. Springer. Dordrecht.

Fowler D., Smith R., Muller J., Hayman G. and Vincent K. 2005.. Changes in the atmospheric deposition of acidifying compounds in the UK between 1986 and 2001. Environ. Pollut. 137: 15–26.

Freeman C., Evans C.D., Monteith D.T., Reynolds B., Fenner N. 2001. Export of organic carbon from peat soils. Nature 412: 785.

Goodman S. This volume. Chapter 3. Geology of Lochnagar and surrounding region. In: Rose N.L. (ed.) 2007. Lochnagar: The natural history of a mountain lake. Springer. Dordrecht.

Helliwell R.C., Lilly A. and Bell J.S. This volume. Chapter 6. The development, distribution and properties of soils in the Lochnagar catchment and their influence on soil water chemistry. In: Rose N.L. (ed.) 2007. Lochnagar: The natural history of a mountain lake. Springer. Dordrecht.

Hustedt F. 1937–39. Systematische und ökologische Untersuchungen üden Diatomeem-Flora von Java, Bali, Sumatra. Arch. Hydrobiol. 15 & 16.

Jansson M.G., Persson G. and Pettersson A. 1986. Phosphorus in acidified lakes: The example of Lake Gardsjön, Sweden. Hydrobiol. 139: 81–96.

Jenkins A. and Cullen J.M. 2001. An assessment of the potential impact of the Gothenburg protocol on surface water chemistry using the dynamic MAGIC model at acid sensitive sites in the UK. Hydrol. Earth Sys. Sci. 5: 529–541.

Jenkins A., Reynard N., Hutchins M., Bonjean M. and Lees M. This volume. Chapter 9. Hydrology and hydrochemistry of Lochnagar. In: Rose N.L. (ed.) 2007. Lochnagar: The natural history of a mountain lake. Springer. Dordrecht.

Jones V.J., Stevenson A.C. and Battarbee R.W. 1989. Acidification of lakes in Galloway, south west Scotland - a diatom and pollen study of the post-glacial history of the Round Loch of Glenhead. J. Ecol. 77: 1–23.

Jones V.J., Flower R.J., Appleby P.G., Natkanski J., Richardson N., Rippey B. Stevenson A.C. and Battarbee R.W. 1993. Palaeolimnological evidence for the acidification and atmospheric contamination of lochs in the Cairngorm and Lochnagar areas of Scotland. J. Ecol. 8: 3–24.

Krug E.C. and Frink C.R. 1983. Acid rain on acid soil: A new perspective. Science 221: 520–525.

Maberly S.C. and Madsen T.V. 2002. Freshwater angiosperm carbon concentrating mechanisms: processes and patterns. Functional Plant Biol. 29: 393–405.

McWilliams P.G. 1983. An investigation of the loss of bound calcium from the gills of brown trout, *Salmo trutta*, in acid media. Comp. Biochem. Physiol. 74A: 107–116.

Mitchell M.J., Driscoll C.T., Kahl J.S., Likens G.E., Murdoch P.S. and Pardo L.H. 1996. Climatic control of nitrate loss from forested watersheds in the Northeast United States. Environ. Sci. Technol. 30: 2609–2612.

Monteith D.T. and Evans C.D. 2005. The United Kingdom Acid Waters Monitoring Network: a review of the first 15 years and introduction to the Special Issue. Environ. Pollut. 137: 3–14.

Monteith D.T. Evans C.D. and Reynolds B., 2000. Are temporal variations in the nitrate content of UK upland freshwaters linked to the North Atlantic Oscillation? Hydrol. Process. 14: 1745–1749.

Monteith D.T., Hildrew A.G., Flower R.J., Raven P.J., Beaumont W.R.B., Collen P., Kreiser A., Shilland E.M. and Winterbottom J.H. 2005. Biological responses to the chemical recovery of acidified fresh waters in the UK. Environ. Pollut. 137: 83–102.

NEGTAP (National Expert Group on Transboundary Air Pollution) 2001. Transboundary air pollution: Acidification, eutrophication and ground-level ozone in the UK. Prepared on behalf of the UK Department for Environment, Food and Rural Affairs (DEFRA) and the Devolved administrations. 314 pp.

NETCEN. 2004. Management and Operation of the UK Acid Deposition Monitoring Network: Data Summary for 2002. A report produced for the Department for Environment, Food and Rural Affairs, the Scottish Executive, the National Assembly of Wales and the Northern Ireland Department of the Environment. AEAT/ENV/R/0740 Issue 1. National Environment Technology Centre.

Olsson H. and Petterson A. 1993. Oligotrophication of acidified lakes – A review of hypotheses. Ambio. 22: 312–317.

Palmer M. A., Bell S. L. and Butterfield I. 1992 A botanical classification of standing waters in Britain: applications for conservation and monitoring. Aq. Cons.: Mar Freshwat. Ecosys. 2: 125–143.

Patrick S., Monteith D.T. and Jenkins A. (eds) 1995. UK Acid Waters Monitoring Network: The first five years. Analysis and interpretation of results, April 1988 - March 1993. ENSIS Ltd., London. 320 pp.

Pearsall W. H. 1920. The development of vegetation in the English Lakes, considered in relation to the general evolution of glacial lakes and rock-basins. Proc. Roy Soc. B 92: 259–284.

Petchey O.W., Downing A.L., Mittelbach G.G., Persson L., Steiner F., Warren P.H. and Woodward G. 2004. Species loss and the structure and functioning of multitrophic aquatic ecosystems. Oikos 104: 467–478.

Potts W.T.W. and McWilliams P.G. 1991. The effects of hydrogen and aluminium ions on fish gills. In: Morris R., Taylor E.W., Brown D.J.A. and Brown J.A (eds.) Acid toxicity and aquatic animals. Society for Experimental Biology. Seminar Series 34. Cambridge University Press, pp. 201–220.

Renberg I. 1990. A 12 000 year perspective of the acidification of Lilla Öresjön, southwest Sweden,. Phil. Trans. Roy. Soc. Lond. Series B 327: 357–361.

Reuss J.O. and Johnson D.W, 1986. Acidic deposition and the acidification of soils and waters. Springer, New York. pp. 1–119.

Reynolds B., Emmett B.A. and Wood C. 1992. Variations in streamwater nitrate concentrations and nitrogen budgets over 10 years in a headwater catchment in mid- Wales. J. Hydrol. 136: 155–175.

Rose N.L. This volume. Chapter 1. An introduction to Lochnagar. In: Rose N.L. (ed.) 2007. Lochnagar: The natural history of a mountain lake. Springer. Dordrecht.

Rose N.L. and Yang H. This volume. Chapter 17. Temporal and spatial patterns of spheroidal carbonaceous particles (SCPs) in sediments, soils and deposition at Lochnagar. In: Rose N.L. (ed.) 2007. Lochnagar: The natural history of a mountain lake. Springer. Dordrecht.

Rose N.L., Metcalfe S.E., Benedictow A.C., Todd M., Nicholson J. This volume. Chapter 13. National, international and global sources of contamination at Lochnagar. In: Rose N.L. (ed.) 2007. Lochnagar: The natural history of a mountain lake. Springer. Dordrecht.

Rosseland B.O. 1986. Ecological effects of acidification on tertiary consumers. Fish population responses. Water Air Soil Pollut. 30: 451–460.

Rosseland B.O., Sevaldrud I., Svalastog D. and Muniz I.P. 1980. Studies on freshwater fish populations - effects of acidification on reproduction, population structure, growth and food selection. In: Tollan A. and Drabløs D. (eds.) Ecological impact of acid precipitation. SNSF Project, Proc. Int. Conf. Ecological Impact of Acid Precipitation, 11–14 March 1980. Sandefjord, Norway. pp 336–337

Rosseland B.O., Eldhurst T.D. and Staurnes M. 1990. Environmental effects of aluminium. Environ. Geochem. Health 12: 17–27.

Rosseland B.O. and Staurnes M. 1994. Physiological mechanisms for toxic effects and resistance to acidic water: An ecophysiological and ecotoxicological approach. In: Steinberg C.E.W. and Wright R.F. (eds.) Acidification of freshwater ecosystems: Implications for the future. John Wiley and Sons Ltd. pp 227–246.

Rosseland B.O., Rognerud S., Collen P., Grimalt J.O., Vives I., Massabuau J-C., Lackner R., Hofer R., Raddum G.G., Fjellheim A., Harriman R. and Piña B. This volume. Chapter 12. Brown trout in Lochnagar: Population and contamination by metals and organic micropollutants. In: Rose N.L. (ed.) 2007. Lochnagar: The natural history of a mountain lake. Springer. Dordrecht.

Skeffington R.A. and Wilson E.J. 1988. Excess nitrogen deposition: issues for consideration. Environ. Pollut. 54: 159–184.

Skjelkvåle B.L., Stoddard J.L., Jeffries D., Tørseth K., Høgåsen T., Bowman J., Mannio J., Monteith D.T., Mosello R., Rogora M., Rzychon D., Vesely J., Wieting J., Wilander A. and Worsztynowicz A. 2005. Regional scale evidence for improvements in surface water chemistry 1990–2001. Environ. Pollut. 137: 165–176.

Smith M. A. 1990. The ecophysiology of epilithic diatom communities of acid lakes in Galloway, southwest Scotland. Phil. Trans. Roy. Soc. London Series - B Biol. Sci. 327: 251–256.

Stevenson A.C., Jones V.J. and Battarbee, R.W. 1990. The cause of peat erosion – a palaeolimnological approach. New Phytol. 114: 727–735.

Stoddard J. 1994. Long term changes in watershed retention of nitrogen. In: Baker L.A. (ed.) Environmental chemistry of lakes and reservoirs. American Chemical Society. pp. 223–284.

Stoddard J.L., Karl J.S., Deviney F.A., DeWalle D.R., Driscoll C.T., Herlihy A.T., Kellogg J.H., Murdoch P.S., Webb J.R. and Webster K.E. 2003. Response of Surface Water Chemistry to the Clean Air Act Amendments of 1990. Report EPA 620/R-03/001, United States Environmental Protection Agency, North Carolina. 92 pp.

Svedäng M.U. 1992. Carbon dioxide as a factor regulating the growth dynamics of *Juncus bulbosus*. Aq. Bot. 42: 231–40.

Tamm C.O. 1992. Nitrogen in terrestrial ecosystems. Ecological studies: 81. Springer, New York. pp. 115.

ter Braak C.J.F. and van Dam H. 1989. Inferring pH from the diatoms: a comparison of old and new calibration methods. Hydrobiol. 178. 209–223.

Thompson R., Kettle H., Monteith D.T. and Rose N.L. This volume. Chapter 5. Lochnagar water-temperatures, climate and weather. In: Rose N.L. (ed.) 2007. Lochnagar: The natural history of a mountain lake. Springer. Dordrecht.

Tipping E. and Hurley M.A., 1988. A model of solid-solution interactions in acid organic soils, based on the complexation properties of humic substances. J. Soil Sci. 39: 505–519.

Twitchen I.D. 1987. The physiological basis of acid resistance in aquatic insect larvae. Surface Water Acidification Programme. Bergen, Norway. June 1987.

Vangenechten J.H.D., Witters H. and Vanderborght O.L.J. 1991. Laboratory studies on invertebrate survival and physiology in acid waters. In: Morris R., Taylor E.W., Brown D.J.A. and Brown J.A. (eds.) Acid toxicity and aquatic animals. Society for Experimental Biology. Seminar Series 34. Cambridge University Press. pp. 153–170.

Vestergaard O. and Sand-Jensen K. 2000. Aquatic macrophyte richness in Danish lakes in relation to alkalinity, transparency, and lake area. Can. J. Fish. Aquat. Sci. 57: 2022–2031.

Woodward G. and Layer K. This volume. Chapter 11. Pattern and process in the Lochnagar food web. In: Rose N.L. (ed.) 2007. Lochnagar: The natural history of a mountain lake. Springer. Dordrecht.

15. TRACE METALS IN THE CATCHMENT, LOCH AND SEDIMENTS OF LOCHNAGAR: MEASUREMENTS AND MODELLING

EDWARD TIPPING (et@ceh.ac.uk)
Centre for Ecology and Hydrology (Lancaster)
Bailrigg
Lancaster LA1 4AP
United Kingdom

HANDONG YANG
Environmental Change Research Centre
University College London
Pearson Building, Gower Street
London WC1E 6BT
United Kingdom

ALAN J. LAWLOR
Centre for Ecology and Hydrology (Lancaster)
Bailrigg
Lancaster LA1 4AP
United Kingdom

NEIL L. ROSE
Environmental Change Research Centre
University College London
Pearson Building, Gower Street
London WC1E 6BT
United Kingdom

and

LAURA SHOTBOLT
Department of Geography,
Queen Mary, University of London
Mile End Road
London E1 4NS
United Kingdom

Key words: atmospheric deposition, biota, lake water, modelling, sediments, soils, trace metals

N.L. Rose (ed.), Lochnagar: The Natural History of a Mountain Lake
Developments in Paleoenvironmental Research, 345–373.
© *2007 Springer.*

Introduction

Trace (or 'heavy') metals emitted to the atmosphere from fossil-fuel combustion, metal smelting, waste incineration and road traffic, are distributed both locally and globally (Nriagu 1990). Here we discuss the impact of long distance atmospheric trace metal deposition onto Lochnagar and its catchment, focusing on metals with potentially adverse effects, principally nickel, copper, zinc, cadmium, mercury and lead.

The concentrations of trace metals in the global environment have been increased by human activity since the domestication of fire (Nriagu 1996). Smelting activity, several thousand years before the present, further increased the anthropogenic contribution; lead has been extracted and smelted for at least 5000 years, copper for around 6500 years (Patterson 1971; Settle and Patterson 1980; Tylecote 1992), and zinc was used in the production of brass probably as early as 4000 BP (Darling 1990). The smelting of lead and copper during Roman times had a hemispherical influence, with detectable increases in deposition found in ice cores from Greenland and Swedish lake sediments from around 2000 BP (Hong et al. 1994, 1996; Boutron et al. 1995; Renberg et al. 2002). An additional source of trace metal pollution, coal burning, was contributing to the atmospheric aerosol by around 1300, becoming a serious urban pollution problem by the early-17th century (Evelyn 1661; Brimblecombe 1987). However, by far the greatest anthropogenic inputs of metals to the atmosphere have occurred since the Industrial Revolution, due to the expansion in manufacturing, industrialisation of the mining and smelting industries, power generation from fossil fuels, and vehicle emissions.

Today, levels of trace metal deposition in the UK have been reduced through a combination of improvements in abatement technology, a switch from coal to natural gas and decreasing energy demands (Baker 2001). They are now low in comparison to peak deposition in the 1960s and early-1970s (Baker 2001; Goodwin et al. 2000). Nevertheless, past heavy metal emissions and deposition have left a legacy of elevated trace metal levels in soils, waters and sediments, even in relatively remote catchments such as Lochnagar.

In this Chapter we review and put into context the available information on trace metals in Lochnagar and its catchment. Most of the information has come from the work of the Environmental Change Research Centre at University College London, especially the PhD (Yang 2000) and post-doctoral research of Handong Yang. This has covered trace metals in deposition, soils, loch water, sediments and biota. We focus on both the historical record of deposition and current stores of metals in the catchment, referring to previously published and new results. The Chapter also includes an account of the application of a dynamic model, CHUM-AM (CHemistry of the Uplands Model – Annual, Metals; Tipping et al. 2005a; b), which has been developed to explore the long-term behaviour of heavy metals in upland UK catchments, to inform the development of the Critical Loads approach for controlling heavy metals emissions within Europe (Ashmore et al. 2004; de Vries et al. 2004). CHUM-AM is an integrated model that describes the accumulation of metals in soil, their transport to the lake in stream water and direct deposition, and their concentrations in the lake's sediments, taking into account historical metal deposition, and the acidification history of the system. This mechanistic, non-steady-state, site-specific approach to modelling metals in lakes contrasts with, and complements, the empirical, steady-state, fate modelling of Rippey et al. (2004).

Methods

Detailed descriptions of the loch and its catchment, soils, vegetation, acidification and hydrology are given by Hughes (Chapter 2: this volume), Helliwell et al. (Chapter 6: this volume), Birks (Chapter 7: this volume), Monteith et al. (Chapter 14: this volume), and Jenkins et al. (Chapter 9: this volume) respectively. Descriptions have been published elsewhere of the analytical methods for deposition chemistry at Lochnagar (Yang et al. 2002a;b) and in Cumbria (Lawlor and Tipping 2003), for soils (Yang et al. 2001; 2002a;b; Tipping et al. 2005b), loch water (Yang et al. 2002a; b), and sediments (Yang et al. 2002a;b;c; Yang and Rose 2003a;b). Further detailed information can be found in Yang (2000) and Rose and Yang (2004). Sampling and analytical details for terrestrial and lake vegetation, zooplankton and macro-invertebrates are given by Rose and Yang (2004). Methodology for sampling and analysis of herbarium mosses are described in Ashmore et al. (2004). Methodology for sediment sampling, in which 17 cores were taken along five transects radiating from the central area of the loch, was described by Yang et al. (2002c). Two previous sediment cores were dated by radiometric techniques (^{210}Pb, ^{137}Cs, ^{241}Am), and the 17 by cross-correlation using spheroidal carbonaceous particles (SCP), after calibration of the SCP profiles.

Trace metals in atmospheric deposition

Concentrations of nickel, copper, zinc, cadmium and lead in atmospheric deposition at Lochnagar have been measured since 1999, and the results are summarised in Table 1. Also shown are data for Banchory, 40 km to the east of Lochnagar, and for a site in Cumbria. The values for the three sites are broadly similar, although those for Lochnagar are generally the lowest. The mean concentrations of mercury in bulk deposition, reported for 1997 to 2003 are somewhat higher than the average value for north temperate systems of 0.016 µg l^{-1} given by Grigal (2002), based on data for 1990-2000. Furthermore, the rainfall volume at Lochnagar is relatively high and so the corresponding annual loads are at the upper end of the range given by Grigal for north temperate systems, and substantially greater than the average of 10 µg m^{-2} yr^{-1}.

A point to be noted, especially in connection with modelling (see below), is that measurements of metals in bulk deposition only refer to relatively labile metals, extractable with dilute HNO$_3$ or, in the case of Hg, HCl. Additional metal is deposited in less-labile or non-labile forms. Furthermore, additional inputs to the catchment and loch occur as dry and occult deposition; for high-altitude wet sites in Cumbria, Lawlor and Tipping (2003) assumed these additional inputs to provide an additional 20% to the values in bulk deposition.

The results for Lochnagar cover only the last few years, too few for trends in metal deposition to be discerned. However, the results of Playford and Baker (2000) for the Banchory site cover periods starting in 1987 (Cu, Zn), 1990 (Cr, Ni, Cd, Pb) or 1991 (As). The Banchory results show decreases in deposition for Ni, Cu, Zn (slight),

Cd and Pb, no change for As, and an increase for Cr. Downward trends have been noted for most metals at three rural sites in England since 1972 (Baker 2001). In the absence of monitored data prior to 1972, analysis of metal concentrations in moss samples can provide a relative record of deposition. Figure 1 shows metal concentrations in herbarium moss samples collected from the Cairngorms between 1870 and 1997, together with concentrations in samples collected as part of the 2000 moss survey (Ashmore et al. 2002). The highest metal levels occur in mosses collected between 1950 and 1975; amounts in mosses collected in 2000 are very low in comparison (particularly for Pb, Cd and As). This supports the observed trend of decreasing metal deposition in the region. Looking further back, levels of As, Cd and Pb (and to a lesser extent Cu and Zn) also appear to have been considerably higher in the latter half of the 19th century and early 20th century than today.

Table 1. Trace metal concentrations ($\mu g\ l^{-1}$) and loads ($\mu g\ m^{-2}\ yr^{-1}$) in bulk deposition at Lochnagar, Banchory and the Duddon valley (Cumbria) for 1997-2003.

Metal	Lochnagar				Banchory[3]		Duddon Valley[4]	
	1998[1]		1999-2003[2]		1997-1999		1998, 2002	
	conc	load	conc	load	conc	load	conc	load
Cr					0.37	362		
Ni			0.14	150	0.22	204	0.39	1200
Cu			0.38	460	1.02	1750	0.40	1200
Zn			4.7	6700	7.03	6850	2.4	7200
As					0.18	167		
Cd			0.043	50	0.12	107	0.02	60
Hg	0.023	28	0.027	31				
Pb			0.72	1300	1.53	1410	0.76	2400

[1]Yang et al. (2002a); [2]Rose and Yang (2004); [3]Playford and Baker (2000); [4]Lawlor and Tipping (2003), Tipping et al. (2005b)

Trace metal inputs from weathering

Aside from atmospheric deposition, the other possible source of trace metals to the Lochnagar system is mineral weathering. Weathering inputs can be estimated very approximately on the basis of rock composition, if certain assumptions are made. Table 2 shows the trace metal contents of the catchment rock, granite, given by Mason (1966), together with the silicon content (see also Chapter 3: Goodman this volume). The mean Si concentration in the loch water over the period 1988 to 1998 was 3.6×10^{-5} M (data from the UK Acid Waters Monitoring Network; Don Monteith, University College London pers. commun.), which is lower than the solubility of quartz (5×10^{-5} M) at the loch temperature (Neal et al. 2005). Moreover, there is not much seasonal variation in Si, indicating that any removal of the element due to diatom formation (see Chapter 5: Thompson et al this volume) plays only a minor role in its dynamics. Therefore, it may be assumed that the rate of mineral dissolution, rather than mineral equilibrium, governs the Si content of the loch water. If we then further assume that the minerals are dissolving congruently, trace metal inputs to the soil-rock system can be estimated by taking ratios of element : Si contents in the rock and multiplying them by the loch water Si concentration (this does not of course imply that the trace metals and silica have the same biogeochemical mobilities). By such a calculation, the input rates shown in Table 2 are obtained. In most cases, they are considerably lower than the values in recent atmospheric deposition (Table 1), but the weathering contributions of Cr and Pb are appreciable (c. 30% and 15% respectively of the atmospheric inputs).

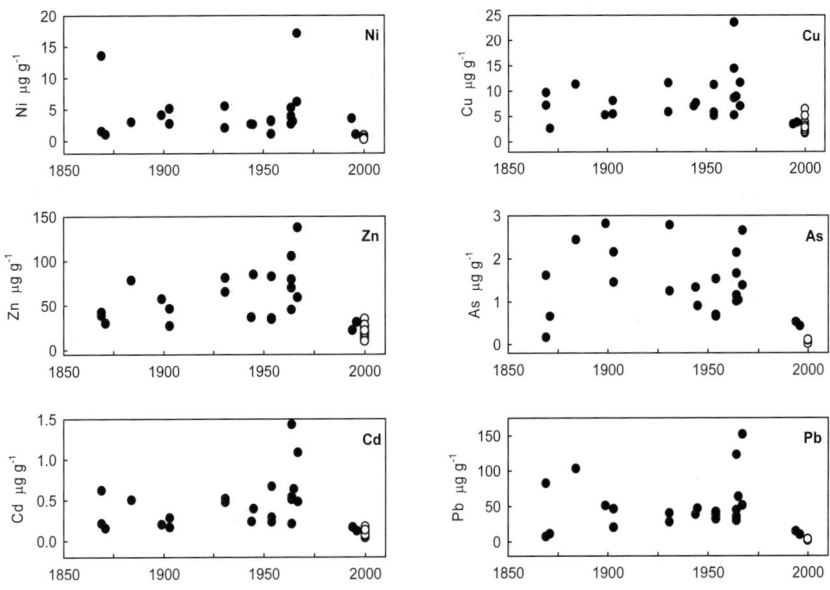

Figure 1. Concentrations of trace metals in Cairngorm moss samples. Closed symbols represent samples from herbarium collections; open symbols are from Ashmore et al. (2002).

Table 2. Estimation of weathering inputs of trace metals to Lochnagar loch water (see text). The values refer to the whole catchment area, and so can be compared directly with the depositional inputs of Table 1.

Element	Granite content $\mu g\ g^{-1}$	Weathering input $\mu g\ m^{-2}\ yr^{-1}$
Si	339600	1.63×10^6
Cr	22	110
Ni	2	9.0
Cu	13	59
Zn	45	200
As	0.8	3.6
Cd	0.06	0.27
Hg	0.2	0.90
Pb	49	220

Trace metals in soil

Yang (2000) and Yang et al. (2001) reported trace metal contents of ten soil cores, representing different soil areas of the Lochnagar catchment. The metal profiles generally show higher metal concentrations in the upper parts of the soil cores, and there are near-surface peaks in some cases. Figure 2 shows two examples. Yang et al. (2001) investigated the use of the results as a historical record of metal deposition, i.e., taking the different depths to represent different times in the past, the chronology being estimated from SCPs (see Chapter 17: Rose and Yang this volume). However, in most of the cores the metal peaks are at depths that correspond to dates in the 19th century, whereas maximum metal deposition almost certainly occurred much later, as evidenced by the sediment core data (see below). Therefore, Yang et al. (2001) concluded that post-depositional movement of metals within the soil had occurred. For Lochnagar, Yang (2000) estimated the increase in the soil metal pool due to anthropogenic contamination, by assuming metal concentrations in the lower soil to represent background values, and found that the anthropogenic metal comprised 93, 72, 87, 77 and 90% respectively of the total pools of Cu, Zn, Cd, Hg and Pb.

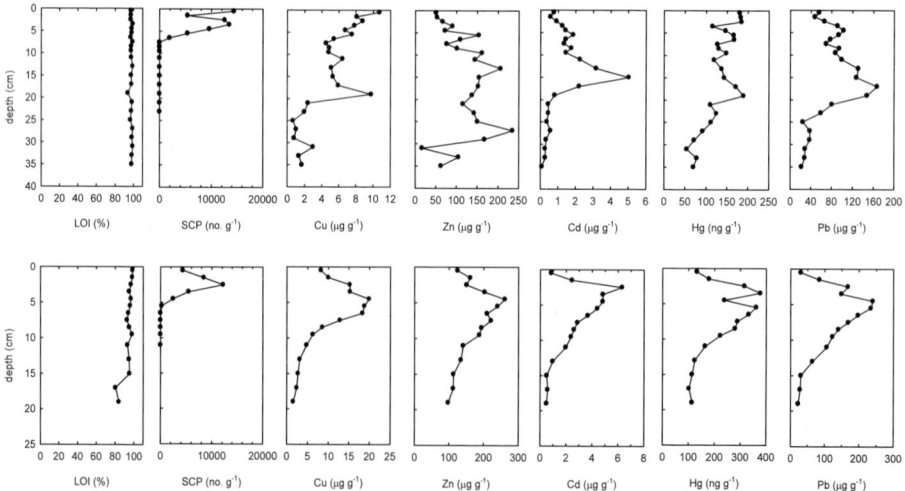

Figure 2. Profiles of percentage loss-on-ignition (LOI) and concentrations of spheroidal carbonaceous particles (SCPs) and trace metals in two organic soil profiles at Lochnagar. From Yang et al. (2002b).

Trace metals in biota

Monitoring of trace metals in biota at Lochnagar began in 1996, the main aim being to employ lake and catchment organisms as passive samplers of trace metals, i.e., to explore their possible use to monitor trace metals in deposition and lake waters at remote sites, rather than to infer toxic damage. Samples were collected annually in late summer. Terrestrial vegetation comprised mosses (*Pleurozium schreberi, Hylocomium splendens*) and ericaceous species (*Calluna vulgaris, Vaccinium myrtillus, Vaccinium vitis-idaea*). Aquatic plants collected were bryophytes (*Nardia compressa, Scapania undulata, Fontinalis antipyretica, Sphagnum auriculatum*) and the macrophyte *Isoetes lacustris*. Zooplankton and, from 2001, aquatic macro-invertebrates were also sampled.

The dataset assembled so far reveals inter-annual variability in the metal contents of biota, and complex temporal trends. This is illustrated by the results for lead (Figure 3). There appears to be a general decline in the Pb content of terrestrial plants, with *Vaccinium myrtillus, V. vitis-idaea* and *Calluna vulgaris* all falling below the limit of detection in the most recent years. Trends for aquatic plants are less clear, and no trends are yet seen for aquatic fauna. Notable observations for other metals include apparent declining contents of Ni in both terrestrial and aquatic species, of Zn in terrestrial species (although all species showed high values in 2001) and of Hg in two aquatic plants (*Sphagnum auriculatum* and *Isoetes lacustris*). The ericaceous terrestrial plants all showed declining Hg from 1996 to 2001, followed by increased levels in 2002 and 2003. The levels of Cd in terrestrial plants showed no clear trends, but there were increases for all the aquatic plants and macroinvertebrates. A detailed discussion of the trends, including possible relationships to deposition trends and lakewater metal concentrations, is given by Rose and Yang (2004). The observations do not yet cover a sufficiently long time span to attempt any statistical work on these data.

Trace metals in loch water

Concentrations in Lochnagar loch water have been determined since 1999. Mean concentrations are shown in Table 3. Where comparisons are possible (Ni, Cu, Zn, Cd, Pb), the values are similar to those reported for three streams in Cumbria (Lawlor and Tipping 2003). Time-series plots for Cr, Zn and Pb (Figure 4) show that, for individual metals, there are quite large ranges of concentrations, and substantial concentration changes between sequential sampling dates. This might indicate a highly dynamic situation, although major ion chemistry is less temporally variable (Chapter 9: Jenkins et al. this volume; Chapter 14: Monteith et al. this volume). It seems unlikely that the concentration variations reflect major changes in loch water metal levels, because the average residence time of the loch (c. nine months) indicates relatively slow water exchange. Some of the high concentration values may therefore reflect (a) the occasional presence in the samples of metal-bearing organic particulates, and (b) difficulties inherent in the sampling, transporting and analysis of low levels of trace metals.

The mean concentrations of Table 3 can be compared with estimates of the values that would be expected just from catchment weathering, calculated from the data of Table 2, if congruent rock dissolution is assumed. Thus, weathering contributes little ($\leq 5\%$) to the loch water concentrations of Ni, Zn, Cd and Hg, but more significantly ($\geq 10\%$) to those of Cr, Cu and Pb. In all cases however, atmospheric deposition makes the major contribution to loch water metal concentrations.

Comparison of the loch water concentrations (Table 2) with those in recent deposition (Table 3) shows them to be quite similar, which might be taken to indicate that the system is in steady state with respect to trace metals. However, this would be to ignore

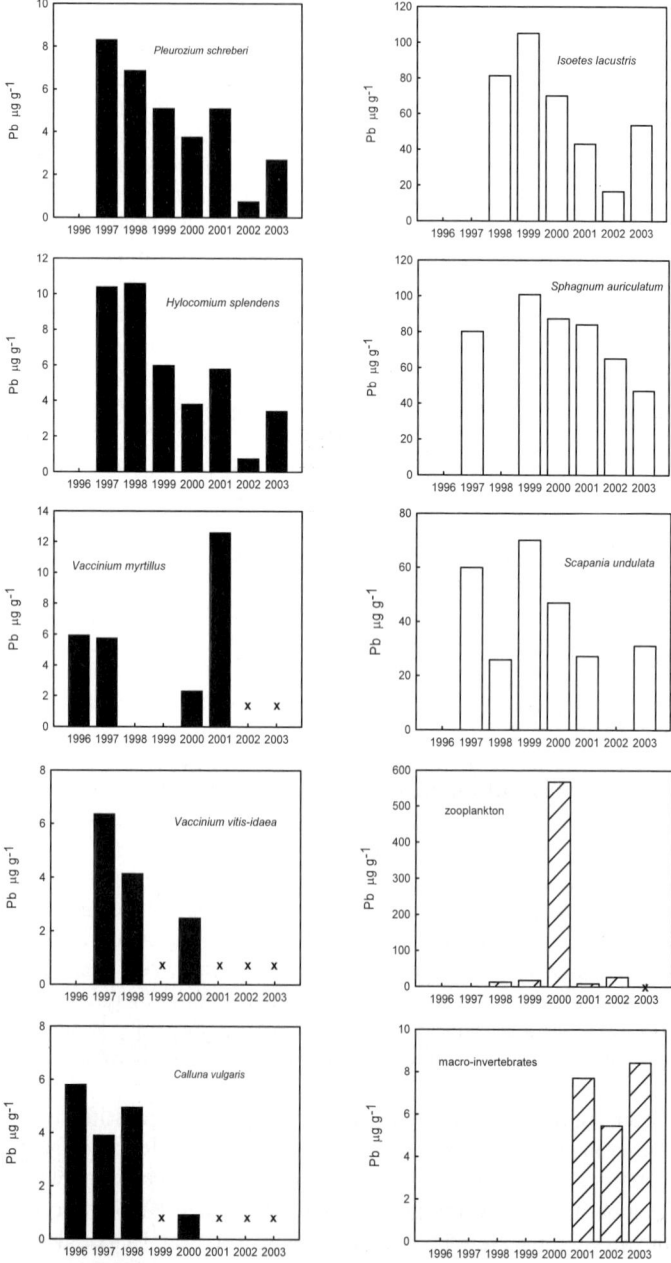

Figure 3. Concentrations of lead in land plants (solid bars), aquatic plants (open) and aquatic animals (hatched) at Lochnagar. Where no data are shown, no sample was taken; "×" indicates values below detection limit.

Table 3. Mean concentrations (μg l⁻¹) of trace metals in Lochnagar lakewater, from 137 data points covering 1999-2004 (14 points in 1998 for Hg), and in three streams in the upper Duddon Valley (Cumbria), from 150 points in 1998 (Lawlor and Tipping 2003; Tipping et al. 2005b).

Metal	Lochnagar	Duddon Valley
Cr	0.22	
Ni	0.11	0.33
Cu	0.27	0.28
Zn	7.4	3.7
Cd	0.055	0.071
Hg	0.015	
Pb	1.2	0.24

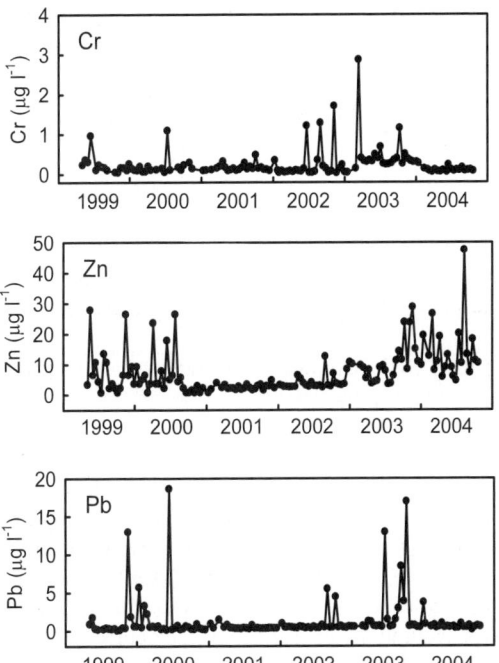

Figure 4. Concentrations of chromium, zinc and lead in Lochnagar loch water.

the large metal pools in the soil (see above). Yang et al. (2002a) used their measured data to compute contemporary annual mass balances for the Lochnagar catchment for Hg and Pb in 1998. They estimated that the loch received 3.5 g of Hg from the atmosphere, but 12.8 g from the catchment, while for Pb the corresponding values were 190g and 1920g. They concluded that the high storage of the two metals in the catchment soils represents a major source of metals to the loch, and would therefore delay restoration based on emission reductions. The role of soils as the main control of metal transfer from catchment to loch is central to the modelling approach (see below).

The loch water metals vary considerably with respect to chemical speciation. Using the Windermere Humic Aqueous Model (WHAM; Tipping 1994) incorporating Humic Ion-Binding Model VI (Tipping 1998), we calculate that for average conditions dissolved Ni, Zn and Cd are essentially all present in inorganic forms, predominantly as free metal ions (Ni^{2+} etc). Twenty-three per cent of the Pb is organically complexed, 78% of the Cu, and essentially 100% of the Hg.

Trace metals in loch sediments

This section is based on Yang (2000) and Yang et al. (2001; 2002a; b; c), and these references should be consulted for detailed information and discussion. Here we will consider five of the 17 sediment cores, the locations of which (Figure 5) form an east-west transect across the loch, with core NAG17 representing the deepest point (water depth 26 m). The trace metal sediment records (Figure 6) show that Lochnagar has been significantly contaminated by Hg, Cd, Pb, Zn and Cu since the 1860s. For both Hg and Pb, the records for NAG14, NAG9 and NAG17 imply that contamination may have commenced even earlier. There are differences in the trends of the trace metal concentration profiles in different cores. For example, in NAG17 and NAG15, Hg concentrations increase in the surface sediments, while in the surface of NAG9 the Hg concentration decreases, and in NAG12 the concentration hardly varies. These data demonstrate that a single metal profile cannot represent the whole accumulating sediment basin.

Yang et al. (2002c) used SCP data and Pb:Ti ratios, calibrated by ^{210}Pb, to identify sediment depths corresponding to the 1860s, i.e., the start of the accumulation of SCPs. Sediment accumulation rates and trace metal inventories in the shallower areas of the loch were inferred to increase with water depth, due to sediment focusing, although the relationships differed between different areas of the loch. Moreover, in the deepwater area, the accumulation rates and inventories were both lower than in other parts of the loch (sediment accumulation rates are discussed in more detail in Chapter 8: Rose this volume). Therefore Yang (2000; Yang et al. 2002a) estimated full basin trace metal accumulation by assigning each of the 17 cores to an area of lake sediment and calculating the dry mass accumulated since 1860. The constant dry mass sedimentation rate model of Oldfield and Appleby (1984) was then used to determine the 14 decadal intervals from 1860 to 1998 for each core, and the trace metal inventory for each decade was calculated. Summing these values for all individual sediment areas yielded decadal fluxes of trace metals for the whole sediment basin, which are compared with modelled values later in the Chapter (Figure 11).

The sediment flux data (Figure 11) fall into two groups. Nickel, copper, mercury and lead all show increases from the 1860s to the present, although Cu, Hg and Pb reach their maxima at about 1950, and are fairly steady thereafter, whereas the Ni flux may only recently have reached a constant value. Zinc and cadmium both show increases to around 1950, followed by a decrease, although it might be concluded that the Cd flux has been roughly constant since 1950. The behaviours of the different metals are considered further in the Discussion.

The surface flux of Hg at Lochnagar (17.6 µg m^{-2} yr^{-1}) is similar in magnitude to, but greater than, values reported for an undisturbed lake in southeastern Hudson Bay (16 µg m^{-2} yr^{-1}; Hermanson 1993), and recent lake sediments from West Greenland (5-10 µg m^{-2} yr^{-1}; Bindler et al. 2001) and the Canadian Arctic (5-8 µg m^{-2} yr^{-1}; Lockhart et al. 1995). The relatively constant flux over the last 40-50 years does not parallel the decrease in UK emissions, a fall from 47.2 tonnes in 1970 to 12.4 tonnes in 1998 (NAEI 2001). Similarly, the relatively constant sediment fluxes of Cu and Pb do not parallel the declines in deposition of these metals over the past several decades.

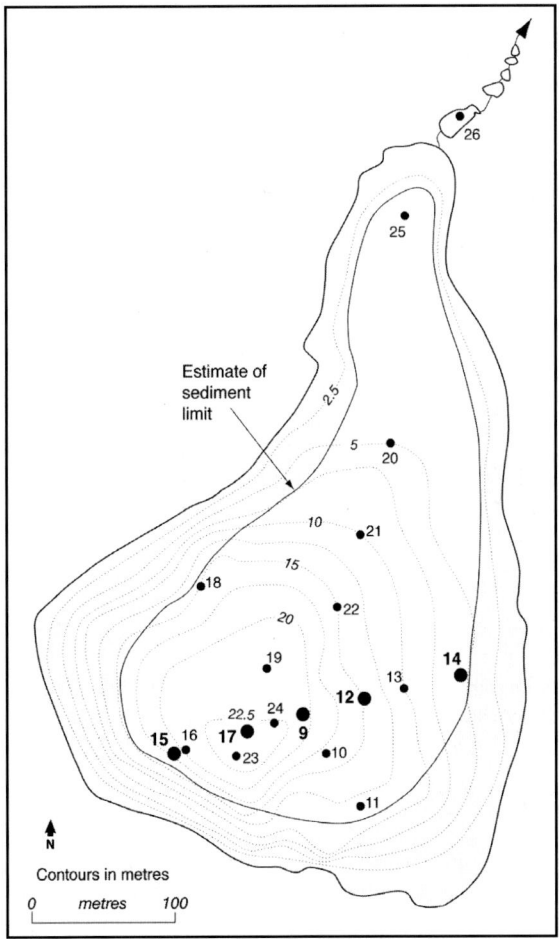

Figure 5. Locations of sediment cores taken from Lochnagar. Larger location points denote those cores featured in Figure 6.

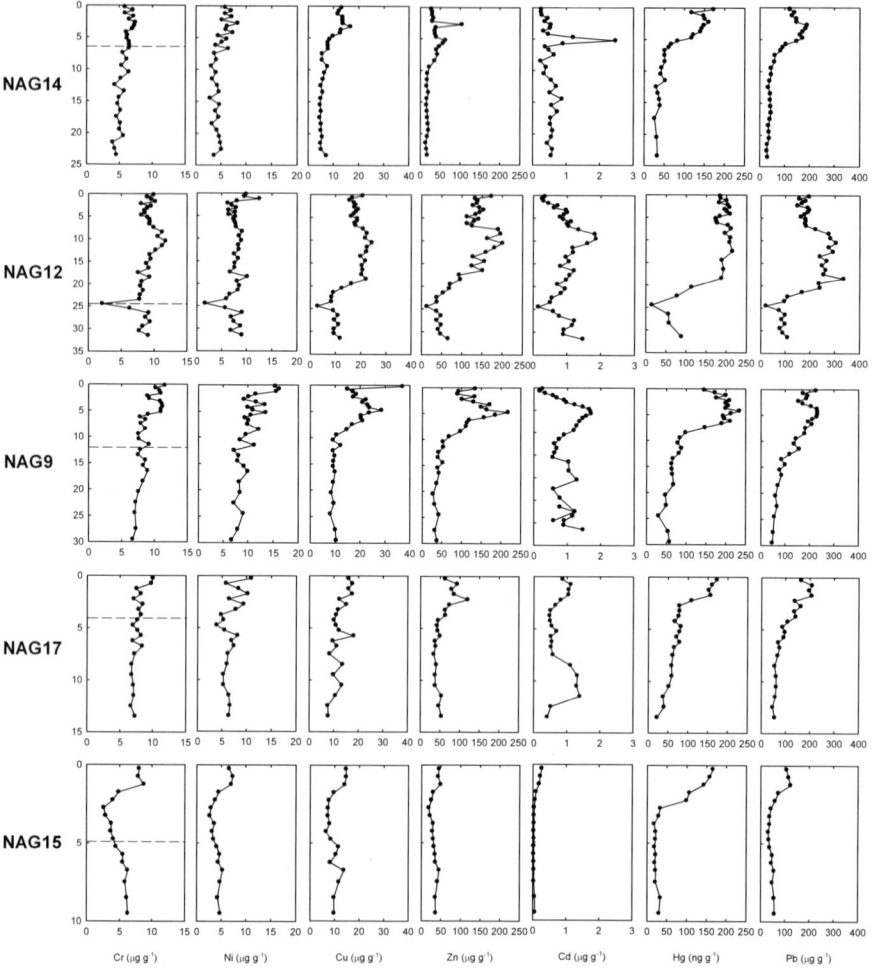

Figure 6. Concentration profiles of trace metals in Lochnagar sediments. The y-axis represents depth in cm. The dashed line indicates the depth corresponding to 1860.

Trace metal toxicity

A simple way to examine whether trace metals are having toxic effects on the Lochnagar ecosystem is by comparing observed metal concentrations with quality standards (or 'Critical Limits'), derived from toxicological experiments. Table 4 makes such comparisons for several different quality standards.

In the case of soils, the most widely used quality standards are expressed in terms of the amount of 'reactive' soil metal in $\mu g \ g^{-1}$; reactivity here usually means that the metal is extractable in dilute (e.g., 0.1 mol l^{-1}) mineral acid. Table 4 compares reactive metal concentrations in the top 5 cm of Lochnagar soils (NAG50 and NAG55; see Figure 2) with 'maximum permissible concentrations' (MPC) used in the Netherlands. Modest exceedances of the MPCs are found for Cd and Pb, but it should be borne in mind that the Dutch standards refer to soil containing only 10% organic matter (and 25% clay); the Lochnagar soil is much richer in organic matter, which will make the metals less bioavailable, and therefore less toxic.

Another way to consider the soil metals is through the use of critical limits expressed in terms of free metal ion concentrations (Lofts et al. 2004). Critical Limit Functions (CLFs) are established, of the form:

$$\log [M^{2+}]_{crit} = a + b \, pH \tag{1}$$

The critical limit at a given pH is considered to protect 95% of soil organisms (plants, invertebrates and microbes) from the effects of a given metal. To estimate free ion concentrations in soil solution, one can apply pedotransfer functions that relate them to pH, soil organic matter content, and total active soil metal. The most appropriate pedotransfer functions for Lochnagar soils are those derived by Tipping et al. (2003). Table 4 shows the results of the application of this approach. We find that $[Zn^{2+}]$ is slightly above the critical limit, while the concentrations of Cu^{2+}, Cd^{2+} and Pb^{2+} are somewhat below the limits.

Current quality standards for lakewater refer in most cases to total metal concentrations. The quoted EA values in Table 4 aim to protect salmonid fish from the toxic effects of Ni, Cu, Zn and Pb, while the values for Cd and Hg just refer to freshwaters in general. The chosen values are for low-hardness waters, which is

Table 4. Comparisons of trace metal concentrations at Lochnagar with Critical Limits (CLim).

	Ni	Cu	Zn	Cd	Hg	Pb
Soil reactive metal ($\mu g \, g^{-1}$)						
obs (NAG50, 55)	4.3	11	120	2.4	0.21	100
CLim[1]	35	36	140	0.8	2.2	85
Soil solution free metal ion (nmol l^{-1})						
calculated[2]		3.5	1300	7.4		4.7
CLim[3]		17	850	15		9.3
Surface water ($\mu g \, l^{-1}$)						
obs	0.11	0.27	7.4	0.055	0.015	1.2
CLim A[4]	50	1	8	5	1	4
CLim B[1]	5.1	1.5	9.4	0.42	0.24	11
Sediment ($\mu g \, g^{-1}$)						
obs (NAG9, 12, 14, 15, 17)	8.0	16	86	0.54	0.15	150
CLim A[1]	44	73	620	30	26	4800
CLim B[5]	22.7	31.6	121	0.99	0.18	35.8

[1]Dutch maximum permissible concentrations (Crommentuijn et al. 1997). [2]Obtained by applying pedotransfer functions; see text. [3]Calculated from soil pH (4.65) and constants given by Ashmore et al. (2004). [4]http://www.environment-agency.gov.uk/yourenv/water [5]Threshold effect concentrations, >5% chance of toxicity (MacDonald et al. 2000).

appropriate for Lochnagar, and they refer to total dissolved metal, except in the case of Zn, for which the total concentration is used. The Dutch standards shown in Table 4 are dissolved MPCs. No exceedances of any of these standards are found for Lochnagar, even though the Lochnagar concentrations are total (as opposed to dissolved) values. Sediment metal concentrations at Lochnagar are all substantially lower than Dutch MPC values, but they are comparable to the threshold effect concentrations for > 5% chance of toxicity given by MacDonald et al. (2000), and for Pb the threshold effect concentration is exceeded (Table 4).

Dynamic modelling

We carried out dynamic modelling to attempt to explain the observations on trace metals at Lochnagar in terms of biogeochemical processes considered likely to control heavy metal behaviour. The aim was to account for contemporary trace metal soil pools and loch water concentrations, and also metal depth profiles in the sediments. CHUM-AM simulates the catchment behaviour of both major and minor solutes, based principally on their interactions with soil organic matter and inputs from chemical weathering.

CHUM-AM

A full description of the soil and streamwater parts of CHUM-AM is provided by Tipping et al. (2005a; b). In brief, a soil-rock profile is considered, consisting of three completely mixed layers (L1, L2, L3) through which water percolates, streamwater being drainage from L3. The model is run with annual average input data (rainfall volume, deposition chemistry, temperature). In L1 and L2, the soil organic matter pool is in steady state, additions of carbon from plant litter being balanced by losses, principally of CO_2, but also of dissolved organic carbon (DOC) and particulate organic carbon (POC) in percolating water. The soil pCO_2 is specified as a constant value. Outgassing of CO_2 occurs in the stream, to a specified partial pressure. Nitrogen and sulphur are taken up by the plant-soil system in L1, and their rates of release, as nitrate and sulphate, depend upon the N:C and S:C ratios. Organic nitrogen and sulphur are transported to and from L2 as dissolved and particulate organic nitrogen and sulphur.

In L1 and L2, solutes bind to the humic acid and fulvic acid fractions of solid phase organic matter, to dissolved organic matter (DOM) and to suspended particulate organic matter (POM). The interactions are described using WHAM / Model VI (Tipping 1998; Tipping et al. 2003), which also takes account of reactions among inorganic species in solution. In addition, precipitation and dissolution of $Al(OH)_3$ and $Fe(OH)_3$ and carbonates may occur. The same reactions occur in L3, together with surface complexation of cations at the surfaces of Al and Fe oxides, described with SCAMP (Lofts and Tipping 1998), and adsorption of DOM, described with the Langmuir equation. In streamwater and lake water, the reactants are the inorganic components, together with DOM and POM. Again, precipitation and dissolution of $Al(OH)_3$ and $Fe(OH)_3$ may occur. Weathering inputs (mol m^{-2} yr^{-1}) of major and trace metals, are given by the expression $k_w\, a_{H+}{}^n$ (Schnoor and Stumm 1986), where a_{H+} is the activity of H^+ and k_w and n are constants.

The loch itself is treated as a constant volume, completely-mixed compartment. Its chemical composition is determined by atmospheric deposition to the water surface, streamwater input, loch water output, particle sedimentation, and sediment-water exchange of solutes. The sediment model focuses on the top layer of sediment, which is considered to be in steady state with respect to particulates. Thus, the input sedimentation flux is equal to the effective downward transfer of material, as sediment accumulates. Solutes enter the sediment adsorbed to settling particles, or by exchange with the sediment pore water. Within the sediment layer, complete reduction of Fe(III), nitrate, and sulphate take place; this assumption follows the field observations of Alfaro-De La Torre and Tessier (2002) on an acid oligotrophic lake in Canada, with a similar major-solute chemistry to that of Lochnagar. The WHAM model is used to compute chemical speciation within the sediment layer, including binding to solid-phase humic acid, which is assumed to represent the active particulate adsorbent. Once material has moved downwards from the active surface layer, its composition is considered to be fixed.

Input and calibration data

Field data characterising the catchment and loch were assembled from various sources (Table 5). The soil properties used for the modelling here are based on the observations of Yang (2000), and differ from the assignments made by Jenkins et al. (2001), although the total carbon (and hence organic matter) contents of the catchment are similar in both cases, and we adopt the N:C ratio reported by Jenkins et al. (2001). All the soil organic matter is assumed accessible to percolating water. Note that not all the soil cores taken by Yang (2000) were sampled completely to the bedrock. Furthermore, whereas Yang (2000) and Yang et al. (2002a) estimated that only 38% of the catchment area is covered by soil, their estimate did not include soil that is currently covered by rocks. Given the uncertainty in the total soil content of the catchment, for modelling we adopted a soil depth based on the sampled cores, and a higher areal coverage of 50%, although the latter value is not a precise estimate.

Historic deposition scenarios were constructed for nitrogen and sulphur by scaling to emissions data (Tipping et al. 2005a), and for metals by using trends based on observations (Baker 2001) and the sediment records of Cumbrian lakes (Hamilton-Taylor 1979; Ochsenbein et al. 1983), combined with contemporary measured values for Lochnagar (Table 1). The assumed metal deposition trends for Lochnagar are shown in Figure 7.

Application to Lochnagar

The first step in calibrating the model was to adjust parameters controlling the soil pools of nitrogen and sulphur and thereby the loch water concentrations of NH_4, NO_3 and SO_4. The results suggest that both nitrogen and sulphur are more strongly retained by the Lochnagar soils than in the streamwater catchments in Cumbria studied by Tipping et al. (2005a). It was possible to match the overall average values, but not the detailed interannual variability (Figure 8). The other major anion, Cl, was assumed to behave conservatively. The second calibration step was to adjust weathering rate constants for major cations (Na, Mg, Al, K, Ca, Fe) and silicon. The weathering

Table 5. Data used for modelling.

Water input and output[1]
Annual mean rainfall = 1144 mm (individual values used for 1987-2003)
Evaporation = 300 mm yr^{-1}

Bulk deposition chemistry (μeq l^{-1})[1]
Mean annual composition (1987-2003): H 34; Na 48; Mg 13; K 2; Ca 7; NH$_4$ 21; Cl 57;
NO$_3$ 25; SO$_4$ 41; F 1

Dry deposition (mmol m^{-2} yr^{-1})[2]
Values for 1992-4: NH$_3$ 7.0; NO$_x$ 1.8; SO$_2$ 6.3

Catchment and lake dimensions
See Hughes (Chapter 2: this volume).

Soil[3]

Mean depth (cm)	29	Humic acid (% of C)	14
Bulk density (g cm^{-3})	0.30	Fulvic acid (% of C)	2.7
C (g g^{-1})	0.38	N:C (mol mol^{-1})	0.027
pH	4.7	S:C (mol mol^{-1})	0.007

Lakewater
Annual mean concentrations of major solutes (1989-2003)[4] ; see Figure 8
Annual mean concentrations of trace metals[5] ; see Figure 9

Sediment
Lake-averaged fluxes of trace metals to sediment[6] ; see Figure 11

---

[1]From data for Lochnagar (1999-2003) and Glen Dye (1987-2003) given by Hayman et al. (2004); evaporation estimated from the difference in rainfall and the outflow of the lake in 1998 (Yang et al. 2002a); rainfall and chemical composition at Lochnagar for years before 1999 estimated from Glen Dye data, scaled using 1999-2003 data. [2]UKRGAR (1997); values for other years were estimated by scaling to emissions. [3]Yang (2000), Jenkins et al. (2001), Tipping et al. (2005a), A.J.Lawlor and E.Tipping, unpublished data. [4]Acid Waters Monitoring Network data provided by D.T.Monteith. [5]Rose and Yang (2004). [6]Yang (2000).

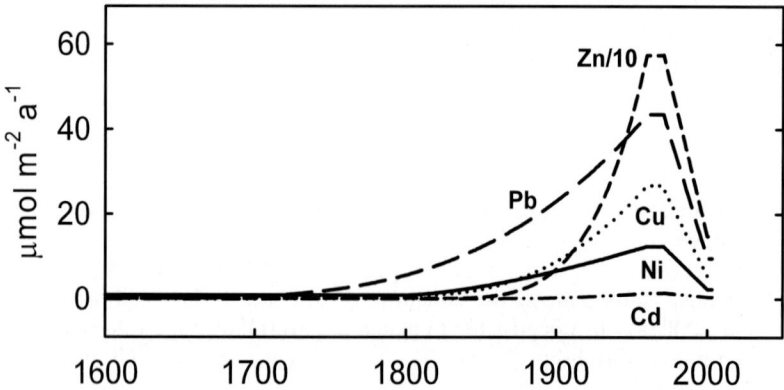

Figure 7. Metal deposition trends for Lochnagar, used for modelling (see text).

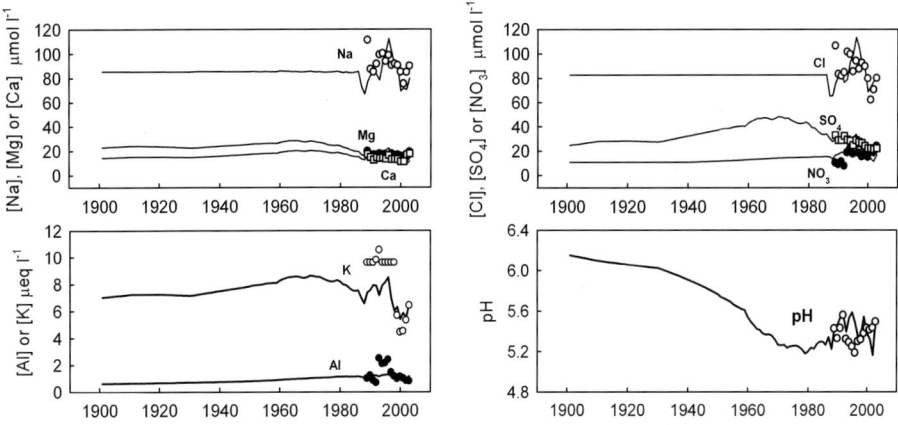

Figure 8. Observed and CHUM-simulated pH and concentrations of major ions in Lochnagar lakewater.

exponent *n* was set to 0.7 for Al and Fe (Stidson et al. 2002) and to zero for the other elements. For simplicity, the values of k_w were assumed to be the same in each soil-rock layer. The calibrations were based on average observed soil and lakewater pH, and lakewater element concentrations. The weathering input rates, expressed in equivalents of charge, of Na, Mg and Ca were similar, and about four times the input rate of potassium. The plots in Figure 8 show that the model also reproduces some of the interannual variations in the data. They also show that lakewater pH is calculated to have fallen from c. 6.4 in the past ('pristine conditions') to 5.2 in around 1980, and to be currently increasing slightly. The acidification development is in rough agreement with a palaeolimnological reconstruction of pH (Patrick et al. 1995; Chapter 14: Monteith et al. this volume), but the diatom-inferred pH values are generally lower than simulated, and they suggest that the major decrease in loch pH had occurred by about 1910, appreciably earlier than suggested by the model.

The calibrated model was used to simulate heavy metal behaviour in soil and loch water, from 1900 to the present, by running it with default WHAM constants, and estimating weathering inputs of metals by taking ratios to Si (cf. Table 2). Figure 9 shows time-series plots (annual means) for the loch water simulations. The model produces less variation than the observations, but there is reasonable agreement in concentration levels, except that the simulated concentration of Ni is too high. Figure 10 compares calculated and observed soil metal pools and streamwater metal concentrations, and also shows results for three Cumbrian stream catchments.

Based on the results of Alfaro-De La Torre and Tessier (2002), a sediment active depth of 2 cm, and a solute exchange coefficient of 1.5 mm day^{-1}, were assumed, and a pCO$_2$ of ten times the atmospheric value was employed. Simulated and observed sediment profiles of trace metals are shown in Figure 11, expressed in terms of annual fluxes to the sediment, following the aggregated fluxes estimated by Yang (2000) and Yang et al. (2002a). The simulated metal contents are the sums of three

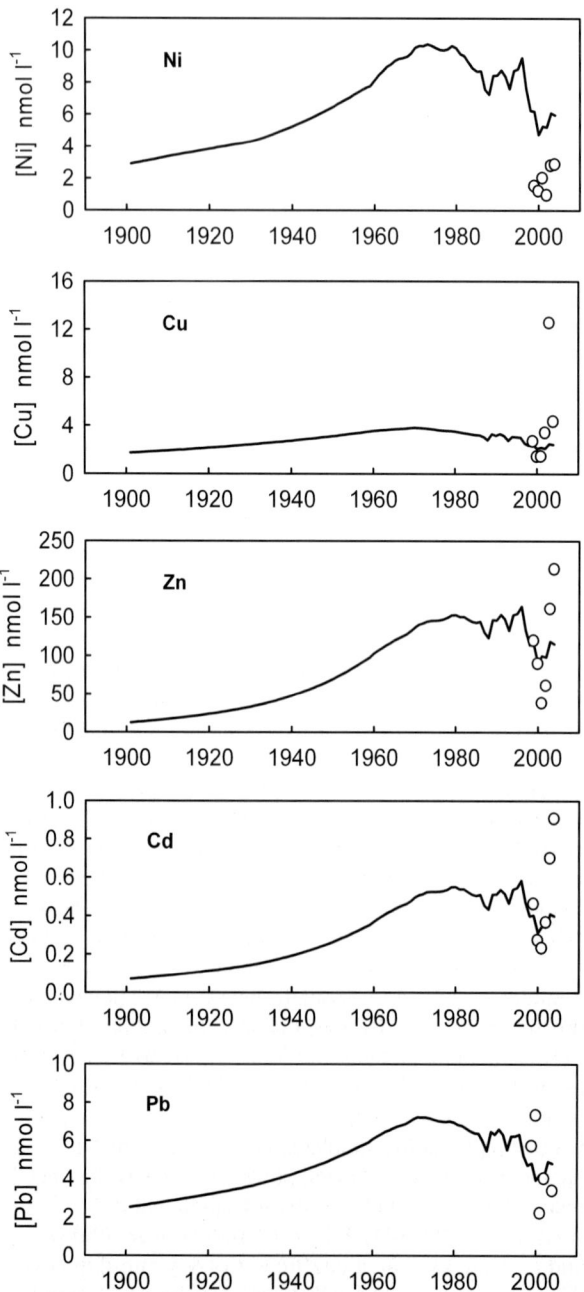

Figure 9. Annual mean concentrations of trace metals in Lochnagar loch water, observed (points) and simulated with CHUM-AM (lines).

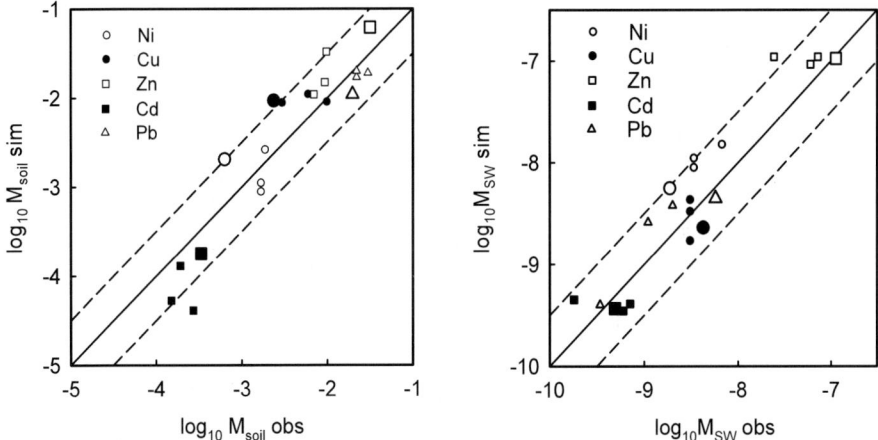

Figure 10. Soil metal pools (mol m^{-2}) and surface water (SW) concentrations (mol l^{-1}) at Lochnagar; comparison of observed and simulated values. Smaller symbols refer to three streams in Cumbria (Tipping et al. 2005b).

Sources – background physical weathering, assumed constant, the transfer of geochemically-active dissolved and particle bound metal from the catchment and directly deposited to the lake surface, and direct inputs of atmospheric 'inert' metals to the lake surface. Physical weathering inputs of metals were estimated from the mineral content of eroded soil and the element content of granite (Table 2). The 'inert' atmospheric depositional input takes into account metal that is not determined by the extraction procedure that measures 'geochemically active' metal in wet deposition, but which might be determined by the more aggressive conditions (100°C; conc. HNO$_3$; 1 hour) used in sediment analysis; it might, for example, include metals held in the matrix of mineral particles. Taking all five trace metals together, the average "inert" metal input was estimated to be 30% of the 'geochemically active' amounts. Note that it is also possible to optimise the 'inert' inputs for the individual metals, and thereby obtain closer agreement with the observations. For example, the observed sediment content of Cu is matched very well by setting the 'inert' input to zero, while good agreement for Pb is achieved by assuming the 'inert' input to be 75% of the 'active' input.

Predicting future metal behaviour

The model was run 100 years into the future to examine possible changes in soil, water and sediment chemistry. In the assumed scenario, pollutant sulphur deposition declines by 50% to 2020, while nitrogen deposition and metal deposition are held at their 2000 values. Soil pH is forecast to increase by 0.2 units, and, in response, the soil pools of Cu and Pb are expected to increase by about 10%, and those of Zn and Cd by about 20%. The soil pool of Ni is expected to decrease by 20% because the assumed input of this metal is now relatively lower than the inputs of the other metals (Figure 8).

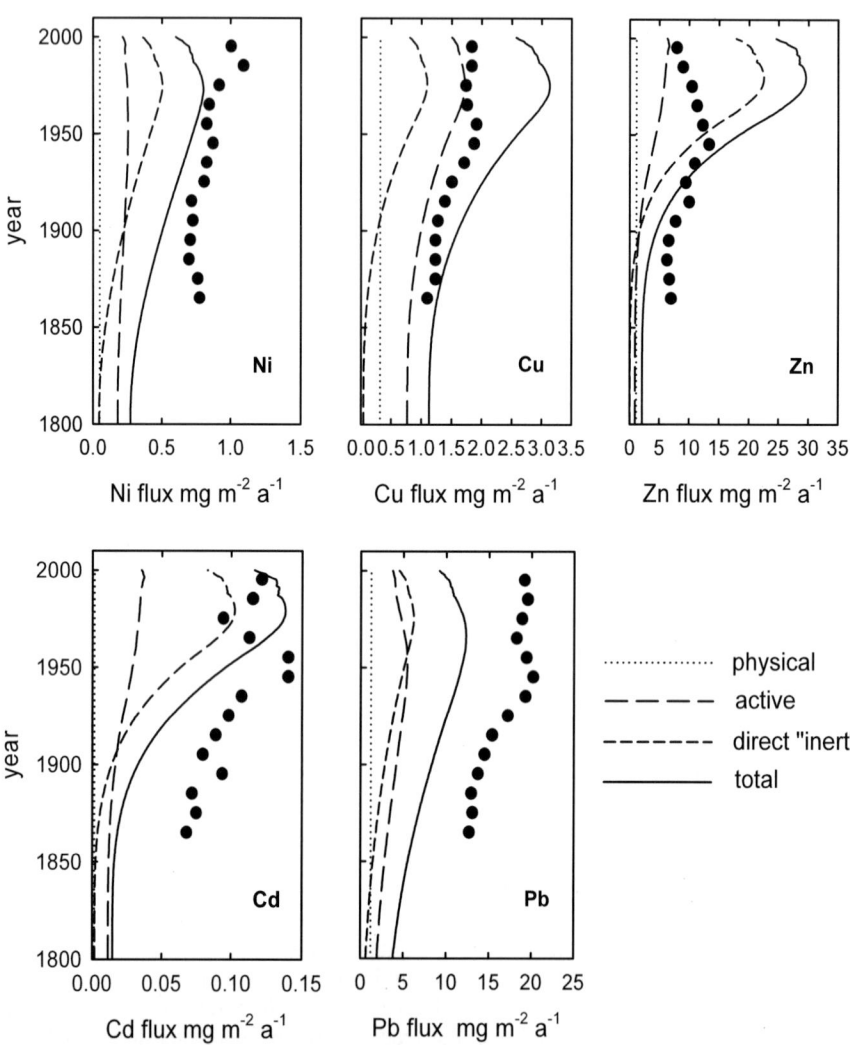

Figure 11. Annual fluxes of metals to the lake sediment, estimated for the whole lake by amalgamating results for different cores (points), and simulated with CHUM-AM (lines). For the simulations, three sources of trace metal are identified (see text).

Loch water pH is assumed to increase, while concentrations of Ni, Cu and Pb will stabilise, and those of Zn and Cd increase slightly. The rates of transfer of geochemically-active trace metals to the sediment will increase, because of stronger binding to sediment particles, while fluxes to the sediment of 'inert' metal will decline. These two factors will lead to initial (50 years) decreases in total fluxes to the sediment, followed by stabilisation, or slight increases for Zn and Cd.

Discussion

Trace metal enrichment of the Lochnagar ecosystem

The soil and sediment trace metal concentration profiles (Figures 2, 6 and 11) demonstrate the recent enrichment of the Lochnagar ecosystem, which can be attributed to increased atmospheric deposition due to human activities. The extents of enrichment, which we express as the ratio of total metal to background metal, can be gauged in several ways (Table 6). From the soil records, Yang (2000) estimated the contributions of anthropogenic metals since 1860 (see section on soils), and obtained values ranging from 3.6 for Zn to 14.3 for Cu. Corresponding ratios calculated from the CHUM-AM model are similar for Cd, but lower for Cu and Pb, and considerably greater for Zn. This suggests that the assumptions used in the modelling, about weathering and background deposition inputs of trace metals, may be inaccurate, although post-depositional transport of metals may have influenced the soil analytical results. The sediment records (Figures 6 and 11) suggest rather constant anthropogenic enrichments, ratios for all metals falling in the range 1.8 to 2.7. The modelled sediment enrichment ratios (cf. Figure 11) are higher, because the background metal fluxes are assumed to be lower than the sediment analyses suggest.

Table 6. Metal enrichment ratios, calculated assuming values in 1860 to represent background metal levels.

	Soil records	Sediment records	Soil model
Ni	-	1.9	2.2
Cu	14.3	1.8	1.6
Zn	3.6	2.6	19.0
Cd	7.7	2.6	6.9
Hg	4.3	2.6	-
Pb	10.0	2.7	3.2

With respect to atmospheric deposition at Lochnagar, the scenarios used in the present work for modelling were based on those for another upland UK location (Cumbria; Tipping et al. 2005b), constructed from measured deposition from the early 1970s to the present, and using local lake sediment records to estimate the start of increased deposition. The deposition trends for the different metals (Figure 7) have similar shapes, with assumed maxima from 1960 to 1970. The shapes of the trends agree quite well with the metal contents of moss samples, collected from 1870 onwards (Figure 1), although the moss data suggest that metal deposition might have been somewhat greater in the 19th century than the scenarios assume. Certainly, both methods of reconstructing past deposition agree that present-day values are much lower than was the case in the past.

The Lochnagar soils, loch water and sediments are contaminated by trace metals from atmospheric deposition, but are these metals having deleterious effects? The data in

Table 4 show slight exceedance of toxic Critical Limits for Cd and Pb in soil, although, as already mentioned, the soil Critical Limits used here are probably too low for the organic-rich Lochnagar soils. There is exceedance for Zn^{2+} in soil solution, and the sediment Pb content exceeds the 5% threshold effect concentration given by MacDonald et al. (2000). The soil and surface water Zn concentrations are also close to the Critical Limits. Given that the exceedances are rather small, it seems unlikely that the trace metals are exerting serious toxic effects. However the metal concentrations in the system are by no means negligible from a toxicological viewpoint.

Modelling soils and waters

CHUM-AM is, of course, a simplification of reality, firstly because water movement is assumed to be constant within a given year, with no account taken of either variation in rainfall or the accumulation and melting of snow. Furthermore, all the water entering the loch is assumed to pass through the soil, ignoring the suggestion of Yang et al. (2002b) that high water concentrations of Hg occur during direct transfer of rainwater into the loch (e.g., during heavy rain events). The simplifications probably cause the dynamic response of the model to be damped in comparison to the real system. The soil itself is represented simply by a single soil column, which allows complete percolation of water. This is neither fully compatible with the presence of zones of permanently sub-oxic soil, which were suggested by Yang et al. (2001) to be sinks for metals, nor with the ready formation of reduced, organically-bound sulphur as a control on the transfer of SO_4 to surface water. It is also assumed that the soil organic matter is in steady state, and that dissolved organic matter (DOM) leaching occurs at a constant rate. DOM export from soil at Lochnagar is increasing, as observed at other upland sites in the UK (Freeman et al. 2001), and this will influence the current and future transport of metals (Cu, Pb) that are significantly organically complexed. A final proviso with respect to trace metals is that their behaviours are predicted after calibrating the model by fitting major solute data. Although CHUM-AM captures the basic chemistry of the loch (Figure 8), there are some clear discrepancies between the simulated and observed data, showing that the model is missing some features, due either to its structure or to inadequate input data.

Bearing the above uncertainties in mind, the simulated results for trace metals in soil and water are generally in reasonable agreement with the available observations (Figures 9 and 10). This supports our picture of an ecosystem contaminated by atmospherically deposited trace metals, and the importance of the organic soil as a significant store of the metals, and the main control on their transfer to the loch. The results in Figure 10 show that the simulated Ni soil pool and loch water concentration are both greater than the observed values, suggesting that the input of Ni to the catchment may be overestimated. The opposite is true of both Cd and Pb, although for each metal the simulated values are only slightly lower than the observed ones. The soil pools of both Cu and Zn are overestimated by the model, but the loch water concentrations are not. For none of the five metals would agreement between simulated and observed values be significantly improved by adjusting the key model parameter, LKMA that regulates metal binding to the soil (Tipping 1998; Tipping et al. 2005a;b). This was also true for four of the metals when CHUM-AM was applied to three stream catchments in Cumbria (Tipping et al. 2005b), but in that study the parameter had to be adjusted to increase the strength of binding of Pb to the soil solids; such an adjustment was not necessary for Lochnagar.

Sediment records

The work on sediment profiles of trace metals at Lochnagar is a rare example of the use of multiple cores to study trace metals, and the variability found across the full accumulating basin is considerable. Rippey et al. (1982) also reported variability in the trace metal profiles of cores taken from Lough Neagh (Northern Ireland). Rose (Chapter 8: this volume) discusses variability in sediment distribution and sediment accumulation rates. The amalgamated sediment core profiles, expressed as overall fluxes in Figure 11, are less well defined than those reported for single cores from other lakes in the UK (Hamilton-Taylor 1979; Ochsenbein et al. 1983). The better definition in these single-core studies may reflect the lack of replication or more favourable conditions for uniform sediment formation. Another possible factor is the low pH of Lochnagar, which reduces the trapping efficiency of metals by the sediment, and may also give rise to greater mobility of metals within the sediment.

The simulated fluxes of metals to the sediment (Figure 11) are very approximately in quantitative agreement with the observations, after adjustment of the fractional deposition rate of 'inert' metals deposited directly onto the loch surface. The variations of the simulated fluxes through time are approximately correct for Ni, Cu and Pb, less good for Cd, and poor for Zn. Taking all the metals together, the 'inert' metal inputs play a major role in determining fluxes of metals to the sediment. This is the case, even though the 'inert' contribution is assumed only to be metal deposited directly to the loch surface, at a level corresponding to only 23% of the total deposition (i.e., it represents a 30% increase above the measured 'geochemically active' deposition). Given the apparent importance of this input, detailed analytical work comparing the different available extraction methods for metals in rainwater, soil and sediment, is warranted.

An alternative approach to the simulation of the sediment metal fluxes is to discount the 'inert' contribution, and seek mechanisms that would give rise to greater sediment binding of geochemically active metals. One possibility is to adjust the values of the key WHAM parameter LKMA (Tipping 1998), i.e., to make sedimentary organic matter a stronger binder of metals than predicted by the default parameters. As already noted, there is no need for stronger binding of Cu. To achieve reasonable simulations of the current sediment fluxes of Ni, Zn, Cd and Pb, the required increases in LKMA are c. 0.3 units, which, since LKMA is expressed in logarithmic units, represents a substantial, but not completely unrealistic, change in binding strength. With the revised values of LKMA, the simulated profile shapes for Cu, Cd and Pb match the observed ones well, but the shapes for Ni and Zn are not improved.

A second possibility is that oxides of Mn and Fe present in the surface sediment boost binding affinity towards the heavy metals. For example, in the sediment model of Boyle (2001), these oxides are considered to be the dominant metal binding phases. However, in studies on Canadian lake sediments, Tessier et al. (1996) and Gallon et al. (2004) concluded that metal binding associated with Fe oxides was mainly due to interactions with humic matter adsorbed to the oxides. Moreover, simple model comparisons suggest that the organic matter would be the dominant binding phase, due to its high binding affinities, compared with Fe oxide, and its abundance, compared with Mn oxide (Tipping 2002). A further possibility is fractionation of granitic minerals during erosion and transport, leading to enrichment of the sediment in fine-grained micas with relatively high metal content, which might explain the tendency of the model to underestimate background metal levels (John Boyle, University of Liverpool pers. commun.).

The ultimate fate of many sedimentary metals is probably sulphide formation (Carignan and Tessier 1985; Gallon et al. 2004), which takes place beneath the zone of initial metal accumulation, modelled here in terms of sorption to organic matter. We examined the possibility that diffusion of dissolved metal from the upper zone to a sulphide sink at depth could increase the overall flux of metals to the sediment, but the effect was small because the dissolved metal concentrations are kept low by the sorption process.

The sediment metal fluxes at Lochnagar have relatively high background values (Figure 11), compared to the values simulated on the basis of physical weathering and background deposition inputs. The modelled physical weathering inputs, based on the assumption that all trace metal associated with sedimented mineral matter is released by the chemical extractant, are probably overestimates. The assumed background atmospheric deposition fluxes for the period before the Industrial Revolution are based on global background data (Nriagu 1989). The values for Lochnagar could conceivably have been higher, since mining and smelting were carried out in Northern Europe for several thousand years before the Industrial Revolution, causing low-level contamination that might be attributed to background levels.

It is clear that there are a number of issues that need to be resolved to make the modelling of metal levels in Lochnagar sediments more convincing. We have discussed issues to do with sorption and source terms. In addition, post-sedimentation diagenesis may influence sediment profiles, and more sophisticated chemical modelling may be required to take this into account (see e.g., Alfaro-De La Torre and Tessier 2002; Gallon et al. 2004). Further work on Lochnagar and other lakes would benefit not only the trace metal fate and impacts modelling, but also the interpretation of sediment records.

The future

Given that CHUM-AM simulates reasonably well the chemical conditions in the soil and loch water at Lochnagar, at least over longer timescales, some insight into future developments has been obtained by running the calibrated model 100 years into the future (see above). The results of the simulation suggest that metal pools and concentrations will change, but quite slowly, with no immediate responses to the substantial decreases in metal deposition seen over the past few decades. The slow response reflects the large buffering capacity of the soil with respect to metal behaviour. This simple preliminary forecast has not considered climate change impacts, the most significant of which might be loss of organic matter from the catchment soils, rendering them less able to retain the trace metals. Climate change may also affect the water balance of the ecosystem, and play a role in the increasing DOC concentrations (see above).

The use of a model with a yearly time step, and which assumes all rainwater to pass through an idealised soil, will tend to produce a relatively damped, or smoothed, response. Many of the observations made for Lochnagar suggest a more dynamic situation, with appreciable variability within and between years (Figures 4 and 9). Interannual variation is also seen in the biota (Figure 3). In a detailed comparison of metal levels in terrestrial and freshwater plants and aquatic organisms, Rose and Yang (2004) reported complex temporal trends with several examples of apparent short-term correlations between levels in biota and concentrations in deposition and loch water. The next phase of such monitoring is crucial, since it will permit statistical analyses to be implemented, and establish the significance or otherwise of relatively short-term trends. The information obtained will complement the longer-term assessments obtained by simulation modelling.

Summary

- Over the last several centuries, the catchment of Lochnagar has been contaminated by atmospherically-deposited trace metals (Cr, Ni, Cu, Zn, As, Cd, Hg, Pb). Monitoring data, analyses of moss sampled over the last 150 years, and lake sediment data, show that trace metal inputs increased from the 19th century to maxima during the mid-to-late 20th century, since when substantial decreases in deposition have taken place.
- Elevated levels of metals in soil, terrestrial plants, loch water, aquatic plants and animals, and sediments have been demonstrated. Enrichment ratios (observed / background) estimated from the soil records range from 3.6 (Zn) to 14.3(Cu), while the sediment records give a smaller range, from 1.8 (Cu) to 2.7 (Pb).
- Some observed Zn, Cd and Pb concentrations in soil, water and sediment are close to, or slightly exceed, metal quality standards or threshold levels.
- Dynamic modelling of Ni, Cu, Zn, Cd and Pb was carried out with CHUM-AM, after calibration of the model for Lochnagar using major solute data. The model provided reasonable predictions of presently-observed soil metal pools and water concentrations, using default parameters to describe the sorption of trace metals by soil organic matter.
- The model forecasts that, over the next 100 years, there will be no major changes ($\leq 20\%$) in trace metal levels at Lochnagar. However, the appreciable interannual variability shown by the observed loch water metal concentrations and the metal contents of biota suggest that faster responses may occur.
- The modelling of metal fluxes to the lake sediments was attempted, assuming the main factors to be (a) background inputs from physical weathering of mineral matter, (b) sorption of metals by organic matter in the upper 2 cm of the sediment, and (c) inputs of geochemically unreactive ('inert') metals that are atmospherically deposited to the water surface, and directly transferred to the sediment. Factors (b) and (c) provide approximate quantitative explanations of sediment metal contents, but faithful simulation of the sediment trace metal profiles probably requires a more sophisticated modelling approach.

Acknowledgements

We thank M.R. Ashmore for helpful discussions, G. Hayman for advice on deposition data, S. Lofts for assistance with modelling, and J. Porter for help with sampling. Thanks are due to the Balmoral Estates Office for facilitating the fieldwork at Lochnagar. This work was financially supported under the EU project MOLAR (1996 - 1999, contract ENV4-CT95-0007), and under the following projects from DEFRA, the Scottish Executive, the National Assembly of Wales, and the Department of the Environment in Northern Ireland; EPG 1/3/117, EPG 1/3/183, EPG 1/3/160, EPG 1/3/188.

References

Alfaro-De La Torre M.C. and Tessier A. 2002. Cadmium deposition and mobility in the sediments of an acidic oligotrophic lake. Geochim. Cosmochim. Acta 66: 3549–3562.

Ashmore M., Bell S., Fowler D., Hill M., Jordan C., Nemitz E., Parry S., Pugh B., Reynolds B. and Williams J. 2002. Survey of UK metal content in mosses 2000. Part 2 of EPG 1/3/144 Final Contract Report.

Ashmore M., Shotbolt L., Hill M., Hall J., Spurgeon D., Svendsen C., Fawehinmi J., Heywood E., Tipping E., Lofts S., Lawlor A.J. and Jordan C. 2004. Further Development of an effects (Critical Loads) based approach for cadmium, copper, lead and zinc. Report to the Department of the Environment, Transport and the Regions, the Scottish Executive, the National Assembly for Wales and the Department of the Environment in Northern Ireland. Contract EPG 1/3/188. University of York. 129 pp.

Baker S.J. 2001. Trace and major elements in the atmosphere at rural locations in the UK: Summary of data for 1999. Report AEAT/R/ENV/0264 to the Department of the Environment, Transport and the Regions, the Scottish Executive, the National Assembly for Wales and the Department of the Environment in Northern Ireland. AEA Technology, Abingdon. 80 pp.

Bindler R., Renberg I., Appleby P.G., Anderson N.J. and Rose N.L. 2001. Mercury accumulation rates and spatial trends in lake sediments from West Greenland. Environ. Sci. Technol. 35: 1736–1741.

Birks H.J.B. This volume. Chapter 7. Flora and vegetation of Lochnagar – past, present and future. In: Rose N.L. (ed.) 2007. Lochnagar: The natural history of a mountain lake. Springer. Dordrecht.

Boutron C. F., Candelone J.-P. and Hong S. 1995. Greenland snow and ice cores: unique archives of large-scale pollution of the troposphere of the Northern Hemisphere by lead and other heavy metals. Sci. Tot. Environ. 160/161: 233–241.

Boyle J. 2001. Redox remobilization and the heavy metal record in lake sediments: A modelling approach. J. Paleolimnol. 26: 423–431.

Brimblecombe P. 1987. The Big Smoke: A History of Air Pollution in London since Medieval Times. Methuen, London. 185 pp.

Carignan R. and Tessier A. 1985. Zinc deposition in acid lakes: The role of diffusion. Science 228: 1524–1526.

Crommentuijn T., Polder M.D. and van de Plassche E.J. 1997. Maximum permissible concentrations and negligible concentrations for metals, taking background concentrations into account. RIVM Report No. 601501 001. National Institute of Public Health and the Environment, Bilthoven. 260 pp.

Darling S.A. 1990. Non ferrous metals. In: McNeil I. (ed.) An encyclopaedia of the history of technology. Routledge, London. pp. 47–115.

de Vries W., Schütze G., Lofts S., Tipping E., Meili M., Römkens P.F.A.M. and Groenenberg J.E. 2004. Calculation of critical loads for cadmium, lead and mercury. Background document to a mapping manual on critical loads of cadmium, lead and mercury. Alterra Report 1104, Wageningen. 143 pp.

Evelyn J. 1661. Fumifugium: Or, the inconveniencie of the aer and smoak of London dissipated. together with some remedies humbly proposed. Printed by W.Godbid for Gabriel Bedel, and Thomas Collins, London. 26 pp.

Freeman C., Evans C.D., Monteith D.T., Reynolds B. and Fenner N. 2001. Export of organic carbon from peat soils. Nature 412: 785.

Gallon C., Tessier A., Gobeil C. and Alfaro-De La Torre M.C. 2004. Modeling diagenesis of lead in sediments of a Canadian Shield lake. Geochim. Cosmochim. Acta 68: 3531–3545.

Goodman S. This volume. Chapter 3. Geology of Lochnagar and surrounding region. In: Rose N.L. (ed.) 2007. Lochnagar: The natural history of a mountain lake. Springer. Dordrecht.

Goodwin J.W.L., Salway A.G., Murrells T.P., Dore C.J. and Eggleston H.S. 2000. UK emissions of air pollutants 1970–1998. National Environmental Technology Centre, AEA Technology, Culham. www.aeat.co.uk/netcen/airqual/naei/annreport/annrep98/naei98.html

Grigal D.F. 2002. Inputs and outputs of mercury from terrestrial watersheds: a review. Environ. Rev. 10: 1–39.

Hamilton-Taylor J. 1979. Enrichments of zinc, lead, and copper in recent sediments of Windermere, England. Environ. Sci. Technol. 13: 693–697.

Hayman G., Vincent K.J., Lawrence H., Smith M., Davies M., Sutton M., Tang Y.S., Dragosits U., Love L., Fowler D., Sansom L. and Kendall M. 2004. Management and operation of the UK Acid Deposition Monitoring Network: Data summary for 2003. Report to the Department for Environment, Food and Rural Affairs and the Devolved Administrations. AEA Technology, Abingdon. 43 pp.

Helliwell R.C., Lilly A. and Bell J.S. This volume. Chapter 6. The development, distribution and properties of soils in the Lochnagar catchment and their influence on soil water chemistry. In: Rose N.L. (ed.) 2007. Lochnagar: The natural history of a mountain lake. Springer. Dordrecht.

Hermanson M.H. 1993. Historical accumulation of atmospherically derived pollutant trace metals in the Arctic as measured in dated sediment cores. Water Sci. Technol. 28: 33–41.

Hong S., Candelone J-P., Patterson C.C. and Boutron C.F. 1994. Greenland ice evidence of hemispheric lead pollution two millennia ago by Greek and Roman civilisations. Science 265: 1841–1843.

Hong S., Candelone J-P., Patterson C.C. and Boutron C.F. 1996. History of ancient copper smelting pollution during Roman and Medieval times recorded in Greenland ice. Science 272: 246–248.

Hughes M. This volume. Chapter 2. Physical characteristics of Lochnagar. In: Rose N.L. (ed.) 2007. Lochnagar: The natural history of a mountain lake. Springer. Dordrecht.

Jenkins A., Ferrier R.C. and Helliwell R.C. 2001. Modelling nitrogen dynamics at Lochnagar, N.E. Scotland. Hydrol. Earth Sys. Sci. 5: 519–527.

Jenkins A., Reynard N., Hutchins M., Bonjean M. and Lees M. This volume. Chapter 9. Hydrology and hydrochemistry of Lochnagar. In: Rose N.L. (ed.) 2007. Lochnagar: The natural history of a mountain lake. Springer. Dordrecht.

Lawlor A.J. and Tipping E. 2003. Metals in bulk deposition and surface waters at two upland locations in northern England. Environ. Pollut. 121: 153–167.

Lockhart W. L., Wilkinson P., Billeck B. N., Hunt R. V., Wagemann R. and Brunskill G. J. 1995. Current and historical inputs of mercury to high-latitude lakes in Canada and to Hudson Bay. Wat. Air Soil Pollut. 80: 603–610.

Lofts S. and Tipping E. 1998. An assemblage model for cation binding by natural particulate matter. Geochim. Cosmochim. Acta 62: 2609–2625.

Lofts S., Spurgeon D.J., Svendsen C and Tipping E. 2004. Deriving soil critical limits for Cu, Zn, Cd and Pb: A method based on free ion concentrations. Environ. Sci. Technol. 38: 3623–3631.

MacDonald D.D., Ingersoll C.G. and Berger T.A. 2000. Development and evaluation of consensus-based sediment quality guidelines for freshwater ecosystems. Arch. Environ. Contam. Toxicol. 39: 20–31.

Mason B. 1966. Principles of Geochemistry. 3rd Ed. Wiley, New York. 329 pp.

Monteith D.T., Evans C.D. and Dalton C. This volume. Chapter 14. Acidification of Lochnagar and prospects for recovery. In: Rose N.L. (ed.) 2007. Lochnagar: The natural history of a mountain lake. Springer. Dordrecht.

NAEI (National Atmospheric Emissions Inventory) 2001. http://www.aeat.co.uk/netcen/ airqual/naei/annreport/annrep98/chap6-3.htn

Neal C., Neal M., Reynolds B., Maberly S.C., May L., Ferrier R.C., Smith J. and Parker J.E. 2005. Silicon concentrations in UK surface waters. J.Hydrol. 304: 75–93.

Nriagu J.O. 1989. A global assessment of natural sources of atmospheric trace metals. Nature 338: 47–49.

Nriagu J.O. 1990. Global metal pollution: Poisoning the biosphere? Environment 32: 7–33.

Nriagu J.O. 1996. A history of global metal pollution. Science 272: 223–224.

Ochsenbein U., Davison W., Hilton J. and Haworth E.Y. 1983. The geochemical record of major cations and trace metals in a productive lake. Arch. Hydrobiol. 98: 463–488.

Oldfield F. and Appleby P.G. 1984. Empirical testing of ^{210}Pb-dating models for lake sediments. In: Haworth E.Y. and Lund J.W.G. (eds.) Lake Sediments and Environmental History, University of Minnesota Press, Minneapolis, pp. 93–124.

Patrick S.T., Monteith D.T. and Jenkins A. 1995. UK Acid Waters Monitoring Network: The first five years. Analysis and interpretation of results, April 1988 – March 1993. ENSIS, London.

Patterson C.C. 1971. Native copper, silver and gold accessible to early metallurgists. American Antiquity 36: 283–321.

Playford K. and Baker S.J. 2000. Atmospheric inputs of heavy metals to the North Sea: Results for 1999. Report AEAT/ENV/336/47044102 to the Department of the Environment, Transport and the Regions, the Scottish Executive, the National Assembly for Wales and the Department of the Environment in Northern Ireland. AEA Technology, Abingdon. 27 pp.

Renberg I., Brannvall M.L., Bindler R. and Emteryd O. 2002. Stable lead isotopes and lake sediments – a useful combination for the study of atmospheric lead pollution history. Sci. Tot. Environ. 292: 45–54.

Rippey B., Murphy J.M. and Kyle S.W. 1982. Anthropogenically derived changes in the sedimentary flux of Mg, Cr, Ni, Cu, Zn, Hg, Pb and P in Lough Neagh, Northern Ireland. Environ. Sci. Technol., 16: 23–30.

Rippey B., Rose C.L. and Douglas R.W. 2004. A model for lead, zinc, and copper in lakes. Limnol. Oceanogr. 49: 2256–2264.

Rose N.L. This volume. Chapter 8. The sediments of Lochnagar: Distribution, accumulation and composition. In: Rose N.L. (ed.) 2007. Lochnagar: The natural history of a mountain lake. Springer. Dordrecht.

Rose N.L. and Yang H. 2004. Work Package 4: Metal deposition and cycling at Lochnagar. In Curtis, C. and Simpson, G.L. (eds.) Summary of research under DEFRA contract "Recovery of Acidified Waters in the UK" EPG/1/3/183. Final Report. Environmental Change Research Centre, University College London, Report No. 98, 362 pp.

Rose N.L. and Yang H. This volume. Chapter 17. Temporal and spatial patterns of spheroidal carbonaceous particles (SCPs) in sediments, soils and deposition at Lochnagar. In: Rose N.L. (ed.) 2007. Lochnagar: The natural history of a mountain lake. Springer. Dordrecht.

Schnoor J. and Stumm W. 1986. The role of chemical weathering in the neutralization of acidic deposition. Schweiz. Z. Hydrol. 48: 171–195.

Settle D.M. and Patterson C.C. 1980. Lead in Albacore: Guide to lead pollution in Americans. Science 207: 1167–1176.

Stidson R.T., Hamilton-Taylor J. and Tipping E. 2002. Laboratory dissolution studies of rocks from the Borrowdale Volcanic Group (English Lake District). Wat. Air Soil Pollut. 138: 335–358.

Tessier A., Fortin D., Belzile N., DeVitre R.R. and Leppard G.G. 1996. Metal sorption to diagenetic iron and manganese oxyhydroxides and associated organic matter: Narrowing the gap between field and laboratory measurements. Geochim. Cosmochim. Acta 60: 387–404.

Thompson R., Kettle H., Monteith D.T. and Rose N.L. This volume. Chapter 5. Lochnagar water-temperatures, climate and weather. In: Rose N.L. (ed.) 2007. Lochnagar: The natural history of a mountain lake. Springer. Dordrecht.

Tipping E. 1994. WHAM - A chemical equilibrium model and computer code for waters, sediments and soils incorporating a discrete site / electrostatic model of ion-binding by humic substances. Comp. Geosci. 20: 973–1023.

Tipping E. 1998. Humic ion-binding model VI: An improved description of the interactions of protons and metal ions with humic substances. Aq. Geochem. 4: 3–48.

Tipping E. 2002. Cation binding by humic substances. Cambridge University Press, Cambridge, 434 pp.

Tipping E., Rieuwerts J, Pan G., Ashmore M.R., Lofts S., Hill M.T.R., Farrago M.E. and Thornton I. 2003. The solid-solution partitioning of heavy metals (Cu, Zn, Cd, Pb) in upland soils of England and Wales. Environ. Pollut. 125: 213–225.

Tipping E., Lawlor A.J. and Lofts S. 2005a. Simulating the long-term chemistry of an upland UK catchment; major solutes and acidification. Environ. Pollut. 141: 151–166.

Tipping E., Lawlor A.J., Lofts S. and Shotbolt L. 2005b. Simulating the long-term chemistry of an upland UK catchment; heavy metals. Environ. Pollut. 141: 139 – 150.

Tylecote R.F. 1992. A history of metallurgy. 2nd Edition. The Institute of Materials, London. 218 pp.

United Kingdom Review Group on Acid Rain 1997. Acid Deposition in the United Kingdom, 4th Report. AEA Technology, Abingdon. 176 pp.

Yang H. 2000. Trace metal storage in lake systems and its relationship with atmospheric deposition with particular reference to Lochnagar, Scotland. Unpublished PhD thesis. University of London. 340 pp.

Yang H. and Rose N.L. 2003a. Arsenic distribution in the UK lake sediments. J. de Physique IV, 107: 1389–1392.

Yang H. and Rose N.L. 2003b. Distribution of mercury in six lake sediment cores across the UK. Sci. Tot. Environ. 304: 391–404.

Yang H, Rose N.L., Boyle J.F. and Battarbee R.W. 2001. Storage and distribution of trace metals and spheroidal carbonaceous particles (SCPs) from atmospheric deposition in the catchment peats of Lochnagar, Scotland. Environ. Pollut. 115: 231–238.

Yang H., Rose N.L., Battarbee R.W. and Boyle J.F. 2002a. Mercury and lead budgets for Lochnagar, a Scottish mountain lake and its catchment. Environ. Sci. Technol. 36: 1383–1388.

Yang H., Rose N.L. and Battarbee R.W. 2002b. Distribution of some trace metals in Lochnagar, a Scottish mountain lake ecosystem and its catchment. Sci. Tot. Environ. 285: 197–208.

Yang H., Rose N.L., Battarbee R.W. and Monteith D.T. 2002c. Trace metal distribution in the sediments of the whole lake basin for Lochnagar, Scotland: A paleolimnological assessment. Hydrobiol. 479: 51–61.

16. PERSISTENT ORGANIC POLLUTANTS IN THE SEDIMENTS OF LOCHNAGAR

DEREK C.G. MUIR (derek.muir@ec.gc.ca)
Water Science and Technology Directorate
Environment Canada
Burlington, Ontario
L7R 4A6, Canada

and

NEIL L. ROSE
Environmental Change Research Centre
University College London
Pearson Building, Gower Street
London WC1E 6BT
United Kingdom

Key words: organochlorines, PBDEs, PCBs, persistent organic pollutants, sediments, toxaphene

Introduction

Persistent organic pollutants (POPs) are characterised by slow rates of environmental degradation, low water solubility, and high sorption to suspended particles. Most POPs have relatively high air-water partition coefficients due to their low water solubility and semi-volatile characteristics. As a result of these properties atmospheric transport can efficiently and rapidly transfer POPs from use areas, which are generally urban and/or agricultural, to remote sites where they are deposited via precipitation and air-water gas exchange. Once these chemicals are dispersed, they are capable of persisting for many years, with environmental half-lives often of the order of years to decades (Mackay et al. 2006). Long environmental half-lives and high particle sorption of POPs also means that lake sediments are natural archives of these chemicals. This chapter will focus only on the measurement of POPs in Lochnagar, with the broader purpose of demonstrating the efficacy of remote lake sediments to record historical and current deposition of legacy POPs and potential new POPs.

In May 2004, the United Nations Environment Programme (UNEP) Stockholm Convention on Persistent Organic Pollutants came into force. It will bring about global

N.L. Rose (ed.), Lochnagar: The Natural History of a Mountain Lake
Developments in Paleoenvironmental Research, 375–402.
© *2007 Springer.*

bans, phase-outs or emission reductions of 12 chemicals (actually groups of substances) (UNEP 2001). Also the United Nations Economic Council for Europe (UNECE) listed 15 chemicals in the long range transport protocol on POPs in 1998. The initial chemicals listed under the Stockholm Convention were the organochlorine pesticides (OCPs) aldrin, chlordane, dieldrin, dichlorodiphenyltrichloroethane (DDT), endrin, heptachlor, mirex, toxaphene, the industrial chemicals, polychlorinated biphenyls (PCBs), and the chlorinated by-products hexachlorobenzene, polychlorinated dibenzo-p-dioxins and dibenzofurans. The UNECE agreement also included hexabromobiphenyl, polyaromatic hydrocarbons (PAHs) and hexachlorocyclohexane (HCH). All these chemicals are thus officially POPs. However, there is a process to add new chemicals to the list and thus additional substances with similar properties such as the brominated diphenyl ethers (PBDEs), endosulfan, methoxychlor, and chlorinated paraffins may well be added to the list in the near future. For detailed information on physical-chemical properties of POPs and potential new POPs useful references are the Handbooks on Physical–Chemical Properties and Environmental Fate of Organic Chemicals by Mackay et al. (2006); the Handbook of Environmental Chemistry on New Types of Persistent Halogenated Compounds (Paasivirta 2000) and the Evaluation of Persistence and Long Range Transport of Organic Chemicals in the Environment resulting from a workshop of the Society of Environmental Toxicology and Chemistry (SETAC)(Kleçka et al. 2000). The chemicals and their physical-chemical properties are listed in Table 1.

While the study of many of the POPs is largely a legacy issue in Europe, Japan and North America, where chlorinated pesticides have been banned since the 1970s, their presence in environmental media can teach us much about the kinds of chemicals in current use that might also be subject to long range transport and deposition. There are more than 100000 chemical substances in commercial use and a significant fraction of these have some properties in common with the chemicals on the POPs list. The ongoing chemical evaluation activities conducted by government environment agencies, for example, in the United States, Canada, the European Union, and Japan are potentially a major source of information from which to develop future 'priority' lists for chemical contaminants. Unfortunately, physical properties and degradation information exist for <1% of these chemicals thus properties must be predicted using Quantitative Structure-Property Relationships (QSPRs). Recent screening of about 11000 chemicals in commerce by Environment Canada found that about 20% had predicted log K_{ow}s > 5, a criterion for bioaccumulation potential under the Stockholm POPs protocol and also an indicator of high sediment sorption. In the same assessment 28% had predicted sediment half-lives of >180 days (Environment Canada 2005). Thus within the list of chemicals in current use there may be other candidates which could be determined in sediments if suitable analytical methods were available. Also it should be kept in mind that these lists of substances in commerce do not include current use pesticides and pharmaceuticals which are exempted on the grounds that they are regulated by other legislation. Although their physical-chemical properties are better known and they are generally much more degradable than POPs, many current use pesticides have similar vapour pressure and water solubility to the POPs, and are known to undergo regional and long range transport (Glotfelty et al. 1990; Majewski and Capel 1995; Muir et al. 2004).

Table 1. Physical-chemical properties of selected individual components of POPs substances.

Common name	Isomer/congener/ abbreviation or chemical name	Formula	Molecular weight	Water[1] solubility, (mmol m^{-3}) (at 25°C)	Log Kow (at 25°C)	Vapor pressure[2], Pa (at 25°C)	Atmospheric half-life (hrs)	References
Organochlorine pesticides								
Aldrin	Aldrin	C$_{12}$H$_8$Cl$_6$	364.9	0.0465	6.5	0.0160	6.1	1,2
Chlordane	cis-CHL	C$_{10}$H$_6$Cl$_8$	409.8	0.137	6.0	0.00040	55	1
	trans-CHL	C$_{10}$H$_6$Cl$_8$	409.8	0.137	6.0	0.00052	55	1
DDT	o,p'-DDE	C$_{14}$H$_8$Cl$_4$	318	0.126	5.8	0.00087	170	1
	o,p'-DDT	C$_{14}$H$_9$Cl$_5$	354.5	0.073	6.0	2.53x10^{-5}	115	1
	p,p'-DDT	C$_{14}$H$_9$Cl$_5$	354.5	0.016	6.2	2.0x10^{-5}	170	1
Dieldrin	dieldrin	C$_{12}$H$_8$Cl$_6$O	380.9	0.446	5.2	0.0005	55	1,2
Endrin	Endrin	C$_{12}$H$_8$Cl$_6$O	380.9	0.656	5.2	0.0004	43	1,2
Endosulfan	α-endosulfan	C$_9$H$_6$Cl$_6$O$_3$S	406.9	6.30	4.94	0.0044	-	3
Heptachlor	Heptachlor	C$_{10}$H$_5$Cl$_7$	373.3	0.482	6.1	0.0530	6.5	1,2
Hexachlorocyclohexane	α-HCH	C$_6$H$_6$Cl$_6$	290.8	3.44	3.81	0.0030	-	1
	β-HCH	C$_6$H$_6$Cl$_6$	290.8	0.344	3.8	4.0x10^{-5}	-	1
	lindane	C$_6$H$_6$Cl$_6$	290.8	25.1	3.7	0.0037	170	1
Mirex	mirex	C$_{10}$Cl$_{12}$	545.5	0.00012	6.9	0.0001	170	1
Toxaphene	technical	C$_{10}$H$_{10}$Cl$_8$	413.8	1.21	5.5	0.0009	170	4
	P26	C$_{10}$H$_{10}$Cl$_8$	414	-	5.5	-	-	4
	P50	C$_{10}$H$_9$Cl$_9$	448	-	5.8	-	-	4
Byproducts & industrial chemicals								
1,3,5-TrCB	1,3,5-trichlorobenzene	C$_6$H$_3$Cl$_3$	181.4	33.1	4.19	32.0	24	2
1,2,4-TrCB	1,2,4-trichlorobenzene	C$_6$H$_3$Cl$_3$	181.4	270	4.02	61.3	30	2
1,2,3-TrCB	1,2,3-trichlorobenzene	C$_6$H$_3$Cl$_3$	181.4	99.2	4.05	28.0	59	2
1,2,3,4-TeCB	1,2,3,4-tetrachlorobenzene	C$_6$H$_2$Cl$_4$	215.9	27.4	4.6	5.2	201	2
PeCB	Pentachlorobenzene	C$_6$HCl$_5$	250.3	20.7	0.831	0.1	286	2
HCB	Hexachlorobenzene	C$_6$Cl$_6$	284.8	0.0176	5.5	0.0023	17000	1
HCBD	Hexachlorobutadiene	C$_4$H$_3$Cl$_6$	260.76	9.78	20	4.78	17	5
Polychlorinated biphenyls	dichlorobiphenyls	C$_{12}$H$_8$Cl$_2$	223.1	0.269-8.96	4.9-5.30	0.0048-0.279	170	1
	trichlorobiphenyls	C$_{12}$H$_7$Cl$_3$	257.5	0.0582-1.55	5.5-5.90	0.0136-0.1430	550	1
	tetrachlorobiphenyls	C$_{12}$H$_6$Cl$_4$	292	0.0147-0.342	5.6-6.50	5.9x10^{-5} - 0.0054	1700	1
	pentachlorobiphenyls	C$_{12}$H$_5$Cl$_5$	326.4	0.0123-0.0613	6.2-6.50	0.0003-0.0093	1700	1
	hexachlorobiphenyls	C$_{12}$H$_4$Cl$_6$	360.9	0.0011-0.002	6.7-7.30	2.0x10^{-5}-0.0015	5500	1
	heptachlorobiphenyls	C$_{12}$H$_3$Cl$_7$	395.3	0.00114-0.0051	6.7-7.0	2.73x10^{-5}	5500	1
	octachlorobiphenyls	C$_{12}$H$_2$Cl$_8$	429.8	0.00047-0.0007	7.1	2.66x10^{-5}	17000	1
	nonachlorobiphenyls	C$_{12}$HCl$_9$	464.2	3.8x10^{-5}-2.4x10^{-4}	7.2-8.16	-	17000	1
Polybrominated diphenyl ethers	BDE 47	C$_{12}$H$_8$Br$_4$O	485.2	0.1950	6.4	2.15x10^{-4}	-	6
	BDE99	C$_{12}$H$_7$Br$_5$O	564.7	0.0688	6.8	3.63x10^{-5}	-	5
	BDE153	C$_{12}$H$_6$Br$_6$O	643.6	0.0259	7.1	8.87x10^{-6}	-	5

1. Mackay et al., 2006; 2. SRC (2002); 3. Shen and Wania (2005); 4. Fisk et al. (1999); 5. LeCloux (2003); 6. Wania and Dugani (2003)

Analysis of dated sediment cores has been used to infer the depositional history of POPs in lakes in the UK (Sanders et al. 1992; Sanders et al. 1993; Gevao et al. 1997), and in western Europe, particularly in alpine regions. A series of studies of remote lakes in Europe has provided detailed information on the sedimentary record of PAHs (Fernández et al. 2000) as well as PCBs and OCPs (Grimalt et al. 2001; Rose et al. 2001; Grimalt et al. 2004) and dioxins (Rose et al. 1997). The use of sediment cores as part of mass balance estimates for persistent organic compounds in remote lakes has a long history starting with work on PAHs (Hites et al. 1977; Laflamme and Hites 1978; Furlong et al. 1987). Substantial work has also been done on deposition of POPs in the Laurentian Great Lakes in North America and nearby inland lakes (Czuczwa and Hites 1984; Swackhamer and Armstrong 1986; McVeety and Hites 1988; Swackhamer and Hites 1988; Eisenreich et al. 1989; Pearson et al. 1997; Pearson et al. 1998), in sub-arctic and arctic lakes (Muir et al. 1996; Malmquist et al. 2003; Stern et al. 2005). These sediment records have provided information on the historical temporal trends and recent fluxes of POPs which generally track known use and emission histories.

Methodology

A single core from Lochnagar has been analysed for POPs and this section focuses on the methods used for that work. For detailed information on analytical methods for POPs in sediment readers should consult Blais and Muir (2001).

Sediment coring and dating

A sediment core (NAG23; for location see Figure 2 in Chapter 8: Rose this volume) was taken from the deep area of the loch (23 m) using a modified Glew gravity corer (Glew 1988). The core was extruded vertically in the field in 0.5cm slices directly into sterile Whirlpak (polyethylene) bags which were immediately sealed and stored cold (4°C) prior to analysis. Homogeneous portions of 18 samples from the core were treated using a variation on the Eakins and Morrison (1978) polonium distillation procedure. Further details on the ^{210}Pb analysis are given in Turner (1998).

Organohalogen analysis

Sediment extraction to determine toxaphene, PCBs, other OCPs and chlorinated by-products as well as PBDEs was performed as described by Muir et al. (1996) with minor modifications. Wet sediments were centrifuged to remove excess water then combined with anhydrous Na_2SO_4 to yield a dry powder. Samples were extracted using a pressurised fluid extractor (Dionex Instruments, ASE) using dichloromethane (DCM). The conditions involved two separate 10 min extractions at elevated temperature (100°C) and pressure (2000 psi). PCB 30 and octachloro-naphthalene (OCN) were added to all sediments prior to extraction and used as internal standards. Combined DCM extracts were mixed with mercury to remove sulphur. Extracts were then transferred into hexane and were chromatographed on an activated silica column into two fractions. Fraction 1 (hexane elution) contained PCBs, hexachlorobenzene and p,p'-DDE. Fraction 2 (1:1 hexane:DCM).contained chlordane, hexachlorocyclohexane

(HCH) isomers, p,p'-DDD and p,p'-DDT, and most toxaphene related compounds (except the octachlorobornane B8-1413 which eluted with hexane). Extracts were then reduced in volume and taken up in isooctane for gas chromatographic analysis. Gas chromatography using electron capture negative ion mass spectrometry (GC-ECNI/MS) was used to quantify toxaphene related compounds and PBDEs in the sediment core extracts. Separation was performed with an HP5 column (30m x 0.25 mm id; film thickness, 0.25 μm) using a temperature program optimised earlier by Glassmeyer et al. (1999) and splitless injection. Further details are given in Rose et al. (2001). Sample extracts were analysed for a suite of 32 PBDE congeners by GC-ECNI/MS and were quantified using an external standard calibration curve.

Quantitative analysis of PCBs, OCPs and by-products was carried out by high resolution GC with 63Ni electron capture detection on a Hewlett Packard 6890 GC equipped with a 30 m x 0.25 mm (0.25 μm film thickness) DB-5 column (J&W Scientific) using H_2 carrier gas and electronic pressure control. This method permitted the quantification of about 104 PCB congeners (including co-eluting congeners) and about 33 OCPs and by-products. An external standard calibration curve was used to quantify these organochlorines. Toxaphene was quantified with external standard of 'Hercules' technical toxaphene (Ultra Scientific). This is referred to as 'total toxaphene' in the text. In addition, 20 individual hexa- to nonachlorobornane congeners were determined using standards from Dr. Ehrenstorfer (Augsburg Germany) or Promochem (Wesel, Germany) of which 12 were detected in one or more slices (B6-923, B7-1001, P25, B8-1413, P31, B8-789, B8-531, B8-1414/1945, B8-809, B8-2229, B9-1679 and B9-715). Further details on these congeners can be found in Rose et al. (2001). Quality assurance included the analysis of laboratory blanks consisting of all reagents. A certified reference sediment (EC5, National Laboratory for Environmental Testing, Burlington ON) was run every six samples. In addition sediment core slices dated to pre-1900 were analysed as a further check on background levels. All silica gel chromatography was conducted in a clean room equipped with carbon and HEPA filtered air. Method detection limits (MDLs) for individual chlorobornane congeners were approximately 0.01 ng g^{-1} (dry wt) and 0.1 ng g^{-1} for total toxaphene and homologue groups. MDLs for PCB congeners ranged from 0.02 to 0.2 ng g^{-1} with the median of 0.04 ng g^{-1}.

Flux estimates, half-lives and doubling times

Concentrations of each POP were expressed as ng g^{-1} dry wt and are reported only on this basis in the results and discussion section. Fluxes (ng m^{-2} yr^{-1}) were calculated by multiplying the sedimentation rate (g m^{-2} yr^{-1}) for each slice estimated from the CRS model by the concentration. Sedimentation rates averaged 120 ± 20 g m^{-2} yr^{-1} and ranged from 105 to 150 g m^{-2} yr^{-1} over the period 1900-1996 (Turner 1998). Doubling times (t2) and half-lives (t½) of degradation products were calculated using the formula, ln(2)/k, where k is the slope of the ln [conc] versus time (CRS date) relationship.

Results and discussion

Historical record of legacy organochlorine pesticides

The historical profiles of OCPs in the Lochnagar core are a good place to start the discussion of the suitability of Lochnagar cores to reflect UK, European and global use and emissions of POPs. Most OCPs have been banned in western Europe for 30 years e.g., DDT, aldrin/dieldrin, chlordane, technical HCH. Others had limited use or no use in the UK or neighbouring countries (e.g., toxaphene), while a few (e.g. endosulfan and lindane) continue in use but use and emissions have been estimated. Global emissions have been estimated for HCH, DDT, endosulfan, and toxaphene (Li 2003) and thus historical profiles in sediment can be compared. However, to our knowledge, there are no specific emission estimates for legacy OCPs available for the UK or western Europe.

DDT
The historical profile of DDT-related compounds in Lochnagar is shown in Figure 1a. Maximum total (Σ)DDT fluxes were found in the sediment horizon dated to 1973. The major DDT-related compounds in the samples were p,p'-DDT, its dechlorination product p,p'-DDD and p,p'-DDE. Also detected was o,p'-DDT which was present at 10-12% of p,p'-DDT in horizons dated to the 1970s and 1980s. The dechlorination product o,p'-DDD was also detected. Similar increases in fluxes were not seen for other legacy OCPs (Figure 1). The ratio of DDD/DDD+DDT generally increased with depth below the top surface slice (Figure 2a). Doubling times in the abundance of o,p'-DDD and p,p'-DDD were 41 ± 5 and 25 ± 4 yr, respectively, in Lochnagar sediments which is an indication of significant reductive dechlorinating capacity of these sediments.

P,p'-DDT exceeded p,p'-DDD only in near surface horizons (Figure 1a). P,p'-DDE was generally about 1.4-fold greater than p,p'-DDT in the slices dated from 1970 to the surface (data not shown). A high flux of p,p'-DDD was found in the surface slice which was unexpected. Sanders et al. (1992) did not see this trend in their core from Esthwaite Water a semi-rural lake in northwest England about 300 km directly south of Lochnagar. Maximum fluxes of ΣDDT in Esthwaite Water were 16000 ng m^{-2} yr^{-1}, about four-times higher than the maximum at Lochnagar, and the maximum occurred much earlier (1958). However, this lake also has municipal waste inputs from a rural community and thus reflects broad use of DDT e.g., for domestic insect control, rather than strictly atmospheric inputs. Fox et al. (2001) reported historical profiles of ΣDDT in cores from salt marsh sediments of the River Mersey estuary. All three cores showed maximum DDT concentrations in the mid-1960s about 5-8 years earlier than in Lochnagar. However, DDT was actually manufactured within the Mersey catchment while at Lochnagar the only route of entry would be via atmospheric deposition. This helps to explain the time delay between the maximum in Lochnagar and the other sites.

Grimalt et al. (2004) studied historical DDT profiles in four alpine lakes, three in the Tatra mountains of eastern Europe and one in the Pyrenees (Spain). One core in particular, from Lake Ladove in the Tatra Mountains, had a temporal resolution comparable to our Lochnagar core. Two maxima for ΣDDT were observed in Lake Ladove, one dated to 1991 and the other to 1971. The later DDT peak may be related to use in Eastern Europe until the early 1990s (Pacyna et al. 2003) while DDT use in Russia and other former countries of Soviet Union, as well as in Spain and Italy

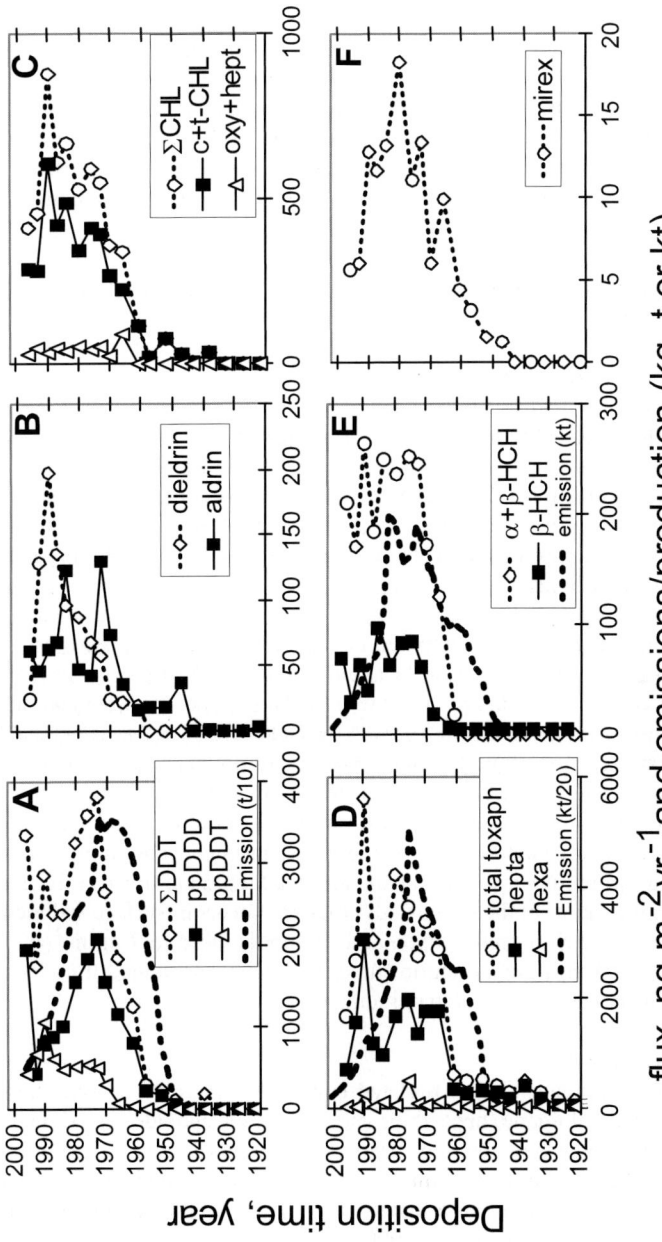

Figure 1. Historical sediment record of six major chlorinated pesticides and selected components or degradation products in a dated core from Lochnagar collected in 1997. (a). ΣDDT, p,p'-DDT and major degradation product p,p'-DDD along with estimated global DDT emissions (Li 2003). (b). Dieldrin and aldrin. (c). Total chlordane related compounds, oxy + hept = oxychlordane + heptachlor epoxide, c+t-CHL = cis- + trans-chlordane. (d). Total toxaphene and hepta and hexa = hepta- and hexachlorobornanes, respectively, along with estimated global toxaphene emissions (Li 2003). (e). Sum of α+β-HCH along with estimated global emissions of technical HCH from Li and Bidleman (2003). (f). Mirex. The same scale is used for fluxes and emissions/production data. The latter have, in some cases, been scaled to fit.

continued at least until 1990 (Pacyna et al. 2003). Unlike Lochnagar, ΣDDT concentrations declined post-1990 in Lake Ladove and two other lakes in the Tatra Mountains. Lake Redon in the Pyrenees also showed a maximum ΣDDT concentration in the 1990s similar to Lochnagar (Grimalt et al. 2004). Taken together these sediment core data suggest continued significant emissions of DDT into the 1990s in western Europe possibly from continued use of old stocks, despite removal from commerce in most countries in the region in the 1970s. Estimated historical global emissions of DDT (Figure 1a) reached a maximum in the period 1965-1970 (Li 2003) and declined by 50% from 1970 to 1985. The decline in Lochnagar ΣDDT flux from 1973 to 1993 was about 55%, in reasonable agreement with the estimated decline in global emissions.

Dieldrin/aldrin
The sediment record of dieldrin and aldrin in our Lochnagar core is shown in Figure 1b. Maximum fluxes of dieldrin were found in horizons dated to 1990 and late-1980s. The maxima for aldrin differ with two peaks, one around 1973 and the other in 1984. The lack of correlation between the two suggests that they have different sources. Aldrin and dieldrin were mainly used as soil insecticides although, following the ban on this use in the 1970s, dieldrin was subsequently used for wood preservation until the 1980s in Europe (to 1981 in the UK) (Pesticide Action Network 2002). Dieldrin is also a degradation product of aldrin, formed by epoxidation, and thus could be formed after atmospheric deposition in surface soils and on plant surfaces within the Lochnagar catchment. Dieldrin is significantly more volatile than aldrin and has a long predicted atmospheric oxidation half-life (Table 1).

Maximum concentrations of dieldrin were 1.4 ng g^{-1} while aldrin peaked at 1.0 ng g^{-1}. These sediment concentrations are low compared to sediments near urbanised areas but higher than remote locations. Fox et al. (2001) found low concentrations in salt marsh sediments of the River Mersey estuary (1-3 ng g^{-1}). This same location had high concentrations and clear historical sediment records for DDT and PCBs, thus indicating that dieldrin use or manufacture within the Mersey basin did not occur. Van Metre and Mahler (2004) reported dieldrin in dated cores from urban reservoirs in the USA as high as 12 ng g^{-1} although a reservoir with little urban development had maximum concentration of 5 ng g^{-1} cores. Muir et al. (1995) found dieldrin concentrations ranging from 0.05 to 3.17 ng g^{-1} in remote lakes in Canada along a latitudinal transect from the mid-latitudes to the high Arctic. In the same study, dieldrin fluxes in surface sediments declined from 40-50 ng m^{-2} yr^{-1} in mid-latitude and sub-arctic lakes to 7 ng m^{-2} yr^{-1} in Lake Hazen at 82 °N. By comparison, dieldrin fluxes in Lochnagar ranged from 24 ng m^{-2} yr^{-1} at the surface to 197 ng m^{-2} yr^{-1} at maximum.

Chlordane related compounds
The sediment record of chlordane (ΣCHL)-related compounds (sum of cis- and trans-CHL, trans-nonachlor, oxychlordane, heptachlor and heptachlor epoxide) are shown in Figure 1c. Cis- and trans-chlordane, the two major components of technical chlordane were the major components in sediment while soil and atmospheric oxidation products, oxychlordane and heptachlor epoxide represented 10-20% of ΣCHL. Maximum fluxes of ΣCHL were found in horizons dated to 1990 and late 1980s, similar to dieldrin. These fluxes of CHL-related compounds (409 ng m^{-2} yr^{-1} at the surface to a maximum of 877 ng m^{-2} yr^{-1}) were higher than dieldrin and all other OCPs except DDT and

toxaphene. These relatively high fluxes are surprising because, according to the recent European survey of POPs (UNEP 2002) chlordane had limited use in Europe. Strandberg et al. (1998) found that fluxes of ΣCHL in the northern Baltic Sea measured by sediment trap (650 - 43000 ng m^{-2} yr^{-1}) were lower than those for dieldrin, ΣHCH and ΣDDT, reflecting its more limited use. Muir et al. (1995) found sum of cis+trans-chlordane fluxes in surface sediments declined from 33-42 ng m^{-2} yr^{-1} in remote mid-latitude and sub-arctic lakes to 3.4 ng m^{-2} yr^{-1} in Lake Hazen at 82 °N. The 10 to 100-fold higher fluxes in Lochnagar suggest much higher use within source regions for the loch i.e., western Europe, notwithstanding the reports of its apparent lack of use.

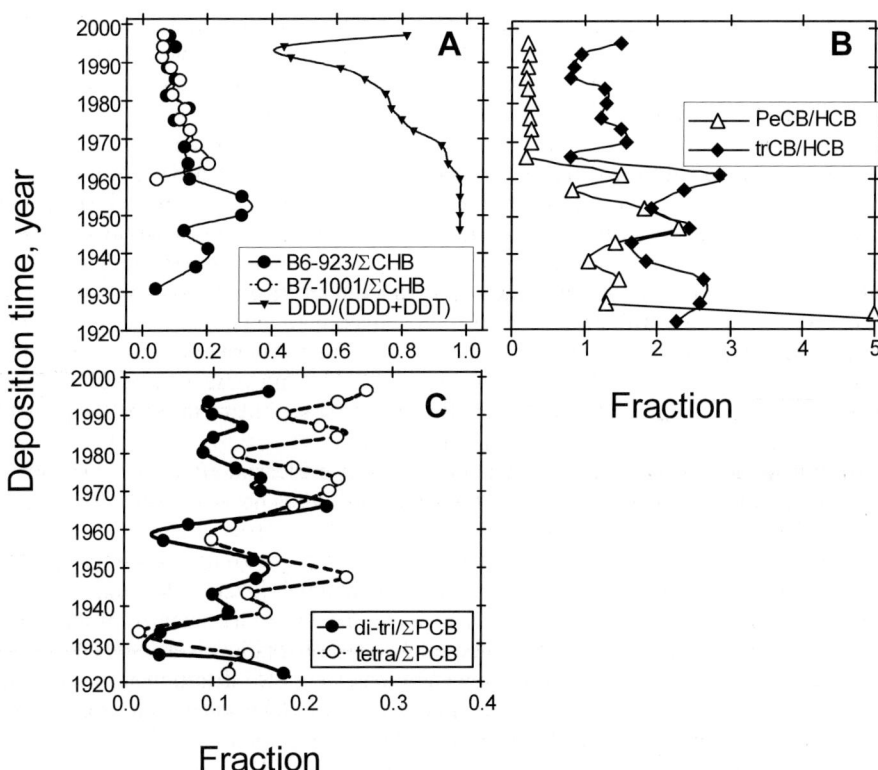

Figure 2. Evidence for dechlorination of OCPs in Lochnagar sediments. A. Historical profiles of the fraction of the dechlorinated products B6-923 and B7-1001 (expressed as a fraction of the sum of 12 chlorobornanes, ΣCHB) are compared with the DDD/DDT+DDD ratios. B. Proportions of PeCB and trichlorobenzenes to HCB. C. Proportions of di-+trichlorobiphenyls and tetrachlorobiphenyls to ΣPCB.

Toxaphene
The historical fluxes of total toxaphene and the hexa- and hepta-chloro-homologue groups are shown in Figure 1d. Toxaphene is a complex mixture of polychlorinated bornanes and camphenes (de Geus et al. 1999) and was widely used as an insecticide following the ban on DDT in the 1970s (Saleh 1991; Li 2001). Total toxaphene concentrations in the Lochnagar sediments are primarily driven by the hepta-group with lesser inputs from the octa-homologues. Hexachloro-bornanes and heptachloro bornanes constituted about 40% of technical toxaphene but comprised 40-68% of chlorobornanes in sediment horizons dated to post-1960. They are formed by dechlorination of higher chlorinated analogues.

Prior to 1960, total toxaphene concentrations were low but uniform at all analysed sediment levels. A dramatic increase in concentration occurred in the 1960s, stabilising for a period between the mid-1960s and the mid-1980s before reaching a peak in 1990 and then declining to the sediment surface. To our knowledge there are no other sediment records of toxaphene from European lakes which can be compared. Toxaphene has been determined in sediment cores from the Laurentian Great Lakes and in a small number of remote lakes (Pearson et al. 1997). Rose et al. (2001) compared concentrations and fluxes of toxaphene in these North American lake cores with results for Lochnagar. Surface sediment concentrations of toxaphene were 14 ng g^{-1} in the Lochnagar core. This is at the upper end of the range previously reported for sites receiving only atmospheric deposition, and is of a similar order to the range of values reported for the Great Lakes where additional riverine sources have been reported (Pearson et al. 1997). Similarly, the maximum total concentration in the Lochnagar core, 40 ng g^{-1}, is far higher than any other reported site receiving only atmospheric deposition and is of the same order as Great Lakes values (Howdeshell and Hites 1996; Pearson et al. 1997) but far below values for sediments in lakes treated with toxaphene for fish removal (Miskimmin et al. 1995).

Toxaphene, in the form of hepta-homologues, was present in all analysed sediment slices, especially for the period 1935-1960 (Figure 1d), thus pre-dating the time of first manufacture in 1945 in the United States (Saleh 1991; de Geus et al. 1999; Li 2001) and 1955 in East Germany (Heinisch et al. 1994). Low toxaphene concentrations were also observed in a slice dated to 1920 (Figure 1d), although levels in laboratory blanks were less than method detection limits. This is similar to results obtained in the Great Lakes where toxaphene was found in Lake Michigan sediment levels dating to 1916 ± 5 yrs (Pearson et al. 1997). Significant concentrations were also determined in sediment levels corresponding to dates prior to the first treatment of lake water with toxaphene at Chatwin Lake, Alberta (Miskimmin et al. 1995). Other studies have suggested that physical mixing, such as bioturbation, rather than molecular diffusion accounts for downward movement of PCBs (Gevao et al. 1997) and toxaphene (Veith and Lee 1971) in lake sediments. At Lochnagar this would require a mixing depth of ca. 2 cm which is unlikely given the well-resolved sediment record of other measured pollutants at the site. Chlorobornanes, the primary compounds in toxaphene, have lower octanol-water partition coefficients (log K_{ow} = 5.2-6.0) than PCBs (Fisk et al. 1999) and hence have lower sediment sorption to sediment organic carbon. Diffusive penetration for heptachlorobornanes and nonachlorobornanes in Chatwin Lake sediments (Miskimmin et al. 1995) was calculated as being between 1.6 and 2.7 cm, depths which would adequately explain the Lochnagar data. Downward smearing of more recent horizons during sampling is also a possible reason for the observations in pre-toxaphene use sediments although precautions were taken against this during the extrusion process.

The historical profile of toxaphene in Lochnagar mirrors the estimated historical global emissions rather well. Global emissions are dominated by large use in the USA which peaked about 1975 (Figure 1d). The later peak at Lochnagar is not reflected in estimated global emissions but may be due to toxaphene usage in Europe. East German production peaked in 1982 at 2630 t and declined to 434 t in 1990 (Heinisch et al. 1994). As late as 1992, toxaphene was still neither banned nor restricted in Spain and Hungary and only restricted in Poland and some other countries of eastern Europe (Voldner and Li 1993). These are areas from which it is known that atmospheric pollutants can impact on this region of the UK (Berge 1997).

Strong evidence for *in situ* dechlorination of chlorobornanes in the core was found when individual chlorobornane congeners were measured. The increase in hexa- and hepta-homologues was primarily accounted for by two congeners, 2-exo,3-endo, 6-exo,8,9,10- hexachlorobornane (B6-923 or Hx-sed) and 2-endo,3-exo,5- endo,6-exo,8,9,10-heptachlorobornane (B7-1001 or Hp-Sed) (Rose et al. 2001). (For additional information on nomenclature see de Geus et al (1999)). Profiles of these congeners in Lochnagar sediments, expressed as an abundance ratio using the sum of the 12 chlorobornane congeners (ΣCHB), are shown in Figure 2a. B6-923 is formed by reductive dechlorination of chlorobornanes with geminal-substituted congeners (2-chlorine atoms at one position), an energetically less favourable configuration (Fingerling et al. 1996). These precursor congeners may include B7-515 and B8-806/B8-809, which are among the major components of the technical toxaphene mixture (de Geus et al. 1999). B6-923 and B7-1001 had doubling times of 17 ± 3 and 12 ± 3 yr, respectively (Figure 2a). Using the same chlorobornane congeners, doubling times for relative abundance of B6-923 and B7-1001 in a core from Jackfish Bay (Lake Superior) in Canada were 8 ± 2 and 27 ± 3 yr, respectively (Muir et al. 2000). The major octa- and nona-chlorobornane congeners in the Lochnagar sediment core were B8-789, B8-2229, B8-1413, B8-1414/ 1945, B8-806/809, B9-715, and B9-1679 (Rose et al. 2001). Maximum concentrations of these congeners occurred in relatively recent sediments, e.g., 6-10 yr of deposition reflecting an increasingly larger proportion degraded in older sediments.

α- and β-HCH

The historical fluxes of sum of α- and β-HCH and α-HCH are presented in Figure 1e. These two isomers are major components of technical HCH which contained 60–70% α-HCH and 5–12% β-HCH (Li 1999). Results for lindane (γ-HCH) are discussed separately below because it replaced technical HCH in Europe and continued to be used during the 1990s (Breivik et al. 1999; UNECE 2004). Technical HCH was banned in Europe as of 1979 (EU 1978). Maximum flux of α- and β-HCH occurred in sediments dated to 1990 although highest concentrations were found in slices dated to 1973. Higher sedimentation rates in the 1980s and early-1990s are the reason for this difference. β-HCH was not correlated with α-HCH or with lindane (p >0.1). However, there was a significant correlation of α-HCH and γ-HCH even in sediments from the 1980s and 1990s when lindane sources might be expected to differ from those of α-HCH because of wide agricultural use in the UK and western Europe. β-HCH is more persistent and much less volatile (Table 1) than the two other isomers (Wu et al. 1997) and may gradually make up a larger proportion of α+β-HCH present in the Lochnagar catchment and in incoming precipitation. Transformation of α- to β-HCH has not been observed under field conditions, however, there is evidence from laboratory studies that lindane (γ-HCH) can be biotransformed into other isomers in soil or sediments. The orientation of chlorine atoms of the γ- isomer makes irreversible transformation into α-HCH the most likely form of isomerisation. Newland et al. (1969) found that 80% of the γ-HCH in a simulated lake impoundment was depleted within three months and that the main metabolites were α-HCH and δ-HCH. Other studies have found a very small percentage of γ-HCH could be transformed into α- or δ-HCH in sewage sludge (Buser and Müller 1995).

Grimalt et al. (2004) determined vertical profiles of α-HCH in a high resolution core from Ladove Lake (Tatra Mountains, Slovakia). The maximum concentration of α-HCH in this core was 1.5 ng g^{-1} which was almost identical to that seen in Lochnagar. The date of maximum deposition in Ladove Lake was about 1992, which was also similar to Lochnagar. Fluxes of α-HCH along a latitudinal gradient in Canada ranged from to 31 to 0.3 ng m^{-2} yr^{-1} in mid-latitude and sub-arctic remote lakes (Muir et al. 1995) although somewhat higher fluxes were found in high latitude lakes (12-25 ng m^{-2} yr^{-1}). These fluxes are about 10 to 100 times lower than maximum fluxes observed in Lochnagar. Overall, the comparison with other locations suggest ongoing emissions of α-HCH occurred well after the time period that it was in use in Europe and that deposition was relatively high compared to background, high latitude sites.

Global emissions of technical HCH are estimated to have peaked in the period 1970-1982 (Li and Bidleman 2003). The historical record in Lochnagar agrees with this rather well although the onset of appearance of HCH isomers is, like the case for DDT and toxaphene, later than the predicted emissions (Figure 1e). This delay may reflect the time needed for significant use and dispersal of HCH within Europe. Unlike DDT, α- and β-HCH fluxes did not decline sharply in parallel to the predicted drop in emissions, reflecting greater redistribution and re-emission of HCH isomers due to their higher volatility and lower sorption to plants and soils.

Mirex
The historical fluxes of mirex in the Lochnagar core are presented in Figure 1f. Mirex is less environmentally mobile than most other OCPs due to its low vapour pressure and strong sorption to aerosols (Scheringer et al. 2000). A maximum flux of 18 ng m^{-2} yr^{-1} was found in the horizon dated to 1980 corresponding to a concentration of 0.12 ng g^{-1}. No other reports for mirex in sediment in European lake waters could be found. Strandberg et al. (1998) did not detect mirex (<0.015 ng g^{-1}) in Gulf of Bothnia (Baltic Sea) surface sediments. Mirex has been determined in sediment cores from Lake Ontario, a lake that has been relatively highly contaminated with mirex due to emissions from upstream production facilities (Oliver et al. 1989; Wong et al. 1995). Wong et al. (1995) reported fluxes of 2000 - 3000 ng m^{-2} yr^{-1} in surface slices dated to the late-1980s and historical trends closely following the time trends of production of mirex in the USA. While originally used as a pesticide, starting in the mid-1950s, largely for the control of ants, it was also used in greater amounts as a fire retardant for plastics, rubber, and electrical goods. Total production in the USA was estimated to be 1.5 kt, (Kaiser 1978), however, global production and emissions have not been estimated. No emissions data for Europe are available (UNEP 2002). Mirex was not

detectable in remote lake sediments along a latitudinal transect in remote sub-arctic and arctic lakes Canada (D. Muir, unpublished results) suggesting much lower fluxes than other OCPs and than observed at Lochnagar. The results from Lochnagar suggest significant past use of mirex within the UK and western Europe.

Current use pesticides

Unlike the banned or 'legacy' organochlorine pesticides discussed in the previous section, endosulfan and lindane have remained in use. Endosulfan increased in use globally in the 1990s while lindane use remained static, at least in Europe (Breivik et al. 1999). There is a pronounced seasonality to the concentrations of endosulfan and lindane in precipitation and air (McConnell et al. 1998; Van Dijk and Guicherit 1999) due to their active use in crop protection as sprays or seed treatments. While input pathways for these current use products to isolated Lochnagar are the same as for the legacy compounds, greater inputs would be expected during the spring and summer months due to use in the UK and in western Europe. The direction of gas exchange may also vary seasonally with absorption into the water column occurring during periods of high air concentrations and volatilisation when concentrations are much lower (McConnell et al. 1998; Rawn and Muir 1999).

Endosulfan
The historical fluxes of α-endosulfan in the Lochnagar core are presented in Figure 3a. α-endosulfan appeared in low concentrations (0.07 ng g^{-1}) in horizons dated to the early- to mid-1950s which is prior to the introduction of this insecticide in 1956. This is similar to observations for chlorobornanes and may be due to diffusion and mixing. Maximum endosulfan fluxes were achieved from the mid-1970s to about 1990 followed by a decline to the surface. Endosulfan has been frequently detected in sediments of agricultural waterbodies (Weston et al. 2004). It is known to degrade in sediment-water systems by hydrolysis and microbial degradation to endosulfan sulphate and endosulfan diol (Peterson and Batley 1993; Navarro et al. 2000). Li et al. (2001) detected α-, β-endosulfan and endosulfan sulphate in a sediment core from the Pearl River estuary in China. The sulphate was the main form of endosulfan detected and low concentrations of α-endosulfan were found in pre-1950s horizons indicating the potential for false positives due to co-eluting interferences.

Stern et al. (2005) reported the detection of α-endosulfan in a laminated core from Devon Island in the Canadian Arctic at low ng g^{-1} concentrations. This core was shown to preserve an accurate record of PCB and toxaphene deposition and thus could be useful for assessing endosulfan time trends as well. The endosulfan concentrations were highest at the sediment surface, and rapidly decreased to below detection limits in core slices dated prior to 1988 (at about 2 cm depth). The authors concluded that the rapid decline in endosulfan with sediment age was due to abiotic and/or biotic degradation. The estimated global emissions of endosulfan reported by Li and Macdonald (2005) and included in Figure 3a, do not coincide well with the historical profile of α-endosulfan flux in Lochnagar. This may be because northern Europe is not a major use area in comparison with south Asia, southern Europe and North America (Li 2003; Li and Macdonald 2005). Endosulfan was first introduced in 1956 and use increased steadily to the early 1990s. However, it was first marketed in Europe and this may explain its appearance in Lochnagar sediments by 1960.

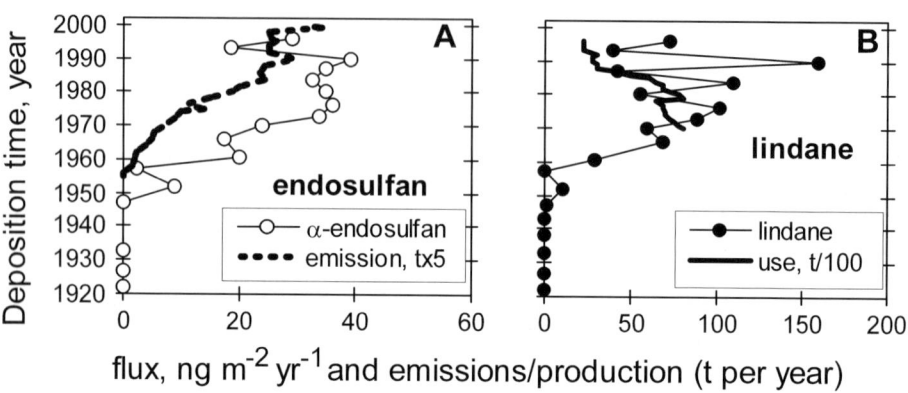

flux, ng m^{-2} yr^{-1} and emissions/production (t per year)

Figure 3. Sediment records of endosulfan and lindane in Lochnagar sediments. (a). The historical estimated global emissions (tonnes) of endosulfan (from Li and Bidleman (2003) and Li(2003)) are compared with α-endosulfan fluxes. (b). The European use data (tonnes) for lindane from Breivik et al. (1999) are included alongside the lindane flux profile.

Lindane

The historical flux of lindane (γ-HCH) is presented in Figure 3b. Lindane appears suddenly in horizons dated to about 1960. An early maximum of lindane was reached in about 1976 after which the profile becomes more erratic. Maximum fluxes were actually found in the horizon dated to 1990 (160 ng m^{-2} yr^{-1}). Fluxes of α-HCH continued to be higher than those of lindane into the 1990s (Figure 1; Figure 3) despite the removal of technical HCH from commerce in 1979 in Europe. Lindane is known to degrade more rapidly in soils and sediments than α-HCH under both aerobic and anaerobic conditions (Willett et al. 1998). However, all three isomers appear to have relatively long half-lives in cold, anaerobic conditions such as sediments and peat bogs judging from monitoring in remote areas. Rapaport and Eisenreich (1988) found that the highest HCH burdens in peat bogs in eastern North America correlated with the lowest mean bog temperature and presumably with the lowest microbial activity. Muir et al. (1995) found much lower concentrations of lindane compared to α- and β-HCH, particularly in mid-latitude remote lakes in central Canada. Fluxes of lindane in these Canadian lakes from the mid-1990s ranged from 0.3-31 ng m^{-2} yr^{-1}, far lower than in Lochnagar during the same time period. Söderström et al. (2002) found median concentrations of lindane of 0.18 ng g^{-1} in surface sediments (0-2 cm) of lakes in southern Sweden and 0.05 ng g^{-1} in northern Sweden. They noted that the α-HCH:lindane ratio generally increased from about 1-2 in the south to 5-7 in northern Swedish lakes. The α-HCH:lindane ratio in Lochnagar 0-2 cm horizon averaged 2.9 similar to the lakes in southern Sweden.

The European usage of lindane was estimated to be 135000 t from 1970 to 1996 when its use in technical HCH was included (Breivik et al. 1999). Lindane, represented 14% of the technical product (UNECE 2004). The overall time trend for use of lindane actually showed a decline from 1970 to 1996 in Europe as illustrated in Figure 3b using data from Breivik et al. (1999). The historical record of lindane fluxes in Lochnagar for the period 1970 to 2000 does not track the usage information very well. The UK and France continued to be the major users of lindane in Europe during the 1990s and thus the relatively high fluxes and variation in the period from the mid-1980s to late-1990s could reflect the close proximity to source areas. A major input pathway to Lochnagar is likely to be via rainfall. Lindane fluxes via precipitation in western Europe have been shown to range from <0.1-50 μg m^{-2} yr^{-1} (Van Dijk and Guicherit 1999) which, even at the low end of that range, would be a very significant contributor to loch inputs. Current use pesticide deposition is known to vary annually depending on timing of cropping and climate factors (Van Dijk and Guicherit 1999). Although banned for spray application to crops in most European countries in the mid-1990s (Breivik et al. 1999) lindane continued to be used for seed and wood treatment and for veterinary and medical purposes in France until 2006 and in the UK until 2002 (UNECE 2004). Modelling of lindane emissions and air concentrations in Europe predicts high concentrations close to the major sources (France) as well as towards Central Europe and Scandinavia (Prevedouros et al. 2004).

Industrial chemicals and by-products

Chlorobenzenes
The chlorobenzenes (tri-, tetra-, penta- and hexa-) were prominent contaminants in Lochnagar sediments. Total (C_{13-6})CBs (ΣCBs) outranked all other groups except ΣDDT and ΣPCBs. Concentrations of ΣCBs reached a maximum of 7 ng g^{-1} in slices dated to 1984 and were above detection limits in slices dated to 1920. The profile shows a major increase in ΣCB fluxes in the 1940s followed by a sharp rise during the 1960s (Figure 4a). Trichlorobenzenes (1,3,5-TrCB; 1,2,4-TrCB; 1,2,3-TrCB) were the predominant CBs detected representing about 36-51% of ΣCBs in post-1940 horizons. The proportion of HCB in ΣCBs ranged from 16-45% and was generally highest in most recent horizons. Maximum fluxes of HCB (399 ng m^{-2} yr^{-1}) were found in slices dated to 1990 (Figure 4b).

Higher ratios of total TrCBs/HCB and PeCB/HCB were found in pre-1960 horizons (Figure 2b). 1,3,5- and 1,2,4-TrCB were especially more prominent in deep sediments. Also detected were 1,2- and 1,4-dichlorobenzene, however, these compounds are quite volatile and the analytical method was not optimised for their recovery, so they are not discussed further. Taken together the results suggest that slow dechlorination of the chlorobenzenes is occurring. Dechlorination of HCB has been observed in anaerobic laboratory sediment cultures with PeCB and 1,3,5-TrCB being the major product (Pavlostathis and Prytula 2000; Chen et al. 2001). This pattern of degradation products is consistent with the appearance of chlorobenzenes in Lochnagar. However, changes in atmospheric inputs over time cannot be ruled out. PeCB as well as trichlorobenzenes and tetrachlorobenzenes were used in combination with PCBs in dielectric fluids (Environment Canada 1993; van de Plassche et al. 2001) and thus their use and emissions may parallel those of PCBs. Greater proportions of trichlorobenzenes in

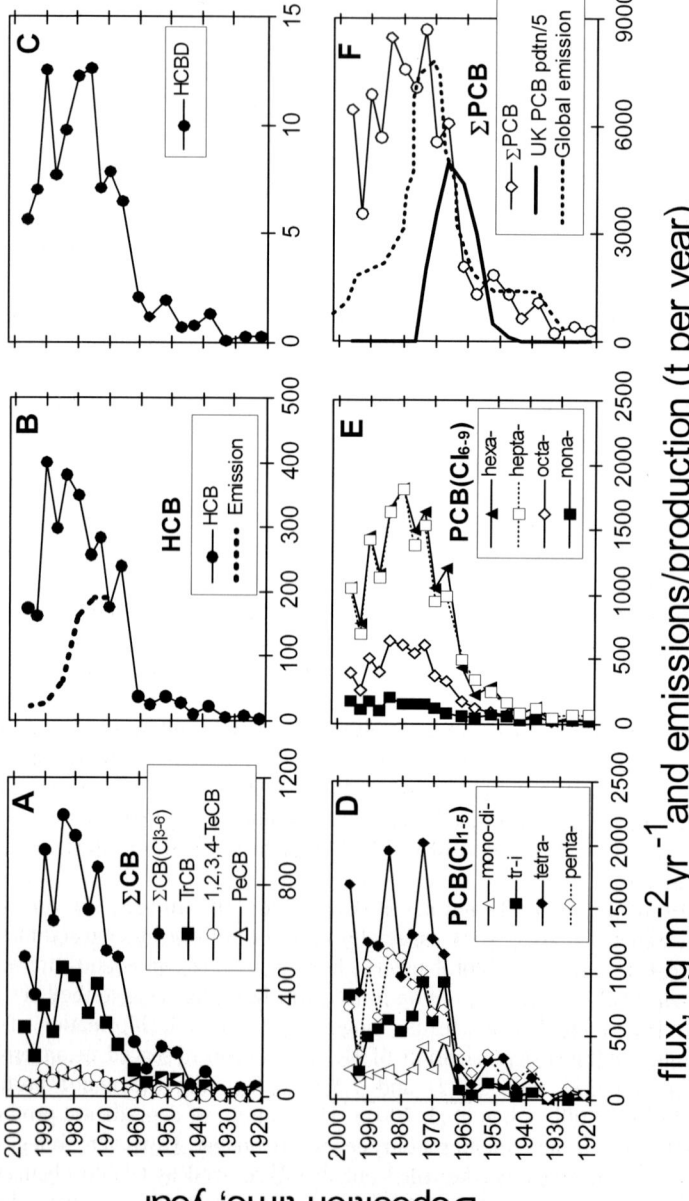

Figure 4. Sediment records of (a) chlorobenzenes (ΣCB, tri-, tetra- and pentachlorobenzene), (b) hexachlorobenzene; (HCB), (c) hexachlorobutadiene (HCBD), (d) mono- pentachloro-biphenyl homologs and (e) hexa- nonachloro-biphenyl homologs, (f) ΣPCBs, in Lochnagar. The HCB profile is compared with HCB emissions for Europe from Pacyna et al. (2003). The PCB profile is compared with global PCB emissions from Breivik et al. (2002) and production in the UK from Sanders et al. (1992); The same scale is used for fluxes and emissions/production data. The latter have, in some cases, been scaled to fit.

deeper horizons would also be consistent with greater diffusion of these compounds which are more water soluble and have lower organic carbon partitioning than HCB (Table 1). PeCB was also a major impurity in the fungicide pentachloronitobenzene (Quintozene) although this chemical is now produced without use of PeCB as an intermediate (van de Plassche et al. 2001). Oliver and Nicol (1982) found higher ratios of PeCB:HCB and TeCB:HCB in pre-1940 sediments in Lake Ontario, however, because of upstream chemical production the authors did not attribute this to dechlorination. They found no evidence for dechlorination of PCBs in the same sediments (Oliver et al. 1989).

HCB has been monitored in European remote lakes. Söderström et al. (2002) found HCB in surface sediments of Swedish rural and remote lakes with higher concentrations in southern (median 0.24 ng g^{-1}) than in Northern Sweden (0.15 ng g^{-1}). Post-1990 surface concentrations of HCB in Lochnagar (1.4-2.8 ng g^{-1}) were generally well above the range for lakes in northern Sweden (0.08-2.0 ng g^{-1}). Grimalt et al. (2004) found inventories of HCB ranging from 120 to 4100 ng m^{-2} in European alpine and remote lakes in post-1978 horizons. This compares with 5300 ng m^{-2} for the HCB inventory in Lochnagar over about the same period.

Fox et al. (2001) found HCB in River Mersey estuary sediment cores at maxima ranging from 9 to 22 ng g^{-1}. These cores were collected near chemical manufacturing facilities where chlorobenzenes, including HCB, were manufactured during the first half of the 20th century. Maximum concentrations occurred as early as 1930-40 at one site and as late as 1965-70 in another. European atmospheric emissions of HCB, estimated by Pacyna et al. (2003), declined from about 190 t in 1970 to 23 t in 1995 (Figure 4b). By comparison, HCB deposition at Lochnagar continued to increase until about 1990 before undergoing a sharp decline. The lag time has been noted for other organochlorines and is discussed further below.

Hexachlorobutadiene

The historical flux of hexachlorobutadiene (HCBD) is presented in Figure 3b. HCBD first appeared above detection limits in horizons dated to the late-1930s. Maximum concentrations of HCBD (0.09 ng g^{-1}) occurred in slices dated to 1976 and generally declined after that although a second maximum was recorded in 1990. Maximum fluxes of 12-13 ng m^{-2} yr^{-1} in the period 1976-1990, (Figure 4c) were much lower than for other organochlorines of similar molecular weight such as HCB and HCH isomers.

There is relatively little comparative information on HCBD in sediments. It has been measured in surface sediments in the German Bight due to emissions from the Elbe River (Schwarzbauer et al. 2000). Mudroch et al. (1992) found that HCBD at concentrations ranging from 0.01 to 0.23 ng g^{-1} at various sediment depths in samples taken from Great Slave Lake in Northern Canada in 1987. A review by Eurochlor (2002) noted a number of studies on marine sediments in western Europe. There seem to be no reports on historical profiles of HCBD for marine or freshwater sediment cores.

HCBD was used as a fumigant for treating grapes against Phylloxera in the former Soviet Union, France, Italy, Greece and Spain but this practice seems to have ended in the 1980s. It was a solvent in the production of rubber and other polymers and also emitted as a byproduct during the production of chlorinated solvents (WHO 1994; Eurochlor 2002). According to Eurochlor (2002) industrial emissions in Europe in 1997 represented 2 kg y^{-1} in air and 100 kg y^{-1} in water, based on a survey of 76 sites from the European chlorine industry. This represents a reduction of 98% of emissions to air and 97% to water between 1985 and 1997 (Eurochlor 2002). This reduced emission may be reflected in declining sediment fluxes of HCBD in Lochnagar from about the mid-1970s onwards.

Polychlorinated biphenyls
The sediment record of total (Σ)PCBs in Lochnagar is shown in Figure 4d-f for the same sediment core analysed for OCPs and by-products. Maximum concentrations ΣPCBs (71 ng g^{-1}) were observed in the horizon dated to 1973 and concentrations generally declined thereafter although with a lot of variability. Fluxes of ΣPCBs also declined post-1973 (Figure 4f) although higher sedimentation rates in Lochnagar in the period 1976-1990 (average 145 g m^{-2} yr^{-1}) compared to 1940-1973 (120 g m^{-2} yr^{-1}) helped fluxes remain elevated into the 1990s. Tetrachlorobiphenyls were the major homologue group in sediments in most horizons (Figure 4d). Hexa- and heptachloro-biphenyls were also very prominent PCB components (Figure 4e). Concentrations and fluxes of the mono-, di-, tri- and tetrachlorobiphenyl groups for the horizons dated to 1966-1986 were generally more variable (RSD 43-60%) than those of the penta- to octachloro- groups (42-48%). The mono- to tetrachlorobiphenyls are more volatile and environmentally mobile than the penta-octa group and may thus follow climate variations, such as shifts in rainfall, more closely. Ratios of di- +tri-, and tetrachlorobiphenyls to ΣPCB do not vary in a consistent pattern with sediment depth (Figure 2c). Unlike DDD:(DDD+DDT) ratios, B6-923- and B7-1001/ΣCHB and PeCB:HCB (Figure 2a;b), there was no indication of dechlorination of PCB congeners in the Lochnagar sediments based on the homologue:ΣPCB ratios (Figure 2c). Gevao et al. (1997) concluded that higher di- and trichlorobiphenyls in pre-1900 horizons of a core from Esthwaite Water could be due to greater diffusion of di- and trichlorobiphenyls as well as possible airborne contamination during extraction and separation steps in the laboratory. This may also explain the presence of PCBs in Lochnagar sediments prior to their first manufacture in the 1930s.

The increase in PCB concentrations starting in the mid-1930s and the large increase during the 1960s coincide well with the known production of PCBs in the UK which reached a maximum in 1968 (Sanders et al. 1992). While UK production ended in 1977, PCB fluxes to Lochnagar continued to be elevated in to the 1990s reflecting European and global emissions. The predicted global PCB emissions from Breivik et al. (2002) for the period 1935-1970 parallel the increase at Lochnagar even more closely than UK production (Figure 4f). However, the predicted decline in global emissions is not reflected closely in horizons dated to the post-1970s. Continued redistribution from shallower sediments as well as inputs of previously deposited PCB residues from the watershed could account for this hysteresis. Jeremiason et al. (1999) concluded that the watersheds of two small, forested lakes were the main source of PCB inputs i.e., from organic material and slow leaching. Rose et al. (2004) concluded that the lack of response in the Lochnagar sediment record to declines in metal deposition was due to a continuing input of previously deposited metals from the catchment.

The historical trend of PCBs in the Lochnagar core agrees well with results from other dated cores in the UK and northern/central Europe although fluxes are generally lower. In a study of PCB deposition to Esthwaite Water maximum ΣPCB fluxes

(60000 ng m^{-2} yr^{-1}) were observed in slices dated to the late-1960s and there was an approximately 50% decline in fluxes between 1970 and 1990 (Gevao et al. 1997). While the timing of onset and decline of PCBs is similar at Lochnagar and Esthwaite Water the fluxes are much higher at the latter site. However, as noted for DDT, this lake also has municipal waste inputs from a rural community. A core from Loch Ness, a large lake to the northwest of Lochnagar, had maximum ΣPCBs in horizons dated to 1968 and a maximum flux of 5500 ng m^{-2} yr^{-1} (Sanders et al. 1993) which is similar to the maximum ΣPCB flux of 6800 ng m^{-2}yr^{-1} in Lochnagar. A peat core from an ombrotrophic bog in a semi-rural location of northwest England (near Liverpool and Manchester) also showed the onset of PCBs in horizons dated to 1932 - 1945 (Sanders et al. 1995) and a maximum flux of 13000 ng m^{-2} yr^{-1} in 1964 probably related to its proximity to urban areas. Two soil cores from undisturbed sites in rural Scotland had ΣPCB inventories of 43 - 46000 ng m^{-2} while soils on the Rothamsted research station near London had similar loads of 35 - 45000 ng m^{-2} (Cousins et al. 1999). By comparison the ΣPCB inventory was 285000 ng m^{-2} in Lochnagar. Of course, the two inventories are not entirely comparable because of particle focusing in Lochnagar which is not a factor for undisturbed soils. Grimalt et al. (2004) also noted 3-20 fold higher concentrations of ΣPCBs in lake sediments compared to soils within the catchments of two alpine lakes. Their results suggested little transfer of soil residues from the catchments to the lakes. Sediment focussing in Lochnagar is discussed further in Chapter 8 (Rose this volume).

ΣPCB concentrations in surface horizons of three remote lakes in northern (68-69° N) Finland ranged from 1.9 - 3.4 ng g^{-1} (Vartiainen et al. 1997), well below, those in the Lochnagar surface (54 ng g^{-1}). The PCB profiles in dated sediment cores collected in the three lakes show post-1950 deposition of PCBs with maxima in the 1970s. Söderström et al. (2002), in their study of 100 Swedish reference lakes, found ΣPCB$_7$ concentrations (sum of seven congeners) were higher in southwest Sweden (10 - 40 ng g^{-1}) than in northern lakes (0.5 - 2.5 ng g^{-1}). ΣPCB$_7$ in the Lochnagar surface horizon was 11.5 ng g^{-1}. Grimalt et al. (2004) determined the same seven congeners in four, high altitude, alpine lakes (three in the Tatra Mountains of Slovakia). Surface ΣPCB$_7$ concentrations ranged from 5.4 to 44 ng g^{-1} in the four lakes, which is similar to that in Lochnagar. Maximum ΣPCB$_7$ in Lake Ladove, the high resolution core, was found in horizons dated to about 1990. Similar to the Lochnagar PCB profile, the trichloro-biphenyl (CB28) and tetrachlorobiphenyl (CB52), showed higher variability in deposition compared with the penta-heptachlorobiphenyls. Sediment inventories of CB153 in eight European high altitude lakes ranged from 170 to 2300 ng m^{-2} (Grimalt et al. 2001) compared to 14500 ng m^{-2} for Lochnagar.

Overall, the onset and maxima for PCBs in Lochnagar are in good agreement with other sediment cores from the UK and northern/central Europe. However, PCB concentrations, fluxes and inventories are generally higher than most other remote locations in continental Europe. Fluxes are lower than a semi-rural peat bog and at Esthwaite Water which may have been impacted by local emissions.

Brominated diphenyl ethers
The sediment record of three major PBDE congeners and total (Σ)PBDEs in Lochnagar is shown in Figure 5 (a and b) for the same core analysed for OCPs and PCBs. Maximum concentrations and fluxes of the three congeners occurred in the horizon

dated to 1990 followed by a sharp decline. BDE47 was the major congener with concentrations ranging from <0.2 ng g^{-1} in horizons dated to 1920 to 21 ng g^{-1} maximum. Low concentrations of BDE66 and BDE153 were also detected in horizons dated to the 1980-90s. Unfortunately the sample for the surface horizon was not available for analysis of PBDEs because the extract was consumed in gas chromatographic determination of PCBs and OCPs. Thus the record only reaches the mid-1990s. BDE47 accounted for 61% of ΣPBDE in the 1990 horizon and up to 100% at lower depths (Figure 5c). A gradual increase in BDE47:ΣPBDE and decline BDE99+100:BDE47 was observed over the period 1990 to 1960. Below this horizon the trend flattens out and is more variable. Concentrations of all three congeners are much lower at these depths the ratios are therefore more uncertain. BDE47, 99 and 100 were found in horizons dated to the 1930s predate the manufacture of PBDEs. All three were found in laboratory blanks at low levels and in extracts of deep sediments (dated to 1890) at concentrations of 0.1-0.2 ng g^{-1}. Results from the lowest horizons were averaged and subtracted from all horizons as an internal blank correction. After correction, BDE99 and 100 were below detection limits (<0.1 ng g^{-1}) in all horizons dated prior to 1960 except for one sample (dated to 1933) which we have to assume was contaminated. BDE47, even after background correction, was present (>0.1 ng g^{-1}) in all horizons except those prior to 1927. This contamination issue has also been noted for PCBs (Alcock et al. 1994) and is an even greater concern for PBDEs because of their use in consumer products like office chairs and computers.

PBDEs have been measured in several dated sediment cores in Europe (Nylund et al. 1992; Zegers et al. 2003), and in remote lakes in Greenland and in North America (Malmquist et al. 2003; Stern et al. 2005). Nylund et al. (1992) analysed a laminated core from the Baltic Sea and found major increases of BDE47 and 99 after 1978 with low concentrations (<0.05-0.24 ng g^{-1}) prior to that. Zegers et al. (2003) detected PBDEs in a core from a small lake in Germany. This core showed maximum BDE47 concentrations in horizons dated in 1979 followed by a slow decline in the 1980-90s. BDE47 and 99 also reached maximum concentrations in horizons of a core from the western Wadden Sea (North Sea) in horizons dated to 1989 and were lower in the mid-1990s (Zegers et al. 2003). A core from Drammenfjord in Norway (a branch of Oslofjord) showed increasing concentrations of BDE47, 99 and 100 to the late 1990s. Overall the results from Nylund et al. and Zegers et al. suggest that tetra- and pentabromo-diphenyl ether deposition peaked in the mid-1990s in western Europe in systems that had significant atmospheric inputs. The PBDE results for Lochnagar are consistent with this trend.

Estimated PBDE production and emissions in the UK also declined in the 1990s (Figure 5b) according to estimates by Alcock et al. (2003). This is likely also the case for western Europe as well since the UK was a major use area and centre for PBDE manufacture. The dates of onset and maximum fluxes of PBDEs in Lochnagar also correspond well to the UK production and emissions curve.

Results for PBDE in North American and Greenland sediment cores generally show maximum PBDE fluxes and concentrations in surface horizons, i.e., dated to the late 1990s (Malmquist et al. 2003; Song et al. 2004; Stern et al. 2005). Arctic sediments studied by Malmquist et al. (2003) in western Greenland and Stern et al. (2005) in the Canadian arctic had mainly BDE47 at very low concentrations 0.001 - 0.25 ng g^{-1}. These concentrations are very near or below limits of detection given background

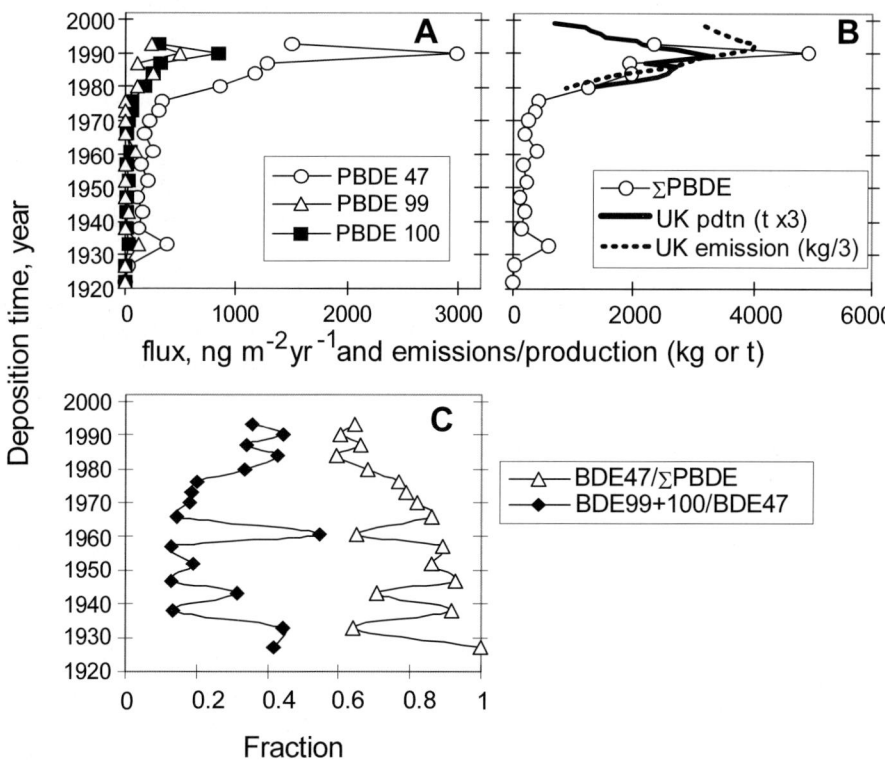

Figure 5. Comparison of sediment records of (a) PBDE congeners 47, 99 and 100 and (b) ΣPBDE fluxes, emissions and production in the UK from Alcock et al. (2003) and (c) Fraction of BDE47 to ΣPBDE and to sum of BDE99 and 100. The same scale is used for fluxes and emissions/production data. The latter have, in some cases, been scaled to fit.

contamination and both studies noted that unexplained PBDE residues were present even after blank correction. Song et al. (2004) measured PBDEs in three cores from central Lake Superior, a system which receives inputs mainly from atmospheric deposition, and found maximum concentrations (6 - 25 ng g^{-1}) in surface horizons dated to the late-1990s. This corresponds well to estimated North American emissions of BDE47 which were predicted to continue to increase through the 1990s (Alcock et al. 2003).

Summary

- The deposition of persistent organohalogen compounds is recorded in Lochnagar sediment just as it is for metals and fly-ash particles. Indeed, Lochnagar appears to be very representative of the regional and global emissions of POPs such as PCBs, DDT, HCB, HCH and toxaphene as well as substances with similar properties to POPs such as the PBDEs.

- Unlike other pollutants, however, most of the deposition has occurred from the mid-1950s onwards paralleling the huge expansion of organohalogen chemical manufacturing and use in that era.
- There is very good agreement between historical trends in POPs contamination, as recorded in the Lochnagar sediment, and predicted UK, European and global emissions or estimated production. The agreement is best for the onset of appearance of the chemicals and maximum fluxes.
- Like the heavy metals (Yang et al. 2002; Rose et al. 2004) the sediments do not show sharp decreases in concentrations or fluxes based on predicted emissions following removal of POPs from use in the 1970s. This is particularly the case for the more volatile compounds, HCB, HCH isomers, dieldrin, and mono-pentachlorobiphenyls and reflects continued recycling of these compounds.
- Sweetman and Jones (2000) noted a significant decline of all PCB congeners in air at their sampling site near Lancaster over the period 1992-1998 while deposition at Lochnagar remained the same or increased over approximately the same time period. Release of previously deposited pollutants stored in the catchment soils and vegetation may explain this. Further work would be needed to develop a mass balance for Lochnagar by measuring concentrations of POPs in water and air, as well as in inflowing waters, to fully address this question.
- As attention turns from legacy chemicals to new substances having properties in common with the chemicals on the POPs list, further analysis of Lochnagar sediments could provide 'ground truthing' of predicted deposition and temporal trends. Alterations of inputs of selected substances, particularly combustion related compounds such as PAHs and chlorinated dioxins/furans due to climate warming or to efforts to reduce CO_2 emissions will also be of interest for future work at Lochnagar.

Acknowledgements

We thank Sean Backus (Environment Canada, Burlington ON) for providing the PBDE data and for his crucial role in all other POPs analysis of the core. We thank L. J. Turner for the sediment core dating, and S. Cagampan (both of Environment Canada, Burlington) for extraction and cleanup of sediments prior to GC-MS analysis, and Heidi Karlsson (now with Biovitrum, Uppsala Sweden) for help with data interpretation and chlorobornane analysis.

References

Alcock R.E., Halsall C.J., Harris C., Johnston A.E., Lead W.A., Sanders G. and Jones K.C. 1994. Contamination of environmental samples prepared for PCB analysis. Environ. Sci. Technol. 28: 1838–1842.

Alcock R.E., Sweetman A.J., Prevedouros K. and Jones K.C. 2003. Understanding levels and trends of BDE-47 in the UK and North America: An assessment of principal reservoirs and source inputs. Environ. Internat. 29: 691–698

Berge E. (ed.) 1997. Transboundary air pollution in Europe. MSC-W Status Report 1997. Part 2; Numerical Addendum to emissions, dispersion and trends of acidifying and eutrophying agents. EMEP/MSC-W Report 1/97; MSC-W, Oslo, Norway. 140 pp.

Blais J.M. and Muir D.C.G. 2001. Paleolimnological methods and applications for persistent organic pollutants. In: Last W.M and Smol J.P. (eds.) Tracking environmental change using lake sediments. Volume 2: Physical and Geochemical methods. Kluwer Academic Publishers, Dordrecht. pp. 271–298.

Blais J.M., Schindler D.W., Muir D.C., Sharp M., Donald D., Lafreniere M., Braekevelt E. and Strachan W.M. 2001. Melting glaciers: a major source of persistent organochlorines to subalpine Bow Lake in Banff National Park, Canada. Ambio 30: 410–5.

Breivik K., Pacyna J.M. and Münch J. 1999. Use of α-, β- and γ-hexachlorocyclohexane in Europe 1970. Sci. Tot. Environ. 239: 151–153.

Breivik K. and Wania F. 2002. Evaluating a model of the historical behavior of two hexachlorocyclohexanes in the Baltic Sea environment. Environ. Sci. Technol. 36: 1014–1023.

Buser H-R. and Müller M.D. 1995. Isomer and enantioselective degradation of hexachlorocyclohexane isomers in sewage sludge under anaerobic conditions. Environ. Sci. Technol. 29: 664–672.

Chen I.M., Chang F.C. and Wang Y.S. 2001. Correlation of gas chromatographic properties of chlorobenzenes and polychlorinated biphenyls with the occurrence of reductive dechlorination by untamed microorganisms. Chemosphere 45: 223–229.

Cousins I.T., Gevao B. and Jones K.C. 1999. Measuring and modelling the vertical distribution of semi-volatile organic compounds in soils. I: PCB and PAH soil core data. Chemosphere 39: 2507–2518.

Czuczwa J.M. and Hites R.A. 1984. Environmental fate of combustion generated polychlorinated dioxins and polychlorinated furans. Environ. Sci. Technol. 18: 444–450.

de Geus H-J., Besselink H., Brouwer A., Klungsøyr J., McHugh B., Nixon E., Rimkus G.G., Wester P.G. and de Boer J. 1999. Environmental occurrence, analysis and toxicology of toxaphene compounds. Environ. Health Perspect. 107: 115–144.

Eakins J.D. and Morrison R.T. 1978. A new procedure for determination of lead-210 in lake and marine sediments. Int. J. Appl. Radiat. Isotope. 29: 531–536.

Eisenreich S.J., Capel P.D., Robbins J.A. and Bourbonniere R. 1989. Accumulation and diagenesis of chlorinated hydrocarbons in lacustrine sediments. Environ. Sci. Technol. 23: 1116–1126.

Environment Canada. 1993. Pentachlorobenzene. Priority substances list assessment report. Environment Canada and Health Canada, Ottawa, ON. 39 pp.

Environment Canada. 2005. Categorization and screening of the domestic substances list. Path forward towards ecological prioritization of substances for assessment. Ottawa ON. pp 18.

EU. 1978. Council Directive 79/117/EEC of 21 December 1978 prohibiting the placing on the market and use of plant protection products containing certain active substances. Official Journal L 033 , 08/02/1979 European Commission, Brussels, p. 36–40.

Eurochlor. 2002. Eurochlor risk assessment for the marine environment. Hexachlorobutadiene (HCBD). OSPARCOM Region – North Sea. Eurochlor, Brussels. Belgium. 35 pp.

Fernández P., Vilanova R.M., Martinez C., Appleby P. and Grimalt J.O. 2000. The historical record of atmospheric pyrolytic pollution over Europe registered in the sedimentary PAH from remote mountain lakes. Environ. Sci. Technol. 34: 1906–1913.

Fingerling G., Hertkorn N. and Parlar H. 1996. Formation and spectroscopic investigation of two hexachlorobornanes from six environmentally relevant toxaphene components by reductive dechlorination in soil under anaerobic conditions. Environ. Sci. Technol. 30: 2984–2992.

Fisk A.T., Cymbalisty C.D., Tomy G.T., Stern G.A. and Muir D.C.G. 1999. Octanol-water partition coefficients of toxaphene congeners. Chemosphere 39: 2549–2562.

Fox W.M., Connor L., Copplestone D., Johnson M.S. and Leah R.T. 2001. The organochlorine contamination history of the Mersey Estuary, UK, revealed by analysis of sediment cores from salt marshes Mar. Environ. Res. 51: 213–227

Furlong E.T., Cessar L.R. and Hites R.A. 1987. Accumulation of polycyclic aromatic hydrocarbons in acid sensitive lakes. Geochim. Cosmochim. Acta 51: 2965–2975.

Gevao B., Hamilton-Taylor J., Murdoch C., Jones K.C., Kelly M. and Tabner B.J. 1997. Depositional time trends and remobilization of PCBs in lake sediments. Environ. Sci. Technol. 31: 3274–3280.

Glassmeyer S.T., Shanks K.E. and Hites R.A. 1999. Automated toxaphene quantitation by GC-MS. Anal. Chem. 71: 1448–1453.

Glew J.R. 1988. A portable extruding device for close interval sectioning of unconsolidated core samples. J. Paleolimnol. 1: 235–239.

Glotfelty D.E., Williams G.H., Freeman H.P. and Leech M.M. 1990. Regional atmospheric transport and deposition of pesticides in Maryland. Long range of transport of pesticides. D. A. Kurtz (ed.) Lewis Publishers. Chelsea, MI. pp 199–221.

Grimalt J.O., Fernández P., Berdie L., Vilanova R.M., Catalan J., Psenner R., Hofer R., Appleby P.G., Rosseland B.O., Lien L., Massabuau J.C. and Battarbee R.W. 2001. Selective trapping of organochlorine compounds in mountain lakes of temperate areas. Environ. Sci. Technol. 35: 2690–2697.

Grimalt J.O., van Drooge B.L., Ribes A., Vilanova R.M., Fernández P. and Appleby P. 2004. Persistent organochlorine compounds in soils and sediments of European high altitude mountain lakes. Chemosphere 54: 1549–1561.

Heinisch E., Kettrup A., Jumar S., Wenzel-Klein S., Stechert J., Hartmann P. and Schaffer P. 1994. Schadstoff Atlas Ost-Europa. Chapter 2.9 (Ed), Ecomet/Landsberg. Lech, Germany pp 39–47.

Hites R.A., Laflamme R.E. and Farrington J.W. 1977. Polycyclic aromatic hydrocarbons in recent sediments: The historical record. Science 198: 829–831.

Howdeshell M.J. and Hites R.A. 1996. Historical input and degradation of toxaphene in Lake Ontario sediment. Environ. Sci. Technol. 30: 220–224.

Jeremiason J.D., Eisenreich S.J., Paterson M.J., Beaty K.G., Hecky R. and Elser J.J. 1999. Biogeochemical cycling of PCBs in lakes of variable trophic status: A paired-lake experiment. Limnol. Oceanogr. 44: 889–902.

Kaiser K.L.E. 1978. Pesticide Report: The rise and fall of mirex. Environ. Sci. Technol. 12: 520–528.

Kleçka G., Boethling R., Franklin J., Grady L., Graham D., Howard P., Kannan K., Larson R., Mackay D., Muir D. and van de Meent D. 2000. (eds). Evaluation of persistence and long-range transport of organic chemicals in the environment. SETAC Books. Pensacola FL 400 pp

Laflamme R.E. and Hites R.A. 1978. The global distribution of polycyclic aromatic hydrocarbons in recent sediments. Geochim. Cosmochim. Acta 42: 289–303.

Lecloux A. 2003. Hexachlorobutadiene. Sources, environmental fate and risk characterisation. Dossier prepared by Envicat Consulting. Euro Chlor. 31 pp.

Li X.D., Mai B.X., Zhang G., Sheng G.Y., Fu J.M., Pan J.M., Wai O.W.H. and Li Y.S. 2001. Distribution of organochlorine pesticides in a sediment profile of the Pearl River estuary. Bull. Environ. Contam. Toxicol. 67: 871–880.

Li Y.F. 1999. Global technical hexachlorocyclohexane usage and its contamination consequences in the environment: From 1948 to 1997. Sci. Total Environ. 232: 121–158.

Li Y.F. 2001. Toxaphene in the United States: 1. Usage gridding. J. Geophys. Res. 106D: 17919–17928.

Li Y.F. 2003. Global organochlorine pesticide emission inventories. Synopsis of research conducted under the 2001–2003 Northern Contaminants Program, Indian and Northern Affairs Canada. Ottawa, ON. pp 172–181.

Li Y.F. and Bidleman T. 2003. Usage and emissions of organochlorine pesticides. Sources, occurrence, trends and pathways in the physical environment In: Bidleman T., Macdonald R. and Stow J. (eds.) Canadian Arctic contaminants assessment report II. Indian and Northern Affairs Canada. Ottawa ON. pp 49–70.

Li Y.F. and Macdonald R.W. 2005. Sources and pathways of selected organochlorine pesticides to the Arctic and the effect of pathway divergence on HCH trends in biota: A review. Sci. Total Environ. 342: 87–106.

Mackay D., Shui W.Y., Ma K.C. and Lee S.C. 2006. Handbook of physical-chemical properties and environmental fate for organic chemicals. 2nd Edition, CRC Press, Boca Raton, FL. 4216 pp.

Majewski H.S. and Capel P.D. (eds). 1995. Pesticides in the atmosphere: Distribution, trends, and governing factors. CRC Press, Boca Raton, FL. 215 pp.

Malmquist C., Bindler R., Marrenberg I., van Bavel B., Karlsson E., Anderson N.J. and Tysklind M. 2003. Time trends of selected persistent organic pollutants in lake sediments from Greenland. Environ. Sci. Technol. 37: 4319–4324.

McConnell L.L., Lenoir J.S., Datta S. and Seiber J.N. 1998. Wet deposition of current-use pesticides in the Sierra Nevada mountain range, California, USA. Environ. Toxicol. Chem. 17: 1908–1916.

McVeety B.D. and Hites R.A. 1988. Atmospheric deposition of polycyclic aromatic hydrocarbons to water surfaces: A mass balance approach. Atmos. Environ. 22: 511–536.

Miskimmin B.M., Muir D.C.G., Schindler D.W., Stern G.A. and Grift N.P. 1995. Chlorobornanes in sediments and fish 30 years after toxaphene treatment of lakes. Environ. Sci. Technol. 29: 2490–2495.

Mudroch A., Allan R.J. and Joshi S.R. 1992. Geochemistry and organic contaminants sediments of Great Slave Lake, Northwest Territories, Canada. Arctic 45: 10–19.

Muir D.C.G., Grift N.P., Lockhart W.L., Wilkinson P., Billeck P.N. and Brunskill G.N. 1995. Spatial trends and historical profiles of organochlorine pesticides in Arctic lake sediments. Sci. Tot. Environ. 160/161: 447–457.

Muir D.C.G., Omelchenko A., Grift N.P., Savoie D.A., Lockhart W.L., Wilkinson P. and Brunskill G.J. 1996. Spatial trends and historical deposition of polychlorinated biphenyls in Canadian mid-latitude and Arctic lake sediments. Environ. Sci. Technol. 30: 3609–3617.

Muir D., Karlsson H., Kohli M., Wang X., Backus S., Lockhart L. and Wikinson. P. 2000. Historical profiles of toxaphene congeners in dated sediment cores collected near two pulp mills. Organohalogen Compd. 47: 256–259.

Muir D.C.G., Teixeira C. and Wania F. 2004. Empirical and modelling evidence of regional atmospheric transport of current-use pesticides. Environ. Toxicol. Chem. 23: 2421–2432.

Navarro S., Barba A., Segura J.C. and Oliva J. 2000. Disappearance of endosulfan residues from seawater and sediment under laboratory conditions. Pest. Manag. Sci. 56: 849–854.

Newland L.W., Chesters G. and Lee G.B. 1969. Degradation of g-BHC in simulated lake impoundments as affected by aeration. J. Water Pollut. Cont. Fed. 41: 174 183.

Nylund K., Asplund L., Jansson B., Jonsson P., Litzen K. and Sellstrom U. 1992. Analysis of some polyhalogenated organic pollutants in sediment and sewage sludge. Chemosphere 24: 1721–1730.

Oliver B.G., Charlton M.N. and Durham R.W. 1989. Distribution, redistribution, and geochronology of polychlorinated biphenyl congeners and other chlorinated hydrocarbons in Lake Ontario sediments. Environ. Sci. Technol. 23: 200–208.

Oliver B.G. and Nicol K.D. 1982. Chlorobenzenes in sediments, water, and selected fish from Lakes Superior, Huron, Erie, and Ontario. Environ. Sci. Technol. 16: 532–536.

Paasivirta J. 2000. Vol. 3K: New types of persistent halogenated compounds. The handbook of environmental chemistry series: Anthropogenic compounds. Paasivirta J. (ed.) Springer. Vol. 3. 359 pp

Pacyna J.M., Breivik K., Münch J. and Fudala J. 2003. European atmospheric emissions of selected persistent organic pollutants, 1970–1995. Atmos. Environ. 37: 119–131.

Pavlostathis S.G. and Prytula M.T. 2000. Kinetics of the sequential microbial reductive dechlorination of hexachlorobenzene. Environ. Sci. Technol. 34: 4001–4009.

Pearson R.F., Swackhamer D.L., Eisenreich S.J. and Long D.T. 1997. Concentrations, accumulations, and inventories of toxaphene in sediments of the Great Lakes. Environ. Sci. Technol. 31: 3523–3529.

Pearson R.F., Swackhamer D.L., Eisenreich S.J. and Long D.T. 1998. Atmospheric inputs of polychlorinated dibenzo-p-dioxins and dibenzofurans to the Great Lakes: Compositional comparison of PCDD and PCDF in sediments. J. Great Lakes Res. 24: 65–82.

Pesticide Action Network. 2002. Pesticides Database. www.pesticideinfo.org/Search _ Chemicals .jsp.

Peterson S.M. and Batley G.E. 1993. The fate of endosulfan in aquatic ecosystems. Environ. Pollut. 82: 143–152.

Prevedouros K., MacLeod M., Jones K.C. and Sweetman A.J. 2004. Modelling the fate of persistent organic pollutants in Europe: parameterisation of a gridded distribution model. Environ. Pollut. 128: 251–261

Rapaport R.A. and Eisenreich S.J. 1988. Historical atmospheric inputs of high molecular weight chlorinated hydrocarbons to eastern North America. Environ. Sci. Technol. 22: 931–941.

Rawn D.F.K. and Muir D.C.G. 1999. Sources of chlorpyrifos and dachtal to a small Canadian prairie watershed. Environ. Sci. Technol. 33: 3317–3323.

Rose C.L., Rose N.L., Harlock S. and Fernandes A. 1997. An historical record of polychlorinated dibenzo-p-dioxin (PCDD) and polychlorinated dibenzofuran (PCDF) deposition to a remote lake site in north-west Scotland, UK. Sci. Total Environ. 198: 161–173.

Rose N.L. This volume. Chapter 8. The sediments of Lochnagar: Distribution, accumulation and composition. In: Rose N.L. (ed.) 2007. Lochnagar: The natural history of a mountain lake. Springer. Dordrecht.

Rose N.L., Backus S., Karlsson H. and Muir D.C.G. 2001. An historical record of toxaphene and its congeners in a remote lake in western Europe. Environ. Sci. Technol. 35: 1312–9.

Rose N.L., Monteith D., Kettle H., Thompson R., Yang H. and Muir D. 2004. A consideration of potential confounding factors limiting chemical and biological recovery at Lochnagar, a remote mountain loch in Scotland. J. Limnol. 63: 63–76.

Saleh M.A. 1991. Chemistry, biochemistry, toxicity and environmental fate. Rev. Environ. Contam. Toxicol. 118: 1–115.

Sanders G., Jones K.C. and Hamilton-Taylor J. 1992. Historical inputs of polychlorinated biphenyls and other organochlorines to a dated lacustrine sediment core in rural England. Environ. Sci. Technol. 26: 1815–1821.

Sanders G., Jones K.C. and Shine A.J. 1993. The use of a sediment core to reconstruct the historical input of contaminants to Loch Ness: PCBs and PAHs. Scot. Natur. 105: 87–111.

Sanders G., Jones K.C., Hamilton-Taylor J. and Dorr H. 1995. PCB and PAH fluxes to a dated UK peat core. Environ. Pollut. 89: 17–24.

Scheringer M., Wegmann F., Fenner K. and Hungerbuhler K. 2000. Investigation of the cold condensation of persistent organic pollutants with a global multimedia fate model. Environ. Sci. Technol. 34: 1842–1850.

Schwarzbauer J., Littke R. and Weigelt V. 2000. Identification of specific organic contaminants for estimating the contribution of the Elbe River to the pollution of the German Bight. Org. Geochem. 31: 1713–1731.

Shen L. and Wania F. 2005. Compilation, evaluation, and selection of physical-chemical property data for organochlorine pesticides. J. Chem. Eng. Data 50: 742–768.

Söderström M., Asplund L., Kylin H. and Sundin P. 2002. Organochlorine contaminants in sediment from 100 Swedish Lakes - relation to lake type and location. In: Local and Global Contaminants in Swedish waters: Studies on PCBs, DDTs, 4,5,6-trichlorolguaiacol and their transformation products in fish and sediments. Department of Environmental Chemistry. Stockholm, Sweden, Stockholm University. Ph.D. thesis.

Song W., Ford J.C., Li A., Mills W.J., Buckley D.R. and Rockne K.J. 2004. Polybrominated diphenyl ethers in the sediments of the great lakes. 1. Lake Superior. Environ. Sci. Technol. 38: 3286–3293.

Stern G.A., Braekevelt E., Helm P.A., Bidleman T.F., Outridge P.M., Lockhart W.L., McNeeley R., Rosenberg B., Ikonomou M.G., Hamilton P., Tomy G.T. and Wilkinson P. 2005. Modern and historical fluxes of halogenated organic contaminants to a lake in the Canadian arctic, as determined from annually laminated sediment cores. Sci. Tot. Environ. 342: 223–243.

Strandberg B., van Bavel B., Bergqvist P-A., Broman D., Ishaq R., Näf C., Pettersen H. and Rappe C. 1998. Occurrence, sedimentation, and spatial variations of organochlorine contaminants in settling particulate matter and sediments in the northern part of the Baltic Sea. Environ. Sci. Technol. 32: 1754–1759.

Swackhamer D.L. and Armstrong D.E. 1986. Estimation of the atmospheric and non-atmospheric contributions and losses of polychlorinated biphenyls for Lake Michigan on the basis of sediment records of remote lakes. Environ. Sci. Technol. 20: 879–883.

Swackhamer D.L. and Hites R.A. 1988. Occurrence and bioaccumulation of organochlorine compounds in fishes from Siskiwit Lake, Isle Royale, Lake Superior. Environ. Sci. Technol. 22: 543–548.

Sweetman A.J. and Jones K.C. 2000. Declining PCB concentrations in the U.K. atmosphere: Evidence and possible causes. Environ. Sci. Technol. 34: 863–869.

Syracuse Research Corporation 2002. Interactive PhysProp Database Demo. http://exc.syrres.com/interkow/physdemo.htm

Turner L.J. 1998. ^{210}Pb dating of lacustrine sediments from Lochnagar, Scotland (Core 207). Report 98-5. National Water Research Institute, Burlington ON. 22 pp.

UNECE. 2004. Technical review report on lindane. Reports on substances scheduled for re-assessments under the UNECE POPs protocol. United Nations Economic Commission for Europe, Geneva, Switzerland. 38 pp.

UNEP. 2001. Final act of the conference of plenipotentiaries on the Stockholm convention on persistent organic pollutants. United Nations Environment Program. Geneva, Switzerland. 44 pp.

UNEP. 2002. Regionally based assessment of persistent toxic substances. Europe regional report. United Nations Environment Programme, UNEP Chemicals, Geneva, Switzerland. 158 pp.

van de Plassche E., Schwegler A., Rasenberg M. and Schouten G. 2001. Pentachlorobenzene. Report available from the UNECE. United Nations Economic Commission for Europe, Geneva, Switzerland. 19 pp.

van Dijk H.F.G. and Guicherit R. 1999. Atmospheric dispersion of current use pesticides. A review of the evidence from monitoring studies. Wat. Air Soil Pollut. 115: 21–70.

van Metre P.C. and Mahler B.J. 2004. Contaminant trends in reservoir sediment cores as records of influent stream quality. Environ. Sci. Technol. 38: 2978–2986.

Vartiainen T., Maanio J., Korhonen M., Kinnunen K. and Strandman T. 1997. Levels of PCDD, PCDF and PCB in dated lake sediments in subarctic Finland. Chemosphere 34: 1341–1350.

Veith G.D. and Lee G.F. 1971. Water chemistry of toxaphene - role of lake sediments. Environ. Sci. Technol. 5: 230–234.

Voldner E.C. and Li Y.F. 1993. Global usage of toxaphene. Chemosphere 27: 2073–2078.

Wania F. and Dugani C.B. 2003. Assessing the long-range transport potential of polybrominated diphenyl ethers: A comparison of four multimedia models. Environ. Toxicol. Chem. 22: 1252–1261.

Weston D.P., You J. and Lydy M.J. 2004. Distribution and toxicity of sediment-associated pesticides in agriculture-dominated water bodies of California's Central Valley. Environ. Sci. Technol. 38: 2752–2759.

WHO. 1994. Environmental health criteria 156 hexachlorobutadiene. International Programme on Chemical Safety (IPCS). World Health Organization Geneva, Switzerland. 115pp

Willett K.L., Ulrich E.M. and Hites R.A. 1998. Differential toxicity and environmental fates of HCH isomers. Environ. Sci. Technol. 32: 2197–2207.

Wong C.S., Sanders G., Engstrom D.R., Long D.T., Swackhamer D.L. and Eisenreich S.J. 1995. Accumulation, inventory, and diagenesis of chlorinated hydrocarbons in Lake Ontario sediments. Environ. Sci. Technol. 29: 2661–2672.

Wu W.Z., Xu Y., Schramm K-W. and Kettrup A. 1997. Study of sorption, biodegradation and isomerization of HCH in stimulated sediment/water system. Chemosphere 9: 1887–1894.

Yang H., Rose N.L. and Battarbee R.W. 2002. Distribution of some trace metals in Lochnagar, a Scottish mountain lake ecosystem and its catchment. Sci. Total Environ. 285: 197–208.

Zegers B.N., Lewis W.E., Booij K., Smittenberg R.H., Boer W., de Boer J. and Boon J.P. 2003. Levels of polybrominated diphenyl ether flame retardants in sediment cores from Western Europe. Environ. Sci. Technol. 37: 3803–3807.

17. TEMPORAL AND SPATIAL PATTERNS OF SPHEROIDAL CARBONACEOUS PARTICLES (SCPs) IN SEDIMENTS, SOILS AND DEPOSITION AT LOCHNAGAR

NEIL L. ROSE (nrose@geog.ucl.ac.uk) and HANDONG YANG
Environmental Change Research Centre
University College London
Pearson Building, Gower Street
London WC1E 6BT
United Kingdom

Key words: atmospheric deposition, combustion products, fly-ash, pollution transport, spheroidal carbonaceous particles

Introduction

Spheroidal carbonaceous particles (SCPs) (Figure 1) are produced by industrial, high temperature combustion of fossil-fuels such as coal and oil. They are a component of fly-ash, the particulate matter emitted to the atmosphere with flue gases, and hence their dispersal and deposition are controlled by the meteorological conditions the dispersing plume encounters. SCPs are not produced by any natural process and consequently their presence in the environment represents an unambiguous indication of atmospherically deposited contamination from sources such as the electricity generation and other industries (Rose 2001). However, whilst there is no direct evidence for SCPs being harmful to biota *per se*, they are considered an important means by which toxic pollutants may be transported and deposited. Trace metals (e.g., Davison et al. 1974; Coles et al. 1979; Seigneur et al. 2005) and persistent organic pollutants (POPs) including polycyclic aromatic hydrocarbons (PAHs) (Griest and Tomkins 1984; Wey et al. 1998; Ghosh et al. 2000), polychlorinated biphenyls (PCBs) (Bucheli and Gustafsson 2003; Persson et al. 2005) and dioxins and furans (PCDD/Fs) (Ohsaki et al. 1995; Persson et al. 2002) may be adsorbed to the surfaces of the particles in the emitted plume or during transport. Such transport mechanisms and coincident sources explain the similarities in spatial and temporal trends between SCPs and these other pollutants in lake sediment studies (e.g., Broman et al. 1990; Boyle et al. 1999; Fernández et al. 2002).

N.L. Rose (ed.), Lochnagar: The Natural History of a Mountain Lake
Developments in Paleoenvironmental Research, 403–423.
© *2007 Springer.*

In environmental research, SCPs have been used in three main ways. First, as a means to identify and quantify trends in atmospherically deposited contamination, for example as concentrations in bulk deposition monitoring (Rose 2001) or as temporal trends in lake sediment (e.g., Renberg and Wik 1985a; Rose et al. 2002; Rose et al. 2003) or soil cores (Yang et al. 2001). Second, as a means of providing a chronology for lake sediment cores (Renberg and Wik 1984; 1985b; Rose et al. 1995; Rose and Appleby 2005) and third, as a method of source apportionment for deposited contamination, either simply as an indicator of fossil-fuel derived pollutants or by using the chemistry of the SCPs themselves to differentiate them into their original fuel sources (Rose et al. 1996; 1999).

This chapter deals with SCPs in the deposition, soils and sediments of Lochnagar as a means to determine temporal trends in the contamination of this remote site; assess the relative storage of SCPs in loch sediments and catchment soils; explore the possibilities for transfer between the catchment and lake; and consider how climate change may affect the movement of SCPs and the toxic pollutants adsorbed to their surfaces within this mountain ecosystem.

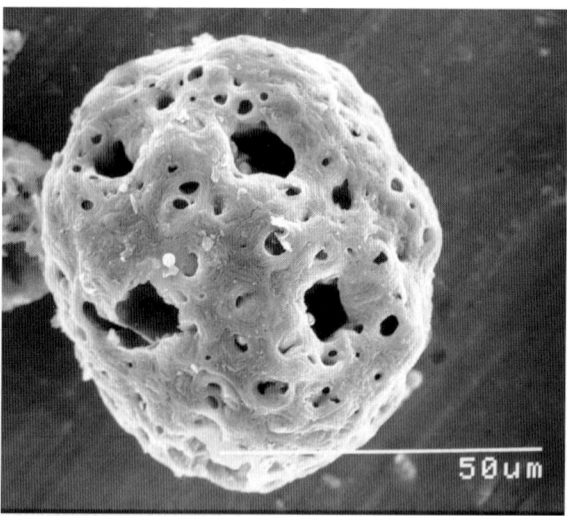

Figure 1. Scanning electron micrograph of a spheroidal carbonaceous particle (SCP) derived from the high temperature combustion of fossil-fuels. (Micrograph: Neil Rose)

Atmospheric deposition of SCPs

Bulk deposition

Sampling of bulk deposition for SCP analysis at Lochnagar began in August 1996 as part of the EU funded MOLAR project and has continued to the present. Data up to the end of 2003 are included in this chapter. NILU (Norwegian Institute of Air Research)-type bulk deposition collectors were used and, where weather conditions permitted, were sampled weekly until November 1997. Since then, samples have been taken every two weeks. The collector is located to the east of the loch near the Automatic Weather Station c. 15m above the loch and 2m above ground level. The position is shown on Figure 7. For SCP analysis, known volumes (as large as possible, but typically c. 2 litres) of the bulk deposition sample were filtered through Whatman GF/C filters. These were then digested and analysed as described on the CARBYNET website (http://www.ecrc.ucl.ac.uk/carbynet/). SCP concentrations can then be expressed as number of SCPs per litre of deposition, or more usefully given the variable length of sampler exposure, transformed to fluxes using site specific rainfall data from the Automatic Weather Station (Chapter 5: Thompson et al this volume).

Figure 2a shows the record of SCP flux in bulk deposition at Lochnagar from 1997 through to the end of 2003. There is considerable variability in the record although there are a higher number of elevated SCP fluxes in the earlier period compared to later in the record. Interestingly, despite the earlier samples being at a higher temporal resolution and thus generally being of lower volume, there are fewer samples below the limit of detection suggesting a more continuous input of SCPs to the site at this time. In the latter half of the record individual high fluxes are separated by periods where SCP

a)

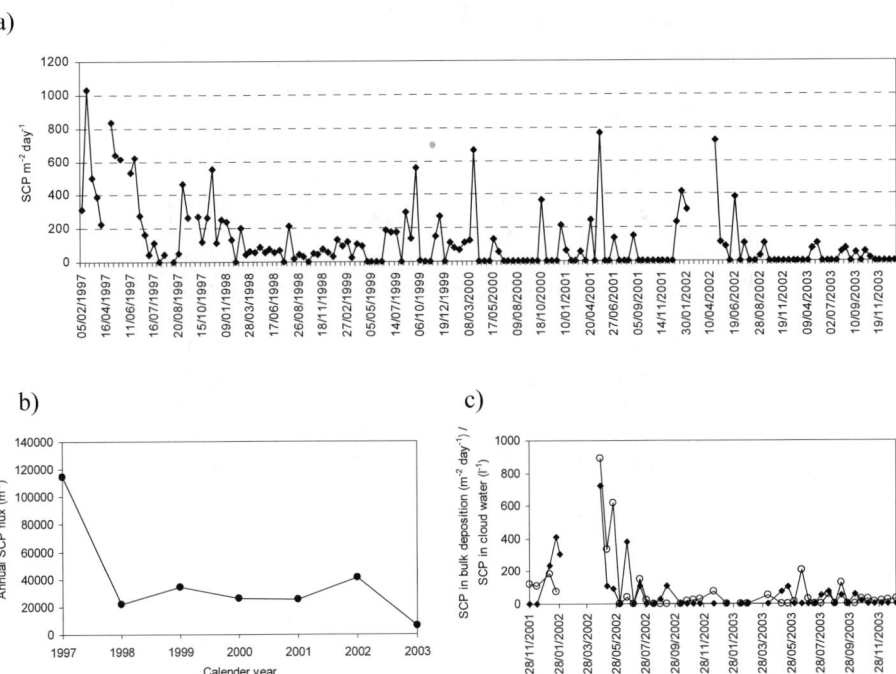

b)   c)

Figure 2. (a) SCP concentrations in bulk rainwater samples (as number of SCP m^{-2} day^{-1}); (b) The same data converted to annual fluxes (m^{-2} yr^{-1}) using rainfall data from the Lochnagar automatic weather station. (c) Expanded latter section of (a) with cloudwater data (○) for the same period (in SCP l^{-1}). The gap represents a period when the collectors were blown over by strong winds.

concentrations, and hence fluxes, fall below the limit of detection suggesting a more episodic input. Converting these data into annual (calendar year) fluxes (Figure 2b) translates the high values from 1997 into a high annual flux. This is followed by a stable period between 1998 and 2002 where annual fluxes vary little before declining again in 2003. Certainly this is in agreement with expected declining trends, although, as 1997 was the first full year of sampling, it is uncertain whether this was an unusually high year or the final year of a previous period of more elevated inputs. Future monitoring will confirm whether the lower flux in 2003 is a further permanent decline in SCP inputs or just a low value in an otherwise stable period since 1998.

Little similar long-term monitoring of SCPs in bulk deposition has been undertaken, either in the UK or elsewhere, although weekly sampling for SCPs over an 18 month period from August 1996 was undertaken at four other European mountain lakes as part of the EU funded MOLAR project (Rose et al. 2002). These four sites represented other studied mountain regions: Øvre Neådalsvatn (mid-Norway); Estany Redo (Spanish Pyrenees); Gossenköllesee (Austrian Alps) and Starolesnianske Pleso (Slovakian Tatras). Comparing the SCP deposition at these sites for the calendar year of 1997 showed that Øvre Neådalsvatn had considerably lower inputs than the other sites (c. 5700 m^{-2}), similar to the lowest Lochnagar annual value for 2003 (Figure 2b). The other sites, in a band across central Europe, showed annual fluxes of between 14000 and 31000 m^{-2}, considerably lower than the figure for Lochnagar over the same period, but comparable with annual Lochnagar fluxes for 1998 – 2002. Unfortunately, monitoring at these four mountain sites was not continued so it is not known whether 1997 was a year of high deposition across the whole of Europe, or whether the reduction in SCP deposition at Lochnagar in 1998 and subsequent years would have shown atmospheric inputs to be equivalent to these other remote European sites.

As SCPs are derived from high temperature fossil-fuel combustion it is interesting to compare trends in SCP deposition with those of deposited acid ions. Rose et al. (2001) undertook this comparison for the mountain lake sites described above, including Lochnagar. Correlations between SCP concentrations and those of the acid ions SO_4^{2-}, NO_3^- and NH_4^+ at Lochnagar were found to be low, but positive, whilst correlations of SCPs with (SO_4^{2-} + NO_3^-) were more positive than for SCPs with (SO_4^{2-} + NO_3^- + NH_4^+) as might be expected given the differing sources of SCPs and the mainly agriculturally derived NH_4^+. The positive correlations between SCPs and NH_4^+ were therefore thought to be an artefact of the high positive correlations of the ions SO_4^{2-} and NO_3^- with NH_4^+, as a result of the formation of the secondary aerosols $(NH_4)_2SO_4$ and $(NH_4)NO_3$, and the positive correlation of SCPs with SO_4^{2-} and NO_3^-.

Occult deposition

As cloudwater in upland UK can demonstrate a tenfold enrichment of fossil-fuel derived elements over bulk rainwater (Wilkinson et al. 1997) a collector was set-up at Lochnagar in October 2002, near the bulk deposition monitoring site, to sample intercepted cloudwater for subsequent chemical and SCP analysis. This cloudwater collector is a passive 'lidded-harp'-type as described in Neal et al. (1997) and shown schematically in Reynolds et al (1996). Briefly, the collector consists of a low density polyethylene-coated frame over which an inverted cone of nylon filaments is stretched. Beneath this cone, a funnel directs the intercepted moisture into a collection bottle. The whole assembly is protected from direct rainfall by a 1.2 m diameter lid mounted on a tubular steel framework, but some rainwater and snow contamination can still occur in windy conditions (Neal et al. 1997). These collectors have been used extensively for cloudwater studies in the UK (e.g., Fowler et al. 1988) and have been used to collect particles in orographic cloud including fly-ash (Crossley 1988). Samples are collected every two weeks, filtered for SCPs and the filters treated as described above for bulk deposition. For the sampling period October 2002 to the end of 2003, cloudwater SCP concentrations were episodic (Figure 2c) with more continuous inputs occurring during autumn and winter and only two samples showing SCP concentrations above detection limit between the end of June and early October 2003. Over the sampling period a cloudwater SCP input of 2970 m^{-2} yr^{-1} was determined (Jennifer Muller and David Fowler, Centre for Ecology and Hydrology, Edinburgh pers. commun.) corresponding to 46.8% of the bulk deposition input for the same period. Therefore, cloud inputs provide significant additional atmospheric deposition of SCPs to Lochnagar and its catchment.

SCPs in sediments and sediment traps

The first sediment core from Lochnagar analysed for SCPs (NAG3) was taken in 1986 using a mini-Mackereth corer (Mackereth 1969) as part of a study on acidification in the Cairngorm region (Jones et al. 1993). Since then, SCPs have been analysed on a number of sediment cores (NAG6, 8 – 23, 25 and 26) and, because of the slow sediment accumulation rate (Chapter 8: Rose this volume), these have all been short sediment cores taken by gravity corer (e.g., Glew 1991). Of these cores, NAG3, 6, 8 and 23 were ^{210}Pb dated.

Simple cylindrical sediment traps are employed for the annual sampling programme of the UK Acid Waters Monitoring Network (UK AWMN). These comprise arrays of three tubes attached to the corners of triangular plastic frames, thus providing triplicate samples. The internal diameter of each tube is 5.2 cm with an aspect ratio (length: diameter) of ≥ 7 to maximise trapping efficiency (Bloesch and Burns 1980; Blomquist and Håkanson 1981). Details of the sampling procedures are given in Rose and Monteith (2005) but at Lochnagar arrays of traps are deployed in the deep water area (> 20m) at c. 1m above the sediment-water interface while an upper trap is located at c. 2m water depth. All SCP analyses of lake sediments and sediment trap material have been undertaken using the method described in Rose (1994) except NAG3 which was analysed using the technique from Renberg and Wik (1984; 1985b).

Temporal trends

The regional diversity of the temporal record of SCPs in lake sediments across the UK is well characterised and now extensively used for providing sediment chronologies (Rose et al. 1995; Rose and Appleby 2005). The historical profile of SCP concentrations from Lochnagar is quite typical for Scotland and shows all the expected features (Figure 3a) associated with documented historical changes in fossil-fuel combustion, industrial development and pollution control (Rose 2001). The first presence of SCPs occurs in the mid-19th century as a result of developments in the industrial use of coal. Initially, this was not for electricity generation as the first public 'power station' in the UK was not commissioned until 1882. However, this early industrial coal combustion was both inefficient and lacking in any emission controls. Thus, the rapid spread of this technology quickly provided numerous point sources and the SCP record begins within a decade or two across the whole of the UK (Rose and Appleby 2005).

The SCP record for the following century at Lochnagar is also typical and shows a slow, but steady increase in concentration resulting from the growth in industrial coal consumption. However, starting in the 1950s, a major increase in SCP concentration is observed at Lochnagar, across the UK and throughout Europe. This is due to two factors. First, a major expansion in the consumption of fossil-fuels at power stations increasing rapidly in size (Laxen 1996) as a result of a dramatic increase in demand for electricity following the end of the Second World War, and second, the availability, for the first time, of cheap fuel oil leading to the development of the first large-scale oil-fired power stations. This rapid increase continued until the late-1970s when an increase in combustion efficiency and the introduction of particle arrestors at source, in addition to the implementation of successively more rigorous control legislation, meant that despite the continued increase in fuel consumption, particulate emissions started to decline. In addition, changes in policy led to a decline in heavy industry which was particularly acute in some areas of central Scotland, emissions from which are known to impact the Lochnagar region (Chapter 13: Rose et al. this volume). This led to a reduction in emissions both directly and indirectly (via a reduced demand for electricity) and was exacerbated by a move away from the use of 'traditional' fuels such as coal and fuel-oil to natural gas. In the lake sediment record, these changes are recorded as a SCP concentration peak and subsequent decline to the sediment surface. This concentration peak occurs in 1978 ± 2 in Lochnagar and is similar throughout northern and central Scotland (Rose and Appleby 2005). Such is the reliability of this temporal record that SCP concentration profiles are widely used for dating lake sediments. In Lochnagar, there is an additional benefit that four cores have been independently dated by ^{210}Pb allowing a reliable calibration. Therefore, there is considerable confidence in the use of SCP profiles to date sediment cores in Lochnagar and this approach has been used to provide dates for cores throughout this volume.

Similarities exist between the historical sediment record of SCPs and that of some atmospherically deposited trace metals such as Hg and Pb (Yang et al. 2002a) and to a lesser extent Cd, Zn and Cu (Yang et al. 2002b). This is most probably due to fossil-fuel combustion also being a major source of these trace metals. However, it is now apparent that metals, previously deposited from the atmosphere and stored in the catchment are being released into the loch such that the sediment basin flux of metals is not declining as would be expected as a result of emission reductions (Rose et al. 2004). It has been hypothesised that this additional catchment input could be due to increased peat erosion or leaching of the metals from the soils. SCPs are not leached but could be input from soils as a result of catchment erosion. The comparison of the SCP and metal mass balance will therefore help to ascertain the relative roles of these processes.

Following the determination of the full historical SCP record in all the UK AWMN sites, including Lochnagar, a programme of annual deep-water sediment trapping was introduced in 1990 in order to monitor the continued input of SCPs as a marker for atmospherically deposited pollutants. These data, although at a greater temporal resolution than the sediment core, show considerable inter-annual variability in SCP flux (Figure 3b Δ) (Rose and Monteith 2005) emphasising the 'smoothing' that occurs once the depositing sediment is incorporated into the full-basin record. However, despite this 'noise' the decline in SCP flux remains evident over the monitoring period and emphasises the remarkable reduction in deposition that has taken place since the peak in the late-1970s. Fluxes to the sediment in the deep-water area of Lochnagar are now at levels previously only observed prior to the 1930s in agreement with other UK AWMN lakes (Rose and Monteith 2005) and with levels in modelled non-marine sulphate deposition (Fowler et al. 2005).

a)   b)

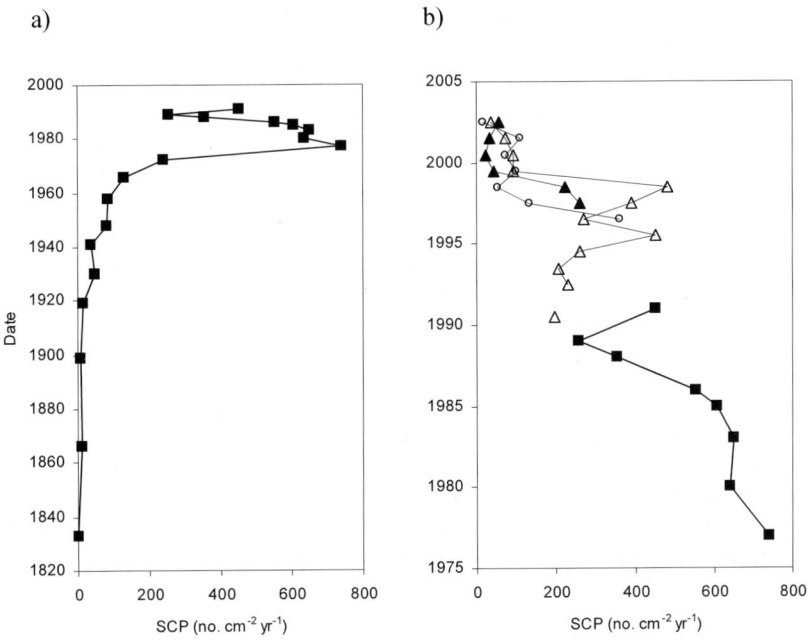

Figure 3. SCP fluxes (in SCP cm^{-2} yr^{-1}) from (a) a dated sediment core (■) from Lochnagar (NAG6) showing the full SCP record since the mid-19th century, and (b) the post-1975 SCP flux record from this sediment core (■) with more recent data from deep (△) and shallow (▲) water sediment traps and bulk deposition (○). The location of NAG6 is shown on Figure 4.

From 1997, an additional sediment trap was installed at 2m water depth in Lochnagar and included in the UK AWMN annual sampling programme. The SCP fluxes from these upper traps are also shown in Figure 3b (▲) and, whilst showing similar trends to those of the deep water traps, the fluxes are lower except for the most recent available data (2002 – 2003) where the upper trap samples show comparable SCP fluxes. The higher SCP flux in the deep water traps over the earlier part of this record may represent sediment 'focussing' or the movement of sediment material, including SCPs, from shallow water to deep water areas as a result of resuspension and transport within the body of the loch (Chapter 8: Rose this volume). However, the continued decline in SCP deposition, evident from both Figure 2b and 3b has led to very low depositional levels and the differences between shallow and deep water sediment trap fluxes are now much reduced.

An alternative hypothesis to explain the difference between upper and lower SCP trap fluxes is that the settling velocity of SCPs through the water column of a lake (6 – 8 cm day^{-1}; Punning et al. 2003) would require half a year for SCPs to reach the deeper trap in Lochnagar, but only a few weeks to reach the upper trap. Under this scenario, the upper trap always represents a more recent sampling interval and over a period of declining deposition, fluxes would always be lower in the more recent samples. Such a temporal disparity would also preclude a straightforward comparison between the two sets of trap data. However, Punning et al. (2003) also suggest that their empirically determined rate was probably "reduced substantially" as the result of the presence of a strong thermocline. As no strong thermocline exists at Lochnagar and the water column is generally well mixed (Chapter 5: Thompson et al. this volume) settling rates are probably higher than 6 – 8 cm day^{-1} reducing the difference in the collection intervals of the upper and lower traps, especially on an annual sampling time-scale.

Spatial distribution

As part of a study into the distribution and storage of trace metals in the sediment basin of Lochnagar (Yang 2000; Yang et al. 2002a; c; Chapter 15: Tipping et al. this volume), 17 sediment cores were taken in 1997 along five transects radiating from a central point (NAG9). These cores were all analysed for the full historical record of SCPs and are shown in Figure 4. It is immediately obvious that there is considerable variability in the sediment record of SCPs across the loch basin. Cores from near the centre of the loch (but not, necessarily from the deepest area) show profiles typical of the 'standard' UK pattern (NAG9, 12, 19, 21, 22) whilst 'marginal' cores near the edge of the area of accumulating sediment (e.g., NAG11, 14, 15, 16, 18, 25) show profiles that are markedly different, having surface concentration maxima, short records or unusual temporal patterns. Cores from locations in the deepest areas (e.g., NAG17, 23) seem quite variable; NAG23 shows many of the usual temporal SCP features in good agreement with ^{210}Pb-derived dates and NAG17 appears to have more in common with the nearby marginal SCP profiles of NAG15 and 16. Therefore, there seems to be a central area of sediment where a full SCP profile can be replicated (marked area, Figure 4) but whilst this area is not over the deepest part of the loch it does seem to be spatially central to the basin. This area is the same as that of maximum post-1850 sediment accumulation (Chapter 8: Rose this volume).

A comparison between the nearest core to the outflow within the main basin (NAG25) and the core from the first outflow pool (NAG26) is also of interest. NAG25 shows a typical marginal profile, with low concentrations and a short record, whereas NAG26 shows higher concentrations and a fuller profile. The sediment within this outflow pool is very organic and composed mainly of eroded peat fragments from the surrounding area. The differences between these two cores may, therefore, result from the erosion of SCPs from recent peats in this part of the catchment, highlighting the role that this process may play in the input of previously deposited pollutants to the loch from catchment soils. The difference in the records of NAG25 and 26 may therefore serve as a record of catchment peat erosion rather than a record of SCPs leaving via the outflow.

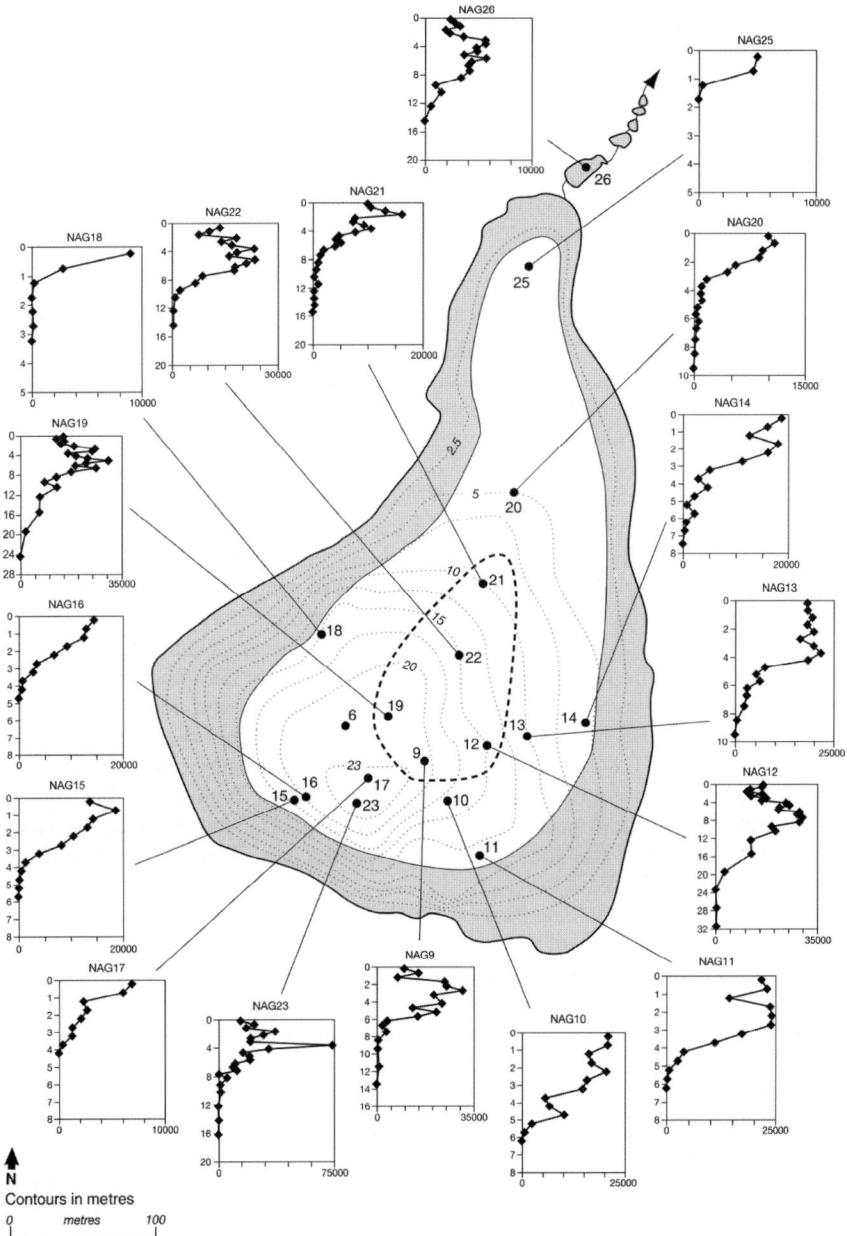

Figure 4. SCP concentration (SCP gDM^{-1}) profiles from sediment cores taken across the Lochnagar basin and from the first outflow pool. Depth contours are in metres. Dotted line denotes region within which maximum sediment accumulation occurs (Chapter 8: Rose this volume). Shaded area represents the empirically estimated region of little or no sediment accumulation.

As well as temporal profiles, SCP concentrations also vary across the loch. Surficial sediment (0– 0.5 cm) concentrations shown in Figure 5a, appear to show little spatial pattern although concentrations in the southeast are slightly higher. However, from Figure 4, this seems to be due to these southeastern cores failing to show the recent decline in SCP concentrations observed in the 'full' profiles seen in more central cores. Furthermore, SCP concentration maxima for individual cores are seen to be highest in this central region. Given these differences, a better spatial comparison, in terms of SCP accumulation within the Lochnagar sediment basin, can be made by converting the concentration profiles from Figure 4 into full-core SCP inventories (Figure 5b). From this it can be clearly seen that SCP accumulation is greatest in the central area and that, in general, inventories increase with depth along the radiating transects. These inventory data also highlight the variability of the record within the deep area (NAG17 and 23) whereas the surface sediment concentrations for these two cores (Figure 5a) were similar. It is uncertain why this is the case, although similar results are obtained for metals data, (Yang 2000) and it maybe that the sediment record in this area of Lochnagar is affected by rockfalls either directly from the backwall or via ice-rafting (Chapter 8: Rose this volume). Similarly, the surface concentrations of NAG25 and the 'outflow pool' core NAG26 show much closer agreement than do their inventories, which show considerably more SCP deposition in the outflow pool, supporting the hypothesis that this additional SCP input is from eroded peats around the outflow area.

The spatial distribution of SCP inventories shown in Figure 5b is close to that observed for inventories of the anthropogenic fraction of trace metals (i.e., 'total' minus 'pre-industrial') over the full industrial period (Yang et al. 2002b) and Figure 6 shows the relationships between SCP inventories with those for the anthropogenic fractions of Hg, Pb and Zn in the Lochnagar cores. The correlations between these are remarkably good with R^2 values of 0.84, 0.91 and 0.63 respectively. This suggests that mechanisms for transport and deposition of trace metals and SCPs within the loch are the same and, given the estimated depth of post-1860 sediment accumulation within the loch (Yang et al. 2002b), are related to the distribution of SCPs and particle-bound metals with the bulk sediment within the basin.

These core inventories can also be used to estimate the number of SCPs stored within the sediment basin. The area of accumulating sediment within Lochnagar can be sub-divided by depth and location so that each analysed core is representative of one of these smaller areas. These are described in Yang (2000) and vary between 1800 and 11000 m^2. The total inventory of SCP within the accumulating sediment basin (I_{tot}) can therefore be calculated from:

$$I_{tot} = \Sigma I_i . A_i \qquad\qquad\qquad (1)$$

where I_i is the SCP inventory for core i and A_i is the area of accumulating basin represented by core i. This results in the estimate of 6.18 x 10^{12} SCPs in the sediment basin of Lochnagar. Rose (2001) estimated that the mass of a 20 μm SCP would be in the region of 4.2 x 10^{-9} g and given this value an estimated 26 kg of SCPs is stored in the Lochnagar sediment basin. This compares with the total masses of anthropogenic Hg and Pb stored in the sediment basin of 71.4 g and 93.8 kg respectively (Yang et al. 2002a).

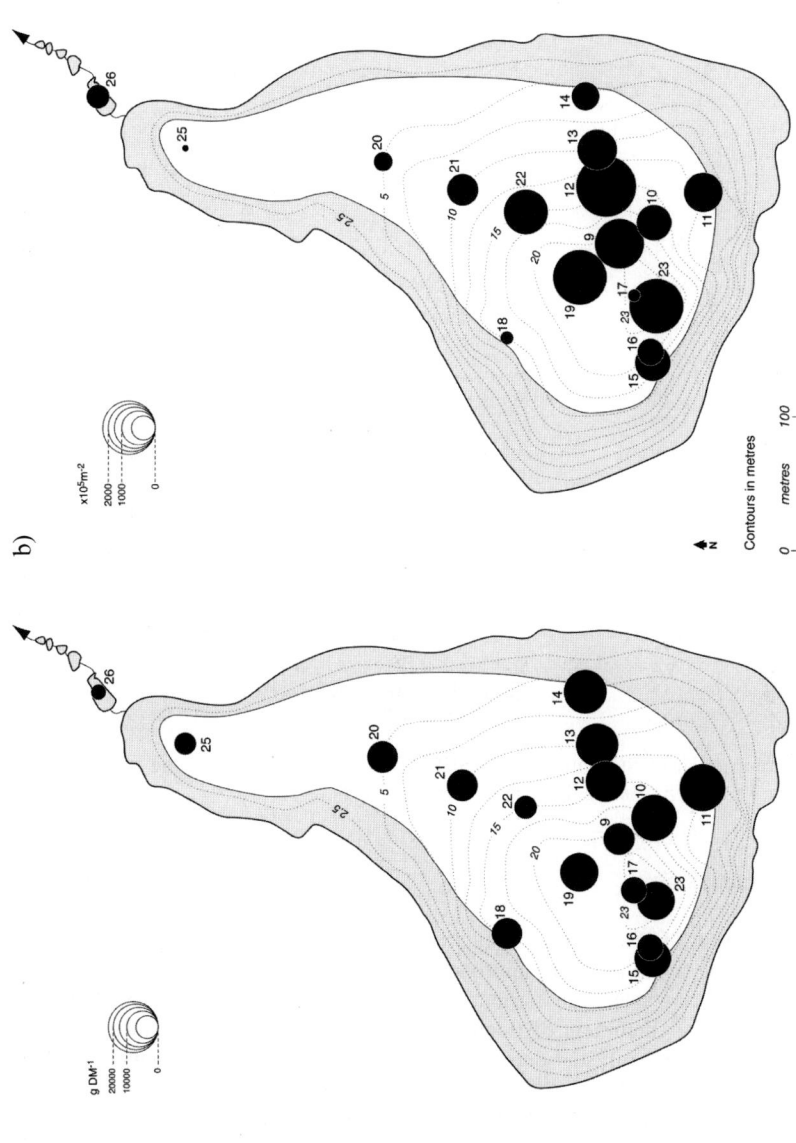

Figure 5. (a) SCP concentrations (SCP gDM⁻¹) in surface sediments and (b) full core SCP inventories (SCP x 10⁵ m⁻²) for cores taken across the Lochnagar basin. Shaded areas represent the empirically estimated region of little or no sediment accumulation.

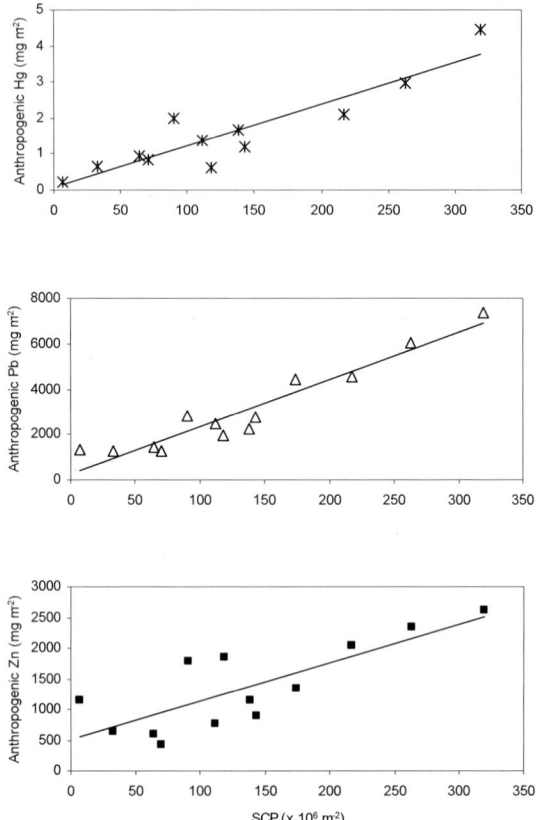

Figure 6. A comparison of SCP inventories (SCP x 10^6 m^{-2}) with those of the anthropogenic fractions of Hg, Pb and Zn from the Lochnagar sediment cores.

SCPs in catchment soils

In the same way that SCP and trace metal storage for the sediment basin of Lochnagar was estimated using a multi-coring approach, so the storage of these contaminants in the catchment soils has also been estimated. Ten soil cores (NAG50-52, 55-61) were taken from around the catchment of Lochnagar and analysed for SCPs as described in (Yang et al. 2001). The soil SCP concentration profiles (Figure 7) are generally short (less than 10 cm) except NAG51 which has a record of 15 cm. SCP profiles from areas of the catchment characterised by shallower slopes and greater accumulation of peats (Chapter 6: Helliwell et al. this volume) appear to show similar features to those of 'typical' SCP lake sediment profiles and as a consequence have been used to apply approximate chronologies to these soil cores (Yang et al. 2001). Maximum concentrations are lower than those of the sediment basin and reach c. 21000 gDM^{-1} on the eastern side of the loch (NAG56; Figure 7).

As with the sediment cores, SCP inventories for each soil core can be calculated. Inventories for the individual cores are lower than those of the sediment cores and range over about an order of magnitude between 6.8×10^6 (NAG52) and 6.7×10^7 m^{-2} (NAG56). In a similar way to the sediment cores, each soil core can be considered representative of an area of the catchment (Yang 2000) and used to estimate the total SCP storage. However, this requires an additional term in equation (1) i.e.

$$I_{tot} = \Sigma I_i . A_i . S_i \tag{2}$$

where S_i is the fraction of area A_i covered by soil. This produces an estimate for the total SCPs in the catchment soils of 9.94×10^{12}. Comparing this value with that for SCPs stored in the sediment basin shows that of the SCPs deposited and accumulated in the loch and catchment over the industrial period, 61.7% are stored in the catchment soils and 38.3% in the loch sediment.

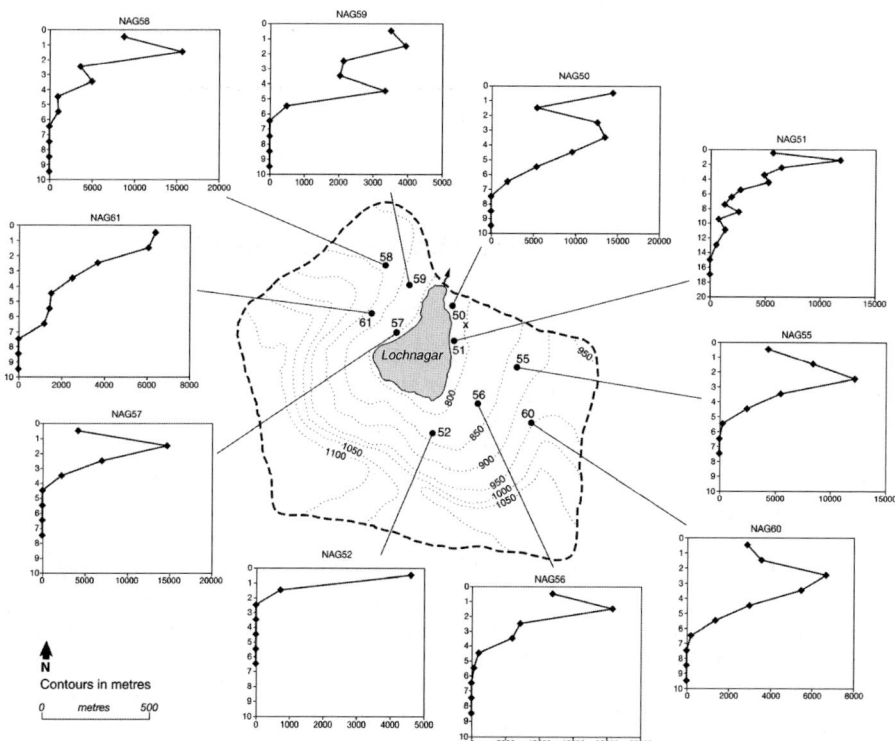

Figure 7. SCP concentration (SCP gDM^{-1}) profiles from soil cores taken across the Lochnagar catchment area. Contours are in metres above sea level. 'X' marks the location of the Automatic Weather Station.

A SCP number balance

The SCP data described above can be used to compile a SCP 'number balance' (cf. 'mass balance' for other pollutants) for Lochnagar. This is done for the calendar year 1997, as there are most data available for this period.

SCP inputs via bulk deposition and cloudwater

Figure 2b shows that, for 1997, SCP deposition calculated from the bulk collectors was 114850 m^{-2}. If it is assumed that this is a reasonable estimate for deposition across the area of the loch and the catchment then the number of SCPs directly deposited to these areas in 1997 are 1.126 x 10^{10} and 1.056 x 10^{11} respectively (Figure 8).

In addition to bulk deposition inputs, cloudwater was estimated to add an additional c. 46% to SCP inputs from the atmosphere for 2002-03. Assuming a similar cloud enhancement for 1997 results in an input of 4.86 x 10^{10} SCP to the catchment from this source. The total SCP input to the Lochnagar catchment area is therefore 1.542 x 10^{11} for the year. These cloudwater estimates were made using a surface roughness appropriate for moorland vegetation and while the use of this same factor for direct deposition to a water surface is probably too high, the influence of the surrounding catchment on small water bodies such as Lochnagar is such that the same factor probably does not over-estimate cloudwater inputs to the Lochnagar surface too much (Jennifer Muller and David Fowler, Centre for Ecology and Hydrology, Edinburgh pers. commun.). Therefore, assuming a similar cloudwater enhancement for the loch as for the catchment results in an additional input of 5.18 x 10^9 SCP (Figure 8) and a total input from the atmosphere to the loch surface of 1.644 x 10^{10} for the year.

SCP loss via the outflow

During 1997, the concentration of SCPs in outflow water was estimated bi-monthly. This was done by filtering large volumes of water ($\geq$40 litres) and analysing the filters for SCPs (see CARBYNET: http://www.ecrc.ucl.ac.uk/carbynet/). The mean and standard deviation of these measurements were 1.6 l^{-1} and 1.46 respectively (N = 6). An estimate of the water volume lost via the outflow in 1997 was made from a calculated hydrological balance where:

$$\text{Outflow} = \text{Precipitation} + \text{Run-off} - \text{Evaporation} \qquad (3)$$

Here, 'Precipitation' is the volume of direct precipitation to the loch surface calculated from the Lochnagar automatic weather station data; 'Run-off' is the volume of precipitation to the catchment area multiplied by a factor of 0.9 (Jenkins et al. 2001) and 'Evaporation' from the loch surface is calculated from the Penman equation (Penman 1948). This provides an estimate of 1.75 x 10^9 litres (Jonathan Tyler, University College London pers. commun.) for the volume of water leaving Lochnagar via the outflow stream in 1997. This is then combined with the mean SCP concentration data to give an estimate for the number of SCPs lost via the outflow in 1997 as 2.795 x 10^9. Further, if it is assumed that the loch water is well mixed, then the number of SCPs in the water column may be estimated from the loch volume. This number is 1.31 x 10^9

SCPs. However, for the purposes of this 1997 number balance calculation it is assumed that storage in the water column does not vary over the year and hence there is no net loss or gain via this compartment.

SCP loss to the sediment

Although annual sediment trap data exist for the years 1996/7 and 1997/8 their location in the deep water area c.1-2 m above the sediment surface may result in an over-estimate of the number of SCPs deposited to the sediment, as a result of focussing. The number of SCPs 'lost' to the sediment in 1997 was therefore estimated from the multiple cores taken from across the sediment basin in that year (NAG9 – 25).

The dry mass of sediment in the uppermost 0.5 cm for each core was estimated from the volume and the dry bulk density of the slice. Up-scaling these masses to the areas of the sediment basin for which each of these cores is representative, and combining these with the surface SCP concentrations for each core, provides an estimate for the number of SCPs in the uppermost 0.5 cm for each area of the basin. Further, the time period covered by this uppermost slice can be estimated from the depth of the SCP peak concentration (taken as 1978) or, where no peak is present (mainly in short cores from the more 'littoral' areas), from the depth of the first presence of SCPs (taken as 1860). The number of SCPs in the uppermost 0.5 cm from each area can then be divided by the number of years represented by this depth of sediment in each area to provide an estimate of SCP deposition to this area for a single year. These 'annual' values can then be summed to provide an estimate for one year across the whole accumulating sediment basin. This summed value can then be used as an estimate for 1997, i.e.

$$\text{SCPs 'lost' to sediment in 1997} = \sum \left(\frac{0.5 A_i.\rho_i.C_i}{Y_i} \right) \qquad (4)$$

where A_i is the area of the accumulating sediment basin represented by core i (in cm^2); ρ_i is the dry bulk density of the $0 - 0.5$ cm slice of core i in area i (in g cm^{-3}); C_i is the concentration of SCPs in the $0 - 0.5$ cm slice of core i (in gDM^{-1}); Y_i is the number of years represented by the $0 - 0.5$ cm slice in core i. This provides an estimate for the number of SCPs deposited to the sediment in 1997 of 5.1×10^{10}.

However, if this estimate is made from the deep-water sediment trap data, then the number of SCPs deposited across the sediment basin is estimated as 1.88×10^{11}, a factor of 3.7 higher than the estimate made using the multiple core approach. This supports the hypothesis that focussing is taking place within the loch and that the sediment traps are representative only of deposition to the deep-water area.

SCP transfer from the catchment

If 2.795×10^9 SCPs were lost from Lochnagar via the outflow in 1997 and 5.1×10^{10} were 'lost' to the accumulating sediment in that year, then the balance for this loss must come from inputs from the atmosphere and indirectly from the catchment. Subtracting direct atmospheric deposition (1.126×10^{10}) and cloudwater input (5.18×10^9) from the total loss thus provides an estimate for the transfer of SCPs from the catchment to the lake in 1997. Figure 8 shows a schematic representation of the loch and catchment system from which this number can be calculated as 3.735×10^{10}, a factor of 3.3 higher than the direct inputs from the atmosphere and a factor similar to that calculated for Hg (3.6; Yang et al 2002a). As the inflow streams are only very minor, these catchment SCPs could be transferred during snow-melt, deposited onto bare rock and transferred directly into the loch, or from SCPs previously deposited and stored in catchment soils and recently eroded. However, unlike Hg and other metals SCPs cannot be leached from the catchment. Figure 8 shows that the number of SCPs directly deposited to the catchment is more than enough to balance the number required in the loch and still allows a surplus of 1.168×10^{11} to be stored in the catchment. Some of these will presumably be in areas available to be transferred to the loch in subsequent years. This calculation suggests that catchment transfer is an important source of SCPs and, by implication, other associated contaminants.

There are, of course, a great many assumptions in this SCP number balance and the errors for each estimated parameter are large. However, it does show that while known inputs from the atmosphere have been, and remain, important as a source of contamination, transfer from the catchment is required for a balance. While the mechanisms described above are all means by which SCPs could be transferred, catchment erosion is thought to be a major source of metals to Lochnagar (Yang et al. 2002b) and SCPs are undoubtedly also eroded into the loch during this process. However, although uncertainty remains as to the relative importance of these processes

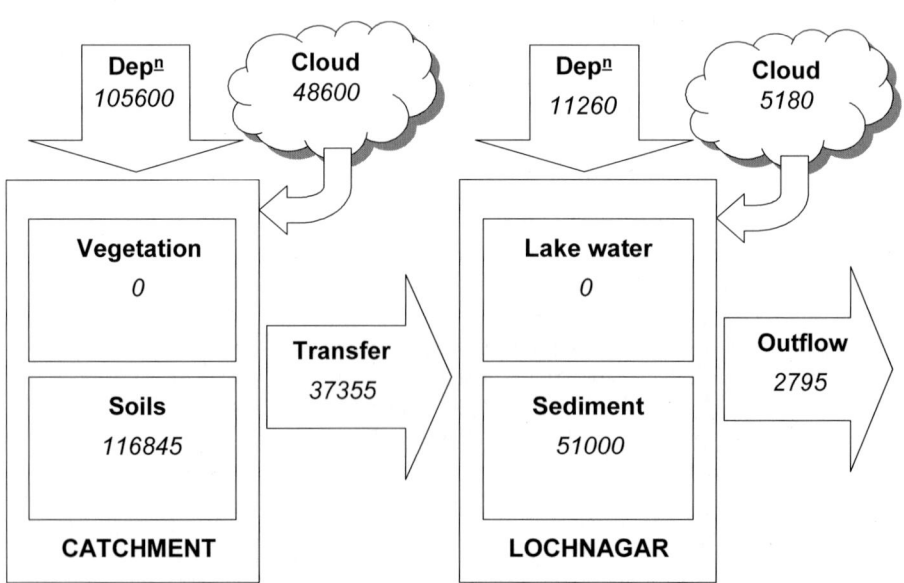

Figure 8. SCP 'number balance' for Lochnagar in 1997 showing: estimated inputs from bulk deposition (Dep^n) and cloudwater (cloud); loss via the outflow and loss via deposition to sediment basin and soils; calculated transfer from the catchment to the loch. All values in millions of SCP (e.g., loss via outflow is 2795 million or 2.795×10^9).

for SCPs compared with that of direct deposition, it is certain that a vast number of SCPs and other previously deposited contaminants remain stored in the catchment soils of Lochnagar with some potential for release as a result of enhanced transfer possibly as a consequence of climate change.

Impacts of climate change

While the SCP number balance raises some interesting questions with regard to the transfer of pollutants from the catchment to the loch, it is currently unknown whether this situation is unique to Lochnagar or typical for upland freshwaters across Scotland. As with many of the detailed studies outlined in this volume, the data for Lochnagar are unique in the UK and further work is required to determine how typical the situation at Lochnagar is. An EU funded project, Euro-limpacs, is currently (2004 – 2009) investigating the role of climate change in enhancing the transfer of contaminants from catchments to waters of upland lakes and it is hoped that the results of this study will help interpret the Lochnagar data still further. Current climate scenarios for Scotland (Chapter 18: Kettle and Thompson this volume) suggest prolonged drier and warmer periods in summers and heavier rainfall in winters with an increased frequency of severe meteorological events. These conditions are favourable for an increase in both erosion and leaching (Noda et al. 2001; Rose et al. 2004) and hence elevated transfer of contaminants, including SCPs, might be expected at Lochnagar as a result.

In terms of the impact of this on the ecosystem, the increased input of SCPs is unlikely to affect biota directly. However, an increase in the catchment input of SCPs to the loch serves as an indication of enhanced inputs of toxic compounds in two ways. First, as a marker for the erosion process thereby indicating that previously deposited toxic metals and POPs stored in catchment soils are similarly, and increasingly, being input via this means. Second, as SCPs bring with them trace metals and POPs adsorbed to the particle surface (e.g., Wey et al. 1998; Ghosh et al. 2000), possibly via complexation with large organic acids which themselves readily adsorb to elemental carbon particles (Seigneur et al. 2005). In this latter case, the porosity of SCPs increases the available adsorption surface making them particularly suitable for the transport of these pollutants via this route. The increase in transfer of SCPs from catchment to the waterbody would therefore indicate an enhanced input of toxic pollutants to Lochnagar regardless of any policy-driven reductions in emissions to the atmosphere.

Summary

- Monitoring of SCPs at Lochnagar has been ongoing since 1990 when the annual sediment trapping programme was initiated. Fortnightly monitoring of SCPs in bulk deposition began in 1996 and a full sediment basin and soil survey was undertaken in 1997.
- Bulk deposition data show a decrease in SCP deposition since 1997 with a stable period between 1998 and 2002. Further monitoring will determine whether a subsequent decrease in 2003 is a temporary reduction or a further decline in deposition.

- Inputs of SCPs from cloudwater at Lochnagar represent an additional c. 46% of bulk deposition inputs.
- Combined sediment trap and dated sediment core data show a continued decline since the late-1970s. SCP fluxes are now at their lowest levels since the 1930s, in agreement with modelled non-marine sulphate data.
- A multi-core survey across the basin of Lochnagar shows replicable temporal trends in an area central to the loch, but not its deepest area. SCP profiles outside this central region are truncated or show unusual trends.
- SCP profiles from the central area of Lochnagar are typical of those across the UK, with the start of the record in 1850 ± 25 and a concentration peak in 1978 ± 2.
- SCP inventory data for Lochnagar shows evidence of sediment focussing in deeper areas in good agreement with trace metals data.
- An SCP number balance for 1997 suggests that inputs can only balance outputs with additional transfer from the catchment. These additional SCPs are either transferred during snowmelt, in-washed from deposition on bare rock areas or enter the loch with eroded catchment peats.
- Predicted climate scenarios imply conditions favourable for increased catchment peat erosion suggesting SCP transfer from this store could increase in the future as a result. While SCPs are not known to be detrimental to freshwater biota, inputs of toxic trace metals and POPs, adsorbed to SCP surfaces, could be enhanced by this mechanism.

Acknowledgements

We would like to thank Jo Porter for his many visits to Lochnagar supporting the monitoring programmes described in this chapter, the Balmoral Estate Office for facilitation of our research efforts and the multifarious members of the ECRC who have helped with field-work over the years. We are also grateful to Ron Harriman and other members of the Freshwater Research Laboratory in Faskally for their help dealing with the many samples and to Jon Tyler (ECRC) for the use of his hydrological estimate. The long-term monitoring and research at Lochnagar has been supported by funding from the Department of the Environment, Food and Rural Affairs (DEFRA) via the UK Acid Waters Monitoring Network (Contract No. EPG 1/3/160) and 'Freshwater Umbrella' (Contract No. EPG 1/3/183) programmes and from the EU via the MOLAR (Measuring and modelling the dynamic response of remote mountain lake ecosystems to environmental change. A programme of mountain lake research, 1996 - 1999) (EU contract ENV4-CT95-0007), EMERGE (European mountain lake ecosystems: Regionalisation, diagnostics and socio-economic evaluation, 2000 - 2003) (EU contract EVK1-CT-1999-00032) and Euro-limpacs (EU project 505540) research projects. Elanor McBay of the UCL Geography Department Cartographic Office produced some of the figures and Tiiu Alliksaar and Jorunn Larsen provided valuable comments on the manuscript.

References

Bloesch J. and Burns N.M. 1980. A critical review of sedimentation trap technique. Schweiz Z. Hydrol. 42: 15–55.

Blomquist S. and Håkanson L. 1981. A review on sediment traps in aquatic environments. Arch. Hydrobiol. 91: 101–132.

Boyle J.F., Rose N.L., Bennion H., Yang H. and Appleby P.G. 1999. Environmental impacts in the Jianghan plain: Evidence from lake sediments. Wat. Air Soil Pollut. 112: 21–40.

Broman D., Nåf C, Wik M. and Renberg I. 1990. The importance of spheroidal carbonaceous particles for the distribution of particulate polycyclic aromatic hydrocarbons in an estuarine-like urban coastal water area. Chemosphere. 21: 69–77.

Bucheli T.D. and Gustafsson Ö. 2003. Soot sorption on non-ortho and ortho-substituted PCBs. Chemosphere 53: 515–522.

Coles D.G., Ragaini R.C., Ondov J.M., Fisher G.L., Silberman D. and Prentice B.A. 1979. Chemical studies of stack fly-ash from a coal-fired power plant. Environ. Sci. Technol. 13: 455–459.

Crossley A. 1988. Particles in orographic cloud and the implications of their transfer to plant surfaces. In: Unsworth M.H. and Fowler D. (eds.) Acid deposition at high elevation sites. Kluwer Academic Publishers. Dordrecht. pp. 453–464.

Davison R.L., Natusch D.F.S., Wallace J.R. and Evans C.A. 1974. Trace elements in fly-ash. Dependence of concentration on particle size. Environ. Sci. Technol. 8: 1107–1113.

Fernández P., Rose N.L., Vilanova R.M. and Grimalt J.O. 2002. Spatial and temporal comparison of polycyclic aromatic hydrocarbons and spheroidal carbonaceous particles in remote European lakes. Wat. Air Soil Pollut.: Focus. 2: 261–274.

Fowler D., Cape J.N., Leith I.D., Choularton T.W., Gay M.J. and Jones A. 1988. Wet deposition and altitude, the role of orographic cloud. In: Unsworth, M.H. and Fowler, D. (eds.) Acid deposition at high elevation sites. Kluwer Academic Publishers. Dordrecht. pp. 231–257.

Fowler D., Smith R.I., Muller J.B.A., Hayman G. and Vincent K.J. 2005. Changes in the atmospheric deposition of acidifying compounds in the UK between 1986 and 2001. Environ. Pollut. 137: 15–25.

Ghosh U., Gillette J.S., Luthy R.G. and Zare R.N. 2000. Microscale location, characterisation and association of polycyclic aromatic hydrocarbons on harbor sediment particles. Environ. Sci. Technol. 34: 1729–1736.

Glew J.R. 1991. Miniature gravity corer for recovering short sediment cores. J. Paleolim. 5: 285–287.

Griest W.H. and Tomkins B.A. 1984. Carbonaceous particles in coal combustion stack ash and their interaction with polycyclic aromatic hydrocarbons. Sci. Tot. Environ. 36: 209–214.

Helliwell R.C., Lilly A. and Bell J.S. This volume. Chapter 6. The development, distribution and properties of soils in the Lochnagar catchment and their influence on soil water chemistry. In: Rose N.L. (ed.) 2007. Lochnagar: The natural history of a mountain lake. Springer. Dordrecht.

Jenkins A., Ferrier R.C. and Helliwell R.C. 2001. Modelling nitrogen dynamics at Lochnagar, N.E. Scotland. Hydrol. Earth Sys. Sci. 5: 519–527.

Jones V.J., Flower R.J., Appleby P.G., Natkanski J., Richardson N., Rippey B., Stevenson A.C. and Battarbee R.W. 1993. Palaeolimnological evidence for the acidification and atmospheric contamination of lochs in the Cairngorm and Lochnagar areas of Scotland. J. Ecol. 81: 3–24.

Kettle H. and Thompson R. This volume. Chapter 18. Future climate predictions for Lochnagar. In: Rose, N.L. (ed.) 2007. Lochnagar: The natural history of a mountain lake. Springer. Dordrecht.

Laxen D. P. H. 1996. Generating emissions? National Power Report. pp 1–74.

Mackereth F.J.H. 1969. A short core sampler for subaqueous deposits. Limnol. Oceanog. 14: 145–151.

Neal C., Wilkinson J., Neal M., Harrow M., Wickham H., Hill L. and Morfitt C. 1997. The hydrochemistry of the headwaters of the River Severn, Plynlimon. Hydrol. Earth Sys. Sci. 1: 583–617.

Noda T., Onodera S., Hirose T. and Naruoka T. 2001. Sediment yield in small catchments with different bedrock in a humid temperate region. Arctic Antarctic Alpine Res. 33: 435–439.

Ohsaki Y., Matsueda T. and Ohno K. 1995. Levels of coplanar PCBs, PCDDS and PCDFs in fly ashes and pond sediments. Int. J. Environ. Anal. Chem. 59: 25–32.

Penman H.L. 1948. Natural evaporation from open water, bare soil and grass. Proc. Roy. Soc. Lond. Series A. 194: 120–145.

Persson N.J., Gustafsson Ö, Bucheli T.D., Ishaq R., Næs K. and Broman D. 2002. Soot-carbon influenced distribution of PCDD/Fs in the marine environment of the Grenlandsfjords, Norway. Environ. Sci. Technol. 36: 4968–4974.

Persson N.J., Gustafsson Ö, Bucheli T.D., Ishaq R., Næs K. and Broman D. 2005. Distribution of PCNs, PCBs and other POPs together with soot and other organic matter in the marine environment of the Grenlandsfjords, Norway. Chemosphere 60: 274–283.

Punning J-M., Terasmaa J., Koff T. and Alliksaar T. 2003. Seasonal fluxes of particulate matter in a small closed lake in northern Estonia. Wat. Air Soil Pollut. 149: 77–92.

Renberg I. and Wik M. 1984. Dating recent lake sediments by soot particle counting. Verh Internat Verein Limnol. 22: 712–718.

Renberg I. and Wik M. 1985a. Carbonaceous particles in lake sediments. Pollutants from fossil fuel combustion. Ambio. 14: 161–163.

Renberg I. and Wik M. 1985b. Soot particle counting in recent lake sediments. An indirect dating method. Ecol. Bull. 37: 53–57.

Reynolds B., Fowler D. and Thomas S. 1996. Chemistry of cloud water at an upland site in mid-Wales. Sci.Tot. Environ. 188: 115–125.

Rose N.L. 1994. A note on further refinements to a procedure for the extraction of carbonaceous fly-ash particles from sediments. J. Paleolim. 11: 201–204.

Rose N.L. 2001. Fly-ash particles. In: Last W.M and Smol J.P. (eds.) Tracking environmental change using lake sediments. Volume 2: Physical and Geochemical methods. Kluwer Academic Publishers, Dordrecht. pp. 319–349.

Rose N.L. This volume. Chapter 8. The sediments of Lochnagar: Distribution, accumulation and composition. In: Rose N.L. (ed.) 2007. Lochnagar: The natural history of a mountain lake. Springer. Dordrecht.

Rose N.L. and Appleby P.G. 2005. Regional applications of lake sediment dating by spheroidal carbonaceous particle analysis I: United Kingdom. J. Paleolim. 34: 349–361.

Rose N.L. and Monteith D.T. 2005. Temporal trends in spheroidal carbonaceous particle deposition derived from annual sediment traps and lake sediment cores and their relationship with non-marine sulphate. Environ. Pollut. 137: 151–163.

Rose N.L., Harlock S., Appleby P.G. and Battarbee R.W. 1995. Dating of recent lake sediments in the United Kingdom and Ireland using spheroidal carbonaceous particle (SCP) concentration profiles. Holocene. 5: 328–335.

Rose N.L., Juggins S. and Watt J. 1996. Fuel-type characterisation of carbonaceous fly-ash particles using EDS-derived surface chemistries and its application to particles extracted from lake sediments. Proc. Roy. Soc. Lond. Series A. 452: 881–907.

Rose N.L., Juggins S. and Watt J. 1999. The characterisation of carbonaceous fly-ash particles from major European fossil-fuel types and applications to environmental samples. Atmos. Environ. 33: 2699–2713.

Rose N.L., Shilland E., Berg T., Hanselmann K., Harriman R., Koinig K., Nickus U., Trad B.S., Stuchlik E., Thies H. and Ventura M. 2001. Relationships between acid ions and carbonaceous fly-ash particles in deposition at European mountain lakes. Wat. Air Soil Pollut. 130: 1703–1708.

Rose N.L., Shilland E., Yang H., Berg T., Camarero L., Harriman R., Koinig, K. Lien L., Nickus U., Stuchlik E., Thies H. and Ventura M. 2002. Deposition and storage of spheroidal carbonaceous fly-ash particles in European mountain lake sediments and catchment soils. Wat. Air Soil Pollut.: Focus. 2: 251–260.

Rose N.L., Flower R.J. and Appleby P.G. 2003. Spheroidal carbonaceous particles (SCPs) as indicators of atmospherically deposited pollutants in North African wetlands of conservation importance. Atmos. Environ. 37: 1655–1663.

Rose N.L., Monteith D.T., Kettle H., Thompson R., Yang H. and Muir D.C.G. 2004. A consideration of potential confounding factors limiting chemical and biological recovery at Lochnagar, a remote mountain loch in Scotland. J. Limnol. 63: 63–76.

Rose N.L., Metcalfe S.M., Benedictow A.C., Todd M., Nicholson J. This volume. Chapter 13. National, international and global sources of contamination at Lochnagar. In: Rose N.L. (ed.) 2007. Lochnagar: The natural history of a mountain lake. Springer. Dordrecht.

Seigneur C., Abeck H., Chia G., Reinhard M., Bloom N.S., Prestbo E. and Saxena P. 2005. Mercury adsorption to elemental carbon (soot) particles and atmospheric particulate matter. Atmos. Environ. 32: 2649–2657.

Thompson R., Kettle H., Monteith D.T. and Rose N.L. This volume. Chapter 5. Lochnagar water-temperatures, climate and weather. In: Rose N.L. (ed.) 2007. Lochnagar: The natural history of a mountain lake. Springer. Dordrecht.

Tipping E., Yang H., Lawlor A.J., Rose N.L. and Shotbolt L. This volume. Chapter 15. Trace metals in the catchment, loch and sediments of Lochnagar: Measurements and modelling. In: Rose N.L. (ed.) 2007. Lochnagar: The natural history of a mountain lake. Springer. Dordrecht.

Wey M-Y., Chao C-Y., Chen J-C. and Yu L-J. 1998. The relationship between the quantity of heavy metal and PAHs in fly-ash. J. Air Waste Man. Assoc. 48: 750–756.

Wilkinson J., Reynolds B., Neal C., Hill S., Neal M. and Harrow M. 1997. Major, minor and trace element composition of cloudwater and rainwater at Plynlimon. Hydrol. Earth Sys. Sci. 1: 557–569.

Yang H. 2000. Trace metal storage in lake systems and its relationship with atmospheric deposition with particular reference to Lochnagar, Scotland. Unpublished PhD thesis. University of London.

Yang H., Rose N.L. and Battarbee R.W. 2001. Dating of recent catchment peats using spheroidal carbonaceous particle (SCP) concentration profiles with particular reference to Lochnagar, Scotland. Holocene. 11: 593–597.

Yang H., Rose N.L., Battarbee R.W. and Boyle J.F. 2002a. Mercury and lead budgets for Lochnagar, a Scottish mountain lake and its catchment. Environ. Sci. Technol. 36: 1383–1388.

Yang H. Rose N.L., Battarbee R.W. and Monteith D.T. 2002b Trace metal distribution in the sediments of the whole lake basin for Lochnagar, Scotland: a palaeolimnological assessment. Hydrobiol. 479: 51–61.

Yang H., Rose N.L. and Battarbee R.W. 2002c. Distribution of some trace metals in Lochnagar, a Scottish mountain lake ecosystem and its catchment. Sci. Tot. Environ. 285: 197–208.

PART IV: FUTURE IMPACTS

18. FUTURE CLIMATE PREDICTIONS FOR LOCHNAGAR

HELEN KETTLE (H.Kettle@ed.ac.uk)
and ROY THOMPSON
School of Geosciences
The University of Edinburgh
Crew Building, Kings Buildings,
West Mains Rd,
Edinburgh EH9 3JN
United Kingdom

Key words: climate, future, IPCC, model, scenarios, stratification, temperature

Our changing climate

The Earth's climate is a result of complex interactions between the atmosphere, oceans, land, ice, snow, and terrestrial and marine biology. The driving energy for the whole system is short-wave radiation from the Sun which is absorbed mainly at the Earth's surface. This energy is then redistributed by the atmosphere and oceans and radiated back to space at longer (infrared) wavelengths. The mean temperature of the Earth is controlled by the balance between incoming solar radiation and outgoing terrestrial radiation – this is known as the 'radiative forcing'. Thus, any factors that alter the amount of incoming or outgoing radiation, or that alter the redistribution of energy within the atmosphere or between the atmosphere, land and ocean, can affect climate. Over geological time scales the Earth has experienced large changes in climate. These glacial and inter-glacial cycles are caused by changes in the amount and distribution of sunlight received, which varies according to the composition of the atmosphere and/or the Earth's position relative to the Sun. There is currently an imbalance in the incoming and outgoing radiation such that the Earth is being driven to a warmer climate by a positive radiative forcing of about 0.85 W m^{-2} (Hansen et al. 2005).

The cause of the warming is generally believed to be the increase in the amount of 'greenhouse gases' in the atmosphere. Most of the atmosphere is made up of nitrogen (78.1%), oxygen (20.9%) and argon (0.9%) which have limited interactions with in-going or out-going radiation. However there are a number of trace gases, such as carbon dioxide (CO_2), methane (CH_4), nitrous oxide (N_2O), ozone (O_3) and water vapour, which absorb and emit infrared radiation (water vapour, CO_2 and O_3 also absorb the

427

N.L. Rose (ed.), Lochnagar: The Natural History of a Mountain Lake
Developments in Paleoenvironmental Research, 427–444.
© 2007 *Springer.*

in-coming solar short-wave radiation). Because these gases absorb the infrared radiation emitted by the Earth and emit infrared radiation upward and downward, they tend to raise the temperature near the Earth's surface. This is known as the 'enhanced greenhouse effect'. Table 1 shows the changes in the concentration of these gases since pre-industrial times and the effect this has had on the radiative forcing. Water vapour, is a natural greenhouse gas (uncontrolled by humans), however it is the dominant one, trapping more of Earth's heat than any other gaseous constituent (Kiehl and Trenberth 1997). Besides these gases, the atmosphere also contains solid and liquid particles (aerosols) and clouds, which interact with both the incoming and outgoing radiation in a complex and spatially variable manner.

Table 1. Pre-industrial (1750) and present (1998) abundance of carbon dioxide, methane and nitrous oxide – the three most important well-mixed greenhouse gases; the radiative forcing due to the change in abundance and their sources (IPCC 2001).

Gas	Abundance (1750)	Abundance (1998)	Radiative forcing (W m^{-2})	Source
CO_2	278 ppm	365 ppm	1.46	Fossil fuel burning, land-use change
CH_4	700 ppb	1745 ppb	0.48	Use of fossil fuels, cattle, rice agriculture, landfills
N_2O	270 ppb	314 ppb	0.15	Agricultural soils, cattle feed, chemical industry

Air bubbles trapped in ice sheets record the chemical composition of the Earth's atmosphere at the time the ice was formed. Thus cores drilled through the ice can provide us with atmospheric data over thousands of years. These data show that for about a thousand years before the Industrial Revolution, atmospheric concentrations of carbon dioxide, nitrous oxide and methane remained relatively constant (e.g., Etheridge et al. 1996; Legrand et al. 1988; Barnola et al. 1987) (see Figure 1). However, since the beginning of the industrial period, the quantities of these gases in the atmosphere have rapidly increased. We know that the increase in atmospheric carbon dioxide is anthropogenic because analysis of the isotopic composition shows that it originates from fossil fuel (e.g., Keeling et al 1995). For over 400000 years before 1800 the amount of carbon dioxide in the atmosphere oscillated between 200 - 280ppm (data from Vostok ice core; Petit et al. 1999) but by the end of 2004 it was 377ppm (Keeling et al, 2005). The amount of carbon dioxide in the atmosphere is increasing at an unprecedented rate, on average by 0.4% per year (IPCC 2001). Carbon dioxide is mainly emitted to the atmosphere by fossil fuel burning and land-use change, whereas methane and nitrous oxide emissions are also related to agricultural practices (Table 1). Human activities have changed the composition of the atmosphere and the character of the Earth's surface and as such, have indirectly affected the climate. Globally, air temperature has risen by about 0.7 °C over the last 100 years (Figure 2). In central England the temperature has risen by almost 1 °C through the 20th century with the 1990s being the warmest decade in central England since records began in the 1660s (Hulme et al. 2002). The UK's thermal growing season for plants is now longer than at any time since the start of the record in 1722, and the average rate of sea-level rise around the UK coastline (after adjustments for natural land movements) is approx 1mm yr^{-1} (Hulme et al. 2002).

Mountain ecosystems may be affected by climate change in two ways; first by the increased concentrations of CO_2 available for photosynthesis (i.e., physiological responses) and second, by the actual changes in climate (i.e., physical responses) (Price and Haslett 1995). These two types of changes are interlinked since photosynthesis and respiration are influenced by temperature and nutrient availability, and in the case of photosynthesis, sunlight. Plant growth is influenced by precipitation/relative humidity (Stampfli and Zeiter 2004) as well as temperature, while phenological (e.g., flowering dates) changes are mostly influenced by temperature (Clark and Thompson 2004). Small changes to maritime climates, such as that of Scotland, can produce large ecosystem responses. Records of flowering dates in Edinburgh since 1850 (Harper et al. 2005) show interannual variations of fifty days due to fluctuations in temperature.

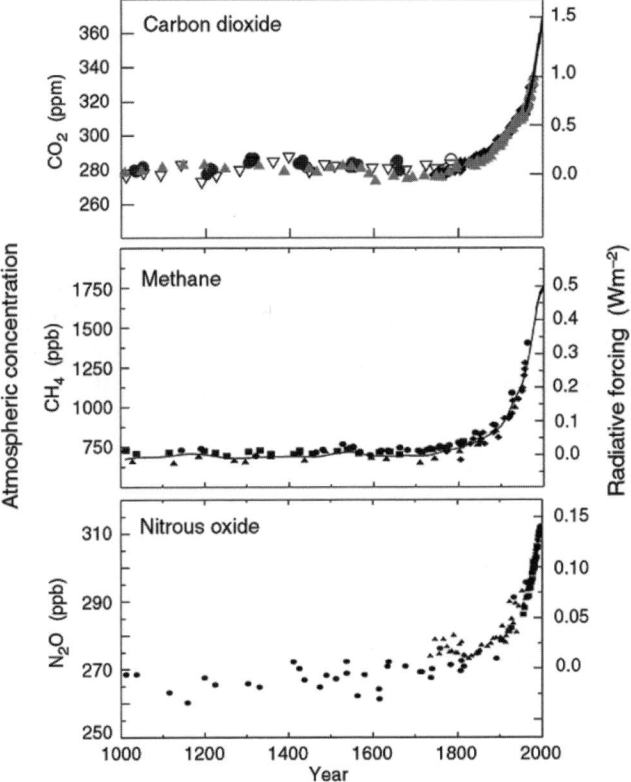

Figure 1. Indicators of the human influence on the atmosphere during the industrial era: carbon dioxide, nitrous oxide, methane from ice core and firn data for several sites in Antarctica and Greenland, supplemented with data from direct atmospheric samples over the past few decades. Right-hand axis shows positive radiative forcing of climate system. (Reproduced with permission from IPCC 2001).

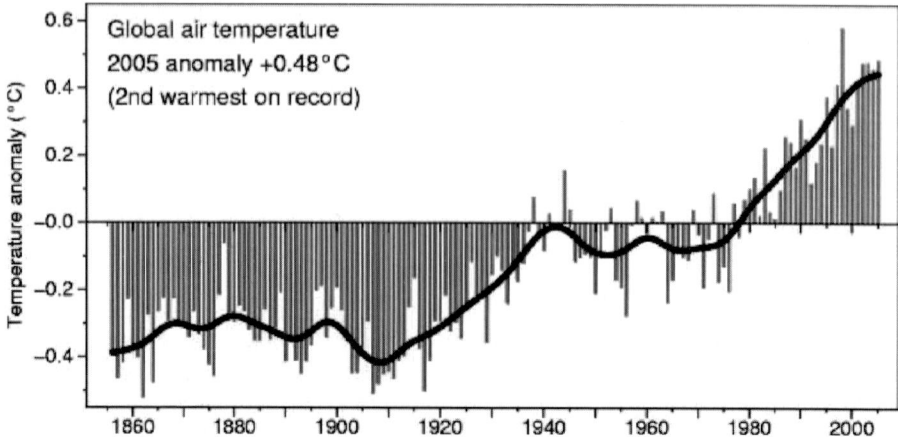

Figure 2. Time series of mean global land and marine surface temperature anomaly (relative to the mean from 1961-1990) from 1856 to 2005. The year 2005 was the second warmest on record, exceeded by 1998. This time series was compiled jointly by the Climatic Research Unit and the UK Meteorological Office Hadley Centre (Jones et al. 1999; Jones and Moberg 2003; www.cru.uea.ac.uk/cru/info/warming/). Reproduced with permission.

Projecting anthropogenic climate change

Climate models

The Earth's climate is a complex non-linear system with many inherent time and space scales. Thus the question of whether it is fundamentally possible to predict the time evolution of the state of this system naturally arises. The work of Lorenz in the 1960s showed that complex non-linear systems have limited predictability even though the time evolution is controlled by perfectly deterministic equations (the 'butterfly effect'). However, there is some evidence to suggest that events caused by the internal chaotic dynamics of the climate system such as El Niño, a disruption of the ocean-atmosphere system in the tropical Pacific, are predictable up to a year ahead (Battisti and Sarachik 1995), and variations due to externally driven forcing such as the annual cycle and short-term variability caused by volcanic eruptions are simulated well by models. Nevertheless it is not possible to predict specific changes far into our future. By contrast we can attempt to project a range of likely future climates under a range of new boundary conditions (such as changes to the composition of the atmosphere). Unlike weather which is the result of short term processes which are 'forgotten' in a couple of weeks, climate is the average weather over a certain time period and region. Significant deviations from the mean climate in a given space and time are referred to as 'climate change'.

In order to provide estimates of the Earth's future climate, general circulation models (GCMs) which solve the governing equations of atmospheric and ocean dynamics have been developed. These equations are non-linear and must be solved numerically on a three-dimensional spatial grid (typically with horizontal resolution of 100 - 300 km and with c. 20 vertical levels in both the atmosphere and ocean, and a time step of 30 minutes). These must represent a very large number of complex processes, interactions and possible feedbacks. Even with the high speed computers available today, the amount of time a model takes to run means it is impossible to explicitly include all of these highly detailed processes. Particularly as in many cases their time and space scales are smaller than the model's grid resolution. Thus many processes are parameterised i.e., approximated in some meaningful way. These parameterisations have to be chosen carefully and in some cases may be highly uncertain.

Quantifying the uncertainty in future climate predictions caused by the chaotic climate variability and model response uncertainty, requires multi-decadal model runs using many different parameter sets. When there are hundreds of parameters each with a range of values this leads to a very large number of computer runs, with each run being potentially the correct run i.e., most representative of the global climate. The climate*prediction*.net project aims to address this problem by using idle CPU on the general public's PCs to run the same climate model each with a different parameter set. This allows thousands of runs to be made and a proper ensemble distribution of possible future climates to be deduced in order to assess the 'most likely' future climate. Of course the actual future climate may not be in the 'most likely' range but may indeed be an outlier of the distribution – it is impossible to say. Nevertheless ensemble modelling can provide a good estimate of the range of possible temperature increases likely to affect our planet. Recent results from the ongoing climate*prediction*.net project show that changes in global air temperature may range between 2 and 11K with 3.4K being the most likely value, for a doubled atmospheric carbon dioxide concentration over the next 45 years (Stainforth et al. 2005).

Emissions scenarios

Greenhouse gas emissions caused by land use, population growth, economics and technological development will change over time. To include this anthropogenic forcing in climate models, 'emission scenarios' are used which are carefully constructed scenarios of socio-economic and demographic developments. The Intergovernmental Panel on Climate Change (IPCC) developed a new set of emission scenarios documented in the Special Report on Emission Scenarios (SRES; Nakicenovic 2000). These scenarios cover a wide range of driving forces of future emissions from demographic to technological and economic developments. The scenarios encompass different future developments that might influence greenhouse gas sources and sinks, such as alternative structures of energy systems and land use changes. However, none of the scenarios includes any future policies that explicitly address additional climate change initiatives (e.g., the Kyoto protocol) although all necessarily encompass various assumed future policies of other types that may directly influence greenhouse gases sources and sinks. The set of SRES emissions scenarios is based on literature assessment, six alternative modelling approaches and participation with many groups of people. The gases covered in emissions scenarios are carbon dioxide, carbon monoxide, hydrochlorofluorocarbons, hydrofluorocarbons, methane, nitrous oxide, nitrogen oxides, non-methane hydrocarbons, perfluorocarbons, sulphur dioxide, sulphur hexafluoride. These future scenarios are highly uncertain, so a very wide range of future emission paths are discussed in the literature. Table 2 (IPCC 2001) gives a detailed overview of SRES scenarios. The main storylines and scenario families are summarised below (IPCC 2001):

- A1: Describes a future world of very rapid economic growth, low population growth, and the rapid introduction of new and more efficient technologies. Major underlying themes are convergence among regions, capacity building, and increased cultural and social interactions, with a substantial reduction in regional differences in *per capita* income. The A1 scenario family develops into four groups that describe alternative directions of technological change in the energy system.

- A2: Describes a very heterogeneous world. The underlying theme is self-reliance and preservation of local identities. Fertility patterns across regions converge very slowly, which results in high population growth. Economic development is primarily regionally oriented and *per capita* economic growth and technological change are more fragmented and slower than in other storylines.

- B1: Describes a convergent world with the same low population growth as in the A1 storyline, but with rapid changes in economic structures toward a service and information economy, with reductions in material intensity, and the introduction of clean and resource-efficient technologies. The emphasis is on global solutions to economic, social, and environmental sustainability, including improved equity, but without additional climate initiatives.

- B2: Describes a world in which the emphasis is on local solutions to economic, social, and environmental sustainability. It is a world with moderate population growth, intermediate levels of economic development, and less rapid and more diverse technological change than in the B1 and A1 storylines. While the scenario is also oriented toward environmental protection and social equity, it focuses on local and regional levels.

Predicting Scotland's future climate

Much of the change in climate that will ensue over the next 30-40 years has already been set in motion by historic emissions because of inertia in the climate system. It is expected that the Gulf Stream (or 'North Atlantic conveyor') will continue to exert a very important influence on UK climate. Changes in the Atlantic conveyor over the past couple of decades are still open to speculation with Bryden et al. (2005) observing a reduction of 30%, and Knight et al. (2005) reporting a speeding up. It is, therefore, presently impossible to conclude with any certainty that there is a persistent weakening of the Gulf Stream. Some computer models indicate that its strength may weaken by up to 25% by 2100, but it is thought unlikely that this will lead to a cooling of UK climate within the next 100 years since the warming from greenhouse gases will more than offset any cooling from a weakening of the Gulf Stream (Hulme et al. 2002). However, we do not at present understand enough about ocean circulation to be completely confident about this prediction, particularly in the longer term.

In 1997 a project called the UK Climate Impacts Programme (UKCIP) was started at the University of Oxford to assess the impacts of climate change at regional and national levels. Later in this chapter we apply this dataset to estimate future climate change at Lochnagar. The UKCIP 2002 (UKCIP02) output data indicates that the

Table 2. Emissions Scenarios (IPCC 2001)

Family	A1				A2	B1	B2
Scenario Group	A1C	A1G	A1B	A1T	A2	B1	B2
Population growth	Low	Low	Low	Low	High	Low	Med.
GDP growth	Very high	Very high	Very high	Very high	Med.	High	Med.
Energy use	Very high	Very high	Very high	High	High	Low	Med.
Land-use changes	Low-Med	Low-Med	Low	Low	Med-High	High	Med.
Resource availability	High	High	Med	Med	Low	Low	Med.
Pace and direction of technological change	Rapid	Rapid	Rapid	Rapid	Slow	Med.	Med.
Favouring	Coal	Oil & gas	Balance	Non-fossils	Regional	Efficiency and dematerialisation	Dynamics as usual

expected scenario for Scotland is that summers will become warmer and drier while the winters will become wetter and warmer (Hulme et al. 2002). To produce these data they used a regional model nested within a GCM. The GCM used was the Hadley Centre coupled ocean-atmosphere global model (HadCM3; c. 300 km grid interval) which was forced with emission scenarios A1F1 (similar to A1B in Table 2), A2, B2 and B1. This model was then used indirectly to drive a higher resolution atmospheric global model (HadAM3H; c.120km grid interval) which in turn was used to drive a high resolution atmospheric regional model for Europe (HadRM3; c. 50km grid interval). These models have a 20 year history of development, have been carefully analysed and evaluated over many model generations, and represent perhaps the most sophisticated set of climate models anywhere in the world. The regional model is very computer intensive and therefore has only been run for the baseline (1961- 1990) and 2080s period and with multi-member ensembles only for the A2 emissions scenarios. Nevertheless by dynamical scaling between the global and regional climate models it has been possible to match the full suite of four scenarios matching the original SRES emissions scenarios. The mean climate for the 30 year intervals, the 2020s (2011-2040), the 2050s (2041-2070) and the 2080s (2071-2100), has then been constructed. These datasets have been used to provide future climate scenarios for the UK and using data representing the grid square containing Lochnagar we have attempted to project possible changes in climate at the loch up to 2100.

Future climate projections at Lochnagar

The UKCIP02 data described above are available at 50 km grid resolution. Predicted changes in air temperature, wind speed, snow and precipitation were extracted for the grid cell containing Lochnagar. The relevant cell is centred at 3.06° W, 56.9° N with an average elevation of 339m. Lochnagar is at 3.23° W; 56.95° N but over 400 m higher (788m). Exactly how the modelled changes at 300 m relate to those 400 m above is uncertain but changes in air temperature for example, can be estimated using present day temperature lapse rates. The UKCIP02 future climate changes are given relative to the baseline modelled climate for 1961-1990, therefore, in order to use these to calculate the absolute possible future climate at Lochnagar we need to know the meteorological characteristics at the loch from 1961-1990. An automatic weather station (AWS) was not installed at the loch until 1996 but we have combined these with long term records of air temperature, sunshine hours and rainfall from the UK Meteorological Office weather station at nearby Braemar (57.01° N; 3.398° W; 339m a.s.l.) to aid in the estimation process. The UKCIP02 data for the monthly mean average air temperature from 1961-1990 at the required grid point matches extremely well with the measured data at the Braemar weather station (coincidentally at the same altitude as the grid cell average), with a mean absolute error of 0.7 °C and a correlation coefficient of 0.99.

Future air temperature at Lochnagar

First, we reconstructed air temperatures at Lochnagar for 1961-1990 using an air temperature lapse rate to transform the monthly mean air temperatures at Braemar (at 339m) to the AWS (at 788m). To relate lowland and upland temperatures a lapse rate of -0.65°C per 100m is often employed (e.g., Agusti-Panareda and Thompson 2002). However, Figure 3 which plots Braemar and AWS air temperatures (1996-2004) shows that significant deviations occur from this relationship (shown by the dashed line) when air temperatures are cold (i.e., when Braemar is below 3 °C) presumably due to atmospheric temperature inversions. Thus, we split the data into two sections and use two different lapse rates:

If $T_{Braemar} > 3°C$ Lapse rate = -0.67°C/100m   MAE=0.52°C, R^2=0.98   (1a)

If $T_{Braemar} \leq 3°C$ Lapse rate = -0.40°C/100m   MAE=0.62°C, R^2=0.66   (1b)

$T_{AWS} = T_{Braemar}$ + Lapse rate x (785-339)/100   (1c)

where MAE is the mean absolute error. Using these formulae and adding on the changes in air temperature predicted by UKCIP02 data we can then predict the possible air temperatures at Lochnagar for 2071-2100 (Figure 4).

Figure 4 shows that the maximum (reconstructed) monthly mean air temperature at Lochnagar from 1961-1990 was 10.0 °C in July. However by 2080 this may increase to between 11.9 °C (low emissions scenario) and 13.7 °C (high emissions scenario). In winter, mean monthly temperatures will rarely fall below freezing whereas before 1999 monthly mean air temperatures were below zero for about 2 months each year. Thus,

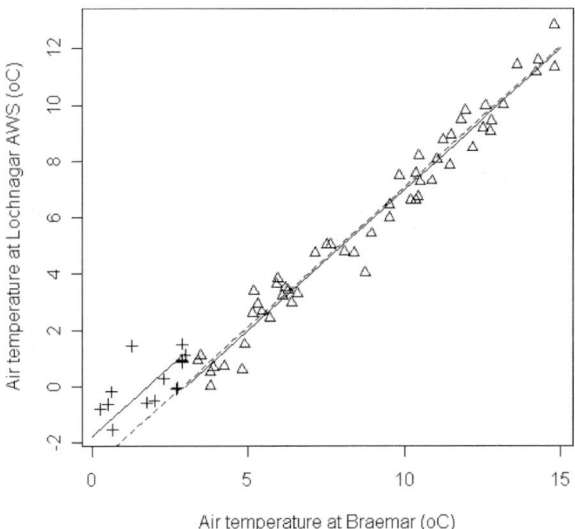

Figure 3. Relating mean monthly air temperatures at the automatic weather station by Lochnagar to the meteorological station at Braemar (1996-2004). When temperatures at Braemar are above 3 °C (triangles) temperatures at the AWS are well predicted by the environmental lapse rate (-0.65K per 100m; dashed line) or by the fitted lapse rate (-0.67K per 100m; solid line). When temperatures at Braemar are below 3 °C (crosses) atmospheric temperature inversions lead to a break down of the usual relationship and a lapse rate of -0.40K per 100m is used (solid line).

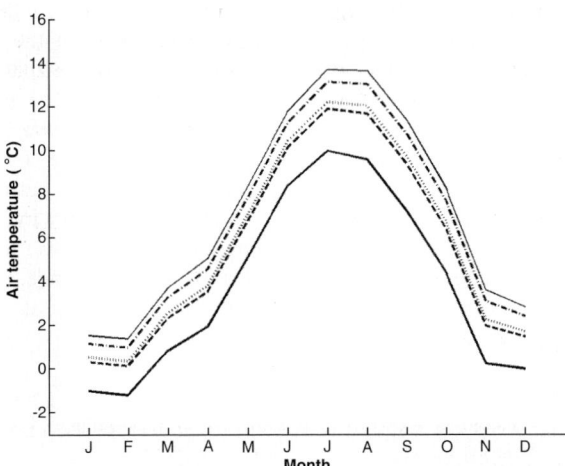

Figure 4. Past and future (2080s) mean monthly air temperatures at Lochnagar. Thick solid line shows temperatures averaged over 1961-1990 (reconstructed from Braemar data using Equations 1a-c); dashed line: future temperatures predicted using the low emissions scenario; dotted line: medium low emissions scenario; dash-dot line: medium high emissions and the thin solid line shows temperatures predicted using the high emissions scenario.

the worst case scenario (emissions wise) sees an almost 4 °C increase in summer air temperature and the 2-3 °C increase in winter air temperatures will mean snow is unlikely to lie in the catchment and ice will not last long on the loch.

Future wind speed at Lochnagar

Unfortunately there are no observed wind speed data at Lochnagar or Braemar for 1961-1990. In any case the predicted future changes in wind speed are very small. By 2080 there will be a small increase in wind speed in winter and spring (less than 1% for all scenarios) and a small decrease in summer and autumn (up to 4% in autumn under the 'high' scenario). This means that the changes in wind speed in winter and spring may be considered negligible and the decrease in summer and autumn (at most 4%) may slightly reduce mixing at the loch although this will be mostly affected by the increase in air temperatures. Lochnagar is windy due to its altitude and it is likely to remain windy.

Future snow and precipitation at Lochnagar

The response to a given weather system may be quite different at high altitudes than in the lowlands. In winter, storm systems from the west may bring rainfall and induce snow melt in the lowlands but cause snowfall in the highlands. When there are temperature inversions there may be snowmelt in the highlands but not in the lowlands due to fog. Thus it is difficult to transform snowfall and precipitation data from a lowland site such as Braemar to a mountain site such as Lochnagar. However, due to the increase in air temperatures the amount of snow predicted to fall in this region is set to decrease rapidly. The baseline model results for 1961-1990 at 339 m a.s.l. give the average snowfall rate to be 24 mm month^{-1} in December-January-February (DJF) and 15 mm/month in March-April-May (MAM). By contrast, even the lowest emission scenario predicts there will be no snow in MAM by 2020. By the 2080s there will be a decrease from the 1961-1990 snowfall average by at least 50% i.e., an annual average of 36mm (low emissions scenario) to 100% i.e., no annual snowfall (high emissions scenario) (see Table 3). The average precipitation rates in winter (DJF) and summer (June-July-August) for 1961-1990 are 300 and 315 cm/month respectively. By the 2080s precipitation at lower elevations is set to increase in winter by 11% (low scenario) to 21% (high) but to decrease in summer by 17 % (low) to 33% (high). In autumn and spring changes are negligible (less than 4% change in all scenarios). However, how these changes in rainfall will manifest themselves at the upland site of Lochnagar is still uncertain.

Table 3. Percentage change in snowfall rate with respect to simulated 1961~90 average, for different emissions scenarios.

	2020	2050	2080
Low	-20	-35	-49
Med-Low	-22	-41	-58
Med-High	-22	-46	-81
High	-23	-55	-96

Predicting water temperatures at Lochnagar

Although it is hard to say with certainty exactly how Lochnagar will be affected by future climate change it seems most likely that air temperatures will be up to 4 °C higher by the 2080s and precipitation will decrease with snowfall becoming a rare event. Using an empirical model for lake surface water temperature (LST) based on air temperature (e.g., as in Kettle et al. 2004) we can also estimate how changes in air temperature will affect the future loch surface temperature at Lochnagar.

To predict future LSTs from future mean monthly air temperatures we derive an empirical model for mean monthly LSTs using air temperature data from the AWS at the loch side and thermistor data from 1.5m depth for 1996-2004. The daily water temperature model described by Thompson et al. (Chapter 5: this volume) requires exponentially smoothed air temperatures. However on a monthly time scale this smoothing is not necessary. Instead we use simple linear regression to find a relationship between air and water temperature (Equation 2). The air temperature coefficient (0.89) is close to 1 which is as expected as a change in air temperature would be expected to produce a similar change in water temperature. It is not equal to 1 however, as changes in air temperature will affect other heat flux processes at the air-water interface such as evaporation and convection. This is supported by a study done in northern USA by Hondzo and Stefan (1993) where they found that epilimnetic temperatures will be higher than air temperature but will increase less than air temperature. Ideally we would like an equation for each month since the intercept probably indicates a solar input which will change through the year but there are not enough data for this. Furthermore, Equation 2 is only applicable to water temperatures above 4°C as the temperature-density relationship reverses at this temperature. Thus monthly mean (mon) of daily mean LST:

$$\text{mon(LST)} = 0.89 \ \text{mon(T}_{\text{AWS}}) + 2.08 \qquad \text{MAE} = 0.62 \ °C \qquad (2)$$

where T_{AWS} is the air temperature at the AWS and mon(LST) represents the monthly mean of daily averaged data. Figure 5 shows a comparison of the model fit (from Equation 2) to observations. We are now in a position to make projections of future summer water temperatures at Lochnagar by inserting the predicted future air temperatures at Lochnagar (computed using the method described in the previous section) into Equation 2. In this way monthly mean LSTs for the high and low emissions scenarios for the 2020s, the 2050s and the 2080s are constructed. These are shown in Figure 6 along with the reconstructed LST for 1961-1990 (the 1970s) based on air temperatures transformed from Braemar (Equations 1a-c). The Figure shows that by the 2080s, under the high emissions scenario, monthly mean water surface temperatures at Lochnagar may be as high as 14.4 °C, and the number of months where the water temperature is above 4 °C may increase from six (in the 1970s) to nine (in the 2080s). This hugely extended growing season has important implications for biological life in the loch.

The empirical model relies on the assumption that the loch will not significantly change its yearly stratification pattern. However, due to the higher air temperatures predicted it is likely that the stratification characteristics of the loch could change in the future. To examine this we also use a 1-d mechanistic model which simulates water

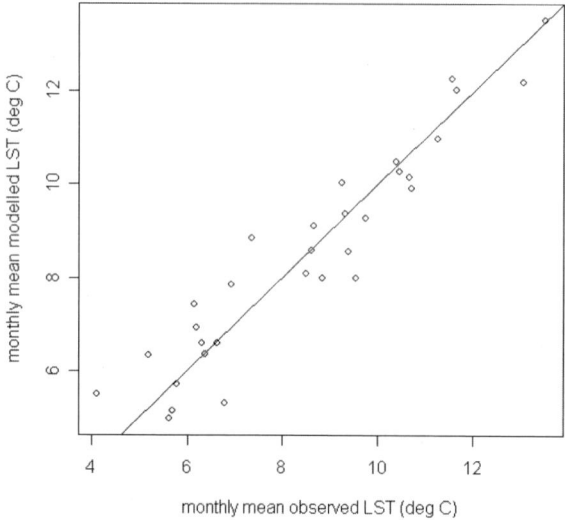

Figure 5. Comparison of modelled monthly mean LST (Equation 2) with monthly mean thermistor data at 1.5m depth.

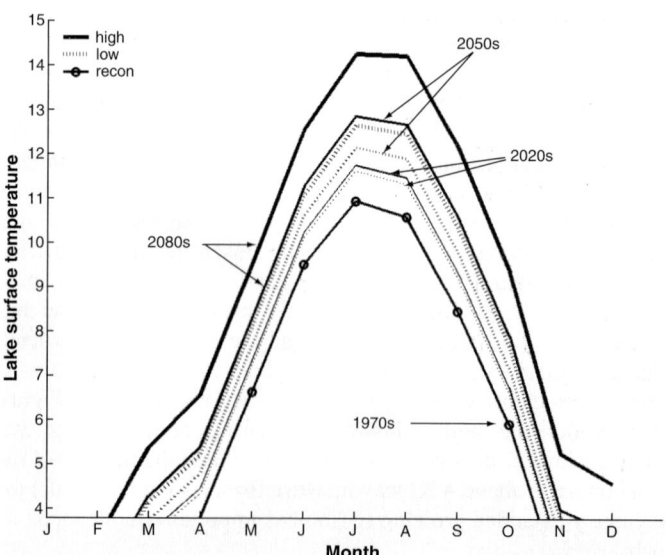

Figure 6. Predicted monthly mean lake surface water temperatures at Lochnagar for the 30 year timeslices 2020s, 2050s and 2080s. The 'recon' curve is constructed from the average monthly mean air temperatures at Braemar from 1961-1990 transformed to the AWS site and then input to the LST model. Note model is not valid for LST<4 °C.

temperatures through the whole depth of the lake, and is driven with wind speed, solar radiation, air temperature, relative humidity and air pressure. The mechanistic model is a general ocean turbulence model (GOTM; www.gotm.net) which is applicable to both oceans and lakes and includes processes such as convective and wind driven mixing through the water column, and sensible, latent and convective heat fluxes at the air-water interface. The model was run using a k-ε scheme to parameterise production (k) and dissipation (ε) of turbulent energy, and is driven using wind speed, solar radiation, relative humidity and air pressure. We used AWS measurements at Lochnagar where possible to drive the model. For solar radiation (which was not measured at the AWS) we used theoretical clear-sky solar radiation, scaled by total cloud cover data from ECMWF reanalysis data (ERA-40). Similarly, for air pressure, the data at the AWS were intermittent, so gaps were filled using ERA-40 pressure data adjusted to the altitude of loch. Figure 7 (a and b) shows a comparison between thermistor chain data and the mechanistic model results for 1999 (the year with the most complete series of AWS measurements). Because the actual solar radiation at the loch is unknown, and because of uncertainties in model parameters the model does not exactly reproduce the data. Nevertheless the model does capture the main characteristics of the loch's stratification and gives generally good results for the time evolution of the surface temperature (although the modelled temperatures are noticeably too warm in autumn). Given these very encouraging overall similarities, a future prediction of loch behaviour was carried out for the 2080s using meteorological conditions produced under the HadCM3 'High' scenario. Changes to relative humidity (-3.7%), air pressure (+1.23 mb), wind speed (<4%), and total cloud cover (-3.2%) by the 2080s are minor. The mean air temperature however, shows large increases. The monthly increases from January to December being 2.6, 2.6, 2.9, 3.1, 3.3, 3.4, 3.7, 4.1, 4.2, 3.9, 3.4 and 2.9 °C. A spline was fitted to interpolate these changes to the hourly driving air temperature data. The mechanistic model was then re-run, the output is displayed in panel c of Figure 7. A comparison of Figure 7b and 7c shows that the summer surface temperature of the loch has risen by about 3 °C. This compares well with the mean monthly predictions given by Equation 2 and shown in Figure 6. Compared with the modelled results for 1999 the stratification of the loch has not changed significantly. The loch is still mixed through winter (at least down to 19m), and only stratifies for about five days earlier in spring. The first stratification event is short-lived and the loch once again mixes about a month later. The second stratification event (which persists through the rest of the summer) lasts about a week longer. Thus, by the 2080s, in our mechanistic model, even though the loch is warmer it is still fully mixed in autumn and winter. The changes in lake behaviour which we find are much more conservative than those found in a similar study using five lakes in Finland (Elo et al. 1998). They found that changes of 30-75 days were predicted in the length of the stratification period by 2090 compared to the present day climate. Our more modest changes are probably related to Lochnagar's maritime climate which damps extreme weather.

Predicting ice cover at Lochnagar

The Lochnagar digicam (http://ecrc.geog.ucl.ac.uk/lochnagar/digcam/) has recorded the growth and loss of ice-cover since it was set up in 2002. Figure 8 in Thompson et al.,

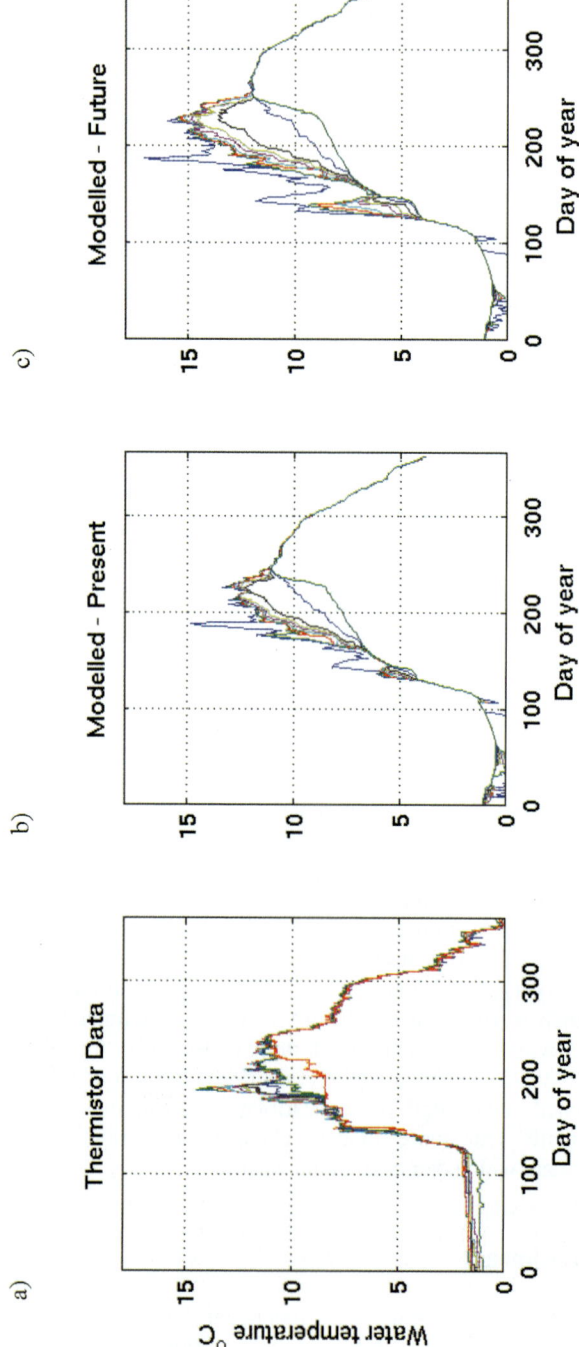

Figure 7. Simulations by mechanistic model. (a) shows thermistor chain data from 1999 at depths: 1.5, 2.5, 3.5, 4.5, 6, 7, 9, 14 and 19m. (b) shows temperatures at the same depths forced by data from the AWS and ECMWF reanalysis data. (c) shows predicted future temperatures.

(Chapter 5: this volume) shows the lake half way through ice-melt on 22nd March 2003. The daily series of such photographs allows the dates of ice-on and ice-off to be determined and so helps in tuning our ice-cover modelling algorithm, which is based on accumulated freezing and melting degree-days. Following the method described by Thompson et al (2005; their Figure 1), a weather generator has been constructed for the Scottish mountains and then run under various climate change scenarios. The ice-cover module was then forced by the synthetic weather series in order to estimate the duration of ice-cover that could be expected, at different elevations, for the various climate scenarios. The ice-cover duration on Scottish lochs turns out to be very sensitive to climate change (Figure 8). In particular Lochnagar at 788m happens to lie exactly within the most sensitive elevation range. Elevations of just under 1000m can be seen in Figure 8 to be expected to witness reductions in ice-cover duration of about 61 days for a 1.5 °C warming, while a 1 °C cooling would increase cover by 66 days. These strong

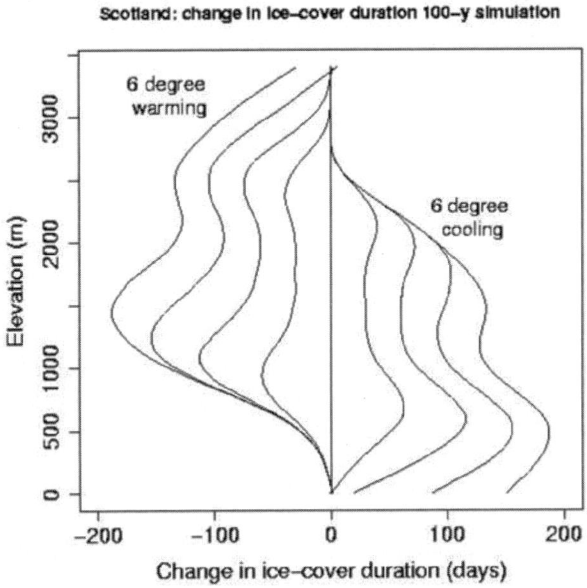

Figure 8. Ice-cover sensitivity versus elevation for Scottish lochs. The left hand part plots the decrease in ice-cover that can be expected for elevations between sea-level and 3000m for temperature increases of 1.5, 3, 4.5 and 6 °C. Although Scottish lochs are not found above 1000m, higher elevations are included as the graph is applicable in other maritime regions, e.g., Norway. At the bottom left of the diagram all the change scenarios go to zero because at sea-level today moderate sized lakes rarely freeze. Under any warming scenario they remain unfrozen and so there is no change to their ice-cover. The response curves show two peaks. The peak at lower elevations is associated with ice forming on lakes which are normally unfrozen. The peak response at higher elevations (around 2000m) is associated with lakes which would normally be frozen throughout the year experiencing an ice break in warmer summers. The right hand part plots the increases in ice-cover that can be expected for temperature decreases of 1.5, 3, 4.5 and 6 °C. Again strong double peaked responses are seen.

responses (42 days per degree) are a direct consequence of the combination of average winter temperatures at Lochnagar being close to freezing along with its maritime situation. Together they set up a threshold effect, whereby a small change in temperature can 'tip the balance' between freezing and melting regimes and hence whether ice forms or not, and consequently whether the ice-albedo effect operates or not. Ice-albedo effects and feedbacks are important climatic phenomena (Cubasch and Cess 1990). Figure 8 illustrates how the response of ice-cover to climate change, at Lochnagar, is twice that normally experienced in marine west coast climates. The normal situation is exemplified by the elevation zone between 1200 and 2200m, where a sensitivity of 30 days per degree pertains.

At present Lochnagar is frozen for about three months in the year. Snow-covered ice reflects most (approximately >70%) of the incident sunlight and so the loch remains frozen for a considerable time. However, a modest warming means that the ice has difficulty in becoming established, incident sunlight penetrates the loch and keeps the water warm. Indeed a three degree warming in climate, as is anticipated this century under many 'business-as-usual' global warming scenarios will cause Lochnagar to become in effect ice-free. Conversely colder winters as encountered in the Little Ice Age would have caused extended ice-cover.

Summary

- Future climate scenarios, as produced by a general circulation model (GCM), are used to estimate future changes to the climate at Lochnagar. The climate scenarios used here are for the UK and were produced by downscaling from a nested regional model within a global GCM. The data are given as changes to the baseline climate from 1961-1990. These meteorological prognoses are in turn used to drive both an empirical and a mechanistic model to estimate future water-temperatures at Lochnagar. Future changes in ice cover at Lochnagar are estimated using a freezing degree-day type model.
- Maximum (reconstructed) monthly mean air temperature at Lochnagar from 1961-1990 was 10.0 °C for July. By 2080 this may increase to between 11.9 °C (low emissions scenario) and 13.7 °C (high emissions scenario). In winter, mean monthly temperatures will rarely fall below freezing whereas before 1999 monthly mean air temperatures were below zero for about two months each year.
- By 2080 there will be a negligible increase in wind speed in winter and spring (less than 1% for all scenarios) and a small decrease in summer and autumn (up to 4% in autumn under the 'high' scenario).
- Due to the increase in air temperatures the amount of snow predicted to fall in this region is set to decrease rapidly. Even the lowest emission scenario predicts there will be no snow in March-April-May by 2020. By the 2080s there will be a decrease from the 1961-1990 annual snowfall average by between 50% and 100% i.e., no annual snowfall. By the 2080s precipitation at lower elevations is set to increase in winter by 11% (low scenario) to 21% (high) but to decrease in summer by 17 % (low) to 33% (high). How these changes in rainfall will manifest themselves at Lochnagar is still uncertain.

- By the 2080s, under the high emissions scenario, monthly mean water surface temperatures at Lochnagar may be as high as 14.4 °C, and the number of months where the water temperature is above 4 °C may increase from six (in the 1970s) to nine (in the 2080s).
- The ice-cover duration on Scottish lochs is sensitive to climate change. In particular Lochnagar at 788m happens to lie exactly within the most sensitive elevation range. Lochs with elevations of just under 1000m can be expected to witness reductions in ice-cover duration of about 61 days for a 1.5 °C warming.
- At present Lochnagar is frozen for about three months in the year. Snow-covered ice reflects approximately >70% of the incident sunlight and so the loch remains frozen for a considerable time. However a modest warming means that the ice has difficulty in becoming established, incident sunlight penetrates the loch and keeps the water warm. A 3 °C warming in climate, as is anticipated this century under many 'business-as-usual' global warming scenarios will cause Lochnagar to become, in effect, ice-free.

Acknowledgements

Many thanks to those who collected the field data used in this work; to UKCIP, ECMWF and CRU for access to climate datasets; and to the creators of GOTM for making it open access. Also thanks to the reviewers for their helpful comments.

References

Agusti-Panareda A. and Thompson R. 2002. Reconstructing air temperature at eleven remote alpine and arctic lakes in Europe from 1781 to 1997 AD. J. Paleolim. 28: 7–23.
Barnola J.M., Raynauld D., Korotkevich Y.S., Lorius C. 1987. Vostok ice core provides 160,000-year record of atmospheric CO_2. Nature 329: 408–414.
Battisti D.S. and Sarachik E.S. 1995. Understanding and predicting ENSO. Rev. Geophys. 33: 1367–1376. Part 2 Supplement S.
Bryden H.L., Longworth H.R. and Cunningham S.A., 2005. Slowing of the Atlantic meridional overturning circulation at 25° N. Nature 438:655–657.
Clark R.M. and Thompson R. 2004. Botanical records reveal changing seasons in a warming world, Australasian Science (October) 37–39.
Cubasch U. and Cess R.D. 1990: Processes and modeling. In: Houghton J.T., Jenkins G.J. and Ephraums J.J. (eds.). Climate Change. The IPCC Scientific Assessment. Cambridge University Press. 365 pp.
Elo A-R., Huttula T., Peltonen A. and Virta J. 1998. The effects of climate change on the temperature conditions of lakes. Boreal Environ. Res. 3: 137–150.
Etheridge D.M., Steele L.P., Langenfelds R.L., Francey R.J., Barnola J-M. and Morgan V.I. 1996. Natural and anthropogenic changes in the atmospheric CO_2 over the last 1000 years from air in Antarctic ice and firn. J. Geophys. Res. 101: 4115–4128.
Hansen J., Nazarenko L., Ruedy R., Mki. S., Willis J., Del Genio A., Koch D., Lacis A., Lo K., Menon S., Tovakov T., Perlwitz Ju., Russell G., Schmidt G.A. and Tausnev N. 2005. Earth's energy imbalance: Confirmation and implications. Science 308: 1431–1435.
Harper G.H., Mann D.G. and Thompson R. 2005. Phenological monitoring at Royal Botanic Garden Edinburgh. Sibbaldia. 2: 33–45.

Hondzo M. and Stefan H.G. 1993. Regional water temperature characteristics of lakes subjected to climate change. Climatic Change 24: 187–211.

Hulme M., Jenkins G.J., Lu X. 2002. Climate Change Scenarios for the United Kingdom: The UKCIP02 Scientific Report, Tyndall Centre for Climate Change Research, School of Environmental Sciences, University of East Anglia, Norwich, UK. 120 pp

IPCC. Houghton J.T., Ding Y., Griggs D.J., Noguer M., van der Linden P. J., Dai X., Maskell K. and Johnson C.A. (eds.). 2001. Climate Change 2001: The Scientific Basis. Contribution of Working Group I to the Third Assessment Report of the Intergovernmental Panel on Climate Change. Cambridge University Press, Cambridge, UK and New York, NY USA. 881pp.

Jones P.D. and Moberg A. 2003. Hemispheric and large-scale surface air temperature variations: An extensive revision and an update to 2001. J. Climate 16: 206–223.

Jones P.D., New M., Parker D.E., Martin S. and Rigor I.G. 1999. Surface air temperature and its changes over the past 150 years. Rev. Geophys. 37: 173–199.

Keeling C.D., Whorf T.P., Whalen M. and van der Plichtt J. 1995. Interannual extremes in the rate and rise of atmospheric carbon dioxide since 1980. Nature. 375: 666–670.

Keeling C.D., Whorf T.P. and the Carbon Dioxide Research Group. 2005. http://cdiac.esd.ornl.gov/ftp/trends/co2/maunaloa.co2

Kettle H., Thompson R., Anderson N.J. and Livingstone D.M. 2004. Empirical modeling of summer lake surface water temperatures in southwest Greenland. Limnol. Oceanog. 49: 271–282.

Kiehl J.T. and Trenberth K.E. 1997. Earth's annual global mean energy budget. Bull. Am. Meteorol. Soc. 78: 197.

Knight J.R., Allan R.J., Folland C.K., Vellinga M. and Mann M.E. 2005. A signature of persistent natural thermohaline circulation cycles in observed climate. Geophys. Res. Lett. 32: 10.1029/2005GL024233.

Legrand M.R., Lorius C., Barkov N.I. and Petrov V.N. 1988. Vostok (Antarctica) ice core: Atmospheric chemistry changes over the last climatic cycle (160,000 years). Atmos. Environ. 22: 317–331.

Nakicenovic N. (ed.) 2000. Special Report on Emissions Scenarios. A special report of working group III of the Intergovernmental Panel on Climate Change. Cambridge University Press, UK. 599 pp.

Petit J.R., Jouzel J., Raynaud D., Barkov N.I., Barnola J.M., Basile I., Bender M., Chappellaz J., Davis M., Delaygue G., Delmotte M., Kotlyakov V.M., Legrand M., Lipenkov V.Y.., Lorius C, Pepin L., Ritz C., Saltzman E and Stievenard M. 1999. Climate and atmospheric history of the past 420,000 years from the Vostok ice core, Antarctica. Nature 399: 429–436.

Price M.F. and Haslett J.R. 1995. Climate Change and Mountain Ecosystems. In Allan N.J.R. (ed). 1995. Mountains at Risk: Current Issues in Environmental Studies. Manohar. New Delhi. pp 73–97.

Stainforth D.A., Aina T., Christensen C., Collins M., Faull N., Frame D.J., Kettleborough J.A., Knight S., Martin A., Murphy J.M., Piani C., Sexton D., Smith L.A., Spicer R.A., Thorpe A.J. and Allen MR. 2005. Uncertainty in predictions of the climate response to rising levels of greenhouse gases. Nature 433: 403–406.

Stampfli, A and Zeiter, M. 2004. Plant regeneration directs changes in grassland composition after extreme drought: A 13-year study in southern Switzerland. J. Ecology. 92: 568–576.

Thompson R, Price D., Cameron N., Jones V., Bigler C., Rosén P., Hall R.I., Catalan J., Garcia J., Weckstrom J. and Korhola A. 2005. Quantitative calibration of remote mountain-lake sediments as climatic recorders of air temperature and ice-cover duration. Arctic Antarctic Alpine Res. 37:626–635.

Thompson R., Kettle H., Monteith D.T. and Rose N.L. This volume. Chapter 5. Lochnagar water-temperatures, climate and weather. In: Rose N.L. (ed.) 2007. Lochnagar: The natural history of a mountain lake. Springer. Dordrecht.

19. PAST AND FUTURE ENVIRONMENTAL CHANGE AT LOCHNAGAR AND THE IMPACTS OF A CHANGING CLIMATE

NEIL L. ROSE (nrose@geog.ucl.ac.uk) and RICHARD W. BATTARBEE
Environmental Change Research Centre
University College London
Pearson Building, Gower Street
London WC1E 6BT
United Kingdom

Key words: acidification, atmospheric deposition, climate change, environmental impacts, eutrophication, toxic pollutants

Introduction

Lochnagar has been part of the UK Acid Waters Monitoring Network (UK AWMN) since its inception in 1988. To date, this has resulted in 18 years of high quality chemical and biological monitoring data (Monteith and Evans 2005) and led to a great deal of further research focussed at the site (Chapter 1: Rose this volume). This research has formed the basis for a large part of this volume. The true value of environmental monitoring lies in such longevity, as progressively refined conclusions based on the interpretation of successive five-year data periods for the UK AWMN have shown (Patrick et al. 1995; Monteith and Evans 2000; Monteith 2005). Such work emphasises the value of long-term monitoring datasets but also shows the need to place contemporary measurements into their correct historical context.

Monitoring cannot provide retrospective data and therefore, in order to obtain records over the longer-term, alternative strategies are required. For many lakes, the sediment record can be used as a natural archive, storing a record of lake and catchment changes as well as that of contaminants deposited from the atmosphere. In the absence of long-term monitoring, lake sediments are one of the few ways by which to provide a record of environmental changes, over hundreds of years or possibly millennia. Contemporary monitoring and palaeolimnological approaches are therefore a complementary and powerful combination. Monitoring provides a means to interpret better the sediment record and the sediment record enables current observations to be placed in context and to be seen as part of longer term environmental change.

445

N.L. Rose (ed.), Lochnagar: The Natural History of a Mountain Lake
Developments in Paleoenvironmental Research, 445–464.
© *2007 Springer.*

At Lochnagar we are fortunate that we have both monitoring and palaeolimnological studies. A wealth of monitoring data now exists for the site from high resolution weather and water temperature data (Chapter 5: Thompson et al this volume) and fortnightly water and deposition chemistry, to annual sampling of macrophytes (Chapter 10: Flower et al this volume), invertebrates (Chapter 11: Woodward and Layer this volume) and fish (Chapter 12: Rosseland et al this volume). Studies on the loch sediments have also covered a wide range of time-scales, from persistent organic compounds whose records are measured on the scale of decades (Chapter 16: Muir and Rose this volume) and post-industrial trends of acidification (Chapter 14: Monteith et al. this volume) and trace metals (Chapter 15: Tipping et al this volume) to natural changes and early anthropogenic impacts measured on a millennial scale (Dalton et al. 2005; Chapter 8: Rose this volume). At the start of the 21st century when unprecedented climatic changes are predicted for at least the next 100 years, these monitoring and sediment data provide an invaluable base-line upon which to gauge future impacts.

Predictive models remain the only means by which to estimate future impacts but, by their very nature, there are considerable uncertainties (Chapter 18: Kettle and Thompson this volume). Therefore the ongoing role of monitoring is critical to determine the accuracy of these predictions and to refine our understanding of how ecosystems respond. The synergy of palaeolimnology, monitoring and modelling is therefore required to fully understand present ecological status and how an ecosystem may respond to future changes whether this is a recovery as a result of declining emissions, or a confounding of that recovery by the influences of a changing climate.

Climate change may be the greatest current anthropogenic threat to mountain lake ecosystems such as that at Lochnagar, but apart from the significant direct impacts there are also a number of potential indirect effects. Until recently, it was thought that ecosystems would respond positively to the removal of a stressor such that a pre-impact 'reference state' would be returned to. However, with global warming, the boundary conditions for ecosystems are changing, and targets for restoration in future may increasingly differ from the reference conditions of the past (Figure 1).

In this chapter we review the conclusions from the preceding chapters to highlight the historical changes determined from the sediment record at Lochnagar and assess its current status. We then discuss the evidence for climate change impacts at Lochnagar and consider the direct impacts that the changes predicted by future climate scenarios will have upon Lochnagar and other mountain lakes. Finally, we consider the indirect effects of climate by considering its effects on the stressors, acidification, eutrophication and toxic pollutant input.

Past environmental change and present status

Natural change

Evidence for environmental change at Lochnagar on a millennial time-scale is indicated by periodic changes in the organic matter content of long sediment cores (Dalton et al 2005). Between 80 and 180 cm in the core NAG27 peaks of c. 30% organic matter occur at 6200 BP, 5000 – 5500 BP and 3750 – 4250 BP, with 'troughs' of c. 15% organic matter in-between (Dalton et al. 2005; Chapter 8: Rose this volume). Above

80cm, however, organic matter increases to c. 30% and remains at 25 – 35% for the remainder of the core. This early period of peaks and troughs in organic matter content therefore covers c. 3000 years and Dalton et al. (2001; 2005) used a range of sediment proxies including C: N ratios, lipid analysis and grain-size analysis to determine that the cause was predominantly driven by catchment-derived sources. These changing inputs at Lochnagar can only have been driven by climate.

Cyclicity in organic matter over millennial time-scales has also been observed in lochs from the adjacent Cairngorm region. Battarbee et al. (2001) show evidence for 18 to 20 organic matter cycles covering c. 4000 years (i.e., each cycle of 200 – 225 years duration) in Lochan Uaine situated at 910 m above sea level on the eastern slopes of Cairn Toul (57° 05'N; 3° 05'W). They suggested that this cycling may be related to the 206 – 208 year ^{14}C cycle attributed by Stuiver and Brazunias (1993) to solar forcing and therefore the influence of naturally varying insolation on lake productivity. While this may explain the organic matter cycle at Lochan Uaine, the frequency of organic matter peaks at Lochnagar are more variable and up to five times lower. However, given that the influences of climate on these two remote lochs, only 35 km apart, must be similar, the reason for these differences must lie at the site-catchment level. The catchment of Lochan Uaine is dominated by steep, bare-rock faces and scree slopes with thin mineral soils. The catchment of Lochnagar is also dominated by a steep back-wall and considerable areas of scree and boulder fields but there are also areas of peat accumulation particularly at the northern end (Chapter 6: Helliwell et al. this volume). Dissolved organic carbon (DOC) and particulate carbon inputs to the loch, derived from these peats, may have added an additional carbon source to the sediments of Lochnagar and obscured the higher frequency organic matter cycles preserved at Lochan Uaine. The sequence of peaks and troughs in the Lochnagar sediment record cease when the organic matter content increases to a consistently elevated level of c. 30% and this increase may be due to the influence of inputs from peat erosion.

Anthropogenic impacts

With no direct impacts and no land-use change apart from possible variations in grazing intensity, the first clear evidence of human impact on Lochnagar comes from the detection of contamination by atmospheric pollutants. Concentrations of Hg are observed to increase from the base of the NAG29 core which, from ^{14}C dating (Dalton et al. 2005), covers a period of around 6000 years (i.e., from c. 4000 BC). While this Hg increase may be linked to an increase in sediment organic content, the ratio of Hg to organic matter (estimated by loss-on-ignition) is also seen to increase suggesting this is not the only cause (Yang and Rose 2003). This early Hg contamination may be due to peat combustion by early people in the same way that early increases in dioxin emissions are thought to have occurred (Meharg and Killham 2003).

Lead isotope data from long sediment cores at Lochnagar also indicate an early anthropogenic impact. Deviations from 'natural' [206]Pb:[207]Pb ratios towards a more anthropogenic Pb signal are observed from Medieval times although these shifts are minor when compared with impacts in the industrial period (Yang et al. in press). Such changes are in line with those detected in other remote regions (e.g., Bindler et al. 2001; Renberg et al. 2000; 2001) and are presumably a result of long distance transport from early industrial regions.

Atmospheric deposition is undoubtedly the greatest source of contamination in remote regions but despite the millennial record of low level impacts from trace metals, significant impacts (i.e., resulting in measurable biological change) as a consequence of atmospherically deposited contamination only occurred at Lochnagar from the mid-19th century with the first effects of acid deposition (Chapter 14: Monteith et al this volume). From c.1850 to 1970 the pH of the water at Lochnagar was reduced from 6.0 to 5.4 and is recorded in the sediments by changes in the diatom and chrysophyte assemblages (Chapter 14: Monteith et al this volume; Chapter 10: Flower et al. this volume). The sediment archive also provides a record of atmospheric deposition of persistent organic pollutants (POPs) (Chapter 16: Muir and Rose this volume), fly-ash particles from fossil-fuel combustion (Chapter 17: Rose and Yang this volume) as well as trace metals (Chapter 15: Tipping et al. this volume). As with non-marine sulphate deposition, and closely related to it by virtue of their sources, trace metals and fly-ash particles both reach a peak in the late-1970s. The temporal record of POP deposition is shorter as, apart from PAHs which have significant natural sources (e.g., forest fires); they were mainly developed, used and banned within the 20th century. Consequently, the POPs record begins and peaks within a matter of decades and many POPs now show significant declines in the uppermost sediments (Chapter 16: Muir and Rose this volume).

Over recent decades the implementation of national 'Clean Air' legislation and international agreements such as the Heavy Metals and POPs Protocols associated with the UNECE Convention on Long-Range Transboundary Air Pollution (http://www.unece.org/env/lrtap/) have led to significant improvements in air quality. The introduction of flue-gas desulphurisation and particle arrestor technology to fossil-fuel combustion sources, the move to unleaded vehicle fuel and a shift in fuels for electricity generation from coal and oil to natural gas have resulted in major reductions in atmospheric emissions and subsequent deposition. Over the last 20 years dry and wet deposition of sulphur in the UK has been reduced by 75% and 47% respectively during a period in which sulphur emissions throughout Europe have declined by around 70% (Fowler et al. 2004). Similarly, UK national trends for trace metals show that Pb, Zn and Cu deposition has been reduced by more than 75%, 55% and 30% respectively since the mid-1970s (Baker 2001) and that Hg emissions have fallen by more than 80% since 1970 (NAEI 1999).

These declining national and international trends are reflected in local measurements. For example, trace metals in deposition at the Banchory monitoring station just 40 km to the east of Lochnagar show considerable declines (Oslo and Paris Commissions 1994; Playford and Baker 2000) while over the period 1986-2001, Fowler et al. (2005) estimate that the non-marine sulphate concentration in wet deposition at Glen Dye declined by 1.46 µeq l^{-1} yr^{-1} and this is reflected in a decline in the concentration of non-marine sulphate in the Lochnagar outflow (Chapter 9: Jenkins et al. this volume).

However, despite the wealth of evidence for significant declines in pollutant deposition, continued monitoring and a new palaeolimnological assessment provide little substantive indication of ecological improvements in the aquatic environment of Lochnagar. The pH reconstruction of a sediment core, taken in 1991, provided tentative signs of a slight pH increase from a minimum of 5.3 in 1960 but since then inter-annual variability in inferred pH has been too large to discern any clear upward trend. The most recent available data show there has been neither increase in loch pH and alkalinity nor decline in labile aluminium concentration, nor a shift to a less acidic epilithic diatom flora relative to a 1988 baseline (Chapter 14: Monteith et al. this volume). Furthermore, fluxes of trace metals to the sediment basin remain elevated, providing no indication of the recent decline detected in deposition (Rose et al. 2004).

The combination of historical data from palaeoecological studies and contemporary monitoring therefore show Lochnagar to be at a critical stage in its history. Anthropogenically-derived contaminants deposited onto the loch and catchment over the industrial period have had a significant impact on the biota including changes in algal assemblages as a result of acid deposition and considerable accumulation of trace metals and POPs in the fish (Chapter 12: Rosseland et al. this volume). More significantly perhaps, where there should now be observable recovery as a result of emissions reductions, there is little or no improvement. The reductions in trace metal emissions have been so significant that the only possible explanation for the lack of reduction in the recent sediment record is that metals previously deposited to the catchment and stored in the soils are now being released. Indeed, for some metals, catchment inputs are now the major source to the loch by some margin (Chapter 15: Tipping et al. this volume; Yang et al. 2002). Two questions therefore remain. First, why is this occurring now and, second, why is this not happening to any significant degree at Lochnagar when recovery from acidification is occurring at other sensitive upland lakes in the UK (Monteith 2005). Hypotheses to explain these observations have been proposed and indicate that the changing climate may be playing a significant role. Therefore, climate change is not only having a direct impact (e.g., through increased air and water temperatures or reduced ice-cover) (Chapter 5: Thompson et al. this volume) but significant indirect impacts by interacting with other environmental stressors. For all these reasons, climate change is currently the greatest anthropogenic threat to mountain ecosystems and its effects are now increasingly observed. These and potential future impacts are discussed in more detail in the following sections.

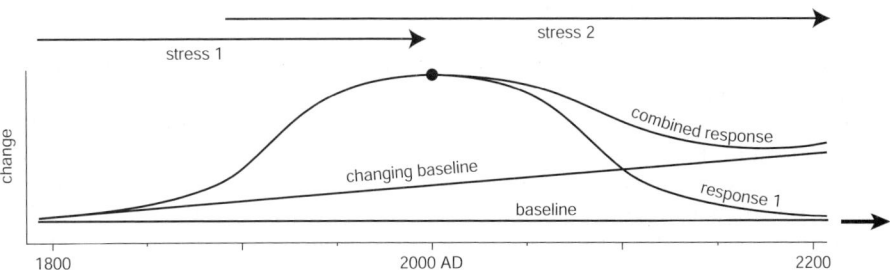

Figure 1. Diagram illustrating the response of a lake system to increasing and decreasing stresses and the potential change in ecosystem recovery (combined response) resulting from a change in boundary conditions (e.g., climate) of the system. (After Battarbee et al. 2005). Reproduced with permission of Blackwell Publishers.

Direct effects of climate

Details of the various climate change scenarios and their effects on air temperature, water temperature, ice and snow cover for Lochnagar are given in Kettle and Thompson (Chapter 18: this volume). The mean (2071 – 2100) maximum monthly air temperature for July is predicted to increase by between 2 °C (low emissions scenario) and 4 °C (high emissions scenario) over the 1961 – 1990 mean (10.7 °C), while winter mean monthly temperatures could increase by 2-3 °C such that they will rarely fall below 0 °C. Under this scenario snow would be unlikely to lie in the catchment and any ice-cover would be rare and short-lived. While precipitation at lower elevations is set to increase in winter by between 11% (low scenario) and 21% (high) by the 2080s and decrease in summer by 17 % (low) to 33% (high) it is still uncertain how these changes will effect upland sites such as Lochnagar. However, the increase in air temperatures would result in a decline in winter (December – February) snowfall by at least 50% on the 1961 – 1990 average by the 2080s while even the low emissions scenario predicts that there will be no snowfall in March-April-May by 2020. Models also suggest that, due to its altitude and maritime position, ice-cover duration at Lochnagar is extremely sensitive to climate change such that reductions in ice duration of around 60 days could be expected for a 1.5 °C warming (Chapter 18: Kettle and Thompson this volume). Therefore, even a modest warming means that ice would have difficulty in becoming established while a 3 °C warming in climate, as anticipated under many 'business-as-usual' warming scenarios will cause Lochnagar to become, in effect, ice free.

However, while predictions for future climate suggest dramatic changes for Lochnagar, there is some evidence, from observational records, for changes having already occurred. Anecdotal evidence suggests the presence of 'permanent' snow patches in the corrie and on the summit in the 19th and early-20th centuries. In the mid-19th century, Backhouse (1849 in Birks 1996) reported 4 feet (1.2 m) of snow in the Black Spout in August while other reports indicate that complete ice-cover between January and April and partial ice-cover between November and June often occurred (Patrick et al. 1991). More recently, personal observations indicate snow patches on the back-wall lasted through to August in the 1970s (H.J.B. Birks personal observation). However, our weekly or two-weekly observations since 1996 have recorded no snow in the catchment in any year after late-June until the following winter (Chapter 5: Thompson et al. this volume). We now rarely observe extended periods of complete ice-cover and ice formation and break-up occurs several times through the winter. The conclusion drawn from these observations is that far from snow and ice-cover reduction being a modern phenomenon, or the result of future climatic change, this is something that has been ongoing for many years. However, acceleration of these effects is likely to occur in the coming decades.

The warming of air and water and the loss of snow and ice-cover will impact on physico-chemical and hydrological processes as well as the biology of mountain lakes. In short, every aspect of mountain lake functioning will be affected, possibly beyond the ranges of anything experienced by these ecosystems since their formation. Details of how these changes are likely to impact Lochnagar specifically are considered in many of the chapters of this volume. Here, we summarise the main points:

- Warmer air and water temperatures may result in a longer growing season for terrestrial and aquatic plants and algae should sufficient nutrients be available. *Isoetes lacustris* is now observed to be growing considerably larger since monitoring began in 1988 and this is considered to be due to a longer growing season and less ice-scouring. The distribution and abundance of *Juncus bulbosus* var. *fluitans* is also now far greater than has been previously recorded (Chapter 10: Flower et al. this volume).
- A decrease in long-lasting snow beds will lead to a decline in chionophilous terrestrial plant species e.g., *Saxifraga rivularis*, as well as a decline in species dependent on snow-melt e.g., *Alopecurus borealis, Carex lachenalii, C. rariflora,* and *Sphagnum lindbergii.* Conversely, both growth and abundances of dwarf-shrubs such as *Calluna vulgaris, Empetrum nigrum* ssp. *hermaphroditum, Vaccinium myrtillus,* and *V. uliginosum* will increase at the expense of fern-rich snow-bed vegetation (Chapter 7: Birks this volume).
- Climatic changes may result in arctic-alpine and sub-arctic species being placed under pressure from 'new' competing species (Bunce et al. 1996) as well as a loss of habitat. In alpine and sub-alpine habitats, a 1.8 °C rise in temperature will be roughly equivalent to a 300m decrease in altitude. This effectively raises the lower altitudinal limit for these habitats reducing the total areal extents of their ranges (Kerr et al. 1999) while the altitude of the tree-line could also increase (McCarthy et al. 2001). In the Lochnagar area this could affect ptarmigan, dotterel and mountain hares as well as a range of rare plant species.
- Reduced snowfall and snow accumulation within the catchment will result in lower peaks in spring flow. Reduced summer rainfall and elevated evaporation will reduce flows through the summer and autumn while rain events following increased dry periods could lead to further acid episodes (Chapter 9: Jenkins et al. this volume).
- Reduced winter ice-cover will lead to a reduction in winter stratification (Chapter 5: Thompson et al this volume). This could lead to greater sediment resuspension resulting in reduced light penetration, more sediment focussing and/or increased loss of sediment material from the outflow.
- Peat erosion is already quite severe in some areas of the Lochnagar catchment (Chapter 6: Helliwell et al. this volume) and increased summer drought and higher winter rainfall could elevate erosion of catchment material into the loch. This could result in reduced light penetration and thereby impact aquatic macrophyte and algal growth. Furthermore, this would increase the amount of allochthonous sedimenting material contributing to an increase in the sediment accumulation rate (Chapter 8: Rose this volume).
- Increased water temperatures will result in a potentially longer growing season for algae (also depending on nutrient availability – see below) leading to higher levels of autochthonous sedimenting material. This would also contribute to an elevation in sediment accumulation rate. The sediment accumulation rate at Lochnagar is known to have doubled in the last 200 years (Chapter 8: Rose this volume).
- Increased algal growth could also affect the water chemistry of the loch. It has been shown that dissolved silica concentrations at Lochnagar decline with

water temperature increases (Chapter 5: Thompson et al. this volume) and it is suggested that this is due to uptake by siliceous algae during growth periods. Greater productivity and longer growing seasons could enhance this effect.

Indirect effects of climate

While the direct effects of increased air and water temperatures and reduced snow and ice-cover will have dramatic impacts at Lochnagar, the changing climate will also interact with other stressors thereby impacting the site in an indirect way. These indirect effects fall under three main categories:

i)   Remote effects, where the impacts of climate elsewhere indirectly causes an effect at Lochnagar. For example, climate change may allow the distributions of 'new' agricultural pests to expand, requiring the application of previously unused pesticides or greater usage of current ones. This results in greater emissions of persistent organic pollutants (POPs) to the atmosphere with subsequent transport to, and deposition in, remote areas.

ii)   Distribution effects, where climate affects the movement of pollutants in the atmosphere from source to sink. For example, warmer temperatures may increase the potential for volatile pollutants (Hg; some POPs) to be transported from temperate and tropical regions to cold areas (either latitudinally or altitudinally) (e.g., Grimalt et al. 2001).

iii)   Storage effects, where previously deposited atmospheric pollutants, stored in catchment soils, are released to freshwaters as a result of climate change.

Acidification

Sulphur deposition is predicted to continue to fall until 2020 when it will be only 7.5% of 1970 emissions (i.e., a reduction of 92.5%) (NEGTAP 2001). In fact with the continued decline of sulphur emissions from terrestrial sources, emissions from international shipping in European waters is expected to surpass the total terrestrial emissions from the 25 EU member states by 2020 and it therefore seems likely that this source will become increasingly important for sulphur deposition at Lochnagar over the next two decades unless more stringent controls on marine emissions are introduced (Chapter 13: Rose et al. this volume).

By comparison, while the emissions of oxidised and reduced nitrogen compounds in the UK have declined by around 45% and 10% respectively over the last ten years, the signal of reduced nitrogen deposition to the UK uplands is not detectable (Fowler et al. 2004; 2005; Chapter 14: Monteith et al. this volume). It is predicted that by 2020 nitrogen deposition will reduce by 61% from its 1988-9 peak of 0.85 Mtonnes, (0.85 terragrams or Tg) but the prospects for future nitrogen deposition in the UK uplands in the absence of further large reductions is for continued accumulation in soils and vegetation and an increase in the probability of nitrate breakthrough to surface waters, possibly leading to renewed acidification (Fowler et al. 2004; Chapter 9: Jenkins et al this volume) and / or eutrophication (see below). With the continuing decline of sulphur inputs to Lochnagar, future impacts from acidification are therefore dependant upon the role of nitrogen. However, while nitrogen dynamics in mountain lake catchments are not yet fully understood, the situation is further complicated by interactions with climate as some aspects aid recovery processes and others delay or confound them. Furthermore, many individual effects may work in opposition to each other. The impact of climate change on acidification at Lochnagar is therefore difficult to determine.

An increase in temperature would elevate the rate of bedrock and soil weathering thereby increasing the supply of base cations which neutralise acidifying processes. Further, evidence for a depression in nitrate concentrations following a period of higher winter temperatures at the start of AWMN monitoring in the late-1980s, suggest that any future increase in winter temperatures may be accompanied by a decline in the leaching of this acidifying anion, thus further assisting the recovery process (Chapter 14: Monteith et al. this volume). Conversely, the predicted increase in annual temperatures may elevate the rate of microbial organic matter decomposition in the catchment soils, potentially increasing the concentrations of DOC, and hence organic acidity, passing into the loch. Although organic acids are often described as 'weak', the net effect of a climatically driven rise in DOC could counter the effects of a decline in acid deposition to an unknown extent. However, while DOC concentrations have been rising across the UK in recent decades (Evans et al. 2005) the primary influence on DOC is now thought to be the reduction in sulphur deposition affecting soil acidity and hence DOC solubility (Evans et al. in press). Most of the increase in DOC currently observed in Lochnagar may, therefore, reflect the process of recovery from acid deposition, with strong mineral acids being partially replaced with weaker organic acids. Hence, the potentially detrimental effects of warming on acidity status may only become apparent once sulphur deposition loads stabilise.

While an increase in precipitation would increase the proportion of sulphur and nitrogen reaching the loch directly and hence a smaller proportion being buffered by the underlying geology, the effects of climate on acidic episodes may be more important (Chapter 9: Jenkins et al. this volume). It is likely that the incidence of sea-salt induced episodes will increase and that large inputs of marine ions (predominantly chloride and sodium) accompanying intense depressions and frontal rainfall may trigger acidic pulses (Chapter 9: Jenkins et al. this volume). This is caused by the large input of sodium being immobilised by the soil and in turn releasing hydrogen and aluminium if the soil is poorly buffered (Wright et al. 1988; Langan 1989; Heath et al. 1992). Furthermore, were the frequency of winter storms to increase, this could lead to elevated flushes of organic matter, known to be correlated with acidic episodes (Easthouse et al. 1992), while the potential for flushes of acidifying sulphate could increase along with the increased prevalence of heavy rainfall following prolonged dry periods. However, the decreasing importance of sulphur could reduce the impact of this particular process. Snowmelt driven episodes, which can also result in acute ecological stress and which are often accompanied by a concentrated release of accumulated dry deposits of acid anions (chloride, sulphate and nitrate) (Helliwell et al. 1998) are likely to decrease in importance as a result of increasing winter temperatures.

Eutrophication

With the recent and predicted further decline of sulphur deposition at Lochnagar, the importance of nitrogen as an acidifier will increase. However, in upland areas and particularly in oligotrophic systems such as Lochnagar, nitrogen deposition also has an important role as a eutrophier (i.e., as a nutrient) (Curtis et al. 2004). Elevated nitrogen deposition is known to be detrimental to terrestrial plants common at Lochnagar including *Vaccinium myrtillus*, *V. vitis-idaea* and the mosses *Hylocomium splendens* and *Pleurozium schreberi* (Nordin et al. 2005) while for the loch itself, nitrogen breakthrough from the catchment is of greatest concern. As deposition continues the terrestrial capacity for nitrogen retention (e.g., in soils and vegetation) is reached and nitrogen saturation is said to have occurred. At this point excess nitrogen can no longer be utilised and will be released to the surface water. This is identified by an increase in nitrate concentration. In the first instance, this is usually observed in winter when biological demand is lower, but, as nitrogen deposition continues, summer demand may also be exceeded and elevated nitrate observed in surface waters all year round. At Lochnagar, the lack of seasonality in nitrate concentrations in loch waters (see Figure 5 in Chapter 14: Monteith et al. this volume) indicates that nitrogen saturation has occurred although the relatively sparse soil cover and vegetation would be unlikely to immobilise a significant fraction of the deposited nitrogen even during the growing season (Chapter 9: Jenkins et al. this volume).

In terms of the impacts of climate on eutrophying processes at Lochnagar, warmer conditions may lead to an increase in the mineralisation rate of organic matter in catchment soils releasing both carbon and nitrogen. The observed increase in surface water DOC since the mid-1990s (see Figure 5d in Chapter 14: Monteith et al. this volume) may support this hypothesis although this may also be related to recovery from acidification. Warmer temperatures would also permit a longer growing season both in the catchment and loch increasing biological demand for nitrogen and possibly leading to phosphorus limitation to growth. However, increased soil temperatures could enhance soil microbial activity leading to higher nitrogen and phosphorus loading from catchment soils while particulate phosphorus input may also be elevated from increased soil erosion. Furthermore, it is known that aluminium, released by acidification, may inactivate phosphorus through complexation (Kopáček et al. 2001). Therefore, recovery from acidification may reduce this complexation resulting in less phosphorus limitation, although the complexation itself may be influenced by the effect of DOC levels on light penetration. If both nitrogen and phosphorus inputs are elevated as a result of these processes and water temperatures also increase, then it is likely that Lochnagar will become more productive, resulting in reduced light penetration, greater autochthonous inputs to the sediment and an elevated sediment accumulation rate. This could also result in the increased scavenging of pollutants from the water column by algae resulting in enhanced trace metals and POPs input to the sediment record (see below).

Finally, there is now some concern over the scattering of cremated human ashes on mountain summits (http://news.bbc.co.uk/1/hi/scotland/4645896.stm) and the impact this may have on nutrient inputs. Both the Mountaineering Council of Scotland and Scottish Natural Heritage are concerned about the possible enrichment this might cause to otherwise impoverished mountain soils due to elevated phosphorus and calcium.

Toxic pollutants

While there have been significant reductions in the emissions and deposition of trace metals over the last 30 years, predicted trends for the future are very uncertain

(NEGTAP 2001). Similarly, while the UK emissions of NM-VOC (non-methane volatile organic compounds) are predicted to have fallen from 2.7 Mtonnes (Tg) in 1989 to 1.4 Mtonnes (Tg) in 2010 (a reduction of 48%) predictions beyond this are also unclear. For upland lakes such as Lochnagar there are three main issues with respect to toxic pollutants under a changing climate:

Increased usage
Previous studies at Lochnagar have shown evidence for the long-range transport of POPs from eastern Europe and from North America where those compounds have neither been produced nor used in the UK (Rose et al. 2001; Chapter 16: Muir and Rose this volume). Thus the elevated use of insecticides and pesticides as a result of the increased need to control agricultural pests due to a changing climate need not be confined to the UK in order to impact Lochnagar. Although it is predicted that species from lower latitudes will colonise the UK as temperatures increase it is unknown to what extent chemical control of these species will be required. However, the same situation will occur across Europe and, in areas where the use of pesticides is less regulated, this may lead to increased usage of currently employed pesticides and / or the development and employment of new ones. Given the potential for increased mobility under a warmer climate this could result in an increase in atmospheric inputs and subsequent transport to remote areas such as Lochnagar. Changes in climate will also inevitably lead to changes in land-use and agricultural practices both in Europe and globally (McCarthy et al. 2001). Increased usage of fertilisers will lead to increased metal inputs which in turn will result in greater trace metal run-off to surface waters. Such a scenario is not expected to impact remote lakes such as Lochnagar although increased cycling of trace metals in the environment may indirectly lead to increased trace metal inputs at remote sites. Conversely, warmer winters may lead to reduced electricity usage for heating which could result in a reduction in coal use. Coal combustion is one of the main sources of metal emissions to the atmosphere and this could help offset increases from recycled agricultural sources.

Distribution and storage
The volatilisation of compounds is a temperature dependent process and therefore in a warmer atmosphere those compounds which volatilise readily (e.g., Hg and some POPs) will be able to travel greater distances prior to deposition. However, in mountain lakes it is the absorption of compounds in water and their retention in organic matter that is important and this process is also temperature dependent. In mountain lakes, because of the lower water temperature, less volatile POPs (or 'semi-volatile' compounds) are preferentially absorbed in water and selectively trapped in sediments and fish. Thus, high altitude lakes are preferentially enriched in less-volatile compounds over more volatile ones. Hence, while mobility within the atmosphere may increase as a result of increasing air temperatures, in mountain lakes there will be less absorption of semi-volatile compounds and possibly even some release from them. As a consequence we may expect a shift in the altitudinal (or latitudinal) gradients of POPs accumulation (e.g., Grimalt et al. 2001; Fernández et al. 1999; 2000; Vives et al. 2004a; b) although the predicted temperature increases are unlikely to affect the least-volatile compounds in any significant way.

Remobilisation of previously deposited pollutants
The catchment area of Lochnagar is almost ten times larger than the lake area and the scale of pollutant deposition to the catchment over the industrial period will be around an order of magnitude larger. Therefore, despite the limited soil coverage in the catchment a considerable store of trace metals and POPs are calculated to be stored there. Indeed, Yang et al (2002) estimate that there is c. 400 times the amount of Pb stored in Lochnagar soils as was deposited from the atmosphere onto the loch and its catchment in 2000. A similar figure is also estimated for Hg. Studies of the whole basin flux of Pb and Hg in Lochnagar (Yang et al. 2002; Chapter 15: Tipping et al. this volume) have shown that while the sediment flux reflects the increase in deposition over the period 1850 – 1950, the subsequent dramatic decline in metal emission and deposition is not observed. As there are no direct sources of contamination the 'additional' metal entering the site and causing this lack of response to depositional changes can only be from metals previously deposited onto the catchment now being transferred to the loch.

There are four hypotheses for the cause of this enhanced catchment transfer (Rose et al. 2004). First, a simple time-lag, i.e., metals deposited onto the Lochnagar catchment take a number of years to pass through. Second, increased erosion of atmospherically contaminated catchment soils is bringing metals into the loch (e.g., Lindeberg et al. 2006). Third, increasing levels of DOC are being transferred from the catchment to which metals are adsorbed (Kolka et al. 2001; Ravichandran 2004). POPs are also known to have a strong affinity for DOC (Gao et al. 1998; Winch et al. 2002) and this mechanism could also be applicable for catchment stored organic contaminants. Fourth, with available nutrients, warmer winters could result in longer growing seasons for algae which could then scavenge metals from the loch water eventually becoming incorporated into the sediment record.

It is likely that the increase in metal inputs from the catchment at Lochnagar is a combination of several of the above effects. However, if the temperature and precipitation driven processes are important then climate scenarios that infer increased summer or winter temperatures and/or increased rainfall and/or increased storminess would point to a continuation of enhanced metal inputs as a result of the second, third and fourth hypotheses. We have already observed that peat erosion is increasing (Chapter 6: Helliwell et al. this volume), DOC is increasing (Rose et al. 2004; Chapter 14: Monteith et al. this volume) and that ice-cover periods are declining as a result of water temperature increases (Chapter 18: Kettle and Thompson this volume). Consequently, any of the latter three causes, or a combination of them could result in the observed lack of response to decreasing metal emissions. While these hypotheses were formulated to explain trace metal transfer (Rose et al. 2004), little data exist for POPs. However, the adsorption of POPs to organics is also strong and the same pathways are applicable. Whichever way these pollutants enter the loch, once mobilised from the catchment they can enter the aquatic food chain in two ways, either via benthic feeders or scavenged from the water column by algae, a process which itself may be enhanced by possible increased productivity.

Other impacts

While the effects of a changing climate will impact on all aspects of the Lochnagar ecosystem and create a diverse set of responses with other environmental stressors, there may be other effects more directly related to human activities in the area. As temperatures increase and the rainfall in the Scottish summer declines (Chapter 18: Kettle and Thompson this volume) the level of recreational activities in mountain areas may also increase. The numbers of walkers in the Lochnagar area is highly seasonal (Figure 2) and so it may be expected that warmer and drier weather may increase the numbers of people walking to the site and extend the period during which high numbers visit the area, although such activities may be tempered should the extent or abundance of the Highland Biting Midge (*Culicoides impunctatus*) also expand.

The number of tracks crossing the catchment is now quite evident at Lochnagar particularly in aerial view (Craig 1995). In areas that are popular with hill walkers and rock climbers such as Lochnagar, the impacts have increased in recent decades (Lance et al. 1991) and damage due to trampling is concentrated along pathways on the main climbing routes and to the summits (Chapter 6: Helliwell et al. this volume). Human trampling can affect soil erosion (Brazier et al. 1996) as the protective vegetation cover is damaged and soil is exposed to erosion by wind and water. Furthermore, Grieve (2001) showed that the effects of trampling in soils similar to those identified in the Lochnagar catchment were intensified where natural cryoturbation processes (disturbance due to the effect of freezing and thawing in soils) were predominant. This increase in footpath erosion cannot be ascribed to other possible causes such as an increase in deer numbers as these have not dramatically increased since the 1960s (Balmoral Estate Office data; Youngson and Stewart 1996) (Figure 3).

Erosion around the loch shore has also increased and paths around the loch have become more evident, as elsewhere in the catchment. In some areas, particularly to the northeast of the loch, deep peat faces are now severely eroded (Chapter 6: Helliwell et al. this volume). Undoubtedly this erosion has, at least in part, contributed to the doubling of the sediment accumulation rate over the last two centuries (Chapter 8: Rose this volume) but it seems unlikely that the erosion of these significant peat faces can be solely due to walkers or deer activity. Climate probably has an influence.

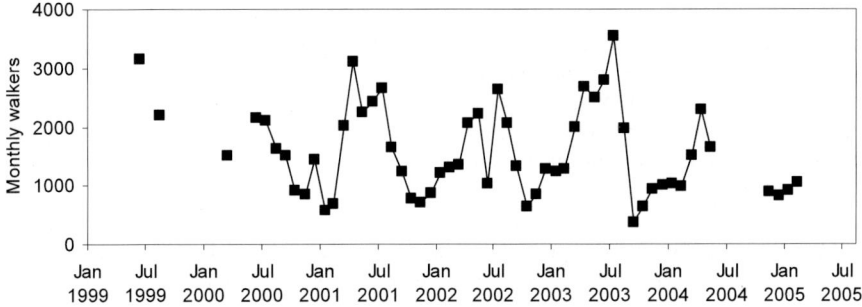

Figure 2. Numbers of walkers on the Lochnagar path 1999 to 2005. (Data from Balmoral Estates Office.)

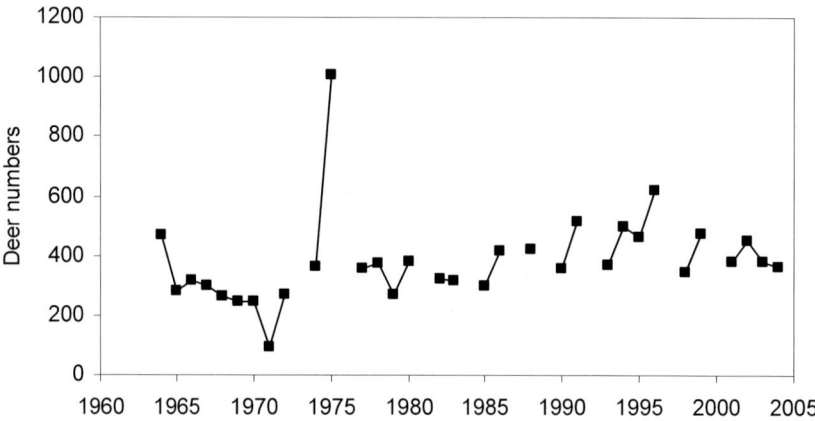

Figure 3. Number of deer in the Glen Gelder beat 1964 to 2004. The Glen Gelder beat includes the whole of the Lochnagar catchment but also areas beyond. Hence, while these data provide an indication of the numbers of deer that are in the Lochnagar area, they do not necessarily represent the number of deer directly affecting the Lochnagar catchment. Data from Balmoral Estates Office.

Conclusions

The sediment record at Lochnagar reveals that environmental change has been occurring at the site for millennia as a result of natural climatic fluctuations. Anthropogenic impacts are detectable for over a thousand years but significant impacts resulting in measurable biological change have only occurred in the last 150 years as a result of acid deposition. While significant progress has been made on the reduction of emissions to the atmosphere of acidifying species, trace metals and POPs, it is current and future climate change which is the greatest threat to the Lochnagar ecosystem and upland and mountain lakes like it across the globe.

While the greatest impacts are yet to come, and to a certain extent are already probably irreversible, the effects of climate change are already observable at Lochnagar in a number of ways: the reduction in ice and snow duration; changes in abundance, distribution and growth patterns of aquatic macrophytes; increased catchment soil erosion; increasing sediment accumulation rate (from both allochthonous and autochthonous sources); and the enhanced input of trace metals from the catchment. Only recently have we begun to realise the extent to which climate has already affected the loch and its catchment and more significantly perhaps, the extent to which all ecological compartments and processes will be impacted both directly and indirectly. The effects of climate change could adversely affect the loch and other upland ecosystems in a way in which no other anthropogenic impact has done in the past. Furthermore, the change could be rapid. In 100 years Lochnagar could have no ice or snow cover, a different aquatic flora and fauna, while the vulnerable species in the catchment dependent on snow beds and melt-water could have been lost and replaced by an increase in dwarf shrubs or new species and possibly even trees from lower altitudes.

Monitoring over the last 20 years combined with palaeolimnological data spanning centuries or millennia have provided a baseline against which to identify the rate and extent of future change. Many of the impacts observed at Lochnagar have not been observed at other sites, and this is probably due to the lack of intense monitoring and, at lower altitudes, due to the effects being masked by other signals. Therefore, mountain lakes provide a means by which an 'early warning' can be given for impacts on both sensitive mountain ecosystems and elsewhere. Maintaining the research and monitoring programmes at these fundamentally important sites is vital, while the expansion of the work being undertaken at Lochnagar to a national and international network of upland and mountain lakes would provide insights into the impacts of climate change on a broader geographical scale.

Summary

- Changes in organic matter content of long sediment cores from Lochnagar indicate a response to natural climatic fluctuations over 3000 years. Shorter cycles (c. 200 years) observable at other mountain lakes are not seen at Lochnagar, possibly due to the effects of peat-derived inputs from the catchment.

- The first clear evidence of human impact on Lochnagar comes from the detection of contamination from atmospheric deposition of Hg possibly as a result of human peat combustion since c. 6000 BP. Lead isotope data also indicate an early anthropogenic impact from Medieval times.

- First significant impacts resulting in measurable biological change occurred at Lochnagar from the mid-19th century with the first effects of acid deposition. The sediment archive also provides a record of atmospheric deposition of persistent organic pollutants (POPs), fly-ash particles from fossil-fuel combustion and trace metals.

- Climate change is currently the greatest anthropogenic threat to mountain ecosystems. At Lochnagar mean monthly air temperatures could increase by 2 – 4 °C by 2100, summer precipitation could decrease and winter precipitation could increase. Snow cover and duration could decline and snowfall vanish completely in March - May. By 2100 the loch could effectively be ice-free.

- Anecdotal evidence suggests that ice and snow cover has already been in decline for many years.

- Other effects of climate change at Lochnagar could include: changes to aquatic macrophyte abundance, distribution and growth; changes in terrestrial plant species dependent on snow cover or snow-melt; the introduction of invasive new species and the possible raising of the tree-line; changes to the hydrology, particularly with respect to spring melt and summer flows; increased catchment peat erosion leading to reduced light penetration and an increase in sediment accumulation rate; and increased growing seasons for aquatic plants and algae.

- Effects of climate on the recovery of Lochnagar from acidification are complex with some processes working to aid recovery whilst others will help confound it. However, the roles of nitrogen and dissolved organic carbon are likely to become increasingly important.
- Nitrogen is also critically important at upland lakes as a nutrient. Nitrogen breakthrough may already be indicating saturation in the Lochnagar catchment. However, some responses to climate may also cause further phosphorus inputs. Increased nitrogen and phosphorus combined with warmer temperatures could lead to much greater productivity in the loch.
- Climate change could influence toxic pollutant (trace metals; POPs) inputs to Lochnagar in three ways: (i) increased usage of pesticides and fertilisers, (ii) changes to long-range transport to, and incorporation within, the loch and (iii) release of pollutants previously deposited and stored in catchment soils.
- Increased mountain recreation in the area could elevate the erosion of soils in the catchment as more people broaden paths, reduce vegetation and expose soil to wind and rain.
- In 100 years Lochnagar could have no ice or snow cover, a different aquatic flora and fauna, while the vulnerable species in the catchment could have been lost and replaced by an increase in dwarf shrubs or new species or possibly even by trees from lower altitudes.
- Current monitoring combined with palaeolimnological data provides a baseline against which to identify the rate and extent of future change. Mountain lakes can provide an 'early warning' and maintaining research and monitoring programmes at these sites is vital. The expansion of the work being undertaken at Lochnagar to a national and international network of upland and mountain lakes would provide insights into the impacts of climate change on a broader geographical scale.

Acknowledgements

We would like to thank the Balmoral Estates Office for providing data on the numbers of people visiting the Lochnagar area and deer numbers in Glen Gelder; chapter authors for the use of their conclusions; Roland Psenner, Jordi Catalan and Don Monteith for their helpful comments on the manuscript.

References

Baker S.J. 2001. Trace and major elements in the atmosphere at rural locations in the UK: Summary data for 1999. National Environmental Technology Centre Report number AEAT/R/ENV/0264. 80pp.

Battarbee R.W., Cameron N.G., Golding P, Brooks S.J., Switsur R., Harkness D., Appleby P., Oldfield F., Thompson R., Monteith D.T. and McGovern A. 2001. Evidence for Holocene climate variability from the sediments of a Scottish remote mountain lake. J. Quat. Sci. 16: 339–346.

Battarbee R.W., Anderson N.J., Jeppesen E. and Leavitt P.R. 2005. Combining palaeolimnological and limnological approaches in assessing lake ecosystem response to nutrient reduction. Freshwat. Biol. 50: 1772–1780.

Bindler R., Renberg I., Anderson N.J., Appleby P.G., Emteryd O. and Boyle J. 2001. Pb isotope ratios of lake sediments in West Greenland: Inferences on pollution sources. Atmos. Environ. 35: 4675–4685.

Birks H.J.B. 1996. Palaeoecological studies in the Cairngorms - Summary and future research needs. Bot. J. Scot. 48: 117–126.

Birks H.J.B. This volume. Chapter 7. Flora and vegetation of Lochnagar– past, present and future. In: Rose N.L. (ed.) 2007. Lochnagar: The natural history of a mountain lake. Springer. Dordrecht.

Brazier V., Gordon J.E., Hubbard A. and Sugden D.E. 1996. The geomorphological evolution of a dynamic landscape: The Cairngorm Mountains, Scotland. Bot. J. Scot. 48: 13–30.

Bunce R.G.H., Watkins J.W., Gillespie M.K. and Howard D.C. 1996. The Cairngorms environment and climate change in the European context. Bot. J. Scot. 48: 127–135.

Craig R. 1995. A Lochnagar Climb. Cairngorm Club J. 20: 294–295

Curtis C.J., Monteith D.T., Evans C.D. and Helliwell R. 2004. Nitrate and future achievement of "good ecological status" in upland waters. In: Battarbee R.W., Curtis C.J. and Binney H.A. (eds.). The future of Britain's upland waters. Proceedings of a meeting held on 21st April 2004. Ensis Publishing. London. 51pp.

Dalton C., Battarbee R.W., Birks H.J.B., Brooks S.J. Cameron N.G., Derrick S., Evershed R.P., Peglar S.M., Scott J.A. and Thompson R. 2001. Holocene lake sediment core sequences from Lochnagar, Cairngorm Mts., Scotland – UK Final report for CHILL-10,000. Environmental Change Research Centre, University College London. Research Report No. 77. 94 pp.

Dalton C., Birks H.J.B., Brooks S.J. Cameron N.G., Evershed R.P., Peglar S.M., Scott J.A. and Thompson R. 2005. A multi-proxy study of lake-development in response to catchment changes during the Holocene at Lochnagar, north-east Scotland. Palaeogeog. Palaeoclimat. Palaeoecol. 221: 175–201.

Easthouse K.B., Mulder J., Christophersen N. and Seip H.M. 1992. Dissolved organic carbon fractions in soil and streamwater during variable hydrological conditions at Birkenes, S. Norway. Water Resour. Res. 28: 1585–1596.

Evans C.D., Monteith D.T and Cooper D.M. 2005. Long-term increases in surface water dissolved organic carbon: Observations, possible causes and environmental impacts. Environ. Pollut. 137: 55–72.

Evans C.D., Chapman P.J., Clark J.M., Monteith D.T. and Cresser M.S. in press. Alternative explanations for rising dissolved organic carbon export from organic soils. Global Change Biology.

Fernández P., Vilanova R.M. and Grimalt J.O. 1999. Sediment fluxes of polycyclic aromatic hydrocarbons in European high altitude mountain lakes. Environ. Sci. Technol. 33: 3716–3722.

Fernández P., Vilanova R.M., Martinez C., Appleby P. and Grimalt J.O. 2000. The historical record of atmospheric pyrolytic pollution over Europe registered in the sedimentary PAH from remote mountain lakes. Environ. Sci. Technol 34: 1906–1913.

Flower R.J., Monteith D.T., Tyler J., Shilland E. and Pla S. This volume. Chapter 10. The aquatic flora of Lochnagar. In: Rose N.L. (ed.) 2007. Lochnagar: The natural history of a mountain lake. Springer. Dordrecht.

Fowler D., Coyle M. and Muller J. 2004. The future chemical climate of the UK uplands. In: Battarbee R.W., Curtis C.J. and Binney H.A. (eds.). The future of Britain's upland waters. Proceedings of a meeting held on 21st April 2004. Ensis Publishing. London. 51pp.

Fowler D., Smith R., Muller J., Hayman G. and Vincent K. 2005. Changes in the atmospheric deposition of acidifying compounds in the UK between 1986 and 2001. Environ. Pollut. 137: 15–26.

Gao J.P., Maguhn J. Spitzauer P. and Kettrup A. 1998. Distribution of polycyclic aromatic hydrocarbons (PAHs) in pore water and sediment of a small aquatic ecosystem. Int. J. Environ. Anal. Chem. 69: 227–242

Grieve I.C. 2001. Human impacts on soil properties and their implications for the sensitivity of soil systems in Scotland. Catena 42: 361–374.

Grimalt J.O., Fernández P., Berdie L., Vilanova R.M., Catalan J., Psenner R., Hofer R., Appleby P.G., Rosseland B.O., Lien L., Massabuau J.C. and Battarbee R.W. 2001. Selective trapping of organochlorine compounds in mountain lakes of temperate areas. Environ. Sci. Technol. 35: 2690–2697

Heath R.H., Kahl J.S., Norton S.A. and Fernandez I.J. 1992. Episodic stream acidification caused by atmospheric deposition of seasalts at Acadia National Park, Maine, US. Water Resour. Res. 28: 1081–1088.

Helliwell R.C., Soulsby C., Ferrier R.C., Jenkins A. and Harriman R. 1998. Influence of snow on the hydrology and hydrochemistry of the Allt a' Mharcaidh. Sci. Total Environ. 217: 59–70.

Helliwell R.C., Lilly A. and Bell J.S. This volume. Chapter 6. The development, distribution and properties of soils in the Lochnagar catchment and their influence on soil water chemistry. In: Rose N.L. (ed.) 2007. Lochnagar: The natural history of a mountain lake. Springer. Dordrecht.

Jenkins A., Reynard N., Hutchins M., Bonjean M. and Lees M. This volume. Chapter 9. Hydrology and hydrochemistry of Lochnagar. In: Rose N.L. (ed.) 2007. Lochnagar: The natural history of a mountain lake. Springer. Dordrecht.

Kerr A. Shackley S., Milne R. and Allen S. 1999. Climate change: Scottish implications scoping study. Scottish Executive Central Research Unit. http://www.scotland.gov.uk/cru/kd01/ccsi-00.htm

Kettle H. and Thompson R. This volume. Chapter 18. Future climate predictions for Lochnagar. In: Rose, N.L. (ed.) 2007. Lochnagar: The natural history of a mountain lake. Springer. Dordrecht.

Kolka R.K., Grigal D.F., Nater E.A. and Verry E.S. 2001. Hydrologic cycling of mercury and organic carbon in a forested upland-bog watershed. Soil Sci. Soc. America J. 65: 897–905.

Kopáček J., Ulrich K-U., Hejzlar J., Borovec J. and Stúchlik E. 2001. Natural inactivation of phosphorus by aluminium in atmospherically acidified water bodies. Wat. Res. 35: 3783–3790.

Lance A., Thaxton R., Watson A. 1991. Recent changes in footpath width in the Cairngorms. Scot. Geogr. Mag. 107: 106–109.

Langan S.J. 1989. Seasalt induced streamwater acidification. Hydrol. Proc. 3: 25–42.

Lindeberg C., Bindler R., Renberg I., Emteryd O., Karlsson E. and Anderson N.J. 2006. Natural fluctuations of mercury and lead in Greenland lake sediments. Environ. Sci. Technol. 40: 90–95.

McCarthy J.J., Canziani O.F., Leary N.A., Dokken D.J. and White K.S. (eds.). Climate change 2001: Impacts, adaptation and vulnerability. Contribution of Working Group II to the Third Assessment report of the Intergovernmental Panel on Climate Change. Published for the IPCC by Cambridge University Press. http://www.grida.no/climate/ipcc_tar/wg2/index.htm

Meharg A.A. and Killham K. 2003. A pre-industrial source of dioxins and furans. Nature. 421: 909–910.

Monteith D.T. (ed.) 2005. UK Acid Waters Monitoring Network: 15 year report. Ensis Publishing, London. http://www.ukawmn.ucl.ac.uk/report/

Monteith D.T. and Evans C.D. (eds.) 2000. UK Acid Waters Monitoring Network: 10 year report. Ensis Publishing, London. 364 pp.

Monteith D.T. and Evans C.D. (eds.) 2005. Recovery from acidification in the UK: Evidence from 15 years of acid waters monitoring. Special Issue of Environmental Pollution. 137: 1–178.

Monteith D.T., Evans C.D. and Dalton C. This volume. Chapter 14. Acidification of Lochnagar and prospects for recovery. In: Rose N.L. (ed.) 2007. Lochnagar: The natural history of a mountain lake. Springer. Dordrecht.

Muir D.C.G. and Rose N.L. This volume. Chapter 16. Persistent organic pollutants in the sediments of Lochnagar. In: Rose N.L. (ed.) 2007. Lochnagar: The natural history of a mountain lake. Springer. Dordrecht.

National Atmospheric Emissions Inventory (NAEI) 1999. UK Emissions of Air Pollutants 1970 to 1999. 13th Annual Report. http://www.aeat.co.uk/netcen/airqual/naei/annreport/annrep99/

NEGTAP 2001. Transboundary air pollution: Acidification, eutrophication and ground-level ozone in the UK. Prepared by the National Expert Group on Transboundary Air Pollution (NEGTAP) on behalf of the UK Department for Environment, Food and Rural Affairs and the Devolved Administrations. 314 pp.

Nordin A., Strengbom J., Witzell J., Näsholm T and Ericson L. 2005. Nitrogen deposition and the biodiversity of boreal forests: Implications for the nitrogen critical load. Ambio 34: 20–24.

Oslo and Paris Commissions 1994. Calculation of atmospheric inputs of contaminants to the North Sea 1987–1992. OSPARCOM. pp 1–27.

Patrick S.T., Juggins S., Waters D., Jenkins A. 1991. The United Kingdom Acid Waters Monitoring Network: Sites and methodologies report. Ensis Publishing, London. 63 pp.

Patrick S.T., Monteith D.T. and Jenkins A. (eds.) 1995. UK Acid Waters Monitoring Network: The first five years. Analysis and interpretation of results, April 1988 – March 1993. Ensis Publishing, London. 320 pp.

Playford K. and Baker S.J. 2000. Atmospheric inputs of heavy metals to the North Sea: Results for 1999. National Environment Technology Centre Report AEAT/ENV/336/47044102. 36 pp.

Ravichandran M. 2004. Interactions between mercury and dissolved organic matter – a review. Chemosphere 55: 319–331.

Renberg I., Bränvall M-L, Bindler R. and Emteryd O. 2000. Atmospheric lead pollution history during four millennia (2000 BC to 2000 AD) in Sweden. Ambio. 29: 150–156.

Renberg I., Bindler R. and Bränvall M-L. 2001. Using the historical atmospheric lead-deposition record as a chronological marker in sediment deposits in Europe. Holocene. 11: 511–516.

Rose N.L. This volume. Chapter 1. An introduction to Lochnagar. In: Rose N.L. (ed.) 2007. Lochnagar: The natural history of a mountain lake. Springer. Dordrecht.

Rose N.L. This volume. Chapter 8. The sediments of Lochnagar: Distribution, accumulation and composition.   In: Rose N.L. (ed.) 2007. Lochnagar: The natural history of a mountain lake. Springer. Dordrecht.

Rose N.L. and Yang H. This volume. Chapter 17. Temporal and spatial patterns of spheroidal carbonaceous particles (SCPs) in sediments, soils and deposition at Lochnagar. In: Rose N.L. (ed.) 2007. Lochnagar: The natural history of a mountain lake. Springer. Dordrecht.

Rose N.L., Backus S., Karlsson H. and Muir D.C.G. 2001. An historical record of toxaphene and its congeners in a remote lake in western Europe. Environ. Sci. Technol. 35: 1312–1319.

Rose N.L., Monteith D.T., Kettle H., Thompson R., Yang H. and Muir D.C.G. 2004. A consideration of potential confounding factors limiting chemical and biological recovery at Lochnagar, a remote mountain loch in Scotland. J. Limnol. 2004. 63: 63–76.

Rose N.L., Metcalfe S.M., Benedictow A.C., Todd M., Nicholson J. This volume. Chapter 13. National, international and global sources of contamination at Lochnagar. In: Rose N.L. (ed.) 2007. Lochnagar: The natural history of a mountain lake. Springer. Dordrecht.

Rosseland B.O., Rognerud S., Collen P., Grimalt J.O., Vives I., Massabuau J-C., Lackner R., Hofer R., Raddum G.G., Fjellheim A., Harriman R. and Piña B. This volume. Chapter 12. Brown trout in Lochnagar: Population and contamination by metals and organic micropollutants. In: Rose N.L. (ed.) 2007. Lochnagar: The natural history of a mountain lake. Springer. Dordrecht.

Stuiver M. and Brazunias T.F. 1993. Sun, ocean, climate and atmospheric $^{14}CO_2$: An evaluation of causal and spectral relationships. Holocene 3: 289–305.

Thompson R., Kettle H., Monteith D.T. and Rose N.L. This volume. Chapter 5. Lochnagar water-temperatures, climate and weather. In: Rose N.L. (ed.) 2007. Lochnagar: The natural history of a mountain lake. Springer. Dordrecht.

Tipping E., Yang H., Lawlor A.J., Rose N.L. and Shotbolt L. This volume. Chapter 15. Trace metals in the catchment, loch and sediments of Lochnagar: Measurements and modelling. In: Rose N.L. (ed.) 2007. Lochnagar: The natural history of a mountain lake. Springer. Dordrecht.

Vives I., Grimalt J.O., Catalan J., Rosseland B.O. and Battarbee R.W. 2004a. Influence of altitude and age in the accumulation of organochlorine compounds in fish from high mountain lakes. Environ. Sci. Technol. 38: 690–698.

Vives I., Grimalt J.O., Lacorte S., Guillamón M., Barceló D, and Rosseland B.O. 2004b. Polybromodiphenyl ether flame retardants in fish from lakes in European high mountains and Greenland. Environ. Sci. Technol. 38, 2338–2344.

Winch S., Ridal J. and Lean D. 2002. Increased metal bioavailability following alteration of freshwater dissolved organic carbon by ultraviolet-B radiation exposure Environ. Toxicol. 17: 267–274.

Woodward G. and Layer K. This volume. Chapter 11. Pattern and process in the Lochnagar food web. In: Rose N.L. (ed.) 2007. Lochnagar: The natural history of a mountain lake. Springer. Dordrecht.

Wright R.F., Norton S.A., Brakke D.F. and Frogner T. 1988. Experimental verification of episodic acidification of freshwaters by seasalts. Nature. 334: 422–424.

Yang H. and Rose N.L. 2003 Distribution of mercury in six lake sediment cores across the UK. Sci. Tot. Environ. 304: 391–404.

Yang H., Rose N.L. Battarbee R.W. and Boyle J.F. 2002. Mercury and lead budgets for Lochnagar, a Scottish mountain lake and its catchment. Environ. Sci. Technol. 36: 1383–1388.

Yang H., Linge K. and Rose N.L. in press. The Pb pollution fingerprint at Lochnagar: The historical record and current status of Pb isotopes. Environ. Pollut.

Youngson R.W. and Stewart L.K. 1996. Trends in Red Deer populations within the Cairngorms core area. Bot. J. Scot. 48: 111–116.

GLOSSARY, ACRONYMS AND ABBREVIATIONS

Abiotic ligand: An atom, ion, or molecule of inorganic origin, that generally donates one or more of its electrons through a coordinate covalent bond to, or shares its electrons through a covalent bond with, one or more central atoms or ions.

Acid phosphatase: A phosphatase enzyme with an acid pH optimum which is used to free attached phosphate groups from other molecules and thereby make them available for cellular uptake

ADN: Acid Deposition Network.

Aerodynamic diameter: The diameter of a unit density sphere (density $= 1$ μg cm^{-3}), which has the same settling velocity as the particle in question.

Air-water partition coefficient: The equilibrium constant for the ratio of concentrations of a single solute in air and water.

Albic: A light coloured sub-surface horizon from which iron oxides and clay have been removed.

Aldrin: An insecticide containing a naphthalene-derived compound, $C_{12}H_8Cl_6$.

Alkalinity: Total alkalinity of a water body refers to its ability to neutralise a strong acid, i.e., its buffering capacity. Carbonate and bicarbonate are often the most important contributors to alkalinity in unacidified freshwaters. Dual endpoint alkalinity is used to determine the ability of an acidic water body to resist change in pH, and is determined as the amount of strong acid required to shift pH from, say, pH 4.5 to 4.2, and in this case is closely related to hydrogen ion concentration.

Allochthonous: From elsewhere, as opposed to autochthonous (produced in situ). Used, for example, for organic matter that may be imported from elsewhere.

AL:PE: Acidification of mountain lakes: Palaeolimnology and Ecology. EU funded research projects (AL:PE 1990 – 1993 and AL:PE2 1993 – 1996).

AMS ^{14}C dating: ^{14}C dating performed by accelerator mass spectrometry.

ANC: Acid Neutralising Capacity.

Anthesis: The time at which a flower comes into full bloom.

Archean: Time during which the oldest rocks on Earth were formed, approximately 3800-2500 million years before present.

Assimilation: Data assimilation involves the use of numerical models to obtain a dynamically and thermodynamically consistent interpolation of data onto a regular grid. The method was developed for Numerical Weather Prediction systems in order to provide an accurate analysis for the initialization of predictions. For climate purposes, however, it also yields global analyses that are geographically comprehensive.

Autochthonous: From within, as opposed to allochthonous which is material from elsewhere. Used, for example, to describe material such as algal cells which are produced within a lake basin.

AWIC: Acid Waters Indicator Community.

AWS: Automatic Weather Station.

Back trajectory analysis: A technique used to identify the pathways of airborne species and particularly in estimating emission source areas.

Basal resource: The energy inputs to the base of the food web (algae, detritus), which are then eaten by the primary consumers (herbivores, detritivores).

Base cation saturation: The degree to which negatively charged exchange soil sites, particularly on clay and organic particles, are occupied by the base cations, calcium, magnesium, sodium and potassium.

BBC: British Broadcasting Corporation.

Benthic: Living attached to or associated with a substrate.

Binding affinity: A measure of the binding tendency of an atom or compound to combine by chemical reaction with atoms or compounds of unlike composition.

Bioaccumulation: Uptake of a substance from the environment into an organism with an increasing concentration (accumulation) according to the concentration in the environment and time of exposure.

Bioconcentration: Uptake of a substance from the environment into an organism, resulting in a higher concentration in the organism than in the environment.

Biofilm: A thin layer of biota that lives on submerged substrates, such as rocks (epilthic biofilm) and plant surfaces (periphytic biofilm), and is often embedded within a polysaccharide matrix. Biofilms in freshwaters typically include algae, fungi, and bacteria, and often form an important element in the diet of a number of macroinvertebrates.

Biogenic: Produced by living organisms or by biological processes.

Biomagnification: Uptake of micropollutants through the food chain. For mercury and many fat-soluble and persistent chemicals (e.g., POPs), biomagnification is the dominant factor.

Bio-reactive metal: A metal ion that can be bound to an organ or a biotic ligand.

Biotic ligand: An atom, ion, or molecule of organic origin, that generally donates one or more of its electrons through a coordinate covalent bond to, or shares its electrons through a covalent bond with, one or more central atoms (like metals) or ions.

Biotope: An area of uniform environmental (physical) conditions providing habitat(s) for a specific assemblage of plant and animals.

Biotransformation: The transformation of chemical compounds in a living system

Blanket-mire: A type of extensive mire (usually bog) that blankets flat or gently sloping upland areas forming peat of varying depth (25-250 cm). Dependent on rainfall for its water and nutrient supply.

BLM: Biotic Ligand Model. Describes the magnitude and effect of metal binding to gills.

Blocked synoptic situation: Atmospheric blocking commonly refers to the situation when the normal zonal flow in the atmosphere is interrupted by a strong and persistent meridional flow. The normal eastward progression of synoptic disturbances is obstructed leading to episodes of prolonged extreme weather conditions. Such persistent weather extremes can last from several days to a few weeks.

Bmax: Binding Capacity. The maximum amount which can be bound to a ligand etc. before saturation.

Boulder lobes: A feature associated with downslope mass movement processes which results in a curved line of boulders on a slope below which is a marked drop similar to a step.

Boulder mantle: A cover of boulders produced by frost weathering and heave of underlying bedrock.

Boundary conditions: The conditions at the edge of the domain over which ordinary or differential equations are applied.

BP: Before Present. Represents the age of a sample or level within a sediment core prior to the date of sampling (i.e., the 'present').

Bryophytes: Non-vascular plants comprising mosses and liverworts.

Calibrated years: Radiocarbon (^{14}C) dates transformed into absolute ages or sidereal years by comparing radiocarbon dates with the radiocarbon years–absolute years calibration curve.

CAM: Crassulacean acid metabolism. A means by which a plant can acquire inorganic carbon for photosynthesis by capturing CO_2 at night (when it is most available) and storing it intra-cellularly as malic acid, before it is decarboxylated during the following day.

Cambisols: Soils that show some degree of alteration from the parent material but do not show significant differences between horizons. The subsurface horizons are uniform in colour and often different from the parent material. These equate with brown earths.

Carotenoids: Any of a class of yellow to red pigments, including the carotenes and the xanthophylls naturally occurring in plants and some other photosynthetic organisms like algae, some types of fungus and some bacteria.

CBs: Chlorobenzenes.

CCCMA: Canadian Centre for Climate Modelling and Analysis.

CCW: Countryside Commission for Wales.

CEH: Centre for Ecology and Hydrology.

CF: Condition Factor. A means for determining the physiological state of a fish. Defined as CF = ((length in cm)^3/(weight in g) x100); where CF = 1 is a normal "well shaped" trout.

Chionophilous: Thriving in snow-covered habitats with a characteristic flora of dwarf shrubs and herbs, mosses, liverworts, and lichens.

CHL: Chlordane. ΣCHL is the sum of chlordane-related compounds (sum of cis- and trans-CHL, trans-nonachlor, oxychlordane, heptachlor and heptachlor epoxide).

Chlordane: A colourless, viscous liquid, $C_{10}H_6Cl_8$, used as an insecticide. It may be toxic to humans and wildlife as a result of its effect on the nervous system.

Chloremia: Low level of chloride (Cl) ions in blood plasma.

Chlorobornane: Various congeners of chlorobornane comprise the pesticide toxaphene.

CHUM-AM: CHemistry of the Uplands Model – Annual, Metals.

Circumneutral: Of, or about, neutral pH (7.0).

CLAM: Critical Loads of Acidity and Metals. A research programme funded by the UK Department for Environment, Food and Rural Affairs (DEFRA).

CLF: Critical Limit Functions.

Climate: Climate is the average of weather conditions in a certain area over a certain period. The exact boundaries of what is climate and what is weather are not well defined and depend on the application.

CLIMEX: Climate Change Experiment; a research project under the EU Framework Programme V.

Co-elution: Whereby two adsorbed substances are extracted at the same time using a chromatographic process.

Cohort: Age group; a group of individuals that were all born at approximately the same time. Cohorts can often be tracked through time, and can be used to estimate secondary production.

Community closure: Theory hypothesizing that the structure and/or dynamics of an existing biological community might 'close' an ecological niche, making it impossible for previously present species to re-colonize.

Competition niche theory: Theory that states that species that are too ecologically similar will be unable to coexist, as one will ultimately out compete the other. Niches are therefore divided among coexisting species within a community, such that there is a degree of complementarity in resource use, which serves to prevent competitive exclusion.

Congener: A form or variety of a chemical substance. For example, PCBs have 209 possible forms or congeners due to the different possible positions of chlorine atoms on the biphenyl.

Conspecifics: Individuals of the same species.

Continental climate: The type of climate found in the interior of the major continents in the middle, or temperate, latitudes. The climate is characterized by a great seasonal variation in temperatures, four distinct seasons, and a relatively small annual precipitation.

Continentality: The tendency for the middle regions of continents to have a wider temperature range than coastal areas. The greater the continentality, the greater the temperature range.

Corixids: Species in the family Corixidae (water boatmen).

Corrie: A steep-walled semicircular basin created by glaciation in high mountains; it may contain a lake. The word is an anglicisation of the Gaelic word coire meaning hollow.

CPU: Central Processing Unit.

Critical Load exceedance: The amount by which the depositional input of a polluting substance exceeds a threshold (the Critical Load) below which there is no long term impact (usually to a named biological class or organism).

CRS: Constant rate of supply. A model used in determining a ^{210}Pb chronology for lake sediments that assumes the rate of deposition of unsupported ^{210}Pb (derived from atmospheric flux) from the atmosphere is constant.

Cryosols: Soils that have a perennially frozen layer (permafrost) within 100cm of the soil surface.

Cryoturbation: Soil water expands as it freezes causing soil particles to be moved and displaced on thawing. This process mixes the soil limiting the development of layering.

Cryptomonads: A group of flagellated algae.

CSIRO: Commonwealth Science and Industrial Research Organisation.

Cultural landscape: Areas created and maintained by human activity.

Dalradian: Name given to a group of sedimentary and volcanic rocks, extending from Shetland Isles in the northeast to Connemara (Ireland) in the southwest.

DCM: Dichloromethane.

DDD: Dichlorodiphenyldichloroethane. A metabolite of DDT.

DDE: Dichlorodiphenyldichloroethene. A metabolite of DDT.

DDT: Dichlorodiphenyltrichloroethane. A powerful insecticide.

DEFRA: Department for Environment, Food and Rural Affairs. A UK Government Department.

Desmids: A group of green algae.

dGPS: Differential Global Positioning System.

Diagenesis: The process of chemical and physical change in deposited sediment.

DIC: Dissolved inorganic carbon.

Dieldrin: A chlorinated hydrocarbon, $C_{12}H_8Cl_6O$, used as an insecticide.

Dielectric fluid: An electrically insulating fluid. Dielectric materials can be made to hold an electrostatic charge, but current cannot flow through them.

Dilution Gauging: A technique for measuring streamflow by injection of a chemical tracer and measuring the time series of downstream change in concentration.

Dinoflagellate: Planktonic algae of the order Dinoflagellata in which the motile stage has two flagella.

Dinophytes: A group of flagellated algae.

Diol: any of a class of alcohols having two hydroxyl groups in each molecule.

Diptera: An order of insects commonly known as the true flies.

DJF: December-January-February. Used to denote data pertaining to the months of winter in the Northern Hemisphere.

DOC: Dissolved Organic Carbon. The dissolved fraction of carbon that remains after filtration through a 0.45 μm filter paper.

DOM: Dissolved Organic Matter.

DR: Dynamic ratio. A metric of lake morphology determined by dividing the lake surface area by the mean depth and then taking the square root of the result.

Drift cover: Superficial deposits that overlie bedrock.

Dwarf shrub: Low growing (< 50 cm high) much branched, woody plants, sometimes growing only as a prostrate mat, commonly in the Ericaceae (heath) family.

Dystrophic: An unproductive water body strongly coloured by peat products.

EA: Environment Agency.

ECMWF: European Centre for Medium-Range Weather Forecasts.

ECRC: Environmental Change Research Centre, University College London.

Efflux: The action or process of flowing or seeming to flow out.

Electrochemical gradient: An electrochemical gradient has two components. First, the electrical component is caused by a charge difference across a lipid membrane. Second, a chemical component is caused by a differential concentration of ions across the membrane. The combination of these two factors determines the thermodynamically favourable direction for an ions movement across a membrane.

Electrofishing: A fishing technique typically used for stream classification surveys and catching brood stock for hatcheries, or making estimates of populations in a body of water. A gated pulse of direct current is used to cause muscular contractions in a fish, called galvanotaxis, causing them to turn towards the source of the electrical current and swim towards it when correct pulse speeds and durations are used, along with correct current.

Elodeids: Plants with elongate, branching growth forms.

EMEP: European Monitoring and Evaluation Programme.

EMERGE: European Mountain lake Ecosystems: ReGionalisation, diagnostics and socio-economic Evaluation. EU funded research project on European mountain lakes 2000 – 2003.

Emergent plants: Those plants which grow underwater but emerge into the atmosphere.

EN: English Nature.

Endemism: The quality of belonging to, or being connected with, only a certain place or region by virtue of origin.

Endocrine disruption: Disruption of the endocrine system (hormones) can occur in various ways. Some chemicals mimic a natural hormone, fooling the body into over-responding to the stimulus, or responding at inappropriate times. Other endocrine disruptors block the effects of a hormone from certain receptors by blocking the receptor site on a cell. Still others directly stimulate or inhibit the endocrine system and cause overproduction or underproduction of hormones.

Endosulfan: A highly toxic insecticide ($C_9H_6Cl_6O_3S$) used in the control of crop insects and mites.

Endrin: A highly toxic chlorinated hydrocarbon, $C_{12}H_8OCl_6$, used as an insecticide.

ENSO: El Niño – Southern Oscillation.

Epilimnetic temperature: Temperature of the upper warm layer in lakes.

Epilimnion: A zone of relatively warm upper water within which mixing occurs as a result of wind and convection.

Epilithon: Biofilm on submerged stones.

Epipelon: Algae attached to mud surfaces.

Epiphyton: Biofilm on submerged plants.

Epipsammon: Algae attached to sand grains.

Epithelium: Membranous tissue composed of one or more layers of cells separated by very little intercellular substance and forming the covering of most internal and external surfaces of the body and its organs.

Epoxidation: A chemical reaction in which an oxygen atom is joined to an unsaturated hydrocarbon molecule (such as ethylene or propylene) to form a cyclic, three-membered ether known as an epoxide.

ERA-40: ECMWF Re-Analysis (1957-2002).

ESRI: Environmental Systems Research Institute, Inc.

Estradiol: Estradiol (17-beta estradiol) (also oestradiol) is a sex hormone. Labelled the "female" hormone but also present in males it represents the major estrogen in humans. Estradiol has not only a critical impact on reproductive and sexual functioning, but also affects other organs including bone structure.

EU: European Union.

Eulittoral: The shoreline between seasonal water level fluctuations.

Eutrophic: Having waters rich in nutrients that promote a proliferation of plant life.

Evapotranspiration: The transfer of water from a soil to the atmosphere by either direct evaporation from the soil surface, water surface or via plants by transpiration.

Exoatmospheric: Originating from outside Earth's atmosphere

Exoskeleton: A hard outer structure, such as the shell of an insect or crustacean, that provides protection or support for an organism.

Extra-regional pollen: Pollen and spores derived by far-distance transport from well outside the region of study.

FAO: Food and Agriculture Organisation of the United Nations.

Fell-field: Type of wind-blasted alpine or arctic tundra vegetation with very sparse dwarfed vegetation and flat, very open and stony soil (= feldmark).

Fetch: The distance traversed by waves without obstruction. The length of this distance measured in the direction of the wind.

FFGs: See Functional Feeding Groups.

FGD: Flue-gas desulphurisation.

FIAM: Free Ion Activity Model. Describes the magnitude and effect of metal binding to gills.

Fissile: Easily split in one direction.

Free atmosphere: The part of the atmosphere that lies above the frictional influence of the Earth's surface.

Free-radical: Radicals (often referred to as free radicals) are atomic or molecular species with unpaired electrons on an otherwise open shell configuration. These unpaired electrons are usually highly reactive, so radicals are likely to take part in chemical reactions. Free radical attack on protein, lipid and nucleic acids leads to a reduction in their respective function, thereby decreasing cell function, then organ function, and finally, organismal function. For most biological structures free radical damage is closely associated with oxidative damage, which can lead to cancer.

Frontal rainfall: Precipitation associated with the passage of weather fronts within cyclonic systems.

Frost heave: The process where the soil surface is raised due to the expansion of soil pore water on freezing.

FRS: Fisheries Research Services. Institute based in Faskally, Perthshire.

Fluted moraine: A ridge of glacial deposits streamlined by ice flow.

Food chain: The number of feeding links between a consumer and a resource species (i.e., if a eats b eats c, chain length is 2). Chains are also sometimes measured as the number of species, rather than links.

Food web: A matrix of consumer and resource species, with nodes representing species (or "trophic elements") and vertices representing feeding links between nodes. Arrows are often used to denote the direction of energy flux (i.e., from resource to consumer).

Functional Feeding Group (FFG): Categorical definitions of a consumer's feeding mode, based on its mouthpart morphology (see Cummins and Klug 1979). FFGs are often related to diet, but the two are not necessarily interchangeable. The main FFGs in Lochnagar are shredders (with tearing/chewing mouthparts, commonly used to process detrital leaf-litter), collectors (usually smaller than shredders, which feed by collecting fine particles of detritus or algae), predators (carnivores and omnivores) and grazers (with scraping mouthparts that are used to remove biofilm from surfaces). Some taxa employ more than one feeding mode, and are often amalgamated (e.g., grazer-collectors).

GC-ECNI/MS: Gas chromatography using electron capture negative ion mass spectrometry.

GCM: General Circulation Model.

GDP: Gross Domestic Product

Geminal substitution: Where two substituents are attached to the same carbon atom.

Gill epithelium: A tissue composed of a layer of cells, of the fish gill.

Gillnet: Fishing net.

GIS: Geographical Information System.

Glaciofluvial: Material transported, sorted and subsequently deposited by rivers flowing from melting ice.

Gley: Also known as a gleysol. A gley soil is one that is waterlogged for long periods where the lack of oxygen in the soil leads to the reduction and transportation of iron compounds which can be locally deposited as orangey coloured mottles where oxygen is present, for example, next to roots or in large pores. When permanently water-logged the iron remains in the reduced state and soil colours are bluish grey.

Gleying: The process of oxidation and reduction in a soil that is periodically waterlogged.

Glycoprotein: A macromolecule composed of a protein and a carbohydrate (an oligosaccharide).

GMT: Greenwich Mean Time.

Gneiss: A coarse-grained, coarsely layered metamorphic rock.

GOTM: General Ocean Turbulence Model.

GPS: Global Positioning System.

GSIM: Gill Surface Interaction Model. Describes the magnitude and effect of metal binding to gills.

ha: Hectare. Unit of area equivalent to 10000 m^2. 100 ha = 1 km^2.

Haemocrit: A measurement of the volume of red blood cells in a blood sample. Low haemocrit suggests anaemia.

Hags: A mosaic of intact remnants of peat, generally capped with a vegetated surface, surrounded by non-vegetated gullies. The gullying is formed primarily by water erosion but wind erosion is also a factor.

Half-life: (Phys) The time required for half the nuclei in a sample of a specific isotopic species to undergo radioactive decay. (Ecol) The time required for the quantity of a substance in a living organism or in another environmental compartment (e.g., a sediment or a soil) to be reduced by half as a result of the processes of metabolisation, degradation and diagenesis.

HARM: Hull Acid Rain Model.

HCB: Hexachlorobenzene.

HCBD: Hexachlorobutadiene. Formerly used as a fumigant for treating grapes against Phylloxera .

HCH: Hexachlorocyclohexane. γ-HCH is also known as lindane.

HEPA filters: High efficiency particulate air filters.

Heptachlor: $C_{10}H_5Cl_7$, formerly used as a pesticide. It is highly toxic to humans and a suspected carcinogen.

Heptachlorobornane: A chlorobornane with seven chlorine atoms.

Hexachlorobornane: A chlorobornane with six chlorine atoms.

Highland Clearances: Massive clearance of whole districts of the Scottish Highlands in the mid 19th century to make way for sheep. The dispossessed people moved south to industrial areas, emigrated to North America or Australia, or were confined to tiny areas on the west coast of Scotland. The inhumanity of the Clearances led to the legal recognition of crofters by the Crofters Act of 1886.

Histic Gleyosol: A gley soil with a peaty surface layer.

Histosol: An internationally recognised term for a peat soil.

Homeostasis: The ability or tendency of an organism or cell to regulate its internal environment to maintain a stable, constant condition, by means of multiple dynamic equilibrium adjustments, controlled by interrelated regulation mechanisms.

HOST: Hydrology of Soil Types is a classification system that groups all soils in the UK into one of 29 classes based on the dominant pathways of water movement through the soil and substrate.

Humification: A name given to the processes by which organic matter decomposes.

Hydrograph: Time series of flow measurements at a point.

Hydrological pathways: The route that water takes through the soil and substrate.

Hyperskeletic leptosol: An extremely stony, poorly developed soil.

Hypoxia: A deficiency in the amount of oxygen.

IBA: Important Bird Area.

Incident solar radiation: Radiation from the sun which reaches the Earth's surface.

Indurated: Hardened due to the interlocking growth of mineral grains.

Insolation: The quantity of solar radiation falling upon a body, especially per unit area.

Intercalated: Interlayered or interbedded.

Ion-regulation: Regulation of ions in the body fluid inside and outside cells.

IPCC: Intergovernmental Panel on Climate Change

IPPC: Integrated Pollution Prevention and Control.

Isoetids: Plants with prostrate, slow growing, rosette form.

Isomer: Any of two or more substances that are composed of the same elements in the same proportions but differ in properties because of differences in the arrangement of atoms.

Isotherm: A line of equal or constant temperature; an isopleth of temperature. Isotherms are commonly used on weather maps to show large-scale temperature distributions.

JJA: June-July-August. Used to denote data pertaining to the months of summer in the Northern Hemisphere.

JNCC: Joint Nature Conservation Committee.

ka: Thousand years.

k cal yr: Thousand calendar years.

k-ε turbulence closure model: A method for closing the Navier-Stokes equation for turbulent flow based on estimating dissipation (k) and production (ε) of turbulent kinetic energy.

K_{OW}: The octanol-water partition coefficient. This is the ratio of the concentration of a chemical in octanol and in water at equilibrium and at a specified temperature. Octanol is used as a surrogate for natural organic matter. K_{OW} is used in environmental studies to help determine the fate of chemicals in the environment e.g., to predict the extent a contaminant will bioaccumulate in fish.

Lagrangian model: The basic concept of Lagrangian dispersion models is the observation of individual air pollutants or any substance in the air. Each substance is associated with certain characteristics and modifications of these characteristics are identified at the same time as the motion of the substance itself. In large scale atmospheric dispersion models the objects of observation become air volumes instead of substances. The precise form of a Lagrangian model is mainly determined by the chosen scale (a few centimetres to some kilometres), affecting for example the type of the simulated turbulence. The models use winds and fluctuation caused by turbulence to predict the pathways of individual substances or air volumes in each time step. Lagrangian trajectory models are appropriate for the description of dispersion in complex meteorological situations or heavily structured orography. Deposition effects can be considered each time a trajectory touches the ground.

LBAP: Local Biodiversity Action Plan.

Lentic: Refers to standing or still waters such as lakes and reservoirs.

Leptosol: Soils with either solid rock within 25 cm of the soil surface or where the soil is comprised mainly of rock or gravel with little evidence of soil development.

LGM: Last Glacial Maximum. The period of maximum extent of the last ice sheet between 28 and 22000 years ago.

Ligand: An atom, ion, or molecule that generally donates one or more of its electrons through a coordinate covalent bond to, or shares its electrons through a covalent bond with one or more central atoms or ions.

Light compensation point: The depth at which photosynthesis and respiration are equal.

Lindane: (see also HCH) $C_6H_6Cl_6$, used chiefly as an agricultural pesticide.

Lipophilicity: Represents the affinity of a molecule, or part of a molecule, for a lipophilic environment i.e., having a tendency to dissolve in fat-like solvents. 'Hydrophobic' is often used interchangeably with lipophilic. Lipophilic or hydrophobic species, or hydrophobes, tend to be electrically neutral and nonpolar, and thus prefer other neutral and nonpolar solvents or molecular environments. Hydrophobic molecules in water often cluster together.

Lithic: Indicates the presence of rock at shallow depths within a soil.

Lithosol: Soils less than 10 cm thick that overly hard, coherent rock.

Littoral: Shallow water zone of a lake that typically supports rooted aquatic macrophyte growth; the region between the eulittoral and the limit of rooted vegetation.

LKMA: Parameter in WHAM describing the binding of metals to carboxylic acid groups.

LOI: Loss-on-ignition. A means by which to estimate the organic content of sediments by combustion of the sample at 550°C.

Lotka-Volterra predator-prey dynamics: The basic building block of many dynamical models of predator-prey and competitive interactions between species. In their simplest form such models include only two species (e.g., the classic snowshoe hare and lynx example quoted in basic ecology textbooks), but they can be adapted to include multispecies assemblages, and such models underpin much of the theoretical exploration of food web dynamics.

LRTAP: Long-Range Transboundary Air Pollution.

Ma: Million years.

Macrophage: White blood cells, more specifically phagocytes, acting in the nonspecific defense as well as the specific defense system of vertebrate animals. Their role is to phagocytise (engulf and then digest) cellular debris and pathogens either as stationary or mobile cells.

Macrophyte: Plants which are visible to the naked eye.

MAE: Mean absolute error.

MAGIC: Model of Acidification of Groundwater in Catchments. A computer simulation model to predict acidification of soils and waters through time.

MAM: March-April-May. Used to denote data pertaining to the months of spring in the Northern Hemisphere.

Maritime climate: A climate strongly influenced by an oceanic environment, found on islands and the windward shores of continents. It is characterised by small daily and yearly temperature ranges and high relative humidity.

MDL: Method detection limit.

Meiofauna: Microscopic aquatic animals (40 to 500µm), including those that live permanently in this size class (for example rotifers, microcrustaceans, nematodes and ciliate protozoa) as well as temporary members, for example early instars of chironomid midge larvae and many of the smaller oligochaete worms, which, as larger individuals form part of the macroinvertebrate community.

Melano-macrophage: Melano-macrophages are macrophages containing melanine granula incorporated from melanophores (pigment cells).

Melanophores: Melanophores are chromatophores, pigment cells, containing eumelanin, a type of melanin that appears black or dark brown due to its light absorbing qualities. It is packaged in vesicles called melanosomes and distributed throughout the cell.

Mesocosm: An experimental enclosure designed to approximate natural conditions – these are commonly deployed in field experiments, but are also suitable for laboratory trials.

Mesopredator release: Concept relating to trophic cascades, whereby the extirpation of higher predators (e.g., fish) from a community leads to increased abundance of the adjacent 'trophic level' represented by the mesopredators (e.g., large-bodied invertebrate predators). This release of mesopredators from predation can, in turn, suppress prey population size.

Methoxychlor: $Cl_3CCH(C_6H_4OCH_3)_2$, used as an insecticide.

Microphyte: Plants which are not visible to the naked eye.

Micropollutants: Term sometimes used to include heavy metals and persistent organic pollutants.

Microsatellite markers: Microsatellites are repetitive stretches of short DNA sequences (usually up to four base pairs). The number of repeats can vary greatly among individuals within a population. This variability makes them useful tools for quantifying genetic diversity and examining gene flow, as inferences can be made about population dynamics and dispersal capacity.

Mineral acidity: Acidity derived from inorganic, sulphur or nitrogen based minerals by chemical reaction as opposed to organic acids.

Mineralisation of nitrogen: The conversion of organic forms of nitrogen to the ammonium ion. Ammonium can then be converted to nitrate by a group of bacteria and become available for the plant.

Mires: Areas of wet ground supporting a characteristic flora and vegetation and developing peat. Includes bogs and fens.

Mirex: A formerly used organochlorine insecticide ($C_{10}Cl_{12}$) and a suspected carcinogen.

ML: Maximum likelihood.

MOLAR: Measuring and modelling the dynamic response of remote mountain lake ecosystems to environmental change. A programme of mountain lake research. EU funded research project 1996 – 1999.

Morphotype: A visually recognisable plant type which may or may not be a true species.

Mottling: The presence of orange spots or blotches (ochreous mottles), within a soil caused by the reduction and subsequent oxidation of iron, under conditions of intermittent waterlogging.

MPC: Maximum Permissible Concentration.

MSCE: Meteorological Synthesizing Centre – East. A part of the European Monitoring and Evaluation Programme (EMEP).

MSCE-Hem-HM: A model developed by the MSCE to describe the hemispherical movement of heavy metals.

MSCE-Hem-POP: A model developed by the MSCE to describe the hemispherical movement of persistent organic pollutants.

Mucous clogging: The clogging and consequent loss of function of gills by the excessive production of gill mucous.

Muir-burning: Deliberate burning of upland moorland ('muir') for management purposes.

MW: Mega-watt.

NaK-ATPase: The sodium-potassium exchanger or $Na^+/K^+ATPase$, is an ezyme which establishes the ionic concentration balance that maintains the cell potential. ATPases are a class of enzymes that catalyse the decomposition of adenosine triphosphate (ATP) into adenosine diphosphate (ADP) and a free phosphate ion. This dephosphorylation reaction releases energy, which the enzyme uses to drive other chemical reactions that would not otherwise occur.

n-alkane: An alkane is an aliphatic hydrocarbon having the chemical formula C_nH_{2n+2}. A normal alkane, or n-alkane is one which does not have a branched carbon backbone.

Nanoplankton: Very small phytoplankton cells, usually less than 5 microns in diameter.

NAO: North Atlantic Oscillation. A large-scale fluctuation in sea-level atmospheric pressure in the Atlantic Ocean between the Azores and Iceland, quantified in the NAO Index.

Natremia: Low level of sodium (Na) ions in blood plasma.

NEGTAP: National Expert Group on Transboundary Air Pollution.

Neoproterozoic: Youngest Era of PreCambrian time, approximately 1000 – 570 million years before present.

NETCEN: National Environmental Technology Centre.

NH-N: Reduced nitrogen.

Nonachlorobornane: A chlorobornane with nine chlorine atoms.

NO_Y: Total oxidised species of nitrogen.

NSA: National Scenic Area.

NWP: Numerical Weather Prediction.

OC: Organochlorine compounds.

Oceanic climate: An oceanic climate (also called marine west coast climate and maritime climate) is the climate typically found along the west coasts at the middle latitudes of all the world's continents, and in southeastern Australia; similar climates are also found at high elevations within the tropics.

Oceanicity: The tendency for the hinterland to oceans, or large lakes, to have a lower temperature range than continental interiors. The greater the oceanicity, the more equable the climate.

OCN: Octachloro-naphthalene.

OCPs: Organochlorine pesticides.

OD: Ordnance Datum.

Oedema: The swelling of any organ or tissue due to accumulation of excess fluid, without an increase of the number of cells in the affected tissue.

Oligotrophic: Pertaining to low trophic (nutrient content) status.

Oligotrophication: A process of a reduction in nutrients in an aquatic system toward levels limiting primary production.

Ontogenetic: Referring to the sequence of an individual's development, as it passes through the different stages of a life cycle (e.g., for chironomid midge larvae the sequence is: egg, instars I-IV, pupa, adult).

Ontogeny: The development of an individual organism from embryo to adult.

Oroarctic: A term used to describe the arctic-like climate of the high tops of the Scottish mountains. Although south of the Arctic Circle, the decrease in temperatures with altitude means that the mountains have a climate similar to that found within the Arctic Circle.

Orographic cloud: A cloud whose form and extent are determined by the effects of mountain relief upon the flow of air.

Orographic enhancement: Mechanism leading to an increase in deposition over mountainous terrain.

OS: Ordnance Survey.

Osmoregulation: The active regulation of the osmotic pressure of bodily fluids to maintain the homeostasis of the body's water content so that the body's fluids will not become too dilute or too concentrated. By regulation of osmotic active particles like ions and small amino acids in body fluids and cells, the cell volume and homeostasis will be maintained.

Otolith: A calcareous concretion in the inner ear of a vertebrate or in the otocyst of an invertebrate that is especially conspicuous in many bony fishes where it forms a hard body.

Overland flow: The movement of water over the soil surface during heavy or prolonged rainfall events.

Oxychlordane: Oxychlordane is the major metabolite of the chlordanes and nonachlors. It is bioaccumulative and is among the most toxic of the chlordane-related contaminants.

QMUL: Queen Mary University of London.

PAH: Polycyclic aromatic hydrocarbon. Group of chemicals with two or more adjacent aromatic rings.

Paludification: Process by which soils become waterlogged and peat formation begins and/or spreads.

PAR: Photosynthetically active radiation.

PBDE: Polybrominated diphenyl ethers. A group of compounds used as flame retardants.

PCA: Principal components analysis.

PCB: Polychlorinated biphenyl. A toxic substance used in plastics and electrical equipment.

ΣPCB_7: Sum of seven PCB congeners.

PCDD/F: Polychlorinated dibenzo-p-dioxins and polychlorinated dibenzofurans.

pCO_2: Partial pressure of carbon dioxide.

PDM: Probability Distributed Moisture model.

PE: Potential evaporation.

PeCBs: Pentachlorobenzenes.

Pedogenesis: The term given to the processes of soil development.

Pedotransfer functions: Equations relating metal concentrations in solution to soil properties (pH, organic matter content, metal content).

Pelagic: Open water area of the lake.

Pelitic: Metamorphosed muddy sediments.

Periglacial: A term given to the processes that are controlled by freezing and thawing of the soil and drift which often lead to distinctive patterns in the landscape. Initially applied to the land surrounding glaciers, it is now used to denote those processes that occur in non-glacial cold environments.

Permafrost: Perennially frozen soil.

Phenol oxidase: An enzyme involved in the breakdown of organic matter and which is inhibited in oxygen poor, waterlogged soils.

Phenolic compounds: Weakly acidic organic compounds found in decaying plant material.

Phenological change: Changes in the timing of recurring natural phenomena e.g., life cycles of plants and animals, freezing dates.

Phenotype: A morphologically recognisable group of organisms.

Phospholipid: Any of various phosphorous-containing lipids, such as lecithin and cephalin that are composed mainly of fatty acids, a phosphate group, and a simple organic molecule. Phospholipids are a major component of all biological membranes, along with glycolipids and cholesterol.

Phylloxera: Any of several small insects of the genus *Phylloxera* that are related to aphids. They are a widely distributed species very destructive to grape crops.

Physiographic: Pertaining to physical topography.

Phytobenthos: Plants living on underwater surfaces.

Phytoplankton: Free floating plants, usually algae, of the open water.

Piscicide: A chemical that kills fish.

Piscivorous: Feeding on fishes.

Plecoptera: An order of insects containing the stone flies.

Pluton: A large-scale mass of granitic rock.

PM_{10}: Particulate matter with an aerodynamic diameter less than 10 μm.

POC: Particulate Organic Carbon.

Podzol: A type of soil where there is an upper light coloured layer formed due to the removal of iron and aluminium sesquioxides and an underlying, brighter, orange-coloured layer where these sesquioxides have been deposited. The process of podzolisation is different from gleying even though both involve the movement of iron.

Pollen-source area: The geographical area from which the major part (50% or more) of a pollen assemblage deposited in a lake or mire originates from.

POM: Particulate Organic Matter.

POPs: Persistent organic pollutants.

Potential evaporation: The maximum flux of water vapour that can evaporate from a surface given optimum conditions of other physical conditions.

Precursor: A substance from which another substance is formed.

Preferential flow paths: Pathways of water movement through the soil or substrate that are active during most rainfall events.

psi: Pounds per square inch. A unit of pressure. 1 psi = 0.06805 atmospheres or 0.07031 kg cm^{-2} or 6892.7 N m^{-2}.

Pteridophyte: Fern or club-moss.

PTWI: Provisional Tolerable Weekly Intake. An amount of food containing a pollutant that can be eaten by a person per week without being harmed by that pollutant.

QSPRs: Quantitative Structure-Property Relationships.

Rain shadow: An area of reduced precipitation on the lee side of a high mountain barrier and characterized by descending air and, as a consequence, a relatively dry climate.

Ranker: A shallow soil less than 50 cm thick and underlain by hard coherent rock.

Reanalysis: Reanalyses, in which a long time series of observations is analyzed with a consistent data assimilation system, produce meteorological products with long-term consistency. The fundamental concept behind reanalysis is the desire to avoid inhomogeneities that result solely from changes to the analysis system. Reanalysis is the same as analysis, except for two important practical differences. First, it is not done in real time, and second, the background field is made by a Numerical Weather Prediction model that does not change over the entire period of the reanalysis.

Recruitment failure: No success of spawning, caused by mortality of new recruits.

Refugia: Usually referred to in a spatial context – i.e., areas where special environmental circumstances have enabled a species or a community to persist. This can also refer to size-refugia, whereby prey species are either too small or too large to be eaten by predators.

Regolith: The loose material that covers bedrock, comprising soil, weathered rock and deposits.

Regosol: Deep, well drained drift such as sands or screes with little soil development.

Reticulate: A surface network patterning.

RGAR: UK Review Group on Acid Rain.

RSD: Relative standard deviation.

RYA: Recombinant Yeast Assay. A means of analysis for endocrine disruption.

SAC: Special Area of Conservation.

SCAMP: Surface Chemistry Assemblage Model for Particles.

Schist: A metamorphic rock characterised by the parallel arrangement of minerals giving a platy texture.

SCP: Spheroidal carbonaceous particle. A component of fly-ash, produced by high temperature combustion of fossil-fuels such as coal and oil.

SD: Standard deviation

SDI: Shoreline Development Index.

Secchi Disc: Named after Pietro Secchi, a Jesuit scientist, who studied the Mediterranean Sea in the mid-1800s. The Secchi disc is a 20cm disc painted with alternating black and white quadrants used to measure water transparency. The disc is lowered into the water until it can no longer be seen. This depth indicates the water transparency and provides a rough estimate of light penetration in the water column. In general, light can penetrate to a depth of two times the Secchi depth.

Secondary lamella: The tissue of the gill arch is divided into a primary and secondary lamella to increase surface area for gas exchange and ion exchange.

Sediment focussing: the resuspension of sediments in shallower zones by waves and water currents with subsequent transport to and deposition in deeper areas of a lake.

Seed-set: Production and dispersal of mature, viable seeds of flowering plants.

SEM-EDS: Scanning Electron Microscopy linked to Energy Dispersive Spectroscopy.

Sesquioxides: A term used to denote oxides and hydroxides of iron and aluminium.

SETAC: Society of Environmental Toxicology and Chemistry.

Sheet joint: A fracture in granite often running parallel to slopes and opened after the removal of overburden and rock expansion.

Skeletic podzols: Podzolic soils that have a very large volume of stones.

SNH: Scottish Natural Heritage.

Solifluction: The downslope movement of soil caused by successive freeze/thaw cycles.

Solifluction lobes: Topographic features formed by the movement of material downslope due to solifluction. Generally curved, there lobes often have a distinct drop on the downslope side giving the appearance of a series of terraces or steps.

Soligenous: Localised ground-water seepage associated with emergent drainage water from springs or flushes, resulting in soligenous (or flush) mires with a characteristic flora and vegetation overlying shallow peat on gentle slopes.

SON: September-October-November. Used to denote data pertaining to the months of autumn in the Northern Hemisphere.

Sorption: The process in which one substance takes up or holds another (by either absorption or adsorption).

SO$_X$: Total oxidised species of sulphur.

Speciose: Rich in numbers of species.

Sphagna: An assemblage of Sphagnum moss species.

SPR: Standard Percentage Runoff, a hydrological index indicating the percentage of rainfall that will run off during a rainfall event.

SR: Source-receptor.

SRES: Special Report on Emission Scenarios.

Sterol: Any of a group of predominantly unsaturated solid alcohols of the steroid group, such as cholesterol and ergosterol, present in the fatty tissues of plants and animals.

Stomatocyst: A resting stage of chrysophyte algae.

Strath: A broad mountain or upland valley in the Scottish Highlands.

Sub-alpine: Areas near the potential limit of woodland and the flora and vegetation that occur in the transition between potential woodland limit and distinct alpine vegetation types.

Subangular stones: Stones that have had their angular edges slightly rounded during transportation processes.

Sub-fossil: Remains of an organism that are essentially unaltered chemically.

Sub-oxic: Condition in which the oxygen partial pressure is lower than that of the atmosphere.

Substratum: An underwater surface often colonized by algae.

Supralittoral: The region immediately above the high water level (eulittoral zone) subject to wave spray.

SWAP: Surface Water Acidification Project.

Synoptic: An adjective derived from the Greek meaning, "to view together, or at the same time". In meteorology, the term refers to the use of meteorological data obtained simultaneously over a wide area for the purpose of obtaining a comprehensive and nearly instantaneous picture of the state of the atmosphere.

Tall herbs: Tall (50-100 cm) herbaceous plants, often with large nutritious leaves and sensitive to grazing and trampling.

Talus: Rock waste.

Teleost: Of or belonging to the Teleostei or Teleostomi, a large group of fishes with bony skeletons, including most common fishes.

Temperature inversion: An increase in temperature with height above the Earth's surface, a reversal of the usual pattern in which temperature decreases with increasing altitude.

Terracettes: A small scale landform feature where the movement of soil downslope results in a step-like or terraced landscape.

TF HTAP: Task Force on Hemispherical Transport of Air Pollution.

Thermistor: The name comes from thermal resistor. A semiconductor comprised of a mixture of certain oxides with finely divided copper, which exhibit rapid and extremely large changes in resistance for relatively small changes in temperature. Because of the known dependence of resistance on temperature, the resistor can be used as a temperature sensor.

Till: Deposit formed directly from glacier ice.

Tipulidae: A family of insects containing the crane-flies.

Toposequence: A sequence of related soils that differ from each other, primarily because topography was the dominant soil-forming factor. For example, soils developed on similar parent material are dry (freely draining) upslope but are wet (poorly draining) at the base of the slope.

Top predator: A species in a food web that is not preyed upon by any other species.

Toxaphene: A complex mixture of chlorinated compounds (polychlorinated bornanes and camphenes) widely used as an insecticide following the ban on DDT. Also used as a piscicide.

Transmissivity: The degree to which (as a proportion) the atmosphere transmits radiation i.e., an atmospheric transmissivity of 0.5 means that 50% of the radiation received at the top of the atmosphere reaches the Earth's surface.

TrCBs: Trichlorobenzenes.

Tree-line: Altitudinal, latitudinal, or artificial human-determined limits of tree occurrence in the landscape. Tree-line strictly refers to the extent of the tree growth-form of individual trees and of isolated tree clusters that extend beyond the forest-limit. Tree-limit is the absolute altitudinal or latitudinal limit of a tree species, often growing in a stunted form.

Tricoptera: An order of insects containing the caddis flies.

Tundra: Vegetation zone lacking trees north of the Arctic Circle and above the tree-line in mountains, consisting of low-growing plants.

Tychoplanktonic: Organisms occasionally carried into the plankton by factors such as turbulence.

UK AWMN: United Kingdom Acid Waters Monitoring Network .

UKCIP: United Kingdom Climate Impacts Programme.

UNECE: United Nations Economic Commission for Europe.

UNEP: United Nations Environment Programme.

UNFCCC: United Nations Framework Convention on Climate Change.

US EPA: United States Environmental Protection Agency.

UTC: Coordinated Universal Time.

VOC: Volatile organic compound.

WA: Weighted averaging.

WAPLS: Weighted averaging – partial least squares.

Weather: Weather is the specific condition of the atmosphere at a particular place and time. It is measured in terms of wind, temperature, humidity, atmospheric pressure, cloudiness, precipitation and so on. A simple way of describing the difference between climate and weather is that climate is what you expect, and weather is what you get.

WFD: Water Framework Directive. A European Union Directive (2000/60/EC) which establishes an integrated approach to the protection, improvement and sustainable use of Europe's rivers, lakes, estuaries, coastal waters and groundwater and aims to achieve good ecological quality in all relevant waters. It requires that biological, hydromorphological and chemical elements of water quality should be based on the degree to which present day conditions deviate from those expected in the absence of significant anthropogenic influence, termed reference conditions.

WHAM: Windermere Humic Aqueous Model.

WHO: World Health Organisation.

Xeric: Pertaining to, characterised by, or adapted to an extremely dry habitat.

Younger Dryas stadial: A well-known and much studied cold period during the last glacial-interglacial transition. It occurred between 10700 – 10000 ^{14}C years BP or between 12650 – 11550 calibrated years BP. Named after the arctic-alpine plant *Dryas octopetala* whose fossil leaves occur in sediments of this age in southern Scandinavia.